D0138768

*AN INTRODUCTION TO
COMMUNICATION SYSTEMS*

ELECTRICAL ENGINEERING, COMMUNICATIONS, AND SIGNAL PROCESSING

Raymond L. Pickholtz, Series Editor

OTHER WORKS OF INTEREST

AN
INTRODUCTION
TO
COMMUNICATION
SYSTEMS

Allan R. Hambley
Michigan Technological University

COMPUTER SCIENCE PRESS

An imprint of W. H. Freeman and Company
New York

Library of Congress Cataloging-in-Publication Data

Hambley, Allan R.
   An introduction to communication systems/Allan R. Hambley.
     p. cm.
   Includes bibliographies and index.
   ISBN 0-7167-8184-0
   1. Digital communications. I. Title.
TK5103.7.H36 1989   88-39459
621.38—dc19   CIP

Printed in the United States of America

Computer Science Press
An imprint of W. H. Freeman and Company
41 Madison Avenue, New York, NY 10010
20 Beaumont Street, Oxford OX1 2NQ, England

1 2 3 4 5 6 7 8 9 0   RRD   8 9

CONTENTS

PREFACE

This book is a complete introduction to the major concerns of communication systems engineering at a suitable level for the typical junior and senior student in electrical engineering. Although traditional analog communication systems are treated, the emphasis of the book is on digital systems. In addition, error control coding, spread-spectrum techniques, and optical fiber systems are covered at an introductory level.

Side issues have been avoided in favor of a clear and explicit discussion of the main issues of communication systems. A large number of complex mathematical tools is required for the full development of the subject. Where possible, discussion of these tools is delayed until they are needed. It is hoped that this approach makes the major points more interesting and accessible to the average student.

Features of the book

The book features the use of design theory, including topics such as antenna characteristics, propagation effects, noise, and intermodulation chacteristics of amplifiers, all of which must be considered in realistic design examples. Theoretical quantities are related to physical measurements, and the units of all quantities are stressed, enabling the reader to make quick and effective use of theory in practical situations.

Features have been included in the book to make the important points stand out from the examples and explanatory material. Important equations and facts are marked with a bullet (●). A summary enumerating the important points and drawing conclusions is given at the end of each chapter, making it easier for students to focus on the major issues when reviewing the material.

The book contains a large number of problems at the end of each chapter, some of which are straightforward drill problems and others that can be used to extend the coverage into many of the side issues avoided by the text and to challenge advanced students. Many problems that illustrate the use of the theory in design situations have been provided.

Prerequisite material

The assumed background includes circuit analysis, Laplace transform techniques, and introductory courses in analog and digital electronics. A signal analysis course including Fourier series and transforms would be helpful but is not necessary if adequate time is alloted to Chapter 2, which reviews this material. Similarly, a previous course in probability is desirable but not necessary because the material is reviewed in Chapter 7.

The content of the book

The first chapter is a broad introduction to the subject, intended to motivate the student and provide a road map to the subject.

A fairly complete review of Fourier methods is given in Chapter 2 to introduce notation, refresh the student's knowledge, and provide a convenient reference. Thus, the book could be used in a course in which the students are meeting Fourier theory for the first time, provided that adequate time and some supplementary material is used by the instructor. In keeping with the philosophy of this book, certain topics, which other texts often treat in an early chapter on Fourier methods, have been avoided in Chapter 2. These include the Hilbert transform, complex envelope representations, questions of the existence of the transform, and so forth. Some of these are treated in later chapters as the need arises or in the problem sets and others are left out entirely.

Chapters 3 and 4 discuss various forms of analog amplitude and angle modulation. Modulators and demodulators are discussed at the block diagram level and, in some cases, at the circuit design level. Phase-locked loops are introduced as demodulators for angle modulated signals and are treated in depth because of their pervasiveness in communication systems.

Those aspects of noise theory required for a discussion of the noise performance of AM and FM systems are developed in Chapter 5. (Additional material on noise signals involving probability and random process theory is included in Chapter 7.) Information of a practical nature, such as the physical sources of noise, noise figures of amplifiers, antenna characteristics, and received power calculations has been included so that the results of the noise analysis can be applied to realistic design problems. The noise analyses of the various modulation types are carried out in parallel fashion to make the differentiating characteristics of the modulation types as clear as possible. In addition, an explicit discussion comparing the various modulation types is provided. Finally, a design example illustrates the system design process.

Chapter 6 considers pulse modulation techniques. Various types of sampling are considered both for baseband and bandpass signals. The concept of time division multiplexing is then introduced. Pulse amplitude modulation is covered and the Nyquist criterion for zero intersymbol interference is discussed. The noise performance and bandwidth requirements of analog pulse amplitude are analyzed and compared to other analog modulation formats. Pulse code modulation is discussed, including the effects of quantization error and bit errors. Finally, delta modulation schemes are briefly considered.

The theory of probability, random variables, and random processes is presented in Chapter 7. The coverage is appropriate for students who have not had prior exposure to this material. It also provides a review and a convenient reference for those who have had a previous course in the subject. Autocorrelation functions of several baseband digital signaling formats are computed and the corresponding power spectral densities are found.

Signal-space concepts are developed at the beginning of Chapter 8. Communication with known binary signals in the presence of white gaussian noise is then discussed and optimum receivers are developed. The discussion relies heavily on signal-space concepts, giving students the clearest possible understanding of the performance of the various digital modulation techniques. Detailed treatments of baseband NRZ signals and phase reversal keying are given. Noncoherent receivers, M-ary signal formats, and timing considerations are then presented in separate sections. The chapter ends with a comparison of the various modulation approaches on the basis of signal to noise ratio required at the receiver for a given bit-error rate and the required channel bandwidth. Performance is also compared to the theoretical limit allowed by the channel capacity equation.

Error control coding is discussed in Chapter 9. Block codes are considered first, followed by a discussion of convolutional codes. Then several examples of coding applications are considered, to illustrate the performance improvements that are made possible by coding. Finally, automatic repeat query methods are considered.

An introduction to information theory is given in Chapter 10. Source coding and channel capacity are discussed for discrete channels. The discrete channel models are related to the digital modulation techniques considered in Chapter 8. A derivation of the channel capacity for the binary symmetric channel is given.

Chapter 11 is an introduction to the basic principles of spread-spectrum communication systems. Most texts intended for undergraduates do not treat this subject. However, the use of these techniques in military communication systems has become so widespread that many new graduates will begin their engineering careers working on these systems. Furthermore, the basic concepts are easily within the grasp of undergraduates when properly presented. The chapter begins with a discussion of maximal-length pseudorandom codes. This is followed by a discussion of direct sequence techniques, including the use of spread spectrum for antijamming, convertness, code division multiplexing, mitigation of multipath problems, and determination of time and position. Next, frequency-hop systems and their applications are considered and compared to direct sequence systems. Finally, code acquisition and tracking techniques are briefly discussed.

The final chapter of the book discusses optical fiber communication systems. This topic has not usually been treated in introductory texts on communication systems. Because of the growing importance of optical fiber systems, this material will be of interest to many communication systems engineers. The advantages of fiber optic systems compared with electrical cable are considered. Ray optics is briefly reviewed and various fiber types are considered. The characteristics of light sources are considered, followed by a discussion of photodetectors. Then the sources of system noise and fundamental performance limits of both coherent and noncoherent systems are considered. Last, the design of optical fiber communication links is considered.

Courses based on the book

This book is suitable for use in either short introductory courses or in a full year sequence of courses for junior/senior students in electrical engineering. It contains enough material so that a selection can be made to emphasize various aspects of communication systems are desired by the instructor.

The book has been used in manuscript form at Michigan Tech in a three quarter sequence of courses. The first quarter is required of all electrical engineering students and covers most of the material in the first six chapters. The students have had previous courses in Fourier theory and linear systems. Students usually take the course in their senior year, though some juniors are present. In the second quarter, most of the students are seniors with a strong interest in communication systems. A course in probability is a prerequisite for the second quarter. The second course reviews and completes the coverage of the first six chapters as well as covering chapter 7 and part of Chapter 8. The third quarter completes the book with the exception of the chapter on fiber optics.

Acknowledgments

I wish to thank the many students who have used this text in manuscript form over the past several years. Their many constructive and supportive comments have improved the book in large measure. I would also like to thank my colleagues who have reviewed parts of the text and engaged in helpful discussions of the topic and its presentation. These include Ashok Ambardar, Robert Bickmore, John Clark, Paul Lewis, and James Rogers. I especially thank Douglas Brumm who carefully reviewed the chapter on fiber optics. Also, I wish to thank my Department Head, Keith Stanek, for his help and encouragement throughout the course of this project.

Finally, I wish to thank my wife, Judy, and son, Tony, for their encouragement and cheerful attitude toward the undertaking which greatly lightened the load.

Allan R. Hambley

AN INTRODUCTION TO
COMMUNICATION SYSTEMS

1

INTRODUCTION

The subject of this book is the engineering principles underlying the design of communication systems. A communication system conveys information from a source to a destination. The information can be in the form either of an analog signal such as a voice waveform or of a digital signal such as a symbol produced by a keyboard or a computer. The emphasis of this book is on digital systems, but analog systems are considered first because they precede digital systems historically and because it is important to have an understanding of analog signals before considering digital systems. Digital systems inherently use analog signals in the process of communicating digital information.

The goal of a communication system engineer is to design systems that provide high quality service for the maximum number of users with the smallest cost and least usage of limited resources. The resources to be conserved include hardware for generating, transmitting, and receiving information signals, the channel bandwidth, and the transmitter power. As we will see, signals can be understood to consist of a sum of sinusoidal components with various frequencies. In many communication systems, signals from different transmitters are designed so that their components have frequencies in different nonoverlapping ranges. Thus, a specific range of frequencies is set aside for each transmitter. Since any physical system for conveying signals is effective for only a limited range of frequencies, it is often important to minimize the width of the band of frequencies employed by each user so that the maximum number of users can be accommodated.

Some obvious examples of communication systems are the telephone network, broadcast radio, and television. Mobile radio, used by police, fire departments, aircraft, and private citizens, is another example. Radio aids to navigation, often used for position determination by ships and aircraft, use principles similar to those of communication systems. Radar systems also use many concepts from communication system theory.

That something operates as a communication system may not be obvious. For example, many readers may be aware that electric typewriters and personal computers radiate unintended signals that can interfere with nearby radio and television receivers. Possibly a communication system engineer, working in the intelligence community, could design a receiver to intercept such signals and to retrieve the information being processed. In this case, a communication system would exist, but the design of the transmitter is not in the control of the engineer who must design the receiver.

Some other examples of less obvious communication systems are audio and video recordings and computer memories. An audio compact disk uses several principles conceived for digital communication systems. (Incidently, though the compact disk stores information in digital form, the original message is an analog waveform that is converted to digital format for storage on the disk.) A compact disk is part of a communication system that conveys information from the recording studio to the listener when the disk is played. A computer memory is a system for conveying digital information from the time of storage to the time it is read from the memory.

Many applications of communication systems with specialized or extreme requirements exist in the military arena. For example, a communication system may be required to operate under the influence of an intentional jammer that is capable of delivering much higher power to the input of the receiver than the power of the desired signal. Another example is a communication system that can reliably communicate without being detected by an enemy receiver that may be much closer to the transmitter than the intended receiver.

In recent years, great strides have been made in the hardware technology in communication systems. Communication satellites are used for transoceanic telephone links, distribution of television signals to cable systems, military communication systems, and interoffice business communications. The development of fiber optic communication links, which transmit information on lightwaves through silica fibers, began in the 1960s and has progressed at an astounding rate, about a tenfold increase in the capacity of optical fibers every three years. Fiber optic links recently installed in telephone systems are capable of transmitting the contents of the *Encyclopaedia Britannica* through a single fiber over a distance of 29 miles between repeaters in less than two seconds. No doubt, in a few years, systems will exceed this feat by an order of magnitude or more. Another area in which technology that is important in communication systems is expanding at a rapid rate is the fabrication of integrated circuits and high bit rate switching circuits.

Advancing technology not only improves our implementation of communication systems, it also increases the demand for communication resources. Advanced digital integrated circuit technology has led to the increased use of computers, which in turn has led to greater demand for communication between computers and other devices, for example, the concept of a local area network, which is a communication system that interconnects computers, printers, and storage devices in office buildings or campuses. New weapon systems require increased information transfer and coordination. The Strategic Defense Initiative (SDI), popularly known as "star wars," will need accurate and rapid communication of very large amounts of information if it is ever implemented.

In this chapter, we will describe the main features of typical analog and digital communication systems, introduce some of the most important concepts, briefly discuss the fundamental limitations inherent to all systems, and discuss how later chapters relate to the overall subject.

1.1
TYPICAL COMMUNICATION SYSTEMS

The block diagram of a typical analog communication system is shown in Figure 1.1. The analog signal to be transmitted can be a voice waveform, a television signal, or any other information-bearing signal. Typically, this message

FIGURE 1.1
Block diagram of a typical analog communication system.

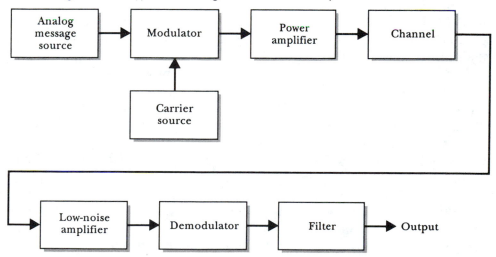

signal must first be filtered to eliminate undesired components and amplified to a suitable level, depending on the source, but we will assume here that the signal is suitable for transmission.

The message signal is often modulated onto a carrier, which can be a sinusoidal signal, a pulse train, or a lightwave. In the modulation process, the signal affects some parameter of the carrier in a predetermined way. For example, if the carrier is a sinusoid, the message amplitude might determine the amplitude of the resulting modulated signal. The result is known as an *amplitude modulated signal*. On the other hand, the message might modulate the frequency of the carrier. A higher message amplitude increases the frequency, and a lower message amplitude lowers the frequency, resulting in frequency modulation.

There are several reasons for using modulation in communication systems. It can be used to change the range of the frequencies of the components of the signal, which is sometimes useful in making the transmitted signal match the characteristics of the system that conveys it. For example, an audio waveform contains frequency components from approximately 20 Hz (hertz) to 15 kHz (kilohertz). This range of frequencies is very difficult to radiate from an antenna but modulation can move the signal to an easily radiated frequency range. Another reason for moving the frequency range of a signal by modulation is to enable several users to transmit signals at the same time. The modulation of users' signals to different frequency ranges allows their separation at the receiving end by filters. Thus, modulation is one method for multiple access to a communication resource. Finally, certain types of modulation enable a transmitter power reduction at the expense of wider bandwidth. We will illustrate how this is possible in the following chapters.

The modulated signal is amplified to a suitable power level and input to the channel, which is the mechanism for conveying the modulated signals to their destination. The channel can consist of a transmitting antenna, the space between the transmitter and the receiver, and the receiving antenna. Other possibilities for the channel are a coaxial cable or an optical fiber connecting the transmitter to the receiver.

Various things can happen to a signal in transmission through the channel. Usually the transmitted signal is greatly attenuated, and random noise, due to thermal agitation or the discrete nature of matter and energy, is added. Signals from other transmitters may be added. The signal may also arrive at the receiver from several paths through the channel; for example, a radio signal may be reflected from buildings and arrive at the receiver from several directions. The "ghosts" in television signals received by antennas in urban environments are due to such *multipath* phenomena. Real communication channels are difficult to analyze, so we will begin by assuming that the channel simply attenuates the signal and adds noise. This will give a convenient basis for comparison of the various analog and digital modulation schemes.

The signal and noise received from the channel are amplified to a suitable level and filtered to eliminate noise and interfering signals that fall outside the frequency range of the desired signal. The result is then demodulated to recover the original message signal. After another filtering, the message waveform is presented to the destination.

Because of the noise added in the channel, the demodulated signal is not identical to the source signal but contains added noise. The primary performance measure of an analog communication system is the ratio of the message signal power to the noise power at the destination. Thus, we will be interested in developing expressions for the signal-to-noise ratio at the destination for various forms of modulation in terms of the system parameters.

A block diagram of a digital communication system is shown in Figure 1.2. Digital systems generally are more complex than analog systems. Both an analog information source and an inherently digital source are shown. Before it is transmitted through the system, the output of the analog source is converted to digital form by a circuit that measures the amplitude of the analog signal periodically and converts these measurements to groups of binary symbols. Thus, the analog signal is sampled and the sample values are converted to digital form by the analog-to-digital converter.

The digitized output of the analog source enters a source coder. The source coding operation attempts to reduce the number of symbols that must be conveyed through the channel, which is possible because many natural sources are highly redundant. For example, a television signal is a waveform that transmits the brightnesses and colors of sample points in each of a sequence of scenes. If the images of moving objects are repeatedly sampled, rapid playback of the successive samples creates the illusion of smooth motion. Television signals are highly redundant because the brightnesses and colors of the points in an image, the picture information, usually changes very little from scan to scan. Furthermore, the brightnesses and colors of points that are close together in scenes are often similar. Source coding techniques efficiently represent this redundant source signal using comparatively few digital symbols. For example, a simple approach to reducing the amount of data that must be conveyed in a television signal is simply to send only the values of the image that have changed since the previous scan. Since most of the points in an image usually have the same color and brightness from scan to scan, this can result in a considerable savings in the quantity of data that must be transmitted. Source coding can also be applied to the outputs of digital sources.

In Figure 1.2, the output of the digital source and the output of the source coder enter a multiplexer in which groups of symbols, called *words*, from the two sources are interleaved in time. This is done so that the information from both sources can be transmitted through the system. The two sources have access to the communication link by dividing time into slots and assigning them alternate time slots. This widely used technique is known as *time division multiple*

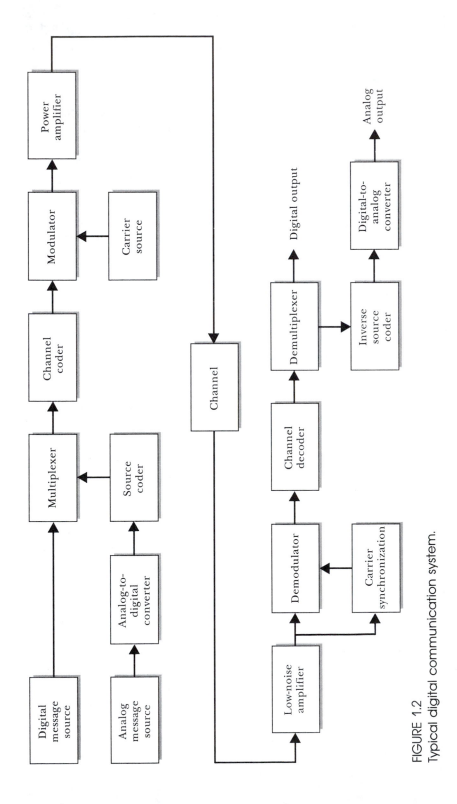

FIGURE 1.2
Typical digital communication system.

access. Telephone networks use this technique on long distance routes in which hundreds of digitized voice signals are interleaved in time. At the receiving end, the data from each source are sorted, or *demultiplexed*, and sent to their destinations.

The output of the multiplexer enters a *channel coder.* Because of signal attenuation and noise added to the signal in the channel, some of the digital symbols will be received in error. The channel coder adds known symbols to the data stream so that errors can be either detected or corrected by the channel decoder at the receiving end of the channel. Correction of channel errors by the decoder is known as *forward error control.* In an alternative technique, known as *automatic repeat query*, the receiver requests the transmitter to repeat erroneous coded data.

The symbols output from the channel coder enter a modulator in which they modulate a carrier. The modulation process in a digital system is similar to modulation in an analog system, differing only in that the modulated signal has a discrete set of amplitudes, frequencies, or phases, depending on which parameter of the carrier is being modulated by the digital stream. Modulation is used to make the signal more suitable for transmission through the channel or to control the location of the frequency components of the modulated signal so that several transmitters can use the channel by using nonoverlapping frequency bands. A large number of digital modulation schemes are in use and they have different performance characteristics with regard to error probability and bandwidth.

At the receiving end of the digital communication system, demodulation, channel decoding, demultiplexing of various data streams, inverse source coding, and digital-to-analog conversion reverse the operations at the source end, resulting in versions of the original analog signals and digital data streams that entered the system. Due to uncorrected errors that have occurred in the channel, some of the digital symbols delivered to the destination may be different than those that entered the system. The primary performance criterion of the system for transmission of digital data is the *error probability.* The system designer tries to find a system that provides the maximum communication capacity with the minimum error probability consistent with cost and use of other resources.

The analog signal delivered to the output of the system contains noise from several sources. First, it is impossible to represent the continuum of values of an analog signal by a finite number of digital words, and therefore, the values recreated by the digital-to-analog converter are simply the closest discrete values to the analog values that entered the system. The resulting error is called *quantizing error*, and it has an effect similar to additive noise. A second source of noise in the output signal is error in the digital words representing the amplitudes.

Another feature of communication systems is the need for synchronization. For example, some of the best performing digital modulation schemes require a version of the unmodulated carrier at the receiving end to demodulate the signal. Thus, a carrier source in the receiver must be synchronized to the frequency and phase of the carrier imbedded in the received signal. Also, in digital systems, a data clock that is synchronized with the boundaries between data symbols is often needed.

Many variations of the typical systems shown in Figures 1.1 and 1.2 are possible. A communication systems engineer must be familiar with many types of modulation, coding, synchronization, and multiplexing schemes to select the system configuration required for a given situation.

1.2
COMPARISON OF DIGITAL AND ANALOG
COMMUNICATION SYSTEMS

The trend in design of new communication systems has been toward increasing use of digital techniques. In general, analog systems are simpler than digital systems, but digital systems can have higher performance and are more versatile. As hardware technology advances, the performance of digital approaches becomes increasingly feasible for systems of reasonable cost.

Digital communication systems have a number of performance advantages compared with analog systems. First, there is a fundamental difference in the way that noise affects a digital signal compared with the way that it affects an analog signal. Figure 1.3a and b show an analog signal and a digital signal before noise is added. The digital signal shown is binary, meaning that it takes only two amplitudes but, in general, a digital signal takes a *discrete set* of values, not necessarily only two. The analog signal contains a continuum of values. The result of adding noise to the signals is shown in Figure 1.3c and d. In the case of the digital signal, if the noise amplitude is less than half of the distance between the levels of the digital signal, the original value can still be determined. However, in the analog signal, not even a small amount of noise can be distinguished from the signal on the basis of amplitude because the message signal takes on a continuous set of values. Thus, it is possible to recover the original digital information from a noisy signal if the noise amplitude is not too large, but usually, not even a small amount of noise and distortion can be completely removed from an analog signal. Often in digital communication systems using cable or optical fiber, regenerative repeaters are spaced through the system to remove the noise and distortion from the signal and regenerate a fresh signal to be sent forward. In an analog system, even though the signal may be amplified at various points, noise and distortion tend to accumulate. Coding techniques can be used with digital transmission of information to further reduce the probability of errors caused by noise.

Of course, when analog signals are digitized, a small amount of quantizing error occurs because a digital word can represent only a discrete set of

8

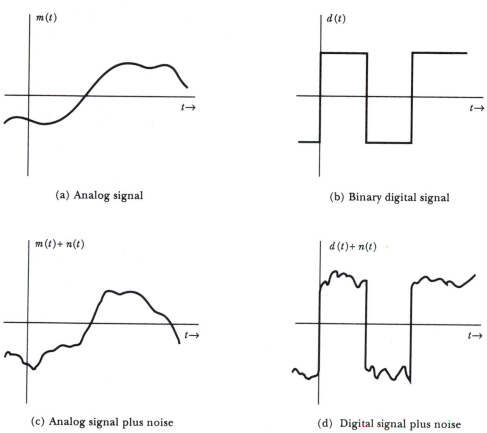

(a) Analog signal   (b) Binary digital signal

(c) Analog signal plus noise   (d) Digital signal plus noise

FIGURE 1.3
Illustration of the effect of noise on analog and digital signals.

values. However, this degradation can be kept very small with currently available hardware. A performance comparison of the compact disk for audio with the older analog records clearly shows some of the advantages of digital techniques.

Another reason for the trend toward digital techniques is that digital systems can be designed to handle many different types of digitized signals. A single communication system can be used to transmit a digitized television signal at one time, a large number of digitized and multiplexed voice signals at another time, and a high-rate data signal from a computer at a third time. Analog transmission systems, on the other hand, tend to be designed for the particular requirements of one type of signal.

The technological advances in digital integrated electronics have reduced the cost of the hardware required for digital techniques to a greater degree than the cost reductions that have occurred for analog systems. This has tended to make digital techniques, which were at one time considered to be too costly, the approach of choice. It is also easier to encrypt digital data for secrecy than analog signals. As business relies more heavily on electronic communication, the need for encryption in civilian communication systems becomes greater.

Despite the many advantages of digital systems, the study of analog approaches is still important. Many existing systems are analog, and the concepts used in analog systems are used in digital systems. After all, the actual signals and noise waveforms in a digital system are often analog in nature. Nearly all of the analog concepts are useful in understanding digital systems.

1.3
FUNDAMENTAL LIMITS OF COMMUNICATION SYSTEMS: THE SHANNON-HARTLEY THEOREM

Communication systems engineers attempt to design communication systems that transmit information at a high rate, with high performance, using the minimum amount of transmitter power and bandwidth. For a digital communication system, high performance means a low error probability, and because we want to minimize transmitter power, the noise added by the channel and the first stages of the receiver is a significant problem. Given these problems, the question of importance is: What is the fundamental limitation of a digital communication system?

The answer is stated by the Shannon-Hartley channel capacity theorem, according to which, digital communication systems that attain as close to zero error probability as desired are theoretically possible, provided that the rate of information transmission is less than the capacity of the channel. This capacity (C) is given by

$$C = B \log_2 \left(1 + \frac{S}{N} \right) \text{ bits/s} \tag{1.1}$$

where B is the bandwidth of the channel in hertz, S is the received signal power, and N is the noise power added to the signal. The logarithm is base two, and a bit is the amount of information that we obtain when we receive one of two equally likely symbols. For example, we receive one bit of information when we learn the result of tossing a coin, provided that "tails" and "heads" are equally likely to occur. Shannon has also shown that it is not possible to design systems that operate at rates in excess of the channel capacity and achieve vanishing error probability.

Unfortunately, the proof of this theorem does not show us how to construct practical systems that achieve the channel capacity. Nevertheless, the theorem provides a level of performance to which the performance of practical systems can be compared to determine how much improvement is possible. From time to time through the book, we will compare the performance of systems to this fundamental limit.

1.4
PREVIEW OF THIS BOOK

Two branches of mathematics, Fourier analysis and probability theory, are particularly important in the study of communication systems. Chapter 2 contains a review of Fourier series and Fourier transform theory, which are used in the book. The theory is concerned with the components of signals as functions of frequency. Fourier theory is important both for analog and digital systems, mainly because bandwidth is limited and because different frequency ranges must be assigned to various systems so that they do not interfere with one another. It is anticipated that most students have studied Fourier theory, and Chapter 2 is intended only as a review of this important material. However, the coverage of the concepts needed in the rest of the book is complete, so the student without previous knowledge should be able to proceed.

Chapters 3 and 4 present amplitude and frequency modulation techniques and related material. This book is mainly concerned with system concepts rather than circuit design, but some circuits are presented to illustrate how the knowledge gained in electronic circuit design courses can be applied to communication systems. Phase-locked loops find application in many places in communication systems and are covered in the last section of Chapter 4.

Chapter 5 presents the information needed for calculating received signal power and noise level in line-of-sight radio communication systems. Noise can have various characteristics, but we will emphasize *white gaussian noise*. An approximation of this type of noise is caused by thermal sources in most communication systems, and it is relatively easy to analyze. The theory of random noise signals needed for predicting the signal-to-noise ratio at the destination for analog modulation systems is presented. Expressions for the signal-to-noise ratio at the destination are obtained for the analog modulation schemes. Finally, a design example is presented that shows how the material presented in the first five chapters is used in the design of an analog communication system.

Chapter 6 begins the study of digital communication systems. Since analog signals must be sampled by measuring their amplitudes periodically before they can be converted to digital form, sampling theory is discussed. Then various aspects of communication of analog signals by digital methods are discussed.

Chapter 7 contains a review of probability, random variables, and random process theory. These topics are needed for a discussion of the error performance of digital communication systems.

Chapter 8 considers a number of digital modulation techniques. The frequency content of the modulated signals, optimum receiver structures, the resulting error probability, and synchronization of reference signals in the receiver are considered for each modulation scheme.

Chapter 9 contains an introduction to coding techniques for error control in digital communication systems.

In Chapter 10 we briefly consider the subject of information theory, which is concerned with the information content of the signals conveyed by a communication system, how to represent this information efficiently, and the fundamental limits of information transfer.

Chapter 11 contains a brief discussion of spread-spectrum techniques. These techniques are particularly useful in military communication systems in which antijamming or covert operation is required, but they are also useful in nonmilitary systems in which protection against multipath and provision for multiple access are desired.

Finally, Chapter 12 presents a brief discussion of lightwave communication through optical fibers.

REFERENCES

The reader who wishes to consult an alternate coverage of topics in the first ten chapters of this book will find the following list of books useful. The *IEEE Communications Magazine* is also an excellent source of articles written at a level appropriate for undergraduate students.

A. B. Carlson. *Communication Systems*, 3d ed. New York: McGraw-Hill, 1986.

L. W. Couch II. *Digital and Analog Communication Systems*, 2d ed. New York: Macmillan, 1987.

S. Haykin. *Communication Systems*, 2d ed. New York: Wiley, 1983.

B. P. Lathi. *Modern Digital and Analog Communication Systems*. New York: Holt, Rinehart and Winston, 1983.

M. Schwartz. *Information Transmission, Modulation, and Noise*, 3d ed. New York: McGraw-Hill, 1980.

F. G. Stremler. *Introduction to Communication Systems*, 2d ed. Reading Massachusetts: Addison-Wesley, 1982.

H. Taub and D. L. Schilling. *Principles of Communication Systems*, 2d ed. New York: McGraw-Hill, 1986.

R. E. Ziemer and W. H. Tranter. *Principles of Communications*, 2d ed. Boston: Houghton Mifflin, 1985.

2

FOURIER METHODS OF SIGNAL ANALYSIS

This chapter is a review of techniques for representing signals of various types as summations of sinusoids whose frequencies range from zero to infinity. The result is a way to represent signals and signal processing operations that is totally different from the time domain approach. We will refer to these representations as the *frequency domain*. The techniques are of great importance in communication system engineering for several reasons:

- Fourier representations give us a viewpoint other than that of the time domain on many issues, enabling us to draw conclusions that would be very difficult or impossible from a purely time domain approach. The frequency domain approach is a powerful theoretical tool.

- The locations of signals in the frequency domain can be easily shifted electronically. Signals that overlap in time but not in the frequency domain can be separated by filtering. As a result, it is possible for several users to transmit signals on the same channel at the same time without mutual interference. This technique, known as frequency division multiplexing (FDM), is used in many communication systems. The use of FDM relies on an understanding of Fourier theory.

- The frequency content of signals in radio communication is carefully regulated by government agencies. The communication engineer must use the frequency domain to understand and comply with these regulations.

2.1
SIGNAL CLASSIFICATION AND DEFINITIONS OF FUNCTIONS

Energy signals versus power signals

If the voltage across a 1Ω (ohm) resistor is denoted as $v(t)$ and the current as $i(t)$, the instantaneous power in watts (W) delivered to the resistor is given by

$$p(t) = \frac{[v(t)]^2}{1\Omega} = [i(t)]^2 \times 1\Omega \text{ W}$$

As a result, it is customary to refer to the square of a signal as its *power*. A more exact term is *normalized instantaneous power* in which *normalized* refers to the assumed resistance. We will mostly be concerned with normalized power in theoretical discussions of communication systems. If, at any time, we should want to find actual power or energy, we only need to divide by the actual resistance when dealing with a voltage signal, or to multiply by resistance when dealing with a current.

If a voltage or current signal is denoted $g(t)$, the total *normalized energy* (E) delivered over all time is, in Joules (J),

$$E = \int_{-\infty}^{\infty} g^2(t)\, dt \text{ J}$$

If this energy turns out to be a finite value, then $g(t)$ is referred to as an *energy signal*. On the other hand, if the energy is infinite, we can calculate the *average normalized power* delivered by the signal as

$$P_{\text{avg}} = \lim_{T \to \infty} \frac{1}{T} \int_{-T/2}^{T/2} [g(t)]^2\, dt \text{ W}$$

If this average power turns out to be greater than zero and less than infinity, then $g(t)$ is referred to as a *power signal*.

Periodic versus aperiodic signals

Periodic signals repeat a fixed sequence of values endlessly. A good example of a periodic signal is a *sinusoid*, which produces the same value each time its argument increases by 2π radians. Thus a periodic function $g_{\text{p}}(t)$ obeys the relation

$$g_{\text{p}}(t) = g_{\text{p}}(t + T) \qquad \text{for all } t$$

If T is selected to be the smallest value for which the signal repeats, then T is referred to as the *period* of the signal. The reciprocal of the period is the *fundamental frequency*, f_0:

$$f_0 = \frac{1}{T} \text{ Hz}$$

We will classify all signals that are not periodic as *aperiodic*.

Deterministic versus random signals

Deterministic signals are those that are completely specified at every point in time, such as a sine wave of known frequency, amplitude, and phase. We can describe a deterministic signal with an equation or a graph that gives its value at every point in time.

A *random signal* is one that requires the use of probability theory in its description. An example is a sine wave whose amplitude is determined by the toss of a coin: five volts for a head and ten volts for a tail. In this case it is not possible to write an equation or draw a graph of the values the signal will take on. Instead, we must discuss the probability that the signal will have a certain value.

Random signals are very important in communication engineering because the signals we want to communicate are almost always of a random nature. In addition, signals are often received in the presence of random noise and interference. Random signals will be discussed in detail in later chapters. We now define some of the deterministic signals we will use.

Unit step function

The unit step function is useful in representing signals that turn on at some time. It is defined as

$$u(t) = 1 \qquad \text{for } t \geq 0$$
$$= 0 \qquad t < 0 \qquad (2.1)$$

A graph of $u(t)$ versus time is shown in Figure 2.1.

Unit impulse function

The unit impulse function, also known as the *Dirac delta function*, is used to represent a concentration of area at $t = 0$. The impulse can be best defined

FIGURE 2.1
The unit step function.

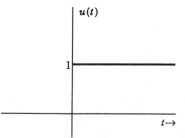

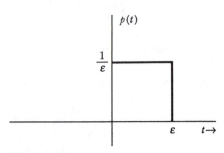

FIGURE 2.2
A unit area pulse $p(t)$ that
becomes a unit impulse function
$\delta(t)$ as ϵ approaches zero.

in terms of its properties:

■
$$\delta(t) = 0 \qquad \text{for all } t \text{ except } t = 0 \tag{2.2}$$

■
$$\int_{-\infty}^{\infty} \delta(t)\, dt = 1 \tag{2.3}$$

   The impulse function can be considered the limiting case of the pulse $p(t)$ shown in Figure 2.2 as ϵ approaches zero. In the limit, the pulse is zero for all t except $t = 0$, and the area under $p(t)$ is 1 for all values of ϵ as ϵ approaches zero. Some additional properties of $\delta(t)$ that can be derived from the defining properties given in Equations 2.2 and 2.3, are

■
$$g(t)\delta(t - t_0) = g(t_0)\delta(t - t_0) \tag{2.4}$$

■
$$\int_{a}^{b} g(t)\delta(t - t_0)\, dt = g(t_0) \qquad \text{if } a < t_0 < b$$
$$= 0 \qquad \text{if } t_0 < a \text{ or if } t_0 > b \tag{2.5}$$

■
$$\delta(at) = \frac{1}{|a|}\delta(t) \tag{2.6}$$

■
$$\int_{-\infty}^{t} \delta(t)\, dt = u(t) \tag{2.7}$$

Rectangular pulse

We define the rectangular pulse function as

■
$$\text{rect }(t) = 1 \qquad \text{for } -\frac{1}{2} < t < \frac{1}{2}$$
$$= 0 \qquad \text{otherwise} \tag{2.8}$$

A graph of rect (t) is shown in Figure 2.3.

16

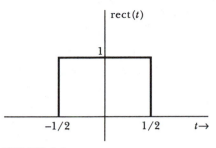

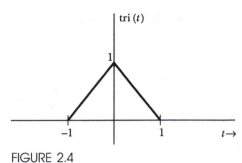

FIGURE 2.3
The rectangular pulse.

FIGURE 2.4
The triangular pulse.

Triangular pulse

The triangular pulse is defined as

$$
\begin{aligned}
\text{tri}\,(t) &= 1 - |t| \qquad \text{for } -1 < t < +1 \\
&= 0 \qquad\qquad \text{otherwise}
\end{aligned}
\tag{2.9}
$$

A graph of tri (t) is shown in Figure 2.4.

Sinc pulse

The sinc pulse is defined as

$$
\text{sinc}\,(t) = \frac{\sin\,(\pi t)}{\pi t}
\tag{2.10}
$$

The definition of the sinc pulse is an indeterminate form for $t = 0$. However, the limit as t approaches zero can be found to be unity by the use of L'Hospital's rule. A graph of the sinc function is shown in Figure 2.5. Some interesting

FIGURE 2.5
The sinc pulse.

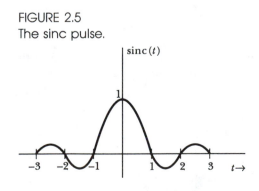

properties of the sinc function are

$$\int_{-\infty}^{\infty} \text{sinc} \ (t) \ dt = 1 \tag{2.11}$$

$$\int_{-\infty}^{\infty} [\text{sinc} \ (t)]^2 \ dt = 1 \tag{2.12}$$

$$\lim_{\epsilon \to 0} \frac{1}{\epsilon} \text{sinc} \left(\frac{t}{\epsilon} \right) = \delta(t) \tag{2.13}$$

$$\int_{-\infty}^{\infty} \text{sinc} \ (t - k) \ \text{sinc} \ (t - n) \ dt = 1 \qquad \text{if } n = k$$
$$= 0 \qquad \text{if } n \text{ is not equal to } k \tag{2.14}$$

where n and k are integers.

2.2
FOURIER SERIES

The exponential Fourier series

The *Fourier series* is a frequency domain technique that is applicable to periodic signals such as square waves and pulse trains. We will assume that the signals are voltages because that is usually the case. In addition, the designation of units is helpful in understanding. The Fourier series for a periodic voltage function $g_p(t)$ is

■
$$g_p(t) = \sum_{n=-\infty}^{\infty} \alpha_n \exp \ (jn2\pi f_0 t) \ \text{V} \tag{2.15}$$

where $f_0 = 1/T$ is the *fundamental frequency* of $g_p(t)$, T is the *period* of $g_p(t)$, and the α_n's are the exponential Fourier series *coefficients* of $g_p(t)$ given by

■
$$\alpha_n = \frac{1}{T} \int_{t_1}^{t_1 + T} g_p(t) \exp \ (-jn2\pi f_0 t) \ dt \ \text{V} \tag{2.16}$$

where the value of t_1 is arbitrary. We will pick t_1 as the value that makes evaluation of the integral most convenient. Note from Equation 2.16 that if $g_p(t)$ is a real-valued function, then

■
$$\alpha_n = \alpha^*_{-n}$$

where ∗ denotes the complex conjugate.

EXAMPLE 2.1

Given that all of the Fourier coefficients of a real function of time are zero except the pair

$$\alpha_k = \alpha^*_{-k} = \frac{C}{2} \exp{(j\phi)}$$

find the corresponding function of time.

SOLUTION

The time function can be found by substituting the values of the coefficients into Equation 2.15. Since all of the coefficients except two are zero, only two terms result, which are given by

$$g_\mathrm{p}(t) = \alpha_k \exp{(j2\pi k f_0 t)} + \alpha_{-k} \exp{(-j2\pi k f_0 t)}$$

Substituting the given values of α_k and α_{-k}, we have

$$g_\mathrm{p}(t) = \frac{C}{2} \exp{(j\phi)} \exp{(j2\pi k f_0 t)} + \frac{C}{2} \exp{(-j\phi)} \exp{(-j2\pi k f_0 t)}$$

which can be combined, using Euler's identity (see Appendix), to obtain

$$g_\mathrm{p}(t) = C \cos{(2\pi k f_0 t + \phi)}$$

From this example, we can see that the Fourier series can also be viewed as a means of representing the periodic function as a sum of sinusoidal terms whose frequencies are multiples of the fundamental frequency f_0. Each exponential term that has a positive index combines with the corresponding negative index term to produce a sinusoidal result. These are called *harmonic components*.

We can also see that the coefficients for positive values of k give the amplitude and phase of the sinusoidal components of the original periodic waveform. The magnitude of the kth coefficient α_k is one-half of the peak amplitude of the kth harmonic, and the angle of the coefficient is the same as the phase angle of the kth harmonic. In steady-state analyses of circuits with sinusoidal sources, phasors are used to represent the sinusoidal currents and voltages. The *phasor* for a sinusoid is a complex number whose magnitude is the peak amplitude of the sinusoid and whose angle is the phase angle of the sinusoid (provided that the sinusoid is written as a cosine function as in Example 2.1). Thus, the exponential Fourier coefficient α_k for positive values of k is one-half of the phasor for the kth harmonic.

The $n = 0$ term of the Fourier series (Equation 2.15) is simply a constant since the exponential is unity when $n = 0$. Thus, α_0 represents the *dc component* or *average value* of the original periodic waveform.

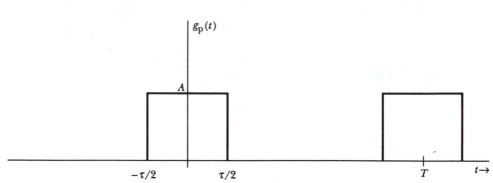

FIGURE 2.6
Periodic pulse train analyzed in Example 2.2.

EXAMPLE 2.2

Find the Fourier series coefficients of the periodic pulse train shown in Figure 2.6.

SOLUTION

The Fourier coefficients are given by Equation 2.16. For convenience we choose t_1 to equal $-T/2$ resulting in

$$\alpha_n = \frac{1}{T} \int_{-T/2}^{T/2} g_p(t) \exp\left(-j2\pi n f_0 t\right) dt$$

Substituting $g_p(t) = A$ when $-\tau/2 < t < \tau/2$, we have

$$\alpha_n = \frac{1}{T} \int_{-\tau/2}^{\tau/2} A \exp\left(-j2\pi n f_0 t\right) dt$$

Integrating, we obtain

$$\alpha_n = \frac{A \exp\left(-j2\pi n f_0 t\right)}{T(-j2\pi n f_0)} \Bigg|_{-\tau/2}^{\tau/2}$$

Evaluating and using Euler's identity for the sine function produces

$$\alpha_n = \frac{A}{n\pi} \sin\left(\frac{n\pi\tau}{T}\right)$$

Now the use of the definition of the sinc function from Equation 2.10 gives

$$\alpha_n = \frac{\tau A}{T} \operatorname{sinc}\left(\frac{n\tau}{T}\right)$$

20

Finally, the Fourier series for the pulse train can be written

$$g_p(t) = \sum_{n=-\infty}^{\infty} \frac{\tau A}{T} \operatorname{sinc}\left(\frac{n\tau}{T}\right) \exp\left(j2\pi n f_0 t\right)$$

Line spectra

The amplitude of the various harmonics contained in a periodic signal can be easily seen if the magnitudes of the Fourier coefficients are plotted versus their corresponding frequency. The resulting plot is called an *amplitude line spectrum*. An example of a line spectrum for the pulse train of Example 2.2 is shown in Figure 2.7a for the special case in which the period is $T_a = 4\tau$. This amplitude spectrum shows that the dc component, or $|\alpha_0|$, is equal to $A/4$. This is exactly what we would expect since the pulse waveform has an amplitude of A for 25% of the time and an amplitude of zero for all other times.

FIGURE 2.7
Amplitude line spectra of the pulse train of Figure 2.6 for two periods.

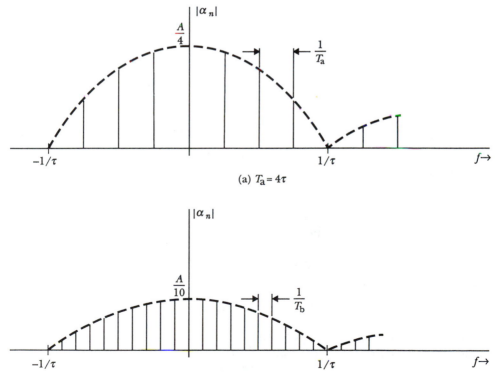

(a) $T_a = 4\tau$

(b) $T_b = 10\tau$

We also see from the line spectrum in Figure 2.7a that the pulse train has no fourth harmonic. This is because α_4 is zero when $T_a = 4\tau$.

Note, by comparing Figure 2.7a with Figure 2.7b, that the shape of the outline of the line spectrum is determined by the pulse width τ and does not depend on the period. However, the spacing between lines is determined by the period, T_a or T_b.

A line spectrum of the phases of the various components of a periodic waveform can be produced by plotting the phase of each coefficient α_k versus its corresponding frequency. In general, we will find less use for the *phase spectrum* than for the amplitude spectrum.

Parseval's theorem

The power in a waveform can be computed from the Fourier coefficients by the use of *Parseval's theorem*, which is given by

$$\blacksquare \qquad P_{\text{avg}} = \frac{1}{T} \int_{t_1}^{t_1 + T} [g_p(t)]^2 \, dt = \sum_{n=-\infty}^{\infty} |\alpha_n|^2 \text{ W} \qquad (2.17)$$

In the summation, $|\alpha_0|^2$ can be interpreted as the power associated with the dc component of the periodic waveform. The sum $|\alpha_{-k}|^2 + |\alpha_k|^2$ is the power associated with the kth component.

EXAMPLE 2.3

Find the power in the dc component, in the fundamental, and in all of the components above 250 Hz for the periodic pulse waveform of Example 2.2 if $T = 10$ milliseconds(ms), $A = 10$ V, and $\tau = T/4 = 2.5$ ms.

SOLUTION
The fundamental frequency is $f_0 = 1/T = 100$ Hz. An expression for the exponential Fourier series coefficients was found in Example 2.2. Substituting the values for T and τ we have

$$\alpha_n = 2.5 \text{ sinc } \frac{n}{4}$$

The power in the dc component is

$$P_{\text{dc}} = |\alpha_0|^2 = 6.25 \text{ W}$$

The power in the fundamental is

$$P_1 = |\alpha_1|^2 + |\alpha_{-1}|^2 = 2|\alpha_1|^2 = 10.13 \text{ W}$$

The total power associated with the waveform is the sum of the dc power and the power in each of the harmonics:

$$P_{\text{avg}} = P_{\text{dc}} + P_1 + P_2 + P_3 + \cdots$$

Thus the power above 250 Hz is given by

$$P_h = P_{avg} - P_{dc} - P_1 - P_2$$

The total average power can be found directly as

$$P_{avg} = \frac{1}{T} \int_{-T/2}^{T/2} [g_p(t)]^2 \, dt = 25 \text{ W}$$

and the power in the second harmonic is

$$P_2 = 2|\alpha_2|^2 = 5.07 \text{ W}$$

Thus, by substitution in the above relation for the power above 250 Hz, we obtain

$$P_h = 3.55 \text{ W}$$

2.3
FOURIER TRANSFORM

The transform and its inverse

The Fourier transform is a frequency domain technique that can be applied to most deterministic signals of interest in communication systems. The *Fourier transform* of a time function $g(t)$ is defined by

$$\blacksquare \qquad G(f) = \int_{-\infty}^{\infty} g(t) \exp{(-j2\pi ft)} \, dt \text{ V/Hz} \qquad (2.18)$$

Since we assume that the units of $g(t)$ are volts, the units of $G(f)$ are volts times seconds (V·s) or, equivalently, V/Hz.

The original time function $g(t)$ can be obtained from $G(f)$ by the *inverse Fourier transform*,

$$\blacksquare \qquad g(t) = \int_{-\infty}^{\infty} G(f) \exp{(j2\pi ft)} \, df \text{ V} \qquad (2.19)$$

The functions $g(t)$ and $G(f)$ are a Fourier transform pair. We will indicate this relationship by the notation

$$g(t) \leftrightarrow G(f)$$

We will also use the notation

$$G(f) = \text{F}[g(t)] \qquad \text{and} \qquad g(t) = \text{F}^{-1}[G(f)]$$

to indicate the Fourier transform operation and its inverse.

Note from the definition of the Fourier transform in Equation 2.18 that if $g(t)$ is a real-valued function, we have

$$G(-f) = G^*(f) \qquad (2.20)$$

As usual, the symbol $*$ denotes the complex conjugate.

EXAMPLE 2.4

Find the time function corresponding to the Fourier transform:

$$G(f) = \frac{A}{2} \exp{(j\theta)}\delta(f - f_0) + \frac{A}{2} \exp{(-j\theta)}\delta(f + f_0)$$

SOLUTION

The time function can be found by substituting the expression given for $G(f)$ into the inverse transform given in Equation 2.19. This results in

$$g(t) = \int_{-\infty}^{\infty} \left[\frac{A}{2} \exp{(j\theta)}\delta(f - f_0) + \frac{A}{2} \exp{(-j\theta)}\delta(f + f_0)\right] \exp{(j2\pi ft)}\, df$$

Using the property of Equation 2.5, we obtain

$$g(t) = \frac{A}{2} \exp{(j\theta)} \exp{(j2\pi f_0 t)} + \frac{A}{2} \exp{(-j\theta)} \exp{(-j2\pi f_0 t)}$$

which can be combined, resulting in

$$g(t) = A \cos{(2\pi f_0 t + \theta)}$$

We can see from this example that, when the Fourier transform is zero everywhere except at the frequencies f_0 and $-f_0$, the corresponding time function is a sinusoid with a frequency of f_0. Also note that the amplitude of the sinusoid is twice the area of the impulse at f_0 and the phase angle is the same as the angle of $G(f)$ at f_0. Since the units of $G(f)$ are V/Hz, we can expect an impulse to appear in the frequency domain whenever the time domain contains a noninfinitesimal sinusoid, as illustrated in Example 2.4.

EXAMPLE 2.5

Find the Fourier transform of the pulse shown in Figure 2.8a which is given by

$$g(t) = A \text{ rect}\left(\frac{t}{\tau}\right)$$

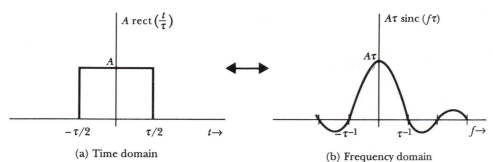

(a) Time domain   (b) Frequency domain

FIGURE 2.8
A rectangular pulse and its Fourier transform.

SOLUTION

The transform can be found by substituting for $g(t)$ in the definition given by Equation 2.18 resulting in

$$G(f) = \int_{-\tau/2}^{\tau/2} A \exp\left(-j2\pi ft\right) dt$$

Integrating, evaluating, and using Euler's formula for the sine function produces

$$G(f) = A\tau \frac{\sin\left(\pi f\tau\right)}{\pi f\tau}$$

$$= A\tau \operatorname{sinc}\left(f\tau\right)$$

This is shown in Figure 2.8b.

In this example, the Fourier transform does not contain impulses. Instead, the transform takes on finite values over a wide range of frequencies. This indicates that the single-pulse waveform is composed of an infinite sum of sinusoids of infinitesimal amplitude.

This conclusion can also be obtained by reexamining the result of Example 2.2. First notice from Figure 2.6 that, as the period T approaches infinity, the periodic pulse train approaches the aperiodic pulse of Figure 2.8a. The line spectrum of the periodic pulse train is shown in Figure 2.7. In Figure 2.7 we can see that, as the period of the pulse train is increased, the lines in its line spectrum become smaller and closer together. In the limit, as the period approaches infinity, the lines approach infinitesimal amplitude and become so close together that there is a component at every frequency. The Fourier transform relations of Equations 2.18 and 2.19 can be derived from the Fourier series by taking the limit as the period approaches infinity.

Amplitude and phase spectra

In Example 2.5, the Fourier transform turned out to be a real-valued function. As a result, it was possible to plot the transform $G(f)$ versus frequency. In general, this is not the case. Instead, the transform may be a complex-valued function. Then if a graphical representation is desired, we plot the magnitude of the transform. This plot of $|G(f)|$ versus f is called the *amplitude spectrum* of the signal since it indicates the relative amplitude of the frequency content of the signal. The *phase spectrum* of a signal is a plot of the angle of $G(f)$ versus f. The phase spectrum portrays the phase angles of the frequency components of the signal.

Fourier transforms of periodic signals

To obtain the Fourier transform of a periodic signal, we will use the following transform pair:

$$\exp(j2\pi f_0 t) \leftrightarrow \delta(f - f_0)$$

It can be demonstrated that this is a valid transform pair by substituting the frequency domain impulse into the inverse Fourier transform given in Equation 2.19. This results in the expression

$$\int_{-\infty}^{\infty} \delta(f - f_0) \exp(j2\pi f t)\, df$$

Use of the integral property of the impulse given in Equation 2.5 results in

$$\int_{-\infty}^{\infty} \delta(f - f_0) \exp(j2\pi f t)\, df = \exp(j2\pi f_0 t)$$

This result can be used to obtain the Fourier transform of a periodic function $g_p(t)$. First, an exponential Fourier series can be found for the function, then the series can be transformed term by term to obtain the pair

$$\blacksquare \quad g_p(t) = \sum_{n=-\infty}^{\infty} \alpha_n \exp(j2\pi n f_0 t) \leftrightarrow G_p(f) = \sum_{n=-\infty}^{\infty} \alpha_n \delta(f - n f_0) \quad (2.21)$$

This reiterates the fact that a periodic signal consists of discrete sinusoidal components that appear in the Fourier transform as impulses at the frequency (as well as the negative frequency) of each harmonic.

Properties of the transform

A number of important properties of the Fourier transform are listed in Table 2.1, and a list of useful transform pairs is given in Table 2.2. We will now discuss and prove some of the properties. For example, the *time delay property* can be proved as follows. Substituting into the definition of the Fourier transform

TABLE 2.1
Properties of the Fourier transform

Property	Time domain	Frequency domain
Linearity	$a_1 g_1(t) + a_2 g_2(t)$	$a_1 G_1(f) + a_2 G_2(f)$
Time shift	$g(t - t_0)$	$G(f) \exp(-j2\pi f t_0)$
Scaling	$g(at)$	$\dfrac{1}{|a|} G(f/a)$
Duality	$G(t)$	$g(-f)$
Modulation	$g(t) \exp(j2\pi f_0 t)$ $g(t) \cos(2\pi f_0 t + \theta)$	$G(f - f_0)$ $\frac{1}{2} \exp(j\theta) G(f - f_0)$ $\quad + \frac{1}{2} \exp(-j\theta) G(f + f_0)$
Differentiation	$\dfrac{dg(t)}{dt}$	$(j2\pi f) G(f)$
Integration	$\displaystyle \int_{-\infty}^{t} g(x)\, dx$	$\dfrac{G(f)}{(j2\pi f)} + \frac{1}{2} G(0)\delta(f)$
Convolution	$g_1(t) \otimes g_2(t)$ $= \displaystyle\int_{-\infty}^{\infty} g_1(x) g_2(t - x)\, dx$ $= \displaystyle\int_{-\infty}^{\infty} g_2(x) g_1(t - x)\, dx$	$G_1(f) G_2(f)$
Parseval's theorem	$E = \displaystyle\int_{-\infty}^{\infty} g^2(t)\, dt$	$E = \displaystyle\int_{-\infty}^{\infty} |G(f)|^2\, df$
Multiplication	$g_1(t) g_2(t)$	$G_1(f) \otimes G_2(f)$ $= \displaystyle\int_{-\infty}^{\infty} G_1(\lambda) G_2(f - \lambda)\, d\lambda$ $= \displaystyle\int_{-\infty}^{\infty} G_2(\lambda) G_1(f - \lambda)\, d\lambda$
Multiplication by t^n	$t^n g(t)$	$\left(\dfrac{j}{2\pi}\right)^n \dfrac{d^n}{df^n} [G(f)]$

given in Equation 2.18 gives

$$g(t - t_0) \leftrightarrow \int_{-\infty}^{\infty} g(t - t_0) \exp(-j2\pi ft)\, dt$$

Now let $t' = t - t_0$ on the right side of the last expression to obtain

$$g(t - t_0) \leftrightarrow \int_{-\infty}^{\infty} g(t') \exp[-j2\pi f(t' + t_0)]\, dt'$$

Factoring the exponential and moving the part that does not contain t' outside the integral, we have

$$g(t - t_0) \leftrightarrow \exp(-j2\pi f t_0) \int_{-\infty}^{\infty} g(t') \exp(-j2\pi f t')\, dt'$$

TABLE 2.2
Selected Fourier transform pairs

Time domain	Frequency domain
$A \operatorname{rect}\left(\dfrac{t}{\tau}\right)$	$A\tau \operatorname{sinc}(f\tau)$

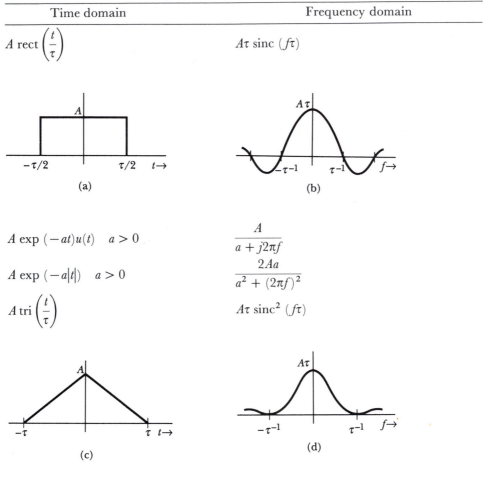

(a)   (b)

Time domain	Frequency domain
$A \exp(-at)u(t) \quad a > 0$	$\dfrac{A}{a + j2\pi f}$
$A \exp(-a|t|) \quad a > 0$	$\dfrac{2Aa}{a^2 + (2\pi f)^2}$
$A \operatorname{tri}\left(\dfrac{t}{\tau}\right)$	$A\tau \operatorname{sinc}^2(f\tau)$

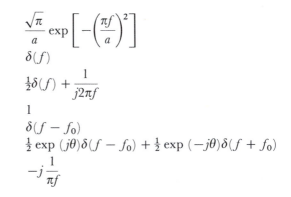

(c)   (d)

Time domain	Frequency domain
$\exp(-a^2 t^2)$	$\dfrac{\sqrt{\pi}}{a} \exp\left[-\left(\dfrac{\pi f}{a}\right)^2\right]$
1	$\delta(f)$
$u(t)$	$\tfrac{1}{2}\delta(f) + \dfrac{1}{j2\pi f}$
$\delta(t)$	1
$\exp(j2\pi f_0 t)$	$\delta(f - f_0)$
$\cos(2\pi f_0 t + \theta)$	$\tfrac{1}{2}\exp(j\theta)\delta(f - f_0) + \tfrac{1}{2}\exp(-j\theta)\delta(f + f_0)$
$\operatorname{sgn}(t) = -1 \quad \text{for } t < 0$ $\qquad\quad\, = +1 \quad \text{for } t > 0$	$-j\dfrac{1}{\pi f}$

Now the integral is the Fourier transform of $g(t)$, so we have the desired result:

∎
$$g(t - t_0) \leftrightarrow \exp(-j2\pi f t_0) G(f)$$

Notice that since

$$\left| \exp(-j2\pi f t_0) G(f) \right| = \left| \exp(-j2\pi f t_0) \right| \left| G(f) \right| = \left| G(f) \right|$$

the amplitudes of the frequency components of $g(t)$ are not affected by a time delay. However, the phase angles of the components are affected since the angle of the transform is changed because of the time delay. We would expect this to be the case since the amplitude of a sinusoidal signal depends on the waveshape rather than the arbitrary choice of the point where $t = 0$. On the other hand, the phase spectrum depends on both the waveshape and the choice of $t = 0$.

The *scaling property*,

∎
$$g(at) \leftrightarrow \frac{1}{|a|} G\left(\frac{f}{a}\right)$$

is very significant to electrical engineers since it shows that if a signal is made to vary more rapidly in the time domain (by choosing a large value for the constant a), the inescapable consequence is that the corresponding frequency domain amplitude spectrum becomes spread over a wider range of frequencies. We are always trying to communicate at higher rates, construct computers to run at higher speeds, and so on. The implication of the scaling property is that these goals will always require circuits capable of transmitting ever wider bandwidths.

Figure 2.9 shows the amplitude spectrums of a wide and a narrow pulse, which illustrates the scaling property by example.

The *duality property* is useful in quickly building a table of Fourier transforms since each transform pair can be used to generate another pair.

EXAMPLE 2.6

Use the duality principle to find the Fourier transform of $A \operatorname{sinc}(t/\tau_1)$.

SOLUTION
From Example 2.5 we have

$$A \operatorname{rect}\left(\frac{t}{\tau}\right) \leftrightarrow A\tau \operatorname{sinc}(f\tau)$$

The duality principle states

$$G(t) \leftrightarrow g(-f)$$

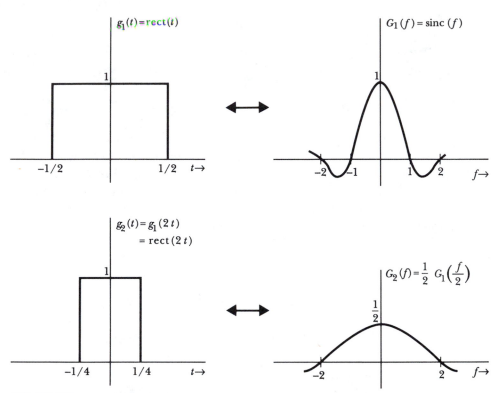

FIGURE 2.9
Example of the scaling property showing that narrower pulses in the time domain result in wider bandwidths in the frequency domain.

which, when applied to the result from Example 2.5, gives

$$A\tau \, \text{sinc} \, (t\tau) \leftrightarrow A \, \text{rect} \left(\frac{-f}{\tau} \right) = A \, \text{rect} \left(\frac{f}{\tau} \right)$$

Substituting $\tau = 1/\tau_1$ and multiplying both sides by τ_1 results in

$$A \, \text{sinc} \left(\frac{t}{\tau_1} \right) \leftrightarrow A\tau_1 \, \text{rect} \, (f\tau_1)$$

The *differentiation property* as given in Table 2.1 applies only if the time function and its derivative are Fourier transformable. The differentiation property can be useful in simplifying the integration process when finding transforms of signals that consist of straight-line segments.

EXAMPLE 2.7

Use the differentiation property to find the Fourier transform of $g(t) = A \text{ tri } (t/\tau)$.

SOLUTION

First differentiate $g(t)$ two times to obtain a function consisting of impulse functions as illustrated in Figure 2.10. The result of the double differentiation is given by

$$g''(t) = \frac{A}{\tau} \delta(t + \tau) - \frac{2A}{\tau} \delta(t) + \frac{A}{\tau} \delta(t - \tau)$$

Taking the Fourier transform of this we obtain

$$g''(t) \leftrightarrow \frac{A}{\tau} [\exp (j2\pi f\tau) - 2 + \exp (-j2\pi f\tau)]$$

From the differentiation property we have

$$g''(t) \leftrightarrow (j2\pi f)^2 G(f)$$

Equating the right-hand sides of the previous two expressions yields

$$(j2\pi f)^2 G(f) = \frac{A}{\tau} [\exp (j2\pi f\tau) - 2 + \exp (-j2\pi f\tau)]$$

$$= \frac{2A}{\tau} [\cos (2\pi f\tau) - 1]$$

$$= -\frac{4A}{\tau} \sin^2 (\pi f\tau)$$

FIGURE 2.10
The triangular pulse and its first two derivatives.

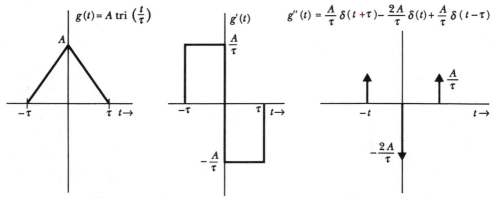

Dividing both sides by $(j2\pi f\tau)^2$ we obtain

$$G(f) = A\tau \frac{\sin^2(\pi f\tau)}{(\pi f\tau)^2}$$

$$G(f) = A\tau \operatorname{sinc}^2(f\tau)$$

Convolution

Convolution is an important concept in electrical engineering because the output signal of a linear system is the convolution of the input signal and the impulse response of the system. Since many of the subsystems in communications are linear, we can often obtain insight into various interrelationships in a system from consideration of the convolution process. It is possibly more important to be able to visualize the process than it is to calculate a convolution. However, it is necessary to go through the process of finding a convolution in detail to gain the understanding required for visualization.

EXAMPLE 2.8

Find the result of convolving the signals shown in Figure 2.11.

SOLUTION
The desired result can be obtained by substituting into the definition for convolution

$$g(t) \otimes h(t) = \int_{-\infty}^{\infty} h(x)g(t-x)\, dx = \int_{-\infty}^{\infty} h(t-x)g(x)\, dx$$

Either integral on the right-hand side of this expression can be used to evaluate the convolution. We use the first integral on the right-hand side for this example. Note that x is a "dummy variable" in the integrals and does not appear in the result. The functions of x in the integrand of the integral are shown for several values of t in Figure 2.12. As indicated in the figure, $g(t-x)$

FIGURE 2.11
Functions that are convolved in Example 2.8.

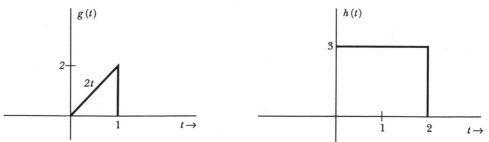

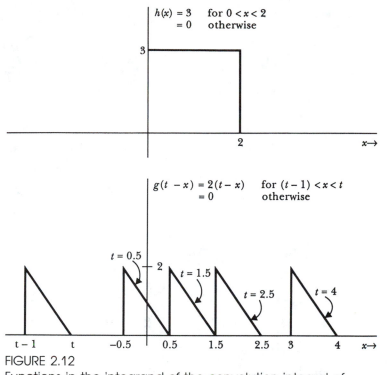

$$h(x) = 3 \quad \text{for } 0 < x < 2$$
$$ = 0 \quad \text{otherwise}$$

3

2

$x \rightarrow$

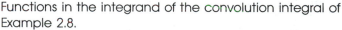

$$g(t - x) = 2(t - x) \quad \text{for } (t-1) < x < t$$
$$ = 0 \quad \text{otherwise}$$

$t = 0.5$

2

$t = 1.5$

$t = 2.5$

$t = 4$

$t - 1 \qquad t \qquad -0.5 \qquad 0.5 \qquad 1.5 \qquad 2.5 \quad 3 \qquad 4 \quad x \rightarrow$

FIGURE 2.12
Functions in the integrand of the convolution integral of
Example 2.8.

slides along the x axis as the value of t varies. Also, note that when $t = 0$, we have $g(-x)$, which is simply a time-reversed version of the original pulse. When t is negative there is no overlap of the nonzero portions of the two functions $g(t - x)$ and $h(x)$. Therefore, the integrand is zero, and we conclude that

$$g(t) \otimes h(t) = 0 \qquad \text{for } t < 0$$

When t is greater than zero but less than one, the right-hand portion of the $g(t - x)$ pulse overlaps the left-hand portion of the $h(x)$ pulse, and we have

$$g(t) \otimes h(t) = \int_0^t 6(t - x)\, dx \qquad \text{for } 0 < t < 1$$

$$= -3(t - x)^2 \Big|_{x=0}^{x=t}$$

$$= 3t^2 \qquad \text{for } 0 < t < 1$$

When t is between 1 and 2, the $g(t - x)$ pulse is totally overlapped by the $h(x)$ pulse, and we have

$$g(t) \otimes h(t) = \int_{t-1}^t 6(t - x)\, dx$$

$$= 3 \qquad \text{for } 1 < t < 2$$

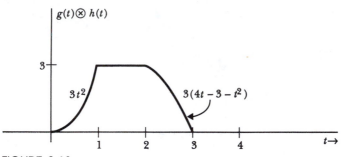

$g(t) \otimes h(t)$

$3t^2$

$3(4t - 3 - t^2)$

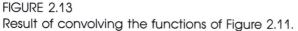

FIGURE 2.13
Result of convolving the functions of Figure 2.11.

Similarly, we have

$$g(t) \otimes h(t) = \int_{t-1}^{2} 6(t - x) \, dx$$

$$g(t) \otimes h(t) = -3(t^2 - 4t + 3) \qquad \text{for } 2 < t < 3$$

and

$$g(t) \otimes h(t) = 0 \qquad \text{for } 3 < t$$

The result of this convolution is shown in Figure 2.13.

You may wish to repeat the convolution of Example 2.8 by use of the integral

$$g(t) \otimes h(t) = \int_{-\infty}^{\infty} g(x)h(t - x) \, dx$$

In other words, compute the convolution by time reversing and shifting the $h(t)$ pulse; the result obtained should be the same as in Example 2.8.

EXAMPLE 2.9

Find the result of convolving the $g(t)$ of Figure 2.11 with the function

$$w(t) = \delta(t) + 2\delta(t - 3)$$

SOLUTION
Substitute into the definition

$$g(t) \otimes w(t) = \int_{-\infty}^{\infty} g(x)w(t - x) \, dx$$

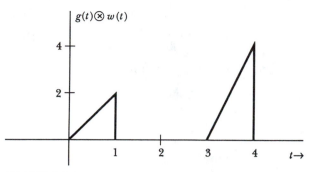

FIGURE 2.14
The result of Example 2.9.

to obtain

$$g(t) \otimes w(t) = \int_{-\infty}^{\infty} g(x)[\delta(t-x) + 2\delta(t-x-3)]\, dx$$

This can be integrated by use of the property of impulse functions given in Equation 2.5 to obtain

$$g(t) \otimes w(t) = g(t) + 2g(t-3)$$

This result is shown in Figure 2.14.

Notice from Example 2.9 that convolving a signal with an impulse results in a shifted and scaled version of the signal.

Parseval's energy theorem

If $g(t)$ is an energy signal, its total energy can be computed either in the time domain or in the frequency domain by use of *Parseval's energy theorem*, which is given by

$$E = \int_{-\infty}^{\infty} g^2(t)\, dt = \int_{-\infty}^{\infty} |G(f)|^2\, df \text{ J} \tag{2.22}$$

Note that (with the usual assumed 1Ω resistance) the units of $|G(f)|^2$ are J/Hz. $|G(f)|^2$ is called the *energy spectral density* of $g(t)$ and shows how the total energy of $g(t)$ is distributed in the frequency domain.

2.4
POWER SPECTRAL DENSITY AND AUTOCORRELATION

The energy spectral density discussed at the end of the previous section does not exist for power signals because the total energy associated with a power signal is infinite. To be able to characterize the frequency distribution of the power in a power signal, we will define the *power spectral density* (PSD). In this section, we will focus on deterministic signals, but we will see in later chapters that this concept can be extended to random signals.

First, a truncated version of the time signal is defined as

$$g_T(t) = g(t) \; \text{rect}\left(\frac{t}{T}\right) \tag{2.23}$$

A graph of a typical signal and its truncated version is shown in Figure 2.15. The Fourier transform of the truncated signal can be found as

$$G_T(f) = \text{F}[g_T(t)] \tag{2.24}$$

When the original signal $g(t)$ has finite power at all times, then the truncated signal $g_T(t)$ has finite total energy, and its energy spectral density can be found as the magnitude-squared of $G_T(f)$:

$$|G_T(f)|^2 \; \text{J/Hz}$$

If the energy delivered over a period of T seconds is divided by T, the average power results. Therefore, we define the PSD as the energy spectral density of the truncated signal divided by T. To obtain a result that is representative of the entire original signal, the limit is taken as T approaches infinity. Thus the PSD of $g(t)$ is

$$\blacksquare \qquad S_g(f) = \lim_{T \to \infty} \left[\frac{1}{T} |G_T(f)|^2 \right] \text{W/Hz} \tag{2.25}$$

If Parseval's energy theorem of Equation 2.22 is applied to the truncated signal $g_T(t)$, we have

$$\int_{-\infty}^{\infty} [g_T(t)]^2 \, dt = \int_{-\infty}^{\infty} |G_T(f)|^2 \, df$$

The integrand in the left integral can be changed to $g(t)$ if the limits are

FIGURE 2.15
An arbitrary signal $g(t)$ and its truncated version $g_T(t)$.

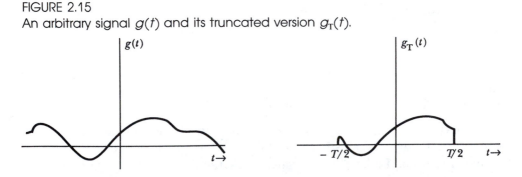

changed to range only over the nonzero part of $g_T(t)$. Then

$$\int_{-T/2}^{T/2} [g(t)]^2\, dt = \int_{-\infty}^{\infty} |G_T(f)|^2\, df$$

If both sides of this expression are divided by T and the limit is taken, we have

$$\lim_{T\to\infty} \frac{1}{T} \int_{-T/2}^{T/2} [g(t)]^2\, dt = \lim_{T\to\infty} \frac{1}{T} \int_{-\infty}^{\infty} |G_T(f)|^2\, df$$

The left side is the total power delivered by $g(t)$ averaged over all time. If the integration and limiting process are interchanged on the right, the integrand becomes the PSD of $g(t)$ as defined in Equation 2.25, and we have

$$\blacksquare \qquad P_{\text{avg}} = \lim_{T\to\infty} \frac{1}{T} \int_{-T/2}^{T/2} [g(t)]^2\, dt = \int_{-\infty}^{\infty} S_g(f)\, df \ \text{W} \qquad (2.26)$$

Relation 2.26 shows that the total average power of a signal can be obtained by integrating the PSD over all frequencies. This suggests that the PSD, $S_g(f)$, shows how the power in $g(t)$ is distributed over frequency. The units of the PSD, which are W/Hz, make the same suggestion. As we proceed we will find that this interpretation is correct; the PSD does show how the power of $g(t)$ is distributed over frequency.

The PSD of a signal can conceptually be computed by the use of Equation 2.25. However, there is an alternative method that is often more useful. This alternative, which we are about to discuss, is particularly important since it can be applied to random signals, whereas the definition of Equation 2.25 cannot. We begin our discussion by defining $R_g(\tau)$ as the inverse Fourier transform of the PSD of the signal $g(t)$:

$$R_g(\tau) = F^{-1}[S_g(f)] = F^{-1}\left[\lim_{T\to\infty} \frac{1}{T} |G_T(f)|^2\right] \qquad (2.27)$$

where τ plays the role of the time variable. Interchanging the limit and the inverse Fourier transform operations, we obtain

$$R_g(\tau) = \lim_{T\to\infty} \frac{1}{T} F^{-1}[|G_T(f)|^2]$$

$$= \lim_{T\to\infty} \frac{1}{T} F^{-1}[G_T(f)G_T^*(f)] \qquad (2.28)$$

Now recall that multiplying Fourier transforms corresponds to convolution of the time functions. Also, it can be easily shown that

$$g_T(-t) \leftrightarrow G_T^*(f)$$

Using these facts, Equation 2.28 becomes

$$R_g(\tau) = \lim_{T\to\infty} \frac{1}{T} [g_T(t) \otimes g_T(-t)]$$

Using the definition of convolution, the last equation can be shown to be the same as

- $$R_g(\tau) = \lim_{T \to \infty} \frac{1}{T} \int_{-T/2}^{T/2} g(t)g(t + \tau)\,dt \qquad \text{(aperiodic signals)} \qquad (2.29)$$

$R_g(\tau)$ is called the *autocorrelation function* of the signal $g(t)$. Notice the similarity of the integral in the autocorrelation to the convolution integral. In convolution, one function was time-reversed and then translated past the other function, and the integral of their product was computed. The autocorrelation integral is the same as convolution except that the time reversal is not performed. We now have a second method for finding the PSD of a signal. We first find the autocorrelation function using Equation 2.29 and then take its Fourier transform.

When $g(t)$ is periodic with period T, the limit in Equation 2.29 can be eliminated. It is only necessary to integrate over a single period and divide by the period. Therefore, the autocorrelation function for a periodic signal is given by

$$R_g(\tau) = \frac{1}{T} \int_{t_1}^{t_1 + T} g(t)g(t + \tau)\,dt \qquad \text{(periodic signals with period } T) \qquad (2.30)$$

As usual, the value of t_1 is picked for convenience. It can be easily shown that the autocorrelation function of a periodic function is also periodic with the same period.

EXAMPLE 2.10

Find the autocorrelation function and PSD of the periodic pulse train shown in Figure 2.16a. Compare the resulting PSD with the results of Example 2.2.

SOLUTION
The autocorrelation function can be found by substituting into Equation 2.30. If we select $t_1 = -T/2$, this results in

$$R_g(\tau) = \frac{1}{T} \int_{-T/2}^{T/2} g(t)g(t + \tau)\,dt$$

Now, if we substitute $g(t) = A$ and $g(t + \tau) = A$, we must change the limits of the integral to extend only over the portion of the interval in which both functions are nonzero. As indicated in Figure 2.16b, the resulting interval of integration depends on the value of τ. The results are given by

$$R_g(\tau) = \frac{1}{T} \int_{-T/8}^{(T/8) - \tau} A^2\,dt \qquad \text{for } 0 < \tau < \frac{T}{4}$$

$$= \frac{A^2}{4}\left(1 - \frac{4\tau}{T}\right)$$

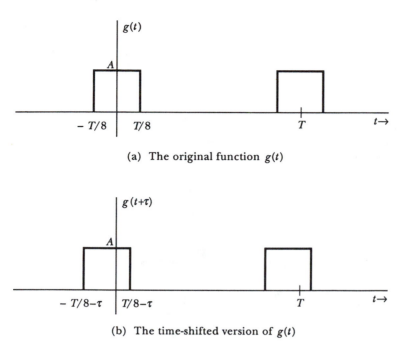

(a) The original function $g(t)$

(b) The time-shifted version of $g(t)$

FIGURE 2.16
Functions in the autocorrelation integral of Example 2.10.

and

$$R_g(\tau) = \frac{1}{T} \int_{-(T/8)-\tau}^{T/8} A^2 \, dt \qquad \text{for } -\frac{T}{4} < \tau < 0$$

$$= \frac{A^2}{4}\left(1 + \frac{4\tau}{T}\right)$$

For $(T/4) < \tau < (3T/4)$, the pulses of the shifted function $g(t + \tau)$ do not overlap the pulses of $g(t)$. Therefore, for τ in this range, the integrand of the autocorrelation integral is zero, and the autocorrelation function is zero. Since the autocorrelation function of a periodic function is itself periodic with the same period, we can construct the entire autocorrelation function from the values found above. The result is shown in Figure 2.17.

To obtain the PSD of the pulse train, we must find the Fourier transform of the autocorrelation function. Since the autocorrelation function is periodic, we will first find the Fourier series representation given by

$$R_g(\tau) = \sum_{n=-\infty}^{\infty} \beta_n \exp\left(jn2\pi f_0 \tau\right)$$

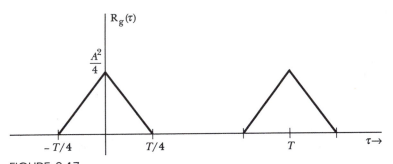

FIGURE 2.17
Autocorrelation function of the periodic pulse train $g(t)$ of Figure 2.16.

where the coefficients can be found by substituting into

$$\beta_n = \frac{1}{T} \int_{-T/2}^{T/2} R_g(\tau) \exp\left(-jn2\pi f_0\tau\right) d\tau$$

Now, if the exponential is written to display its real and imaginary parts, we have

$$\beta_n = \frac{1}{T} \int_{-T/2}^{T/2} R_g(\tau) [\cos\left(n2\pi f_0\tau\right) - j\sin\left(n2\pi f_0\tau\right)] d\tau$$

Note that $R_g(\tau)$ is an even function, the sine function is odd, and the product of an even function and an odd function is odd. Therefore, the sine term can be dropped since the integral of an odd function over a symmetrical interval is zero. The remaining integrand is even, and the limits can be changed to cover only the positive half of the interval if the result is doubled. This is because the integral over the negative half of a symmetrical interval is equal to the integral over the positive half for an even function. These changes result in

$$\beta_n = \frac{2}{T} \int_0^{T/2} R_g(\tau) \cos\left(n2\pi f_0\tau\right) d\tau$$

Substituting the expressions for $R_g(\tau)$ in the range of the integral results in

$$\beta_n = \frac{2}{T} \int_0^{T/4} \left(\frac{A^2}{4}\right)\left(1 - \frac{4\tau}{T}\right) \cos\left(n2\pi f_0\tau\right) d\tau$$

This can be integrated and evaluated by straightforward but tedious methods to produce

$$\beta_n = \frac{A^2}{16} \operatorname{sinc}^2 \frac{n}{4}$$

Using these coefficients, we obtain the Fourier series for $R_g(\tau)$ given by

$$R_g(\tau) = \sum_{n=-\infty}^{\infty} \frac{A^2}{16} \operatorname{sinc}^2 \frac{n}{4} \exp\left(jn2\pi f_0\tau\right)$$

Taking the Fourier transform of this term by term, we obtain the desired PSD, given by

$$S_g(f) = \sum_{n=-\infty}^{\infty} \frac{A^2}{16} \operatorname{sinc}^2 \frac{n}{4} \delta(f - nf_0)$$

A plot of $S_g(f)$ is shown in Figure 2.18. As we may have expected, the power of the periodic pulse train is concentrated at dc and at multiples of the fundamental frequency f_0. In fact, we can verify that the power at each frequency is correct by use of the Fourier series of the pulse train, which we found in Example 2.2. The Fourier series coefficients of the pulse train were found in Example 2.2 as (with $T = 4\tau$)

$$\alpha_n = \frac{A}{4} \operatorname{sinc} \frac{n}{4}$$

Recall that the coefficient α_n is half of the phasor for the nth harmonic component of the pulse train. Thus, the peak value of the nth harmonic is $2|\alpha_n|$, and the root mean square (rms) or effective value is the peak value divided by the square root of two. The resulting rms value of the nth harmonic is

$$V_n = 2^{1/2}|\alpha_n|$$

$$V_n = \frac{A\sqrt{2}}{4} \operatorname{sinc} \frac{n}{4}$$

FIGURE 2.18
Power spectral density for the pulse train of Figure 2.16.

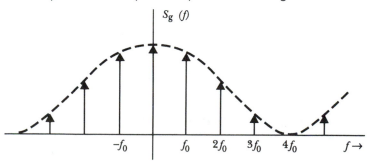

The power delivered to a 1Ω resistor by this voltage is given by

$$P_n = \frac{V_n^2}{1\Omega}$$

$$= \frac{A^2}{8} \mathrm{sinc}^2 \frac{n}{4}$$

The power in the nth harmonic appears in the PSD split equally between the impulses at nf_0 and $-nf_0$. It can be easily verified that this sum agrees with the power in the nth harmonic that we have just calculated directly from the Fourier coefficients of the pulse train.

The PSD of the periodic pulse train in Example 2.10 could have been determined without finding the autocorrelation function. In fact, the use of the autocorrelation function is an unnecessary complication of the theory when we are dealing with deterministic functions. However, when we need to deal with random signals, the autocorrelation function is indispensable. Autocorrelation functions can be found for random signals whereas Fourier series and transforms do not exist for many of the random signals of interest to us in later chapters. We will return to the subject of autocorrelation functions for random signals in Chapter 7.

Some important properties of autocorrelation functions are now given. The autocorrelation function is an even function

- $$\mathrm{R_g}(\tau) = \mathrm{R_g}(-\tau) \qquad \text{for all } \tau \tag{2.31}$$

The value of the autocorrelation at $\tau = 0$ is its largest magnitude

- $$\mathrm{R_g}(0) \geq \left| \mathrm{R_g}(\tau) \right| \qquad \text{for all } \tau \tag{2.32}$$

The value of the autocorrelation function at $\tau = 0$ is the total average power in the signal

- $$P_{\mathrm{avg}} = \lim_{T \to \infty} \frac{1}{T} \int_{-T/2}^{T/2} \left| g(t) \right|^2 dt = \mathrm{R_g}(0) \tag{2.33}$$

2.5
LINEAR SYSTEMS

System classification

A communication system often consists of an interconnection of many subsystems. The various subsystems can be placed into several classes that we will now describe. A subsystem is often shown as a block with an input signal $g(t)$ and an output signal $y(t)$, as indicated in Figure 2.19. We will indicate that

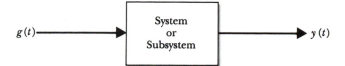

FIGURE 2.19
System block showing the input signal $g(t)$ and the output signal $y(t)$.

the input $g(t)$ gives rise to the output $y(t)$ by the notation

$$g(t) \rightarrow y(t)$$

If, for any two input signals $g_1(t)$ and $g_2(t)$, we have

$$g_1(t) \rightarrow y_1(t) \qquad \text{and} \qquad g_2(t) \rightarrow y_2(t)$$

and if a weighted sum of the inputs results in the same weighted sum of the outputs:

$$\blacksquare \qquad [a_1 g_1(t) + a_2 g_2(t)] \rightarrow [a_1 y_1(t) + a_2 y_2(t)] \qquad (2.34)$$

then the system is classified as being *linear*. Thus, we say that *superposition* holds for a linear system.

All systems that do not behave according to Equation 2.34 are classified as *nonlinear*. An example of a nonlinear system is one in which the output is the square of the input:

$$y(t) = g^2(t)$$

In a *time-invariant* system, a time shift of the input signal results in a corresponding time shift of the output signal. For a time-invariant system, if $g(t) \rightarrow y(t)$, then $g(t - t_0) \rightarrow y(t - t_0)$. Any system that does not obey this property is considered a time-varying system.

In a *causal* system, the impulse response is zero for all negative values of time. It is not physically possible to construct a noncausal system since such a system would have to anticipate the input signal. Nevertheless, we often find it convenient to ignore this fact in theoretical considerations.

Impulse response and transfer functions

Linear time-invariant systems can be completely characterized in terms of their *impulse response* $h(t)$. Provided that there is no energy initially stored in the system, the impulse response is the output of the system when the input is a unit impulse

$$\delta(t) \rightarrow h(t)$$

It can be shown that the output signal of a linear time-invariant system can be found for any input signal by convolving the input signal with the impulse response of the system.

$$\blacksquare \quad y(t) = g(t) \otimes h(t) = \int_{-\infty}^{\infty} g(x)h(t-x)\,dx = \int_{-\infty}^{\infty} g(t-x)h(x)\,dx \quad (2.35)$$

In the frequency domain, convolution corresponds to multiplication of the Fourier transforms of the input signal and the impulse response.

$$\blacksquare \qquad\qquad Y(f) = G(f)H(f) \qquad\qquad\qquad (2.36)$$

The transform of the impulse response $H(f)$ is called the *transfer function* of the system. At any given frequency, it is a complex number whose magnitude shows the amplitude gain of the system for a sinusoidal component of the input at that frequency. The angle of $H(f)$ at any frequency is the phase shift that the system produces for an input at that frequency. For example, if

$$H(f_1) = K_1\underline{/\theta_1}$$

then the steady-state output due to a sinusoidal input at the frequency f_1 is K_1 times as large as the input and shifted in phase by θ_1:

$$A\cos\left(2\pi f_1 t + \phi\right) \rightarrow K_1 A \cos\left(2\pi f_1 t + \phi + \theta_1\right)$$

The term *steady-state* means that we are interested in the output only after the input has been applied long enough for the decay of any transients that may have been caused by the sudden application of the input or by energy stored in the system initially.

When an energy signal is applied to a linear time-invariant system, the energy spectral density of the output signal is given by

$$\blacksquare \qquad |Y(f)|^2 = |H(f)G(f)|^2 = |H(f)|^2|G(f)|^2 \qquad (2.37)$$

If a power signal is applied to a linear time-invariant system, the PSD of the output signal is equal to the squared magnitude of the transfer function times the input PSD.

$$\blacksquare \qquad\qquad S_y(f) = |H(f)|^2 S_g(f) \qquad\qquad (2.38)$$

This is intuitively reasonable since $|H(f)|$ is the factor by which the amplitudes of the frequency components of the input signal are multiplied. When the amplitude of a signal is increased by a factor such as $|H(f)|$, the power level of the signal is increased by the square of that factor. This is exactly the meaning of Equation 2.38.

Distortionless transmission

The requirement for an ideal analog communication system is that the output signal should be a scaled and time-delayed version of the input signal. (Time delay is often unavoidable due to transportation lag.) In equation form, this system requirement is

$$y(t) = Kg(t - t_0) \tag{2.39}$$

Using the time delay property of Fourier transforms from Table 2.1, we find that the frequency domain expression corresponding to Equation 2.39 is

$$Y(f) = K \exp{(-j2\pi f t_0)}G(f) \tag{2.40}$$

Therefore, the ideal analog communication system is a linear time-invariant system with a transfer function given by

$$H(f) = K \exp{(-j2\pi f t_0)} = K\underline{/-2\pi f t_0} \tag{2.41}$$

As indicated in Figure 2.20 this means that the magnitude of the transfer function of an ideal system should be constant over the frequency range in which

FIGURE 2.20
The magnitude and phase of the overall transfer function of an ideal analog communication system versus frequency.

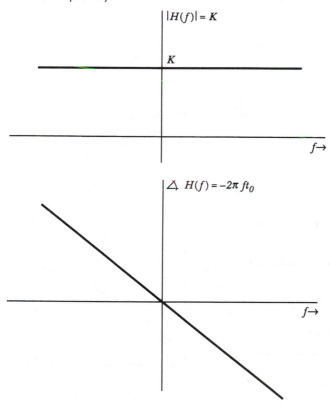

the input signal has nonzero frequency content. In addition, the phase response should be a linear function of frequency.

Relationships between rise time and bandwidth

When a step function is applied to a lowpass filter, the output of the filter eventually reaches a constant steady-state value. This is illustrated in Figure 2.21 for a simple lowpass resistance capacitance RC filter. The *rise time* of the transient response of the filter is usually defined as the time interval between the point at which the response reaches 10% of its final value and the point at which it reaches 90% of the final value. The rise time t_r of the RC filter response is indicated in Figure 2.21. It can be shown that the rise time of the simple RC lowpass filter is given by

$$t_r = RC \ln (9) \tag{2.42}$$

The half-power bandwidth of the lowpass RC filter is given by

$$B = \frac{1}{2\pi RC} \tag{2.43}$$

Combining Equations 2.42 and 2.43, we obtain the following relationship between the rise time and the half-power bandwidth.

$$\blacksquare \qquad t_r = \frac{\ln (9)}{2\pi B} \simeq \frac{0.350}{B} \qquad \text{(lowpass filters)} \tag{2.44}$$

Equation 2.44 is exact only for the simple lowpass RC filter but it is often used by communications engineers to estimate the rise time of other types of lowpass filters. For many filters, the estimate is quite close to the actual value.

FIGURE 2.21
Rise time for a simple RC lowpass filter.

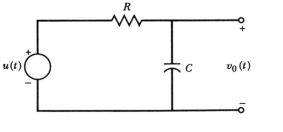
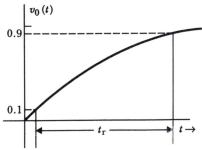

When a sinusoidal signal is suddenly applied to the input of a bandpass filter whose transfer function is centered at the frequency of the applied signal, the response eventually settles to a constant amplitude sinewave. Figure 2.22 shows the stepped sinusoidal input signal applied to a bandpass filter, the transfer function of the filter, and the resulting output transient. The rise time of the bandpass filter is the interval between the point at which the envelope of the response reaches 10% of the final value and the point at which the envelope is 90%. This is indicated in Figure 2.22. A convenient rule of thumb for estimating the rise time of bandpass filters in terms of the half-power bandwidth B is given by

$$t_r \simeq \frac{0.700}{B} \qquad \text{(bandpass filters)} \qquad (2.45)$$

FIGURE 2.22
Bandwidth and rise time of a bandpass filter.

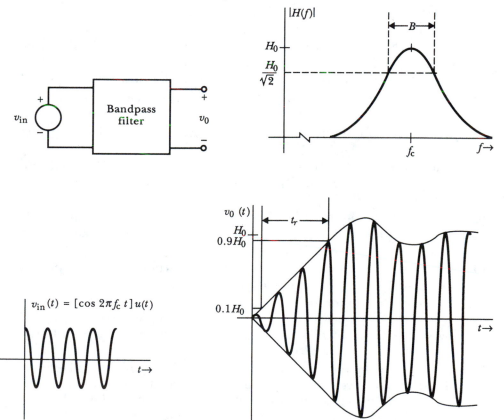

SUMMARY

Many of the important ideas in this chapter are contained in equations. The important equations are marked with a bullet (■). These equations should be reviewed with the summary points given below.

1. Fourier analysis is important as a theoretical tool and in the design of FDM systems. It is also the basis for legal regulation of the emissions of radio transmitters.

2. Signals can be classified as power versus energy signals, periodic versus aperiodic, and deterministic versus random.

3. Various functions are useful in describing signals in communication systems. Some of those defined in this chapter are the step, impulse, rectangular, triangular, and sinc pulse functions.

4. Fourier series is a technique that applies to periodic signals.

5. Periodic signals are composed of a sum of sinusoidal components whose frequencies are multiples of the fundamental frequency.

6. The coefficients of the exponential Fourier series α_n for positive values of n are one-half of the phasor representation of the sinusoidal component at the nth harmonic.

7. α_0 is the dc component of the periodic waveform.

8. The amplitude line spectrum of a periodic signal is a plot of the magnitudes of the Fourier series coefficients versus frequency. The amplitude line spectrum shows the relative amplitudes of the harmonic components.

9. Parseval's theorem relates the total average power in a periodic waveform to the Fourier coefficients.

10. The Fourier transform applies to deterministic signals, including periodic signals.

11. The Fourier transform of a voltage signal has units of V/Hz and shows the relative amplitude and phase of the components of the signal versus frequency.

12. If a signal contains a noninfinitesimal sinusoidal component at some frequency, then the Fourier transform contains an impulse at that frequency and the corresponding negative frequency.

13. The Fourier transform has many useful properties that are summarized in Table 2.1. (The table should be carefully reviewed.)

14. The Fourier transform of a periodic signal consists of impulses at multiples of the fundamental frequency. The Fourier transform of a periodic signal is obtained by first finding the Fourier series and then transforming each term.

15. Convolution in the time domain corresponds to multiplying Fourier transforms. Convolution of two signals is carried out by time reversing and shifting one signal and then multiplying by the second signal followed by integration to find the area of the product. This process is carried out for each possible value of time shift.

16. The ability to visualize the convolution process is important in signal processing and communication engineering.

17. The magnitude squared of the Fourier transform of an energy signal is the energy spectral density of the signal. The energy spectral density shows how the signal's energy is distributed in the frequency domain. The integral of the energy spectral density over all frequencies is the total energy in the signal.

18. The PSD of a power signal shows how the total power is distributed in the frequency domain. The integral of the PSD over all frequencies is the total power in the signal.

19. The PSD of a signal is the Fourier transform of the autocorrelation function of the signal. Autocorrelation is a means of determining the PSD that will be most important when random signals are considered in later chapters.

20. Systems or parts of systems can be classified as being linear versus non-linear, time-invariant versus time-varying, or causal versus noncausal.

21. Linear time-invariant systems are the easiest type to deal with on a theoretical basis. The output signal can be found for any input signal by convolving the input with the impulse response of the system.

22. The Fourier transform of the impulse response is the transfer function of the system. The transform of the output signal is obtained by multiplying the transform of the input signal by the transfer function.

23. The PSD of the output signal of a linear time-invariant system is the product of the magnitude squared of the transfer function and the PSD of the input signal.

24. Nonlinear or time-varying subsystems are of importance in communications, but a general theory is difficult to construct for them. They will be handled on a case-by-case basis as they are encountered in the following chapters.

25. An ideal analog communication system has an overall transfer function that has a constant magnitude (as a function of frequency). The phase shift is linear with frequency over the frequency range of the message signal.

26. The rise time of a filter can be estimated from the bandwidth of the filter.

REFERENCES

R. N. Bracewell. *The Fourier Transform and Its Applications*, 2d ed. New York: McGraw-Hill, 1978.

E. A. Guillemin. *The Mathematics of Circuit Analysis*. New York: John Wiley, 1949.

A. Papoulis. *The Fourier Integral and Its Applications*. New York: McGraw-Hill, 1961.

R. E. Ziemer, W. H. Tranter, and D. R. Fannin. *Signals and Systems: Continuous and Discrete*. New York: Macmillan, 1983.

PROBLEMS

1. Classify the following signals as power signals, energy signals, or those that fall in neither category. Find the total average power of the power signals and the total energy of the energy signals.
   (a) $3 \cos (377t + 30°)$ (b) $\exp (-t)u(t)$
   (c) rect (t) (d) t^2
   (e) $u(t)$

2. Find the period and fundamental frequency for the signal $g_p(t) = 2 \cos (20\pi t) + \cos (30\pi t)$.

3. Use the properties of the unit impulse function to evaluate the following integrals.

   (a) $\displaystyle\int_{-\infty}^{\infty} t^2 \delta(t - 2) \, dt$ (b) $\displaystyle\int_{0}^{\infty} t^2 [\delta(t - 2) + 3\delta(t + 3)] \, dt$

   (c) $\displaystyle\int_{-5}^{10} \exp (-t)\delta(2t) \, dt$ (d) $\displaystyle\int_{-\infty}^{t} \delta(x - 2) \, dx$

4. Prove the properties of the impulse function given in Equation 2.4 through Equation 2.7.

5. Find the coefficients and write the exponential Fourier series for the periodic waveforms shown in Figure P2.1.

FIGURE P2.1

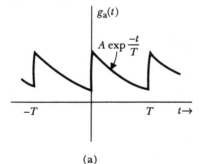

(a)

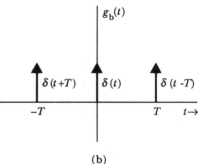

(b)

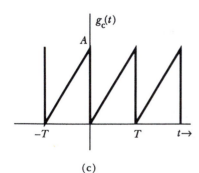

(c)

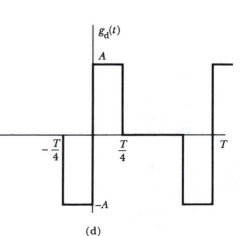

(d)

6. The kth derivative of a periodic signal is also periodic with the same period. Suppose that the exponential Fourier coefficients of $g(t)$ are known and denoted as α_n and the coefficients of the kth derivative are denoted as β_n. Find β_n in terms of α_n.

7. Find the total average power for the signal of Figure P2.1a. What percentage of this power is supplied by the dc component? By the fundamental? What percentage of the total power is contained in the first five harmonics and dc (i.e., the terms from $n = -5$ through $n = 5$)? Repeat for the signals of Figure P2.1c and d.

8. Find the power associated with the dc component, the fundamental component, and the nth component for the signal of Figure P2.1b. What is the total power contained in this signal? Is this signal a power signal?

9. Use the definition of the Fourier transform, Equation 2.18, to find the transforms of the following signals. Use Table 2.2 to check your answers.
   (a) $A \exp{(-at)}u(t)$ (b) $A \exp{(-a|t|)}$
   (c) $\delta(t)$

10. Sketch, to scale, the amplitude and phase spectra of the following signals.
    (a)   rect $(100t)$ (b)   rect $(100[t - 0.01])$
    (c) $\exp{(-20t)}u(t)$ (d) $\delta(t - 2)$

11. Use the transforms of Table 2.2 and the properties of Table 2.1 to find the Fourier transforms of these signals. Sketch the resulting amplitude spectra to scale versus frequency.
    (a) $\exp{(-at)} \cos{(2\pi f_0 t)}u(t)$ (b) 2 sinc$^2 (3t)$
    (c) $\dfrac{2}{t^2 + 9}$ (d)   rect $(t) \otimes$ rect $(2t)$
    (e)   rect $(100t) \sin{(2000\pi t)}$ (f) $\delta(t) + 2\delta(t - 3)$

12. If $g(t) \leftrightarrow G(f)$, find expressions for the transforms of (a) $[\cos{(t)}][dg/dt]$, (b) $tg(2t)$, (c) $g(2t - 1)$, (d) $g(2 - t)$, (e) $G(t - 2)$, and (f) $g(t)G(2t)$.

13. Find the results of the following operations by direct application of the convolution integral. Sketch the resulting functions to scale.
    (a) $[\exp{(-at)}u(t)] \otimes [u(t)]$
    (b) $[\text{rect } (t)] \otimes [\exp{(-t)}u(t)]$
    (c)   rect $(100t) \otimes$ rect $(200t)$

14. Find the transform of $g_1(t) \otimes g_2(t)$ by first finding the result of the convolution operation in the time domain and then transforming. Check by transforming $g_1(t)$ and $g_2(t)$ separately and then taking the product of the transforms.
    (a) $g_1(t) = $ rect (t) (b) $g_1(t) = \exp{(-at)}u(t)$
          $g_2(t) = $ rect (t) $g_2(t) = \exp{(at)}u(-t)$

15. Construct proofs of the following properties of the Fourier transform given in Table 2.1.
    (a)   Linearity   (b)   Duality
    (c)   Modulation   (d)   Time shift
    (e)   Differentiation   (f)   Scaling

16. Use Parseval's energy theorem to verify Equation 2.12.

17. Find and sketch the energy spectral density of the signals given by (a) $\exp{(-at)}u(t)$, (b) rect $(100t)$, and (c) $\exp{(-a|t|)}$.

18. A certain noise signal has an autocorrelation function given by $R_g(\tau) = [\mathcal{N}_0/2]\delta(\tau)$. Find the PSD, $S_g(f)$, of the signal. Can you explain why this type of noise is often called "white noise"?

19. Find and sketch to scale the autocorrelation functions of the periodic signal of Figure P2.1d.

20. Find the autocorrelation function of $A \cos (2\pi f_0 t + \theta)$. Now take the Fourier transform to find the PSD. Notice that the answer does not depend on θ. Why is this fact a reasonable result?

21. A certain filter has an impulse response given by

$$h(t) = 1 \qquad \text{for } 0 < t < T$$
$$= 0 \qquad \text{otherwise}$$

    (a) Find the transfer function of the filter and sketch its magnitude and phase characteristics to scale versus frequency. What type of filter is this? (bandpass, high-pass, etc.)
    (b) Find the approximate half-power bandwidth in terms of T.
    (c) Suppose that a unit step is applied to the filter. Find and sketch the output signal to scale. Find the 10% to 90% rise time in terms of T.
    (d) Use the results of (b) and (c) to find the relationship between rise time and bandwidth for this filter type. Compare this result to the rule of thumb given in the text in Equation 2.44. This type of filter is known as an "integrate and dump" or as a "moving average filter."

22. A white noise voltage signal $x(t)$ with a PSD given by $S_x(f) = N_0/2$ is applied as the input to a simple RC lowpass filter as shown in Figure P2.2. Find the PSD and autocorrelation function of the output signal $y(t)$. Find the total power associated with the input signal and the output signal.

23. Under what conditions (i.e. for signals confined to what frequency range) does the filter of Problem 21 meet the requirements for distortionless transmission? Repeat for the filter of Figure P2.2.

24. Suppose that a (noncausal) filter has an impulse response given by $h(t) = \operatorname{sinc} [100(t - 0.02)]$. Find and sketch the magnitude and phase characteristics of the filter to scale versus frequency. Under what conditions does this filter meet the requirements for distortionless transmission? If the input to the filter is $x_1(t) = \operatorname{sinc} (50t)$, find the output signal. Has distortionless transmission occurred? Repeat for an input signal given by $x_2(t) = \operatorname{sinc} (200t)$.

25. Use the multiplication property of the Fourier transform to find the amplitude spectra of the outputs of the nonlinear systems of Figure P2.3 and sketch the results to scale versus frequency.

FIGURE P2.2

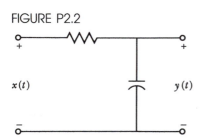

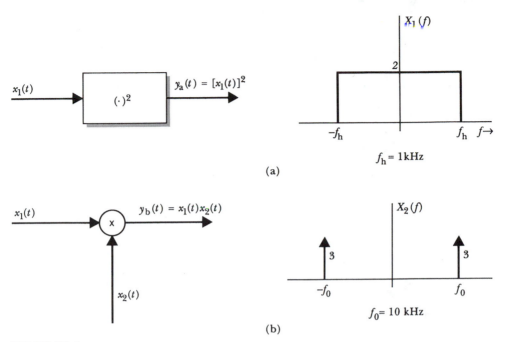

$x_1(t)$ → $(\cdot)^2$ → $y_a(t) = [x_1(t)]^2$

$|X_1(f)|$

2

$-f_h$ f_h $f\rightarrow$

$f_h = 1\,\text{kHz}$

(a)

$x_1(t)$ → ⊗ → $y_b(t) = x_1(t)x_2(t)$

$x_2(t)$

$X_2(f)$

3   3

$-f_0$ f_0

$f_0 = 10\,\text{kHz}$

(b)

FIGURE P2.3

26. For the linear system shown in Figure P2.4, (a) find the impulse response $h(t)$, and (b) find the transfer function $H(f)$ and sketch its magnitude and phase to scale versus frequency,
   (c) If the input signal is $5\cos(20\pi t + 30°)$, find the output signal under steady-state conditions.
   (d) If the input signal is white noise with a PSD given by $S_x(f) = N_0/2$, find the PSD of the output signal and sketch it to scale.
27. Derive Equation 2.42 for the rise time of the RC lowpass filter.

FIGURE P2.4

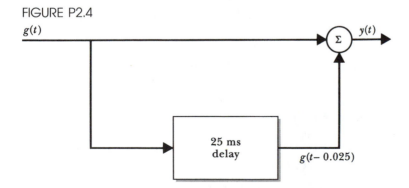

$g(t)$

Σ → $y(t)$

25 ms delay

$g(t - 0.025)$

3

AMPLITUDE
MODULATION

Modulation is the process of causing some parameter of a *carrier* to vary in relation to a message signal. When the carrier is a sinusoidal waveform, given by

$$a(t) \cos \left[2\pi f_c t + \theta(t) \right]$$

several parameters can be changed in response to the message signal. For example, we can cause the amplitude $a(t)$ of the sinusoidal carrier to depend on the message signal, resulting in *amplitude modulation*. This can be done in several ways resulting in various forms of amplitude modulation that will be discussed in this chapter. When the phase angle $\theta(t)$ depends on the message signal, *angle modulation* results, which will be discussed in Chapter 4. Schemes in which the frequency f_c of the carrier is varied in accordance with the message could also be used, but we will see that this effect can be achieved by varying the phase angle $\theta(t)$. Carrier waveforms other than sinusoids will be considered in Chapter 6 where modulation of the parameters of pulse trains is considered. Lightwave carriers are used in optical fiber communication systems, which are are considered in Chapter 12.

Modulation is useful for several reasons. It allows freedom in locating the frequency range of the signal power, which is useful because we can select a frequency that's easy to radiate from an antenna of reasonable size. For example, an audio signal in the frequency range from 20 Hz to 15 kHz would be very difficult to radiate from any antenna of practical size. By using modulation, we can move such a signal up to a frequency range at which it can be radiated from a small antenna.

Modulation enables more than one user to communicate over a channel at one time by selecting the modulation technique and carrier frequency so that the power spectral densities of the signals do not overlap in the frequency domain. The signals can be separated at the receiving end by filters that pass only the desired signal. This is known as either *frequency division multiplexing* (FDM) or as *frequency division multiple access* (FDMA).

In later chapters, we will also see that certain forms of modulation enable the communication system designer to expand bandwidth to gain either lower transmitter power or an improvement in the output signal-to-noise ratio (SNR) of the system.

3.1
DOUBLE-SIDEBAND–SUPPRESSED-CARRIER MODULATION

Definition and spectral characteristics

The *double-sideband–suppressed-carrier* (DSB-SC) signal is described by

$$g_{sc}(t) = A_c m(t) \cos(2\pi f_c t) \qquad (3.1)$$

where $m(t)$ is the message signal, A_c is an amplitude scale factor, and f_c is the carrier frequency. Figure 3.1a and b shows a typical message waveform and the corresponding DSB-SC signal. Notice, as indicated in the figure, that the phase of the modulated signal becomes inverted when the message is negative.

An important characteristic of DSB-SC modulation is that it is a *linear* process in the sense that superposition holds. In other words, if the DSB-SC signal, produced by message $m_1(t)$, is added to the DSB-SC signal produced by $m_2(t)$, the sum is exactly the same as the DSB-SC signal produced by $m_1(t) + m_2(t)$. One of the consequences of linearity is that, if we can show that a linear circuit can produce DSB-SC modulation for a single sinusoidal message, then it will do so for each frequency component in a general message waveform. Angle modulation is a nonlinear process.

The Fourier transform of the DSB-SC modulated signal can be found in terms of the transform of the message signal $M(f)$ by application of the modulation property (see Table 2.1) as

$$G_{sc}(f) = \frac{1}{2} A_c M(f - f_c) + \frac{1}{2} A_c M(f + f_c) \qquad (3.2)$$

The amplitude density spectra of a message and the corresponding modulated signal are shown in Figure 3.1c and d. Notice that the effect of the modulation process is simply to shift the original message spectrum so that it appears centered around the carrier frequency (and the negative of the carrier frequency).

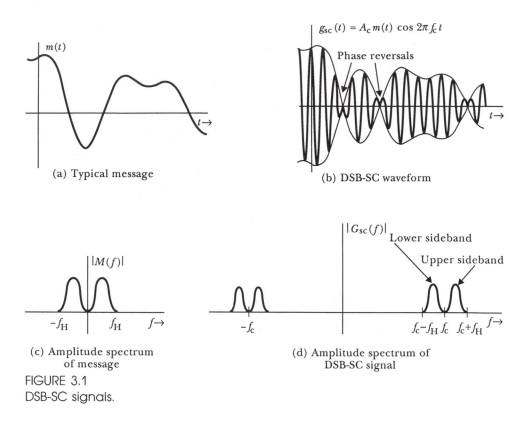

(a) Typical message

$g_{sc}(t) = A_c m(t) \cos 2\pi f_c t$

Phase reversals

(b) DSB-SC waveform

$|M(f)|$

$-f_H$ f_H $f \rightarrow$

(c) Amplitude spectrum
of message

$|G_{sc}(f)|$ Lower sideband

Upper sideband

$-f_c$ $f_c-f_H \; f_c \; f_c+f_H$ $f \rightarrow$

(d) Amplitude spectrum of
DSB-SC signal

FIGURE 3.1
DSB-SC signals.

The part of the spectrum that lies above the carrier frequency is called the *upper sideband* and the part below the carrier frequency is called the *lower sideband*, as indicated in the figure.

The message signal shown in Figure 3.1 occupies the frequency domain from approximately zero frequency up to an upper limit f_H. This type of signal is often called a *baseband signal*. The modulated signal, on the other hand, is concentrated in a band of frequencies from $f_c - f_H$ to $f_c + f_H$. The modulated signal is an example of a *bandpass signal*. Thus, the modulation process has converted a baseband message signal into a bandpass signal. We can adjust the location of the modulated signal in the frequency domain by our choice of the carrier frequency. Also, notice that the bandwidth occupied by the DSB-SC signal is twice the highest message frequency. When several signals are to be transmitted over one channel, we must consider the bandwidth of each signal, when selecting carrier frequencies, in order to avoid overlap.

When the message signal does not contain a dc component, the DSB-SC signal does not contain a component at the carrier frequency. That is the case illustrated in Figure 3.1. However, if the message contains a dc component, the message spectrum will have an impulse at zero frequency and the DSB-SC

3 / AMPLITUDE MODULATION

signal will contain a component at the carrier frequency. In this case, the designation *suppressed-carrier* is incorrect, though common usage still refers to the modulated signal as DSB-SC. When a small dc component is added to the message as an aid in carrier recovery, as will be discussed later in this section, we have a *residual carrier*. When a large dc component is added to the message, *amplitude modulation* (AM) results. AM is discussed in the next section.

In principle, all that is required to produce a DSB-SC signal is a circuit that can produce an output signal that is the product of two input signals. This is indicated in the modulator portion of the system diagram shown in Figure 3.2. *Demodulation*, which is the process of recovering the message signal from the modulated signal, can also be accomplished by multiplication by the original sinusoidal carrier. This product is given by

$$g_{sc}(t) \cos (2\pi f_c t) = A_c m(t) [\cos (2\pi f_c t)]^2$$

Using the identity for cosine-squared, this becomes

$$g_{sc}(t) \cos (2\pi f_c t) = \frac{1}{2} A_c m(t) + \frac{1}{2} A_c m(t) \cos (4\pi f_c t)$$

Notice that the first term on the right-hand side is the desired demodulated message signal multiplied by a scale factor. The second term on the right side is a DSB-SC signal with a carrier frequency of $2f_c$. This second term is undesired in the output from the demodulator and can be eliminated by a lowpass filter, which passes the message frequencies but eliminates everything above $2f_c - f_H$. (Notice that $2f_c - f_H$ is the lowest frequency present in the second term on the right in the expression.) Thus, the demodulator consists of a multiplier followed by a lowpass filter. This type of demodulator is often called a *coherent demodulator* because it uses a carrier signal that is synchronized in phase and frequency with the transmitter's carrier. The modulator, demodulator, and typical waveforms are shown in Figure 3.2.

Modulators

Because the message signal and the DSB-SC signal lie in different bands in the frequency domain, they can be separated by filters. This fact is often used in the design of modulator and demodulator circuits. For example, consider the modulator shown in Figure 3.3. In this circuit, known as a *switching modulator*, the switch opens and closes periodically at the desired carrier frequency. We assume that the bandpass filter has been designed so that it does not load the circuit to its left (i.e., the input impedance of the filter is much higher than resistance R). When the switch is open, the message waveform appears at the input to the filter. When the switch is closed, the input waveform $v_{in}(t)$ is zero. This is indicated in Figure 3.3b, which shows the input waveform $v_{in}(t)$ to the filter.

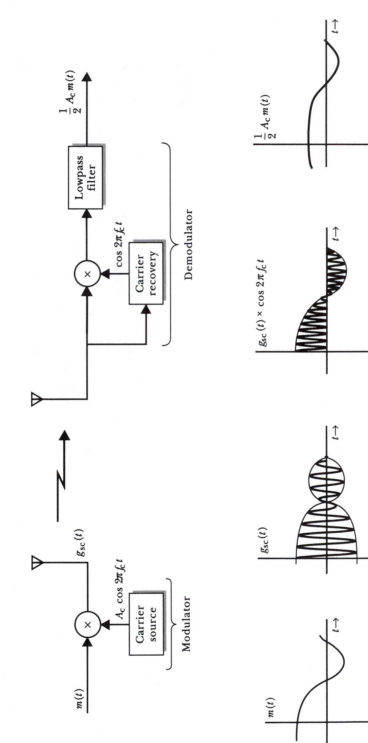

FIGURE 3.2
DSB-SC communication system.

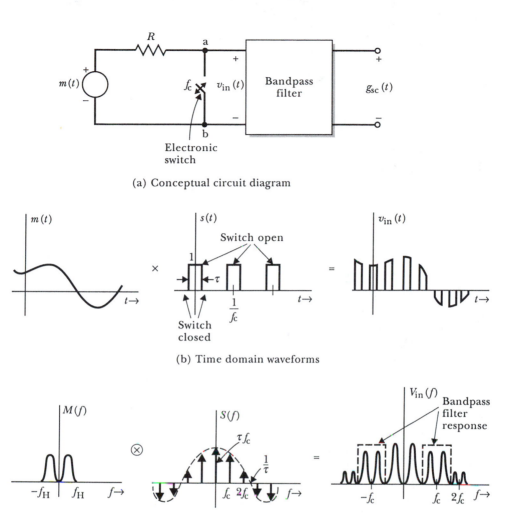

(a) Conceptual circuit diagram

(b) Time domain waveforms

(c) Corresponding frequency domain representations

FIGURE 3.3
DSB-SC modulator and waveforms.

The input waveform to the filter is given mathematically by the product of the message and a switching waveform $s(t)$, which is shown in Figure 3.3b. As indicated in the figure, $s(t)$ is unity when the switch is open, and zero when the switch is closed. The switching waveform is periodic and can be represented by a Fourier series. In fact, this is the waveform analyzed in Example 2.2, in which (with a change of notation to fit the present case) we obtained

$$s(t) = \sum_{n=-\infty}^{\infty} \tau f_c \text{ sinc } (n\tau f_c) \exp (j2\pi n f_c t)$$

If this expression for $s(t)$ is multiplied by $m(t)$ and the result is Fourier transformed by use of the modulation property, we obtain the transform of the input to the filter, given by

$$V_{\text{in}}(f) = \sum_{n=-\infty}^{\infty} \tau f_c \text{ sinc } (n\tau f_c) M(f - n f_c)$$

The amplitude spectrum of the input signal to the filter is the magnitude of this expression, shown in Figure 3.3c. Notice from the amplitude spectrum of v_{in} that the input signal to the filter contains the original message signal, the desired DSB-SC modulated signal centered at the carrier frequency, and additional DSB-SC signals at multiples of the carrier frequency f_c. If these components do not overlap in the frequency domain, the desired DSB-SC signal can be separated by the bandpass filter. Often, the carrier frequency is much higher than the highest message frequency, so the bandpass filter does not require a rapid cutoff characteristic. Thus, the filter is not difficult to implement.

Notice that, since the input signal to the filter in Figure 3.3 also contains DSB-SC modulated signals at multiples of the carrier frequency, it would be possible to obtain these as output signals by simply moving the passband of the bandpass filter. In this case, the effective carrier frequency would be a multiple of the rate at which the switch opens and closes.

It is rather surprising that a message signal can be so perfectly multiplied by a sinusoidal carrier with the use of a simple switch opening and closing at the carrier frequency. This results because the effect of the on-off switching action is to multiply the message by a pulse train. The periodic pulse train consists of a dc component and sinusoidal components at multiples of the carrier frequency. The product of each of these components and the message falls into a different range in the frequency domain, so we can select the desired product (i.e., the product of the message and the sinusoidal component at the carrier frequency) by the use of a filter.

It is not the intention of this book to give a full treatment of the circuits used in communication systems, but a few implementations will be included for illustration. The reader is referred to the electronics texts in the biblio-

FIGURE 3.4
One implementation of the electronic switch of Figure 3.3.

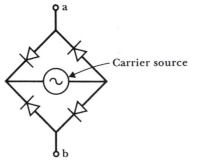

a

Carrier source

b

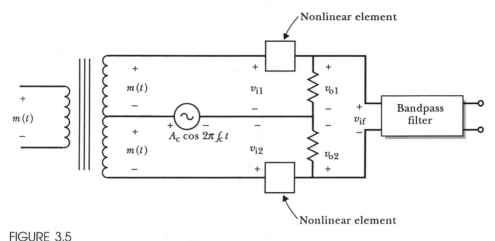

FIGURE 3.5
Balanced modulator. Note that the circuit is symmetrical for the carrier input.

graphy for more information. One method for obtaining the *electronic switch* required for the modulator of Figure 3.3 is shown in Figure 3.4. When the carrier signal is positive on the right, the diodes are turned on and, if the diode forward voltages are equal, point *a* is at the same potential as point *b*. Thus, point *a* is electrically shorted to point *b*. On the other hand, when the carrier source is positive on the left in Figure 3.4, the diodes are reverse biased. Then no current can flow between *a* and *b*, resulting in an open circuit between *a* and *b*.

Another circuit for producing DSB-SC modulation is shown in Figure 3.5. If the nonlinear circuit elements are identical, the circuit is symmetrical about a horizontal line drawn through the middle of the circuit. As a result, it is not possible for the carrier source, acting by itself, to produce an input signal to the bandpass filter. In other words, the circuit acts as a balanced bridge for the carrier input. This is why it is called a *balanced modulator*.

The nonlinear circuit element, typically a diode, and the resistor in the upper half of the circuit in Figure 3.5 act as a voltage divider for the input voltage v_{i1}. When the output of this nonlinear voltage divider at a given instant depends only on the input at that instant, it is said to be a *memoryless nonlinearity*. Such a memoryless nonlinear transfer function can be mathematically described by the Maclaurin series from calculus. Thus, the output voltage for the upper half of the circuit is

$$v_{o1} = K_1 v_{i1} + K_2 [v_{i1}]^2 + K_3 [v_{i1}]^3 + \cdots \tag{3.3}$$

where the second, third, and higher power terms are due to the nonlinearity. The output of the lower half of the circuit v_{o2} is given in terms of its input v_{i2} by an identical equation, provided that the two halves of the circuit are matched. Note from Figure 3.5 that the input voltages to the two halves of the circuit are given by

$$v_{i1} = A_c \cos 2\pi f_c t + m(t) \tag{3.4}$$

$$v_{i2} = A_c \cos 2\pi f_c t - m(t) \tag{3.5}$$

Also, note that the input voltage to the filter is given by

$$v_{if} = v_{o1} - v_{o2} \tag{3.6}$$

Combining Equations 3.3–3.6 results in

$$v_{if} = 2K_1 m(t) + 4K_2 m(t) A_c \cos 2\pi f_c t + 2K_3 m^3(t)$$
$$+ 6K_3 m(t)[A_c \cos 2\pi f_c t]^2 + \cdots \tag{3.7}$$

The first term on the right-hand side of Equation 3.7 is the original message signal, which will be eliminated by the bandpass filter. The second term on the right-hand side is the desired DSB-SC signal, which is passed by the bandpass filter. Usually the carrier frequency is high enough so that the two frequency bands are widely separated, and the filter is easy to implement.

The third term on the right-hand side of Equation 3.7 is the cube of the message signal. Its Fourier transform is the triple convolution of the message spectrum with itself, due to the multiplication property of the Fourier transform. This is indicated by

$$m^3(t) \leftrightarrow M(f) \otimes M(f) \otimes M(f)$$

The highest frequency present in the cube of the message is three times the highest frequency in the original message. This can be verified by visualizing the result of the triple convolution of the message spectrum with itself. If the carrier frequency is high enough, the message spectrum does not overlap the desired DSB-SC signal in the frequency domain and is therefore eliminated by the bandpass filter.

The last term on the right-hand side of Equation 3.7 can be expanded by the use of the identity for cosine-squared. It then consists of the original message plus a DSB-SC signal at twice the carrier frequency. By choosing a carrier frequency high enough and by properly selecting the bandpass filter characteristics, these terms can be eliminated by the filter.

If the nonlinearity requires a fourth-power term in its description, output terms that overlap the desired signal in the frequency domain occur. Since these terms are not eliminated by the filter, some distortion occurs in the output. Therefore, in designing the circuit, the nonlinear elements and drive levels should be selected to minimize the higher order terms of the Maclaurin series. Nonlinearities that can be described by a few low order terms are called *soft nonlinearities* whereas sharp nonlinearities that require higher order terms are called *hard nonlinearities*. Thus, to avoid serious distortion of the output, soft nonlinearities are called for in the design of modulators using the principle of Figure 3.5.

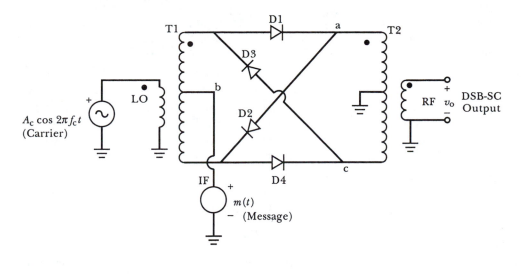

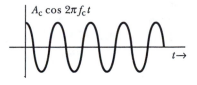

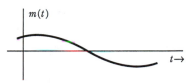

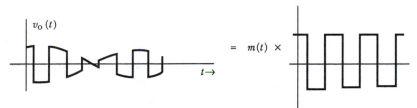

FIGURE 3.6
Double balanced mixer used as a DSB-SC modulator.

This circuit is called a *single balanced* circuit because it is symmetrical for the carrier input. As a result, the carrier component does not appear by itself at the input to the final filter. The circuit is not balanced for the message input, so the message appears at the input to the filter, which must reject it. A circuit that is balanced for both the carrier and the message signals is shown in Figure 3.6.

The *double balanced mixer* shown in Figure 3.6 is carefully constructed so that the diode characteristics match and the transformer taps are precisely at the midpoints of the windings. The carrier input should be much larger than the message signal input, so the on-off states of the diodes are controlled by the carrier rather than by the message signal. When the carrier is positive, diodes D1 and D2 are biased on. Then point *a* and point *b* are at the same potential because the diodes D1 and D2, as well as the two halves of the secondary of the transformer T1, form a balanced bridge circuit with equal voltages across each arm. Since point *b* is driven from the message source, the message signal appears at point *a* during the positive half of the carrier waveform. In the same way, the diodes D3 and D4 are *on* during the negative half cycle of the carrier, resulting in the message signal being connected to point *c*. Thus, the output of transformer T2 is the same as the message signal during the positive part of the carrier cycle. The output is the negative of the message signal during the negative half carrier cycle as shown in the examples of waveforms in Figure 3.6.

As shown in Figure 3.6, the output waveform of the circuit is the original message waveform multiplied by a symmetrical square wave at the carrier frequency. Fourier analysis of the symmetrical square wave shows that it consists of a fundamental component plus components at each of the odd harmonics. Neither dc nor even harmonics are present in a symmetrical square wave. Thus, the output signal contains a DSB-SC signal centered at the carrier frequency and each odd harmonic. The desired DSB-SC signal can be selected by the use of a bandpass filter following the double balanced mixer.

The three ports of the double balanced mixer of Figure 3.6 are labeled LO *for local oscillator*, RF for *radio frequency*, and IF *for intermediate frequency*. These port designations are often used by manufacturers because the circuits are used in radio receivers, in which these labels have meaning. This will become clear later. The LO port of the double balanced mixer is usually driven by a strong sinusoidal signal that controls the states of the diodes, as discussed above. Another signal that we want to multiply by a square wave at the frequency supplied to the LO port can be connected to either the IF or the RF port and the desired product appears at the other port (RF or IF, respectively).

The LO and RF ports of the double balanced mixer of Figure 3.6 are said to be *ac coupled* because the transformers at these ports do not couple dc signals into or out of the circuit. The IF port, on the other hand, is *dc coupled*. That is why the message signal, which may have components at dc or very low frequencies, is connected to the IF port. The DSB-SC signal, which is a bandpass signal far from dc, is taken from the ac coupled RF port. Although the transformers are not effective all the way down to dc, double balanced mixers are available that perform well for over three decades of frequency.

Notice in Figure 3.6 that the diodes are connected end to end in a ring. As a result, the circuit is sometimes called a *ring modulator*. Double balanced mixers find use as demodulators, frequency converters, and phase detectors as we will see later in the text.

Demodulators

As we have noted earlier, DSB-SC signals can be demodulated by coherent demodulation (i.e., multiplying by the original sinusoidal carrier followed by lowpass filtering). Many demodulator circuits are similar to the modulators discussed above. For example, the circuit of Figure 3.3a can be used as a demodulator by replacing the message signal $m(t)$ with the DSB-SC signal to be demodulated. The bandpass filter is replaced by a lowpass filter. The switching action now acts to multiply the incoming DSB-SC signal by the switching signal $s(t)$ just as before. Since the switching signal has a sinusoidal component at the carrier frequency, the desired product is contained in the input to the filter and the demodulated message is passed by the filter. The remaining components of the switching waveform (i.e., dc and harmonics of the carrier) produce products with the DSB-SC input signal that are not passed by the lowpass filter.

The double balanced mixer of Figure 3.6 can also be used to demodulate DSB-SC signals. The carrier is connected to the LO port and is selected to have a large enough amplitude to control the on-off states of the diodes as before. The DSB-SC signal to be demodulated is connected to the RF port and the IF port is input to a lowpass filter. The desired demodulated message appears at the output of the filter. Note that it is necessary to take the demodulated output from the dc-coupled IF port of Figure 3.6, since the low frequency message signal is not usually effectively coupled by the transformers.

A simple demodulator for DSB-SC signals simply adds the signal to a carrier and inputs the sum to a peak rectifier circuit as shown in Figure 3.7. In this circuit, the input to the peak rectifier $v_i(t)$ is

$$v_i(t) = A_c m(t) \cos 2\pi f_c t + A_{lo} \cos 2\pi f_c t$$
$$= [A_c m(t) + A_{lo}] \cos 2\pi f_c t$$

The output of the peak rectifier is the upper envelope of the input to the rectifier as indicated in Figure 3.7. If the local oscillator signal amplitude A_{lo} is larger than the peak amplitude of the DSB-SC signal, the output of the peak rectifier is

$$v_o(t) = A_c m(t) + A_{lo}$$

This is clearly the desired message signal plus a dc component. If the message does not contain a dc component, the dc component of the peak rectifier output can be discarded by ac coupling.

The time constant of the RC network of Figure 3.7 should be longer than a carrier cycle so that the voltage is maintained at the output between peaks of the input signal. However, if the time constant is too long, the capacitor will not be able to discharge rapidly enough to follow the negative excursions of the message waveform, as indicated by the dotted line on the waveform sketches in Figure 3.7.

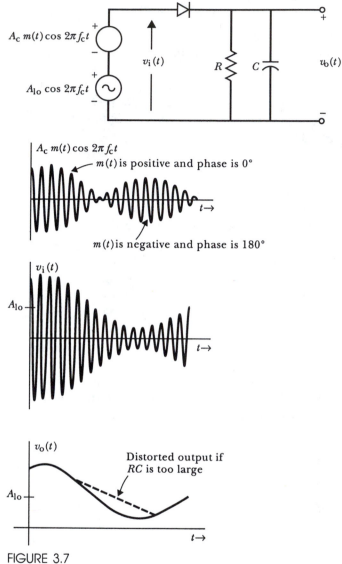

FIGURE 3.7
Simple DSB-SC demodulator.

Carrier recovery techniques

In all of the demodulator circuits for DSB-SC, a carrier reference signal of exactly the same frequency and phase as the carrier used in the modulator is required. The exact frequency and phase of the received DSB-SC signal is not known in advance. This may be due to carrier source drifts or unknown distance and velocity between the transmitter and receiver. Therefore, it is not

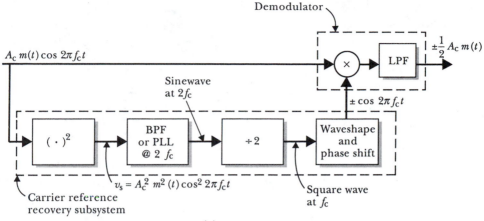

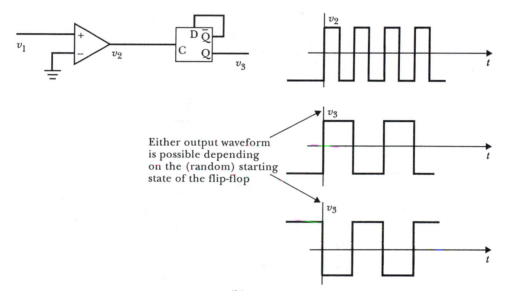

FIGURE 3.8
(a) Block diagram of the squaring method for carrier recovery for DSB-SC; (b) one implementation for the divide-by-two block with example waveforms to illustrate the phase ambiguity.

possible to generate the carrier reference with a free-running oscillator. Instead, it is necessary either to generate the reference directly from the received signal or to derive a control signal to continuously adjust the phase and frequency of a local source.

A scheme for obtaining the carrier reference from the received DSB-SC signal is shown in Figure 3.8a. The received signal is split and applied as the input to two paths: a carrier recovery subsystem and a demodulator. The carrier recovery subsystem first passes the received DSB-SC signal through a square-law circuit, which produces an output equal to the square of the input given by

$$v_s = (A_c)^2 m^2(t) \cos^2 2\pi f_c t$$

If we substitute the identity for cosine-squared, this becomes

$$v_s = \frac{1}{2}(A_c)^2 m^2(t) \cos 4\pi f_c t + \frac{1}{2}(A_c)^2 m^2(t)$$

The first term on the right in the last expression can be recognized as a DSB-SC signal in which the carrier frequency is $2f_c$ and the modulating signal is the square of the original message signal. Since the square of the message signal is always positive, it contains a significant dc component. As a result, the output of the squaring block of Figure 3.8 contains a line component at $2f_c$. The output of the squaring block is the input to a narrow bandpass filter centered at $2f_c$. The bandpass filter passes the twice-carrier component and should be narrow enough to reject the sidebands contained in the last expression. Thus, the output of the bandpass filter is a sine wave at twice the desired carrier frequency. The twice-carrier component is then applied to a frequency divider circuit that produces an output waveform at the desired carrier frequency. The carrier frequency waveform produced by the divide-by-two block is passed to a wave-shaping and phase-shifting circuit that produces a sine wave of the required phase at its output. This is the recovered carrier reference, which is passed to the demodulator.

One possible implementation for the divide-by-two circuit is a comparator to convert the sine wave at $2f_c$ into a square wave followed by a flip-flop connected as a divide-by-two counter. This is shown in Figure 3.8b. The comparator produces a logic-high output when the input is positive and a logic-low output when the input is negative, resulting in an output square wave at twice the carrier frequency. The square wave is the clock input to the D flip-flop. At each positive-going transition of the square wave, the Q output of the flip-flop switches to the logic level present at the D input just before the transition. Since the not-Q output of the flip-flop is the opposite of the Q output, connection of the not-Q output to the D input causes the output state of the flip-flop to reverse at each positive transition of the clock input. This results in an output square wave at half of the frequency of the input square wave.

The output phase of the divide-by-two block depends on the initial state of the flip-flop, shown in the waveforms of Figure 3.8b. Therefore, the phase of the recovered carrier can be either the same as the original carrier or exactly 180° out of phase. As a result, the demodulator output can be either the original message or an inverted version of the message. When the message is a voice signal, the inversion is not important, since an inverted voice signal sounds the same as the original. With other kinds of messages, such as television or digital data, it may be necessary to recover the message with the correct polarity. Various mechanisms are used for detecting the inversion and correcting the internal state of the divide-by-two block to produce the proper message polarity. This problem of *phase ambiguity* is inherent in the DSB-SC modulation technique, and is not confined to the system of Figure 3.8.

A second method, known as a *Costas phase-locked loop*, for recovering the suppressed carrier and demodulating DSB-SC signals is shown in Figure 3.9. In this system, the recovered carrier is produced by a *voltage-controlled oscillator* (VCO). A VCO is an oscillator circuit whose frequency depends on its input

FIGURE 3.9
Costas phase-locked carrier recovery and demodulator for DSB-SC.

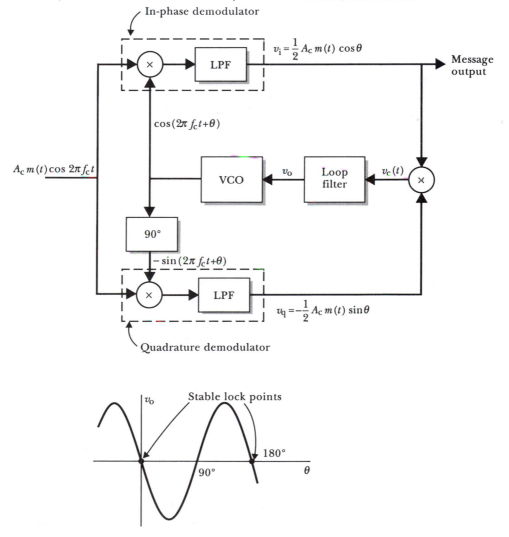

control voltage. The input to the VCO in Figure 3.9 is a control signal $v_o(t)$ that adjusts the frequency and phase of the VCO output to match the original carrier. When the control input is zero, the VCO in Figure 3.9 has a nominal frequency equal to the desired carrier frequency. The VCO frequency changes from its nominal value in proportion to the input control voltage.

The Costas loop contains two demodulators acting on the received DSB-SC signal. The *in-phase demodulator* uses the output of the VCO as its carrier reference, and the *quadrature demodulator* is supplied with a 90° phase-shifted version of the VCO output. The output of the in-phase demodulator is given by

$$v_i(t) = LP\{A_c m(t)[\cos 2\pi f_c t][\cos (2\pi f_c t + \theta)]\}$$

where LP indicates that only the lowpass part of the expression in braces should be retained. Applying the identity for the product of cosines yields

$$v_i(t) = LP\left\{\frac{1}{2} A_c m(t) \cos (4\pi f_c t + \theta) + \frac{1}{2} A_c m(t) \cos \theta\right\}$$

The first term inside the braces of the last expression is a DSB-SC signal at twice the original carrier frequency and is rejected by the lowpass filter. The second term inside the braces is a baseband signal and is passed by the lowpass filter. The resulting output of the in-phase demodulator is

$$v_i(t) = \frac{1}{2} A_c m(t) \cos \theta$$

In a similar fashion, the output of the quadrature demodulator is given by

$$v_q(t) = -\frac{1}{2} A_c m(t) \sin \theta$$

The control signal at the input to the loop filter is the product of the outputs of the two demodulators as shown in Figure 3.9. This control signal is given by

$$
\begin{aligned}
v_c(t) &= v_i(t) v_q(t) \\
&= -\frac{1}{4} (A_c)^2 m^2(t) \sin \theta \cos \theta \\
&= -\frac{1}{8} (A_c)^2 m^2(t) \sin 2\theta
\end{aligned}
$$

The loop filter is designed so that the VCO responds to the time-average of the input to the loop filter. Therefore, the effective input to the VCO is

$$v_o = -K \sin 2\theta$$

This input to the VCO is plotted versus θ in Figure 3.9.

Now the phase correction action of the system can be described. Assume that the VCO is operating with $\theta = 0$ and then θ begins to drift positive because the VCO frequency is slightly too high. The slightly positive value of θ results in a negative input to the VCO as indicated by the plot of v_o versus θ shown in Figure 3.9. This lowers the VCO frequency until θ is returned to zero. In this way, a control is continuously developed that maintains θ at 0. Such a feedback control system is described as being *phase-locked*. Since the system locks at $\theta = 0$, the output of the in-phase demodulator is the demodulated message signal.

Also, notice that if the phase of the VCO starts from 180°, the control action maintains the phase at 180°. As a result, the loop can lock at 180°. Then the in-phase demodulator produces an inverted version of the original message. This is the same phase ambiguity that we encountered earlier with the squaring method of carrier recovery. If message polarity is important, some mechanism must be provided to insure that the loop locks at the desired phase.

The Costas loop is just one example of the *phase-locked loop* (PLL) technique. PLLs are used in many ways in communication systems. We will encounter another example in the next chapter and discuss their design in more detail at that time.

Other methods for carrier recovery may sometimes be useful. For example, if the message signal is known always to contain a nonzero dc component of suitable amplitude, the DSB-SC signal spectrum contains an impulse at the carrier frequency. In this case, the system of Figure 3.10 can be used to recover the carrier. If the message does not contain a dc component, it may be possible to add a small dc component that can be discarded at the receiver so that the system of Figure 3.10 can be used. This approach is called a *pilot carrier system*.

A final possibility for carrier regeneration is to transmit an auxiliary signal that can be used to generate the carrier. This technique is used by the stereo multiplexing system of commercial FM broadcast stations.

FIGURE 3.10
Alternate carrier recovery system useful when $m(t)$ contains a nonzero dc component.

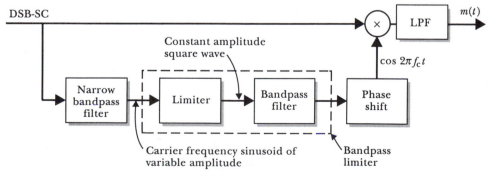

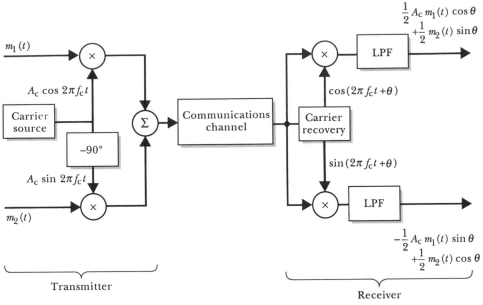

FIGURE 3.11
Quadrature multiplexed DSB-SC communication system.

Quadrature multiplexing

It is possible, using DSB-SC modulation to transmit two independent messages in the same frequency band without mutual interference. This is accomplished by using carriers that are phased 90° apart as shown in the block diagram of Figure 3.11. The quadrature multiplexed DSB-SC signal produced by the transmitter is given by

$$g_{\text{msc}}(t) = A_c m_1(t) \cos 2\pi f_c t + A_c m_2(t) \sin 2\pi f_c t$$

where $m_1(t)$ is one message signal and $m_2(t)$ is the other.

The signals resulting at the outputs of the in-phase and quadrature demodulators are shown in Figure 3.11. Notice that, when the phase error θ of the carrier recovery circuits is zero, the output of the in-phase demodulator contains only $m_1(t)$ and the second message appears at the output of the quadrature demodulator. Note that any residual phase error causes crosstalk between the two message paths, which may be undesirable in some applications. Static phase error is critical when it is important to avoid crosstalk in quadrature multiplexed systems. Quadrature multiplexing is used for the color signals in color television.

Spectrum analyzers

We will discuss the design of spectrum analyzers, which are commonly used by communication engineers, as an illustration of another way in which mod-

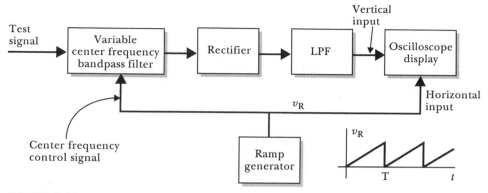

FIGURE 3.12
Simplified conceptual block diagram of a spectrum analyzer.

ulation techniques can be useful. A spectrum analyzer is a laboratory instrument designed to produce a frequency domain display of a signal much as an oscilloscope produces a time domain display of a signal.

In principle, a spectrum analyzer could be designed according to the block diagram shown in Figure 3.12. The input test signal is applied to the input of a narrow bandpass filter whose center frequency moves over the frequency range of interest in response to an internally generated ramp signal. As the center frequency is slowly varied, the rms output of the bandpass filter varies in proportion to the amplitude spectrum of the test signal. This output signal is then processed by a rectifier circuit to produce an output equal to the rms value of the test signal in the passband of the filter. As shown in Figure 3.12, the rectified output of the bandpass filter is applied to the vertical channel of a display. The ramp, which sweeps the center frequency of the filter, is applied to the horizontal channel.

The resulting display is almost identical to the amplitude density spectrum of the test signal if the parameters of the system are selected appropriately. The display approaches the true amplitude density when the filter bandwidth is made very narrow, so only the components of the test signal in a very narrow range of frequencies are passed at each point in the sweep. The sweep rate should be very slow so that the bandpass filter reaches steady state at each frequency. Finally, the lowpass filter should have a long time constant so that the displayed value corresponds to a representative sample of the test signal. In practice, something less than perfect results must be accepted due either to hardware limitations or the desire for a short measurement time.

Although the block diagram of Figure 3.12 works in concept, it is difficult to design the variable center frequency bandpass filter required. Practical spectrum analyzers often modulate a carrier with the test signal and then sweep the carrier frequency in order to move the signal spectrum past a fixed tuned bandpass filter. A typical block diagram using this idea is shown in Figure

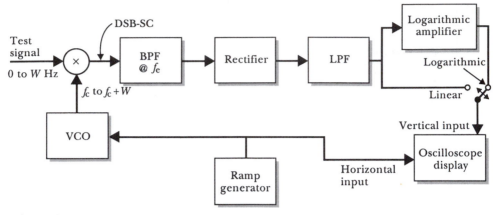

FIGURE 3.13
Block diagram of a practical spectrum analyzer.

3.13. This is similar in operation to the previous system except for the addition of the VCO and the DSB-SC modulator. The VCO sweeps from a frequency, denoted by f_c (which is the same as the center frequency of the bandpass filter), to a frequency of $f_c + W$. (W is assumed to be the highest frequency of interest in the test signal.) Notice that when the VCO frequency is $f_c + f_t$, a component of the test signal at f_t produces a lower sideband component at f_c. This component passes through the bandpass filter and produces a proportional vertical deflection on the display. This system turns out to be more practical because the VCO and modulator are easier to implement than the variable bandpass filter of Figure 3.12.

An additional feature of Figure 3.13 is the *logarithmic amplifier*, which takes the logarithm of the signal amplitude before it is displayed. With this feature in place, the vertical display can be calibrated in decibels (dB) relative to some reference level. When the logarithmic amplifier is in use, the display can be considered either as an amplitude density spectrum or as a power density spectrum depending on how one chooses to interpret the results. This is due to the fact that a voltage level in decibels relative to a voltage reference is the same as the power level of the signal in decibels when it is referred to an equivalent power level.

For a true logarithmic display with no limitations due to hardware, we would expect the display to indicate minus infinity when no input signal is present. In fact, when no input signal is present, real spectrum analyzers indicate a level that is typically only 60–90 dB below the largest rated signals. This is due either to internally generated noise or to imperfections in the logarithmic amplifier. The level indicated on a spectrum analyzer when no input signal is present is called the *noise floor*. The distance between the largest rated signal and the noise floor is called the *dynamic range* of the analyzer.

Typical spectrum analyzer displays obtained for a pulse train are shown in Figure 3.14 for both a linear and a logarithmic display. The advantage of

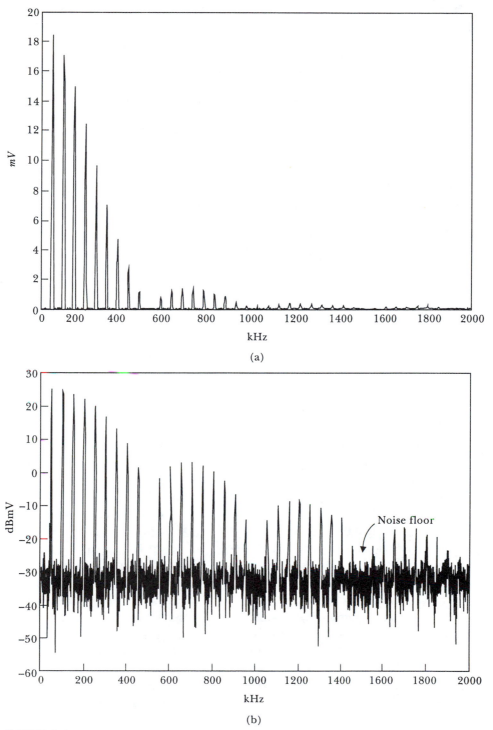

FIGURE 3.14
Spectrum analyzer displays of a filtered pulse train.

a logarithmic display is that it is possible to detect very small signals in the presence of very large signals. With a linear display, either the large signals are saturated or the small signals are too close to zero to be evident.

The spectrum analyzer of Figure 3.13 does not give a true indication of the dc component of the test signal, due to the fact that carrier signals leak past the modulator and result in a deflection of the display at zero frequency even when no dc component is present in the test signal.

3.2
AM WITH CARRIER

Definition and spectral characteristics

The AM signal with carrier is given by

$$g_{am}(t) = A_c[1 + m(t)] \cos (2\pi f_c t)$$
$$= \underbrace{A_c \cos (2\pi f_c t)}_{\text{carrier}} + \underbrace{A_c m(t) \cos (2\pi f_c t)}_{\text{sidebands}} \qquad (3.8)$$

where $m(t)$ is the message signal. Notice that this is the same as the DSB-SC signal given in Equation 3.1 except that a carrier component $A_c \cos (2\pi f_c t)$ has been added. Figure 3.15 shows a typical message signal and the corresponding AM signal. Often, when AM is used, the message signal is constrained to be less than unity in magnitude. Then the resulting AM signal has an upper envelope that is identical to the message waveform as shown in the figure. This is important because the AM signal is often demodulated with the peak rectifier circuit of Figure 3.16. Notice that if $m(t)$ becomes less than -1, the upper envelope of the AM signal will no longer be identical to the message and the use of the peak rectifier will cause distortion.

The amplitude spectra of the message and the AM signal are also shown in Figure 3.15. Notice that the spectrum of the AM signal consists of an upper and lower set of sidebands plus a discrete component at the carrier frequency. This is due to the fact that the AM signal is identical to the DSB-SC signal plus a carrier component.

The term *amplitude modulation* or AM is used in two ways by many communication systems engineers. In one case, it is a generic term referring to the entire collection of modulation schemes discussed in this chapter. The term is also used to refer to the specific modulation format of Equation 3.8.

The AM signal of Equation 3.8 is the format used by commercial AM broadcast stations. This signal format was chosen because it is possible to demodulate with the simple peak rectifier of Figure 3.16. The chief advantage of this is that the complex carrier recovery subsystem needed for DSB-SC is not needed for AM because the carrier is transmitted along with the sidebands. At the time that the commercial AM broadcast system was being designed,

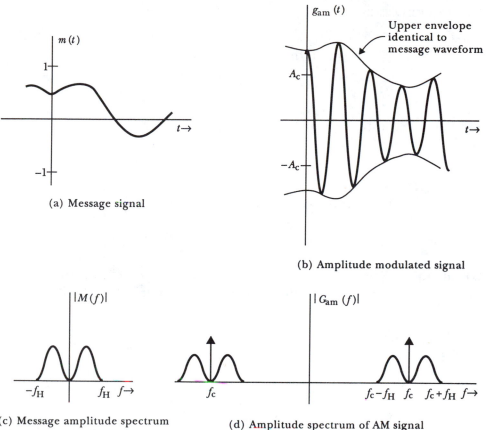

(a) Message signal

(b) Amplitude modulated signal

(c) Message amplitude spectrum

(d) Amplitude spectrum of AM signal

FIGURE 3.15
A typical message waveform and AM signal.

electronics was in its infancy and the system was selected so that simple receivers could be used. If the system were being designed today, some other modulation scheme would probably be used because it is now possible to manufacture complex receiver circuits at relatively low cost.

FIGURE 3.16
Simple AM demodulator.

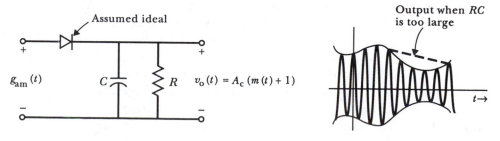

Modulation efficiency

The power in the AM signal is partly in the carrier component and partly in the sidebands. The power in the carrier is

$$P_c = \frac{1}{2} (A_c)^2$$

and the power in the sidebands is

$$P_{sb} = \lim_{T \to \infty} \frac{1}{T} \int_{-T/2}^{T/2} [A_c m(t) \cos (2\pi f_c t)]^2 \, dt$$

Using the identity for cosine-squared, this becomes

$$P_{sb} = \lim_{T \to \infty} \frac{1}{T} \int_{-T/2}^{T/2} [A_c m(t)]^2 \left[\frac{1}{2} + \frac{1}{2} \cos (4\pi f_c t) \right] dt$$

Usually the message signal is nearly constant over a cycle of twice the carrier frequency; so the term in the integral involving the cosine is negligible because each cycle of that term has zero net area. Thus, the sideband power is given by

$$P_{sb} = \frac{1}{2} (A_c)^2 P_m$$

where P_m is the power in the message signal given by

$$P_m = \lim_{T \to \infty} \frac{1}{T} \int_{-T/2}^{T/2} [m(t)]^2 \, dt \qquad (3.9)$$

We can now define the *modulation efficiency* of the AM signal as the ratio of the power in the sidebands to the total signal power. This efficiency is given by

$$\eta = \frac{P_{sb}}{P_{sb} + P_c} = \frac{P_m}{P_m + 1} \qquad (3.10)$$

Demodulation with envelope detectors

Recall that for the AM signal to have an upper envelope with the same shape as the message signal, it is necessary for the peak negative excursion of $m(t)$ to stay above -1. Since the fluctuations of audio signals are symmetrical around zero, this means that the message is bounded by ± 1. Many audio signals have the characteristic that the peak value is much higher than the typical amplitude. As a result, the amplitude is much less than one most of the time. Therefore, the power in $m(t)$ is much less than one, and the modulation efficiency of the AM signal is quite small. The bulk of the power is in the carrier component, which does not convey information. We will see in Chapter 5 that the carrier power is not useful in providing a higher signal-to-noise ratio (SNR) at the receiver output. However, the presence of the carrier com-

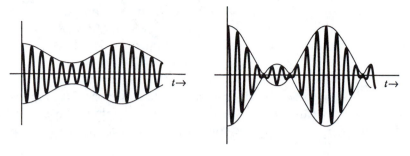

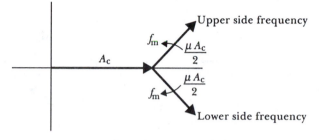

(a) $\mu = 50\%$ (b) $\mu = 150\%$

(c) Phasor diagram

FIGURE 3.17
AM signals for a single tone message.

ponent does make the use of the simple envelope detector possible in the receiver, as we have noted earlier.

Tone modulation

It is instructive to study the AM signal when the message signal is replaced by a single sinusoidal tone. The AM signal can then be written as

$$g_{am}(t) = A_c[1 + \mu \cos (2\pi f_m t)] \cos (2\pi f_c t) \qquad (3.11)$$

where μ is called the *modulation index* and f_m is the frequency of the message sinusoid. A sketch of this waveform is shown in Figure 3.17a and b for two values of the modulation index. Note again that the upper envelope is not the same as the message waveform when the modulation index is over 100%.

Equation 3.11 can be expanded by the use of the identity for the product of two cosine functions to yield

$$g_{am}(t) = A_c \cos (2\pi f_c t) + A_c \frac{\mu}{2} \cos [2\pi (f_c - f_m)t]$$

$$+ A_c \frac{\mu}{2} \cos [2\pi (f_c + f_m)t] \qquad (3.12)$$

The first term on the right is the carrier component, the second term is the lower side frequency, and the last term is the upper side frequency. A phasor diagram illustrating these three terms is shown in Figure 3.17c. The phasor diagram is drawn so that the carrier phasor is stationary at zero degrees. The upper side frequency is higher than the carrier frequency; so it rotates counterclockwise at f_m revolutions per second. Similarly, the lower side frequency phasor rotates clockwise because it is of a lower frequency than the carrier and therefore lags the carrier component by an amount that increases with time.

Modulators

The methods presented in Section 3.1 for generating DSB-SC signals can be modified to produce AM signals. One possibility is to add the carrier to the DSB-SC signals produced by those circuits. A second possibility is to add a dc component to the message before the modulator so the DSB-SC modulator produces the carrier component in the output.

Actually, simpler circuits can be used to generate AM since it is not necessary to suppress the carrier in the output for AM. Thus, it is not necessary to use modulator circuits that are balanced for the carrier signal as was the case for DSB-SC. A simple unbalanced modulator is shown in Figure 3.18. In this circuit, the sum of the carrier and the message is applied as the input to a nonlinear voltage divider circuit. The output of the nonlinear circuit is passed through a bandpass filter that passes only the components of the desired AM signal. The input to the filter is related to the input of the nonlinear circuit by the Maclaurin series, given by

$$v_{\text{if}} = K_1 v_i + K_2 v_i^2 + K_3 v_i^3 + \cdots \tag{3.13}$$

The input to the nonlinearity is

$$v_i = m(t) + A_c \cos (2\pi f_c t)$$

Substituting and expanding, we obtain

$$v_i = K_1 m(t) + K_1 A_c \cos (2\pi f_c t) + K_2 m^2(t) + 2K_2 m(t) \cos (2\pi f_c t)$$
$$+ K_2 \cos^2 (2\pi f_c t) + \text{higher terms}$$

The first term on the right is the message, which is eliminated by the bandpass filter. The next term is the carrier, which is part of the desired AM signal. The next term is the square of the message, which has a highest frequency equal to twice the highest frequency in the message signal and is eliminated by the bandpass filter if the carrier frequency is high enough. The next term is the product of the message and the carrier, which is the desired sidebands of the AM signal. The square of the carrier will be seen to consist of a dc component and a component at twice the carrier frequency if the identity for cosine-squared is used. These terms are eliminated by the filter. Thus, the

3 / AMPLITUDE MODULATION

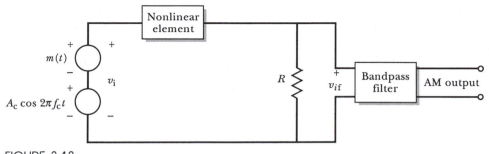

FIGURE 3.18
Simple AM modulator.

desired AM signal is produced by the first- and second-power terms of the nonlinearity.

Notice that higher terms in the nonlinearity of Equation 3.13 result in terms in which powers of the message signal are multiplied by the carrier. These terms are passed by the bandpass filter and cause a distorted output. Thus, it is necessary in designing a modulator using the principle of Figure 3.18 to minimize the higher-order terms of the nonlinearity.

3.3
SINGLE-SIDEBAND MODULATION

Generation by filtering

Single-sideband (SSB) modulation can be produced by passing a DSB-SC signal through a bandpass filter that eliminates one of the sidebands as shown in Figure 3.19. Two versions of the SSB signal are possible. Either the upper or the lower sideband may be passed by the filter.

When the carrier frequency is high, the filter of Figure 3.19 may have very extreme requirements since it is required to pass one of the sidebands and

FIGURE 3.19
Filter method for producing SSB.

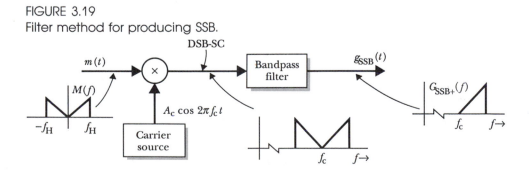

reject the other. This can mean that the filter must have a very sharp cutoff characteristic. One approach to reducing these extreme requirements is to cascade two of the systems shown in Figure 3.19. The first SSB modulator works at a relatively low carrier frequency, so the transition region of the filter is a larger percentage of its center frequency and is therefore easier to realize. The output of the first SSB modulator is the input to the second modulator. The carrier frequency of this second modulator is selected so that the resulting SSB signal is in the desired frequency band. Because the message spectrum has been moved up away from zero frequency by the first modulator, the DSB-SC signal at the input to the filter in the second SSB modulator has widely separated sidebands. Therefore, the cutoff characteristic of the filter does not have to be as sharp as when only a single-stage modulator is employed.

Generation by the phase-shift method

A more complete understanding of the SSB signal can be gained by considering the SSB modulator shown in Figure 3.20. This modulator is known as a *phase-shift modulator*. It works by producing two DSB-SC signals in which the lower sidebands are in phase and the upper sidebands are out of phase. Adding or subtracting these two DSB-SC signals causes one of the sidebands to cancel and the other to add, resulting in a SSB output. To simplify the discussion of the phase-shift modulator, we will assume that the message signal is a single tone given by

$$m(t) = \cos\left(2\pi f_m t + \theta\right)$$

As shown in Figure 3.20, the message signal is split into two paths. In one path, the message is applied to a DSB-SC modulator. This modulator is also

FIGURE 3.20
Phasing method for producing SSB.

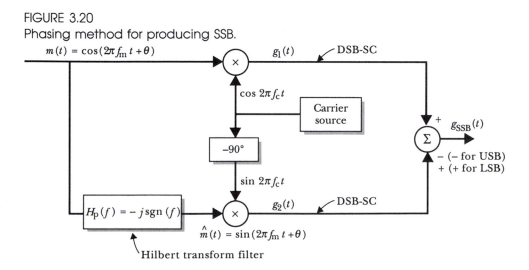

driven by a cosine carrier signal, and its output is given by

$$g_1(t) = \cos\,(2\pi f_m t + \theta)\,\cos\,(2\pi f_c t)$$

Applying the identity for the product of cosines, we obtain

$$g_1(t) = \frac{1}{2}\cos\,[2\pi(f_c + f_m)t + \theta] + \frac{1}{2}\cos\,[2\pi(f_c - f_m)t - \theta]$$

In the lower path of Figure 3.20, the message signal is applied to a filter that phase shifts all of the frequency components of the message by $-90°$ without changing their amplitudes. The transfer function of this filter is given by

$$H_p(f) = -j\,\text{sgn}\,(f). \tag{3.14}$$

For the single tone message under consideration, the output of the filter becomes

$$\hat{m}(t) = \cos\,(2\pi f_m t + \theta - 90°)$$
$$= \sin\,(2\pi f_m t + \theta)$$

This filter output is applied to the input of a DSB-SC modulator that is also supplied with a sine carrier component, as shown in Figure 3.20. The output of this modulator is given by

$$g_2(t) = \sin\,(2\pi f_m t + \theta)\,\sin\,(2\pi f_c t)$$
$$= -\frac{1}{2}\cos\,[2\pi(f_c + f_m)t + \theta] + \frac{1}{2}\cos\,[2\pi(f_c - f_m)t - \theta]$$

Now if the outputs $g_1(t)$ and $g_2(t)$ of these DSB-SC modulators are added, the upper side frequencies cancel and the *lower sideband version of SSB* results. This is given by

$$g_{ssb-}(t) = g_1(t) + g_2(t)$$

For a general message signal $m(t)$, this becomes

$$g_{ssb-}(t) = m(t)\,\cos\,(2\pi f_c t) + \hat{m}(t)\,\sin\,(2\pi f_c t) \tag{3.15}$$

For the special case of a single tone message signal this becomes

$$g_{ssb-}(t) = \cos\,[2\pi(f_c - f_m)t - \theta]$$

Notice that the upper side frequency has canceled, and we are left with only the lower side frequency. On the other hand, if the modulator outputs are subtracted, the upper side frequency remains. Thus, the *upper sideband version of SSB* is given by

$$g_{ssb+}(t) = m(t)\,\cos\,(2\pi f_c t) - \hat{m}(t)\,\sin\,(2\pi f_c t) \tag{3.16}$$

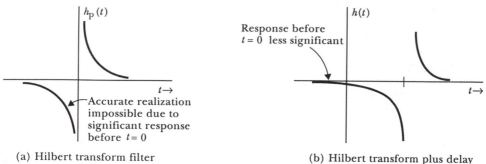

(a) Hilbert transform filter   (b) Hilbert transform plus delay

FIGURE 3.21

The phase-shifted version of the message produced at the output of the filter in Figure 3.20 is known as the *Hilbert transform* of the message signal. The impulse response of the Hilbert transform filter can be found by taking the inverse Fourier transform of the transfer function of the filter given in Equation 3.14. The result is

$$h_{\mathrm{p}}(t) = \frac{1}{\pi t} \tag{3.17}$$

This impulse response is shown in Figure 3.21a. Notice that this filter is not realizable since its impulse response is not zero prior to $t = 0$. However, the Hilbert transform filter can be approximated if a time delay is included. The impulse response of the Hilbert transform filter plus a time delay is shown in Figure 3.21b. This combination can be accurately approximated by a hardware realization, because the portion of the impulse response that comes before $t = 0$ is less significant. This suggests that the system of Figure 3.20 can be accurately realized by including a time delay in both the upper and lower message paths. In practice, this is what is done except that the design focuses on achieving a precise 90° phase difference between the inputs to the DSB-SC modulators in the upper and lower paths rather than accurately implementing the Hilbert transform plus delay block. This is done because a precise 90° phase difference is required for good cancellation of the undesired sideband. The interested reader will find more details in the literature. Another method known as the *third method* or *Weaver's method* for generating the SSB signal can also be found in the problems at the end of the chapter.

Demodulation

The SSB signal can be demodulated by the system shown in the block diagram of Figure 3.22. We will assume in this discussion that the signal to be demodulated is the upper sideband version. The modifications for the lower side-

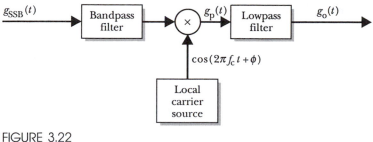

FIGURE 3.22
SSB demodulator.

band version will be obvious. The received SSB signal is first passed through a bandpass filter to eliminate any noise or interference in adjacent frequency bands. This prefiltering is necessary in the case of SSB because any undesired signals below the carrier frequency would be converted to frequencies in the desired message band by the mixer, resulting in interference that cannot be eliminated by lowpass filtering. This is in contrast to the DSB-SC demodulator of Figure 3.2, in which prefiltering is not as necessary. We will assume that the desired SSB signal passes through this prefilter undistorted.

The filter output is applied to the input of a multiplier circuit as shown in Figure 3.22. The other input to the product device is a locally generated version of the carrier. The output signal from the mixer is given by

$$g_p(t) = g_{ssb+}(t) \cos\,(2\pi f_c t + \phi)$$

Substituting the expression for the SSB signal from Equation 3.16 results in

$$g_p(t) = [m(t) \cos\,(2\pi f_c t) - \hat{m}(t) \sin\,(2\pi f_c t)] \cos\,(2\pi f_c t + \phi)$$

Applying the identities for the products of sinusoids results in some terms at baseband and some at twice the carrier frequency. These double frequency terms are rejected by the lowpass filter. The output of the lowpass filter is the baseband terms given by

$$g_o(t) = \frac{1}{2}\,[m(t) \cos\,\phi - \hat{m}(t) \sin\,\phi]$$

Thus, when the phase of the locally generated carrier is zero (i.e., the same as the phase of the carrier in the received signal), the output of the demodulator is the original message signal. However, when ϕ is not zero, the output contains both the message and its Hilbert transform. When it is necessary to preserve both the amplitude and phase relationships of the message waveform, the Hilbert transform signal represents distortion because its phase relationships have been changed from those of the original signal. When the message signal is voice, the phase is not important for intelligibility, and therefore it is not necessary to control the value of ϕ. This is a significant advantage for SSB as compared to DSB-SC because the carrier can be generated by a *free-running local oscillator* rather than by a complex carrier recovery subsystem.

3.4
VESTIGIAL SIDEBAND AM

If the message signal contains frequency components nearly all the way down to dc, it becomes difficult to realize the filters needed for producing SSB. *Vestigial sideband* (VSB) modulation is a method for gaining most of the bandwidth advantage of SSB without filters with difficult specifications. In VSB, all of one of the sidebands except for a small part or vestige is removed by filtering, and the other sideband is reduced slightly to compensate for the contributions of the vestigial sideband at the demodulator. Typically, this filtering is done partly in the transmitter and partly in the receiver.

Figure 3.23 shows the block diagram of a VSB system as well as the frequency response of the filters used to produce the VSB effect. The message signal is multiplied by the carrier to produce either DSB-SC or AM (if a dc component is added to the message before the modulator). This signal is then passed through a bandpass filter, which eliminates most of one of the sidebands. In the receiver, the VSB signal is passed through a prefilter that has a gradual cutoff characteristic before reaching the demodulator. The overall response of these filters is

$$\blacksquare \qquad H(f) = H_{\mathrm{t}}(f)H_{\mathrm{r}}(f) \qquad\qquad (3.18)$$

Ideally, the roll-off characteristic of the overall transfer function magnitude is symmetrical about the carrier frequency. In equation form this symmetry requirement is given by

$$\blacksquare \qquad H(f_{\mathrm{c}} - f_{\mathrm{m}}) + H(f_{\mathrm{c}} + f_{\mathrm{m}}) = C \qquad\qquad (3.19)$$

where C is a constant and f_{m} is the frequency of a message component.

We have assumed that the phase of the transfer function is zero. Of course, if the phase is a linear function of frequency, the received signal is simply time delayed. As long as the carrier recovery circuit tracks the phase of the received carrier, the message is properly demodulated. As usual, a nonlinear phase response results in waveform distortion.

For a simple demonstration of the demodulation process, assume that the message consists of a single frequency component given by

$$m(t) = \cos\,(2\pi f_{\mathrm{m}}t)$$

In this case, the DSB-SC signal produced in the modulator is

$$g_{\mathrm{sc}}(t) = 2\cos\,(2\pi f_{\mathrm{m}}t)\,\cos\,(2\pi f_{\mathrm{c}}t)$$
$$= \cos\,[2\pi(f_{\mathrm{c}} - f_{\mathrm{m}})t] + \cos\,[2\pi(f_{\mathrm{c}} + f_{\mathrm{m}})t]$$

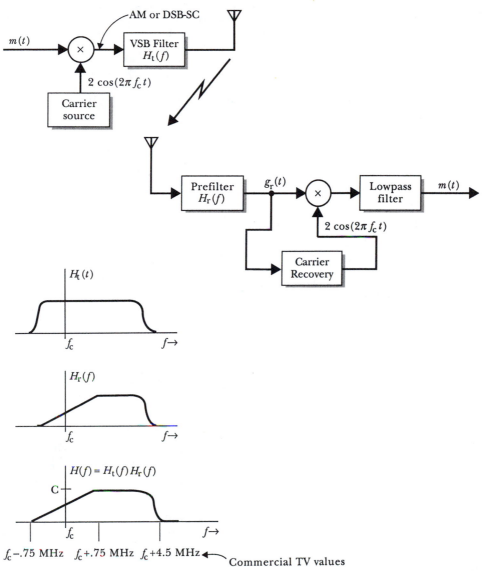

FIGURE 3.23
Typical VSB communication system.

The factor of two is due to the assumed carrier amplitude in Figure 3.23. The resulting received signal at the input to the demodulator is given by

$$g_r(t) = H(f_c - f_m) \cos [2\pi(f_c - f_m)t] + H(f_c + f_m) \cos [2\pi(f_c + f_m)t]$$

The demodulator output is given by the low frequency terms of

$$g_r(t) [2 \cos (2\pi f_c t)]$$

Substituting the expression for $g_r(t)$, using the identity for the product of cosines, and discarding the terms in the vicinity of twice the carrier frequency results in

$$[H(f_c - f_m) + H(f_c + f_m)] \cos 2\pi f_m t$$

Using the symmetry requirement of Equation 3.19 for the overall transfer function of the system filters, this reduces to

$$C \cos 2\pi f_m t$$

This is identical to the original message except for the scale factor C. Since the overall system is linear for the message signal, superposition holds. Thus, a scaled version of each frequency component of the message signal is reproduced at the demodulator output.

VSB is used for commercial television broadcast in the United States. The numerical bandwidths indicated in Figure 3.23 are the nominal values for television transmission. One of the chief reasons for using VSB rather than a double-sided signal is the large savings of bandwidth that is possible. A double-sided TV signal would occupy nine megahertz whereas the VSB signal occupies only six megahertz including guardbands between adjacent channels. This saves three megahertz. Given the demands for the radio spectrum, this is a substantial savings.

We have demonstrated that the VSB signal can be demodulated by the use of a product demodulator with a recovered carrier. At the time that the TV standards were being developed, this demodulator approach was deemed to be too expensive for use in the large number of receivers to be manufactured. As a result, the standards were developed so that the signal can be demodulated by the use of an envelope detector. To achieve this, a dc component is added to the message before modulation. As a result, the modulated signal before filtering is actually an AM signal whose envelope is identical to the message waveform. The largest part of the power in a TV message signal is close to dc. Therefore, the corresponding side frequencies are close to the carrier and appear at the demodulator nearly equal in amplitude so the VSB signal to be demodulated closely resembles an AM signal. As a result, an envelope demodulator can be used with only a small amount of distortion.

3.5
FREQUENCY DIVISION MULTIPLEXING

Situations arise in communication systems in which it is necessary to transmit a number of message signals simultaneously over one channel. One method for avoiding mutual interference is a technique known as *frequency division multiplexing* (FDM).

A block diagram illustrating this technique is shown in Figure 3.24. Two message signals $m_1(t)$ and $m_2(t)$ are transmitted to their destinations over a common channel. The message signals are first used to modulate two carrier frequencies, chosen far enough apart so the modulated signals do not overlap in the frequency domain. The modulated signals are then summed and transmitted over the channel. The spectra of the messages and the sum of the modulated carriers are indicated in the figure. DSB-SC modulation is used in illustrating the spectra of Figure 3.24. However, any type of modulation can be used in FDM as long as the carrier spacing is sufficient to avoid spectral overlap. At the receiving end of the channel, the various modulated signals are separated by bandpass filters that pass only the desired signal and reject all of the others. Once a signal has been selected by the bandpass filter, it is passed to a demodulator, in which the message signal is recovered.

Examples of FDM can be found in commercial AM and FM broadcast, television, and in many other radio transmissions. Each transmitter modulates its message signal onto its assigned carrier frequency and radiates the signal into space where it is summed with the transmissions of all the other transmitters. Each receiver obtains the sum of many signals from its antenna and selects the desired signal by filtering, followed by demodulation.

Commercial AM broadcast stations use carrier frequencies spaced 10 kHz apart in the frequency range from 540 to 1600 kHz. This separation is not sufficient to avoid spectral overlap for amplitude modulation with reasonably high fidelity (say 50 Hz to 10 kHz) audio message signals. Therefore, AM stations on adjacent carrier frequencies are placed geographically far apart to minimize interference. Commercial FM broadcast uses carrier frequencies spaced 200 kHz apart. This larger separation is necessary to accommodate the wider bandwidth associated with frequency modulation. As we will see in later chapters, FM uses more space in the radio spectrum but is able to achieve a higher SNR at the demodulator output.

Another example of the use of FDM on a large scale is in long distance transmission of telephone message signals. In this case, each voice signal is limited to the frequency range from 200 Hz to 3.2 kHz. Up to 600 or more such signals are transmitted over coaxial cable or microwave radio links using SSB modulation with carrier frequencies spaced 4 kHz apart. To avoid the necessity for SSB modulators working at 600 separate frequencies, the message signals are placed in groups of twelve each. Each message signal in a group is SSB modulated on a separate carrier. The carriers for each group are spaced 4 kHz apart from 64 to 108 kHz. The lower sideband version of SSB is used so the message spectra are "stacked" in 4 kHz intervals from 60 to 108 kHz. In the next step of the multiplexing process, the composite signals from each of five groups are combined to produce a "supergroup" by again using SSB modulation and different carrier frequencies. Finally, up to ten supergroups can be multiplexed for transmission over the channel.

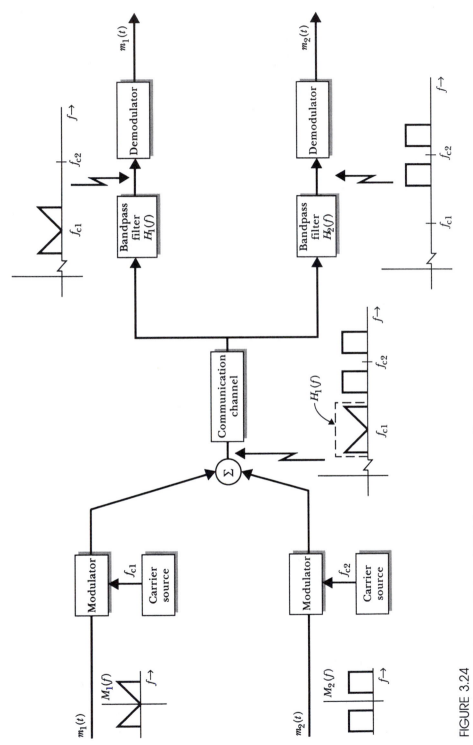

FIGURE 3.24
Frequency division multiplexed (FDM) communication system.

3.6
RADIO RECEIVER DESIGN CONSIDERATIONS
Superheterodyne receivers

A block diagram of the commonly used *superheterodyne architecture* for radio receivers is shown in Figure 3.25. The desired input signal is obtained from a receiving antenna and could be any of a variety of signals such as ordinary AM broadcast, commercial FM, television, and so on. In addition to the desired signal, a number of undesired signals as well as noise and interference are present at the terminals of a typical antenna. The function of the receiver is to reject as much of the noise and interfering signal as possible and then to demodulate the desired signal. The resulting message signal is finally delivered to the output terminals or to an appropriate transducer or display device. In this section, we will describe the function of each of the blocks and address the problem of interfering signals. In Chapter 5, the effects of in-band noise on the demodulation process will be developed.

The received signal is first passed through a bandpass amplifier, known as a *radio frequency* (RF) *amplifier*, to increase its amplitude so that noise added in the *mixer* and following stages is insignificant and some of the incoming noise and interfering signals are rejected. The primary design considerations for the RF amplifier are its noise figure, frequency response, and nonlinear characteristics when handling very strong signals.

The noise performance of an amplifier is given by its *noise figure* which will be defined and discussed in more detail in Chapter 5. The noise added to the signal by the RF amplifier should be small compared to the noise present at the antenna terminals. As we will see later, this often (but not always) implies that the RF amplifier should have a "good" noise figure.

The nonlinear characteristics of the amplifier are important because the RF amplifier is exposed to undesired signals that may be much stronger than the desired signal. This is because the transmitter for the desired signal may

FIGURE 3.25
Typical superheterodyne radio receiver block diagram.

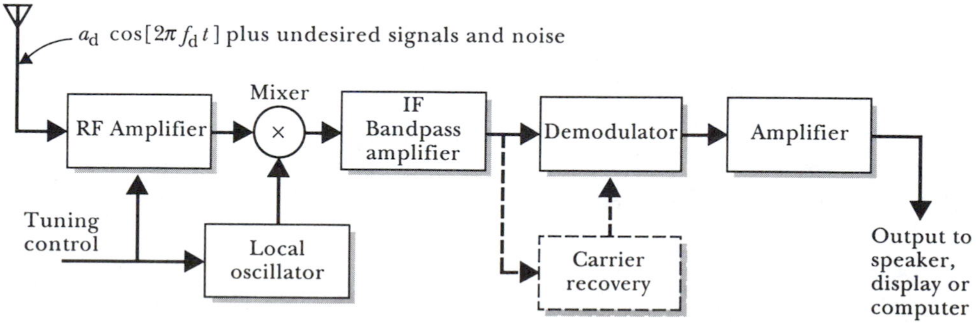

$a_d \cos[2\pi f_d t]$ plus undesired signals and noise

be much farther from the receiver than that of the undesired signal. The output of any memoryless nonlinear device can be written as a power series of the input signal. This is the familiar Maclaurin series from calculus. In equation form, the output of the RF amplifier v_o can be expressed in terms of the input v_{in} as

$$\blacksquare \quad v_o = K_0 + K_1 v_{in} + K_2(v_{in})^2 + K_3(v_{in})^3 + \cdots \quad (3.20)$$

Now suppose that the input signal consists of a desired signal plus an undesired term given by

$$v_{in} = a_d \cos (2\pi f_d t) + a_u \cos (2\pi f_u t)$$

The subscripts indicate the desired (d) and the undesired (u) signal. When this expression is substituted into Equation 3.20, many terms result. The cubic term of Equation 3.20 produces one of the most undesirable terms at the output of the RF amplifier. This troublesome term is

$$3K_3[a_d \cos (2\pi f_d t)][a_u \cos (2\pi f_u t)]^2 = \frac{3}{2} K_3 a_d (a_u)^2 \cos (2\pi f_d t)$$

$$+ \frac{3}{2} K_3 a_d (a_u)^2 \cos (2\pi f_d t) \cos (4\pi f_u t)$$

The first term on the right-hand side of the expression is the one that causes difficulty. This is due to the fact that it represents an AM signal on the desired carrier, but the modulation is due to the square of the amplitude of the undesired signal times the desired signal amplitude. This term represents distortion that overlaps the desired signal in the frequency domain. Therefore, it cannot be eliminated by filtering. This type of distortion is known as *cross modulation* or *intermodulation* because the amplitude of one signal appears modulated on the carrier of another signal. It is also called *third-order distortion* because it arises from the cubic term of the nonlinearity. However, still higher order terms can produce a similar effect.

Detailed expansion of the other terms of Equation 3.20 shows that only sum and difference frequencies of two signal components occur from the second-order term. Thus, a nonlinear device that contains only a second-order term does not produce cross modulation components from signals close to the desired frequency that cannot be removed by filtering. Field-effect devices have a square law input-output characteristic and are often preferred in RF amplifiers for this reason. If strong signals that are widely separated from the desired frequency are allowed to reach the second-order device, it is possible for their sum and difference frequencies to fall in the desired band so that some bandpass filtering ahead of the device is necessary.

One of the objectives in the design of the frequency response of the RF amplifier is to prevent strong undesired signals from reaching the mixer in which cross modulation can also occur. This may be difficult to achieve because highly selective filter characteristics coupled with a requirement for the receiver to be tunable to various incoming frequencies may be difficult to achieve at RF. In general, highly selective filters require many components that must change value in a precisely controlled manner when the center frequency changes. Thus, a high degree of selectivity in the RF amplifier is often not practical. Rejection of adjacent signals must be accomplished later in the *intermediate frequency* (IF) filter.

The primary objective of the mixer is to convert the frequency range of the desired signal so that it lies in the passband of the IF filter. The mixer forms the product of the output signal from the RF amplifier and the *local oscillator* (LO) signal. The term in this product due to the desired signal is given by

$$[a_d \cos(2\pi f_d t)][a_{lo} \cos(2\pi f_{lo} t)] = \frac{1}{2} a_d a_{lo} \{\cos[2\pi(f_{lo} - f_d)t]$$

$$+ \cos[2\pi(f_{lo} + f_d)t]\}$$

One of the terms on the right-hand side of this expression is selected by the IF filter and the other is rejected. The effective carrier frequency of the term selected is the intermediate frequency or IF. Usually, the LO signal is constant in amplitude and phase so any amplitude or angle modulation present on the incoming desired signal is also present on the IF signal. Thus, the local oscillator, mixer, and IF filter combination act to change the carrier frequency of the desired signal up or down to the IF. The combination is called an *up converter* or a *down converter*, depending on whether the resulting IF is higher or lower than the input carrier frequency. The IF is given by

$$f_{if} = f_{lo} + f_d \qquad \text{or by} \qquad f_{if} = f_{lo} - f_d \qquad (3.21)$$

Note that, in the case in which the difference frequency is selected, if the result is a negative frequency, the minus sign should be dropped because the cosine of a negative angle is the same as for a positive angle. In many receivers, the IF is lower than the carrier frequency because it is usually easier to obtain a highly selective filter characteristic when the center frequency is lower. This fact tends to cause a designer to choose a low IF.

Besides the desired signal, it is possible for a signal at a second frequency to produce an output from the mixer that lies at the IF. For example, if the desired signal is lower than the local oscillator frequency, then a signal that is higher than the LO frequency by the IF results in an output from the mixer at the IF. The frequency that produces this second undesired IF signal is called the *image frequency*. The relationship between the LO frequency, the image frequency, and the IF are illustrated in Figure 3.26 for the case in which the LO is above the frequency of the desired signal. The LO frequency and image frequency trade places in the figure for the case in which the LO frequency

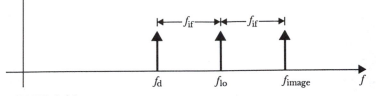

FIGURE 3.26
The relationships between the frequencies in a superheterodyne receiver when the desired frequency is below the local oscillator. When the desired frequency is above the local oscillator, f_d and f_{image} trade places.

is below the desired frequency. Notice that

$$f_{image} = f_d + 2f_{if}$$

when the LO frequency is above f_d and

$$f_{image} = f_d - 2f_{if}$$

when the LO frequency is below f_d.

Once a signal at the image frequency has reached the mixer, it is converted to the IF in the same manner as the desired signal, and it is impossible to stop it from causing undesirable interference at the receiver output. Therefore, one of the prime objectives of the RF filter is to eliminate any signal that may be present at the image frequency. Since the distance in the frequency domain between the desired signal and the image is twice the IF, it is desirable in the design of the RF filter to have a high IF. As we have noted above, a low IF is often desirable when designing the IF filter for a high degree of selectivity, so a compromise is necessary. In extreme cases, it may not be possible to find a suitable compromise and a more complex receiver architecture with two frequency converters in cascade may be necessary. The first IF is selected high to allow the RF filter to reject the image. The second IF is selected low to ease the design of the second IF filter, in which the filtering to eliminate adjacent undesired signals is accomplished. Note that images are possible at each mixer, and the offending image signals must be eliminated in the filters ahead of each mixer.

The output signal of the IF amplifier consists of the desired signal plus any noise that lies in the same frequency range. This is passed to a demodulator that is appropriate for the type of modulation being used. If a suppressed carrier modulation is in use, it is necessary to have a carrier recovery subsystem as indicated in dotted lines in Figure 3.25. Actually the "carrier" recovered is at the IF.

Another consideration in receiver design is to avoid unintentional radiation or leakage of signals from the IF amplifier or local oscillator. These leakage signals can cause interference by entering the circuitry at unintended locations or with nearby receivers by entering through the antenna. Considerable effort may be necessary in the selection of a frequency plan for the receiver and in the physical construction of the receiver to avoid this type of problem. Sometimes receivers are designed without RF amplifiers or filters. In addition to reduced noise performance and poor image rejection, this can lead to greatly increased leakage of the LO signal. Thus, an additional function of the RF amplifier is the provision of a buffer between the antenna and the mixer to avoid interference with other receivers.

Tuned RF receivers

An alternate but rarely used receiver architecture known as the *tuned RF* approach is shown in Figure 3.27. In this type of receiver, no frequency conversion is employed before the demodulator. The bulk of the amplification and interference rejection is accomplished in the RF amplifier and filter.

The tuned RF approach has some serious disadvantages compared to the superheterodyne architecture. If the receiver is to be tuned over a band of frequencies, it is often difficult to design a highly selective filter with a variable center frequency for the tuned RF receiver. (The selectivity of the superheterodyne receiver is provided mainly by the fixed-tuned IF filter.) Another problem of the tuned RF approach is that it is often difficult to design an amplifier with the required gain without having it oscillate due to signals from the output reentering the amplifier input by unintended paths. Oscillation requires the net loop gain of the amplifier and the unintentional path to be greater than unity, which is more likely if the amplifier gain is very high. The superheterodyne receiver has the required gain segmented with part of the gain at RF and part at IF.

On the other hand, the tuned RF approach does not have problems with images or with LO leakage.

FIGURE 3.27
Tuned RF receiver architecture.

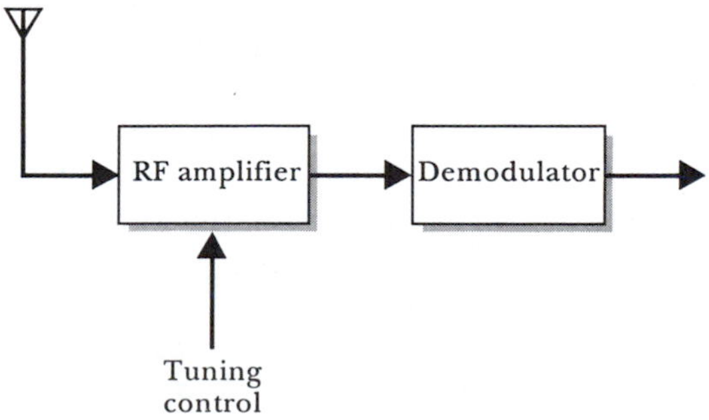

SUMMARY

As usual, the important equations in the chapter are marked with a bullet (■). These equations should be included in a review of the material.

1. Modulation is the process of causing some parameter of a carrier to vary in relation to a message signal. It is used for the purpose of placing the signal in a more advantageous location in the frequency domain or to allow several messages to be carried on a single channel using FDM. Certain forms of modulation allow bandwidth to be traded for improved signal-to-noise ratio at the system output. Various forms of amplitude modulation result from allowing the message signal to control the amplitude of a sinusoidal carrier.

2. DSB-SC modulation results when the message signal multiplies a sinusoidal carrier. Its spectrum consists of both an upper and a lower sideband. A carrier component is present only if the message signal contains a dc component. The bandwidth of the DSB-SC signal is twice the highest message frequency. Demodulation requires a coherent carrier that is synchronized in phase and frequency with the carrier at the transmitter. The modulation efficiency of DSB-SC is 100%.

3. AM results if a carrier component is added to a DSB-SC signal. Usually, the carrier amplitude is larger than the peak amplitude of the sidebands so the upper envelope of the AM signal is the same as the message signal plus dc. The spectrum consists of a discrete carrier component plus upper and lower sidebands. The bandwidth of the AM signal is twice the highest message frequency. Demodulation can be accomplished with a simple envelope detector provided that the carrier component is large enough. Modulation efficiency is low, particularly for voice signals, which have a high ratio of peak amplitude to rms amplitude.

4. SSB results from removing one of the sidebands of DSB-SC signal. It can be generated either by filtering or by the phasing method. The spectrum consists of either the upper or the lower sideband. The bandwidth of the SSB signal is the difference between the highest and the lowest message frequencies. Demodulation can be accomplished by multiplication by the carrier. When phase relationships of the message do not need to be preserved, it is not necessary to synchronize the phase of the local carrier. In this case, a simple free-running oscillator closely matched to the transmitter frequency can be used. Modulation efficiency is 100%. SSB is not suitable for messages containing a dc component.

5. VSB results from removing all but a vestige of one of the sidebands from either a DSB-SC or an AM signal. The bandwidth is more than the highest message frequency but considerably less than twice the highest message frequency. VSB can be demodulated by multiplicatiion with a local carrier followed by lowpass filtering. VSB is used in television

broadcasting, in which it results from filtering an AM signal. In television receivers, demodulation is accomplished, with some distortion, by the use of an envelope detector. Modulation efficiency depends on the starting modulation format.

6. All forms of amplitude modulation we have considered result from linear operations on the message signal. Thus, superposition applies (excluding the carrier component of an AM signal). Therefore, these modulation formats are known collectively as *linear modulation*.

7. Modulation and demodulation is often accomplished by multiplying by the sinusoidal carrier. Multiplication can be accomplished by switching the signal on and off at the carrier rate followed by filtering to select the desired components of the product. Another approach is to add the signals to be multiplied and then pass the sum through a nonlinear device, again followed by filtering to select the desired product terms. The double balanced mixer or ring modulator is an important and widely used device in communication systems. It is used as a modulator, demodulator, mixer in superheterodyne receivers and, as we will see later, a phase detector in phase-locked loops.

8. Carrier recovery is necessary for demodulation of DSB-SC. It can be accomplished by squaring the received DSB-SC signal followed by filtering and a divide-by-two frequency divider to retreive the carrier component. Carrier recovery can also be accomplished by use of the Costas phase-locked demodulator.

9. Two message signals can be multiplexed using DSB-SC with quadrature carriers providing a doubling of the usefulness of the spectrum. However, very tight control of the carrier phase is necessary in the receiver to avoid crosstalk. FDM with SSB modulation is usually a more practical way to gain the equivalent spectral efficiency.

10. Spectrum analyzers use modulation techniques to produce a frequency domain display of the amplitude spectrum of a test signal. By sweeping the carrier frequency, the sideband produced by the test signal is moved past a fixed tuned narrowband filter. The rectified filter output is then displayed versus frequency.

11. If all of the frequency components of a signal are phase shifted by $-90°$, the Hilbert transform of the signal results. The Hilbert transform is useful in giving an analytic description of the SSB signal.

12. FDM is an important technique in communications engineering and finds wide application. Modulation using different carrier frequencies is used to place all signals in nonoverlapping frequency ranges so that they can be separated by filtering.

13. Most radio receivers use the superheterodyne principle. The desired signal is first amplified and filtered in an RF amplifier. Then it is up or down converted to a fixed IF in a mixer supplied with a LO signal. The signal at the IF is filtered, amplified, and then either converted to a second IF or immediately demodulated. The functions of the RF amplifier and filter

are to amplify the desired signal without degrading the signal-to-noise ratio, to avoid crossmodulation, to remove strong undesired signals that could overload the mixer, to strongly reject signals at the image frequency, and to provide a buffer preventing leakage of the LO signal. The mixer produces the product of the RF amplifier output and the LO signal. This results in versions of the desired signal, including its modulation, at the sum and difference frequencies. One of these components is the desired IF. The main functions of the IF filter and amplifier are to remove adjacent signals and provide the bulk of the required gain. The chief advantage of the superheterodyne architecture is that it allows a tunable receiver with a minimum of critical adjustable components. In addition, it allows the large gain to be segmented and placed at different frequencies.

REFERENCES

H. L. Krauss, C. W. Bostian, and F. H. Raab. *Solid State Radio Engineering.* New York: Wiley, 1980.

J. Smith. *Modern Communication Circuits.* New York: McGraw-Hill, 1986.

PROBLEMS

1. A 100 kHz sinusoidal carrier is DSB-SC modulated by a 5-kHz sinusoidal message waveform, and the peak amplitude of the modulated waveform is 5 V.
   (a) Sketch the modulated waveform to scale indicating any phase reversals.
   (b) Find the power in both of the frequency components and the total power in the modulated waveform.
2. The 1 kHz symmetrical square-wave message signal shown in Figure P3.1 is used to DSB-SC modulate a 100 kHz sinusoidal carrier. The average (normalized) power of the resulting modulated signal is 50 W.
   (a) Sketch the waveform of the modulated signal to scale versus time.
   (b) Sketch the amplitude spectrum to scale versus frequency.
   (c) Find the power associated with the 101 kHz component.

FIGURE P3.1

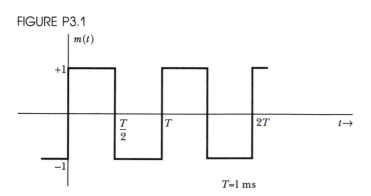

$T = 1$ ms

3 / AMPLITUDE MODULATION

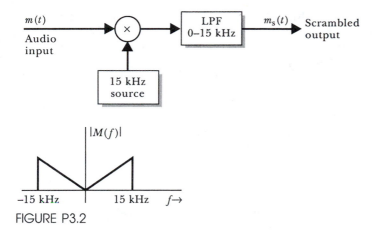

FIGURE P3.2

3. The system shown in Figure P3.2 is sometimes used for scrambling audio signals.
   (a) For the message spectrum shown, sketch the spectrum of the scrambled signal at the output of the filter.
   (b) Draw the block diagram of a descrambler for this signal.
   (c) If the multiplier is implemented with a circuit that is not well balanced for the message signal, the scrambling is partially defeated because the output contains a message component in unaltered form. Design an equivalent system in block diagram form that will be able to use unbalanced modulators but avoid having message feedthrough.

4. Two sinusoidal signals at frequencies f_1 and f_2 are added together and passed through an amplifier having a nonlinear input-output characteristic. Assuming that the nonlinearity is described by first-, second-, and third-power terms, list the frequencies of all of the components present in the output signal. If f_1 is close to f_2 (i.e., $f_1 - f_2$ is much less than either f_1 or f_2), which term in the nonlinearity is responsible for new frequency components that fall close to f_1 and f_2? If both of the input components are increased in amplitude by 3 dB, by how many decibels do the output components, due to each term of the nonlinearity, increase?

5. A message signal $m(t)$ has the Fourier transform shown in Figure P3.3.
   (a) Find the spectrum of $m^2(t)$ and sketch to scale versus frequency. What is the highest frequency present in $m^2(t)$?
   (b) Repeat for $m^3(t)$.
   (c) What is the highest frequency present in $m^n(t)$?

FIGURE P3.3

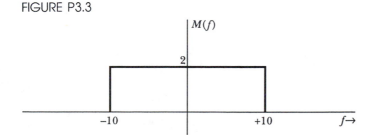

6. Consider the quadrature multiplexed system of Figure 3.11. Assuming that the two message signals have the same power, how large can the phase error θ of the recovered carrier become before the crosstalk term is 40 dB below the desired signal at the system output? What is the effect of this amount of phase error on the demodulation of a DSB-SC signal when quadrature multiplexing is not in use? From the results of this problem you should conclude that phase errors are much more serious when quadrature multiplexing is in use.

7. For a message signal given by $m(t) = \cos(2000\pi t)$ and a 100-kHz carrier, (a) write an expression for an AM signal with 50 W of power in the carrier component and a modulation index of 50%, (b) sketch the signal of part (a) to scale versus time, (c) sketch the amplitude spectrum of the signal of part (a) to scale, and (d) find the power in each of the side frequency components for the signal of part (a).

8. Derive an expression for the modulation efficiency of an AM signal with a single tone sinusoidal message signal in terms of the modulation index.

9. The AM signal $g_{am}(t) = A_c[1 + m(t)] \cos(2\pi f_c t)$ is to be demodulated by the envelope detector shown in Figure P3.4a. Assume that the carrier frequency is very high so the capacitor discharges only a very small amount during a carrier cycle.
   (a) Sketch the output waveform to scale for the message waveform of Figure P3.4b when RC is 0.1 ms.
   (b) Repeat part (a) if RC is 1.0 ms.
   (c) Find the maximum allowed value of the RC time constant as a function of k; so the message of Figure P3.4c can be demodulated without distortion.
   The distortion caused by too long a time constant in this circuit is called *diagonal clipping*.

FIGURE P3.4

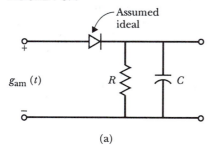

(a)

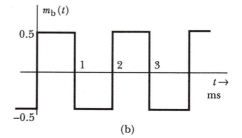

(b)

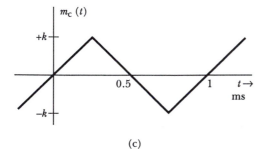

(c)

10. A load resistance equal in value to the resistance R of the envelope detector is connected to the output terminals of the envelope detector of Figure P3.4a through a very large coupling capacitor. The AM signal $g_{am}(t) = A_c[1 + m(t)] \cos (2\pi f_c t)$, with the message signal of Figure P3.4c is to be demodulated by this circuit. Assume that the parallel smoothing capacitor C is large enough so that it discharges a negligible amount during the (very short) carrier cycle but that C is small enough so the message frequency current through it is negligible. Sketch the output waveform to scale for various values of k. Notice that distortion will occur due to the fact that the current supplied by the diode averaged over a carrier cycle cannot be negative. This type of distortion is called *bottom clipping*.

11. One method for evaluating AM modulators is to produce the *trapezoidal pattern* on an oscilloscope by applying the AM signal to the vertical input and the message signal to the horizontal input. A typical resulting pattern is shown in Figure P3.5.
    (a) If the message signal is a single tone, so that the resulting AM signal is $g_{am}(t) = A_c[1 + \mu \cos (2\pi f_m t)] \cos (2\pi f_c t)$, find the modulation index μ in terms of the measurements A and B taken from the pattern.
    (b) How would you judge the linearity of the modulator by observing the pattern?
    (c) Sketch the pattern for a DSB-SC modulator.

12. The message signal $m(t) = \cos (2000\pi t)$ is used to modulate a 10-kHz sinusoidal carrier using the lower sideband version of SSB. The power in the modulated signal is 50 W. Write an expression for the modulated signal and sketch its waveform to scale versus time.

13. Find the Hilbert transform of the signal $m(t) = 3 \cos (2000\pi t) + 5 \sin (6000\pi t)$ by considering the frequency response of the Hilbert transform filter given in Equation 3.14.

14. The message signal $m(t) = \cos (2000\pi t) + \cos (4000\pi t)$ is used to modulate a 100 kHz carrier using the lower sideband version of SSB as given by Equation 3.15.
    (a) The modulated signal is then demodulated using the system of Figure 3.22 with $\phi = 0$ so the original message is recovered. Sketch the demodulated signal to scale versus time.
    (b) Sketch the demodulated waveform when $\phi = 90°$. Notice that the waveform is different from that in part (a). However, if it is an audio signal it would sound the same because the ear is insensitive to phase differences of the frequency components.

15. If the Hilbert transform is taken of the Hilbert transform of a signal, show that the result is the negative of the original signal.

FIGURE P3.5
Trapezoidal pattern.

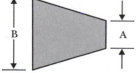

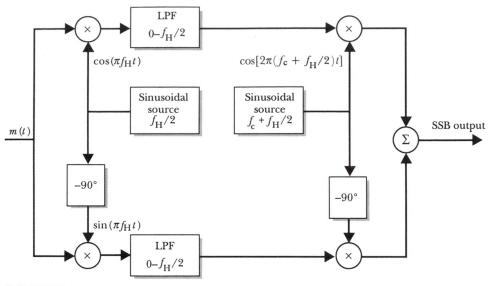

FIGURE P3.6
Weaver's, or the third method, for generating SSB.

16. A 1000 Hz message tone is used to modulate a carrier. The modulated signal is then demodulated in a product demodulator using a carrier that is 0.2 Hz higher than the original. Find an expression for the demodulator output and describe how it would sound as compared to the original message if the modulation is (a) DSB-SC, and (b) the lower sideband version of SSB. (c) What do you conclude about the necessity for recovery of the exact carrier frequency for each modulation format when audio signals are being transmitted and only intelligibility is required?

17. Show that if a large carrier is added to a SSB signal, demodulation can be accomplished with an envelope detector. (It may help to assume a single tone message signal and make use of a phasor diagram such as Figure 3.17c.) What is the consequence of using a carrier only slightly larger than the SSB signal to be demodulated?

18. The block diagram of the system for producing SSB by the third method, also known as Weaver's method, is shown in Figure P3.6. Assume that the message is a single sinusoidal tone and show that the output is in fact the appropriate SSB signal. What argument would you use to show that the circuit produces SSB for a general message signal?

19. Show that the demodulator of Figure P3.7 responds to tones on one side of the carrier frequency but not on the other side. In principle, this approach can be used to eliminate interfering signals on the other side of the carrier from a desired SSB signal. Thus, the system of Figure P3.7 is an alternate approach to using a bandpass filter ahead of the product detector.

20. Consider an AM signal that has been modulated with a single tone message signal with a modulation index of 100%. This signal is then passed through a VSB

3 / AMPLITUDE MODULATION

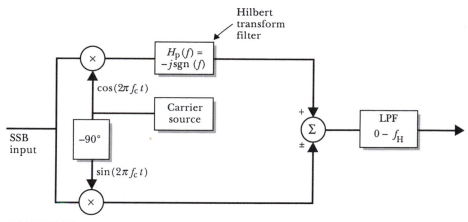

FIGURE P3.7

filter that eliminates the lower side frequency. Note that the carrier component is also reduced.

(a) Write an expression for the resulting VSB signal and sketch its envelope to scale. Hint: Consider the phasor representation of the AM signal shown in Figure 3.17c. Notice that the envelope is a distorted version of the original message. Thus, we expect an envelope detector to have a distorted output.

(b) Would the distortion be as severe if the modulation index is only 10%?

(c) Show that multiplication by the original carrier followed by lowpass filtering recovers the message without distortion.

21. Consider a superheterodyne AM radio with an IF of 455 kHz that must tune from 550 to 1600 kHz.

(a) What are the two possible ranges of the LO frequency? What is the ratio of the highest LO frequency to the lowest LO frequency in each case? Which choice do you think would lead to the easiest design for the LO? Why?

(b) If the radio is tuned to 1000 kHz, what are the two choices for the LO frequency? What are the corresponding image frequencies?

(c) When such a radio is turned to 1500 kHz, it produces a distorted output due to the presence of a strong signal at 750 kHz. What is the probable explanation?

(d) When tuning such a radio through a station at 910 kHz, a tone is often heard that first decreases in frequency to zero and then increases as the receiver is tuned past the 910-kHz station. What is the probable explanation?

(e) What would be the probable effect of having two very strong received signals exactly 455 kHz apart?

Hint for parts c, d, and e: These and many other undesirable effects are due to nonlinear interaction of signals in the receiver amplifiers and due to signals entering or reentering the circuitry at unintended points.

22. Recall from Section 2.5 that a rule of thumb for estimating the risetime of the output of a bandpass filter due to a stepped sinusoidal input is $0.7/\mathrm{BW}$ where BW is the 3-dB bandwidth of the filter. Thus, we can estimate the time to reach steady state as perhaps $2/\mathrm{BW}$. Use this to derive an expression for the approximate minimum allowed sweep time (i.e., the period of the ramp) in terms of the bandwidth BW of the bandpass filter and the sweep range W of the spectrum analyzer of Figure 3.13. Neglect the response time of the lowpass filter.

4

FREQUENCY
AND PHASE
MODULATION

In the last chapter, we introduced the concept of modulation as the process of causing some parameter of a carrier to vary in relation to a message signal. A sinusoidal carrier, given by

$$a(t) \cos \left[2\pi f_c t + \theta(t)\right]$$

has two parameters that can be modulated by the message: the amplitude $a(t)$ and the phase angle $\theta(t)$. We also considered the various forms of amplitude modulation, in which the message varies $a(t)$ while $\theta(t)$ is constant. In this chapter, we consider angle modulation, in which the message modulates $\theta(t)$ while the amplitude $a(t)$ is constant. We consider two closely related forms of angle modulation: *frequency modulation* (FM) and *phase modulation* (PM).

Unlike amplitude modulation, angle modulation turns out to be a *nonlinear* process, and so it is not as easy to give a theoretical treatment for the general case. Therefore, we resort to studying special situations and quoting some empirical rules for estimating the bandwidth of the angle modulated signal.

In the first section of this chapter, we define FM and PM, find the frequency domain representation of a special case known as narrowband (NB), present the block diagram of a modulator for NBFM and NBPM, and present a phasor representation for the case in which the message is a single tone.

The second section presents wideband FM and PM, including modulators and demodulators. One demodulator for FM and PM is the *phase-locked loop* (PLL), which is treated in a separate section of the chapter because of its importance and widespread use in communications.

In Chapter 5, we analyze the effect of in-band noise on various modulation formats and find that wideband angle modulation achieves a higher SNR at the receiver output at the cost of increased channel bandwidth. Later, we will find that digital methods provide another way to make this trade.

4.1
NARROWBAND FM AND PM

Definitions of FM and PM

To define frequency modulation, we must first define the *instantaneous frequency* of a sinusoidal signal. The instantaneous angular frequency of a sinusoid is the rate of change, or time derivative, of its argument. Thus, the instantaneous frequency of the angle modulated sinusoidal signal

$$g(t) = A_c \cos\left[2\pi f_c t + \theta(t)\right] \tag{4.1}$$

is given by

$$f_i(t) = \frac{1}{2\pi}\frac{d}{dt}\left[2\pi f_c t + \theta(t)\right]$$

$$f_i(t) = f_c + \frac{1}{2\pi}\frac{d\theta(t)}{dt} \tag{4.2}$$

Frequency modulation results if the instantaneous frequency of a sinusoidal carrier is varied in proportion to the message signal. In equation form, we have

$$f_i(t) = f_c + k_f m(t) \qquad \text{for FM} \tag{4.3}$$

As usual, f_c is the carrier frequency and $m(t)$ is the message signal. The constant k_f determines how much the instantaneous frequency deviates from the carrier frequency per unit of message signal. Equating the right-hand sides of the last two expressions for $f_i(t)$, we obtain

$$\frac{1}{2\pi}\frac{d\theta(t)}{dt} = k_f m(t)$$

Integrating this expression and multiplying by 2π gives

$$\theta(t) = 2\pi k_f \int_{t_0}^{t} m(x)\, dx + \theta(t_0)$$

The value of the integral at the lower limit and $\theta(t_0)$ simply contribute a fixed phase angle that has little real significance since it can be changed by a different selection of the instant at which $t = 0$. Therefore, we drop the lower limit and the constant to obtain

∎
$$\theta(t) = 2\pi k_f \int^t m(x)\, dx \qquad \text{for FM} \tag{4.4}$$

Phase modulation results when the phase angle of the sinusoidal carrier is proportional to the message signal. In equation form, we have

∎
$$\theta(t) = k_p m(t) \qquad \text{for PM} \tag{4.5}$$

where k_p is a constant that determines the number of radians of phase shift per unit of the message signal.

The FM and PM signals are found by substituting Equations 4.4 and 4.5 into 4.1. The results are

∎
$$g_{fm}(t) = A_c \cos\left[2\pi f_c t + 2\pi k_f \int^t m(x)\, dx \right] \qquad \text{for FM} \tag{4.6}$$

and

∎
$$g_{pm}(t) = A_c \cos\left[2\pi f_c t + k_p m(t) \right] \qquad \text{for PM} \tag{4.7}$$

FM and PM waveforms for a typical message signal are shown in Figure 4.1.

Spectra of narrowband angle modulated signals

Now we consider the special case, known as *narrowband angle modulation*, in which $\theta(t)$ is very small compared to 1 radian. If the general expression for an angle modulated signal given in Equation 4.1 is expanded using the formula for the cosine of the sum of two angles, we obtain

$$g(t) = A_c \cos\left[2\pi f_c t + \theta(t) \right]$$
$$= A_c[\cos \theta(t) \cos (2\pi f_c t) - \sin \theta(t) \sin (2\pi f_c t)]$$

Now, if $\theta(t)$ is much smaller than 1 radian, we have $\cos \theta(t) \simeq 1$ and $\sin \theta(t) \simeq \theta(t)$. This produces the approximate result for NB angle modulation:

∎
$$g(t) \simeq A_c \cos (2\pi f_c t) - A_c \theta(t) \sin (2\pi f_c t) \tag{4.8}$$

The approximate spectrum of the angle modulated signal can be found by taking the Fourier transform of Equation 4.8 by using the modulation property from Table 2.1. This results in

$$G(f) \simeq \frac{A_c}{2}\left[\delta(f - f_c) + \delta(f + f_c) + j\Theta(f - f_c) - j\Theta(f + f_c) \right] \tag{4.9}$$

where $\Theta(f)$ is the Fourier transform of $\theta(t)$, which can be found in terms of the message signal by taking the transforms of Equations 4.4 and 4.5. This

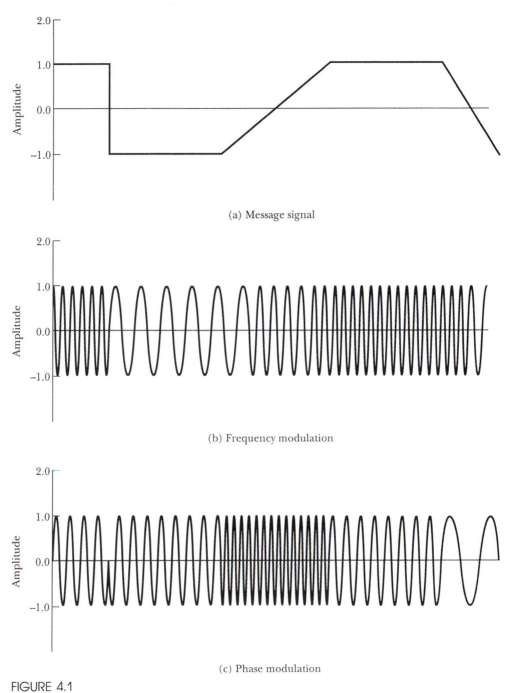

(a) Message signal

(b) Frequency modulation

(c) Phase modulation

FIGURE 4.1
Message waveform and corresponding FM and PM waveforms.

procedure results in

$$\Theta(f) = 2\pi k_f \frac{M(f)}{j2\pi f} \qquad \text{for FM} \qquad (4.10)$$

and

$$\Theta(f) = k_p M(f) \qquad \text{for PM} \qquad (4.11)$$

where $M(f)$ is the Fourier transform of the message signal. (We have assumed that the dc component of the message is zero, or $M(0) = 0$.) Substituting Equations 4.10 and 4.11 into Equation 4.9, we obtain expressions for the Fourier transforms of the NBFM and NBPM signals as

$$\blacksquare \quad G_{pm}(f) \simeq \frac{A_c}{2} [\delta(f - f_c) + \delta(f + f_c)$$
$$+ jk_p M(f - f_c) - jk_p M(f + f_c)] \qquad \text{(NBPM)} \qquad (4.12)$$

FIGURE 4.2
(a) Typical message spectrum, (b) corresponding NBFM spectrum, and (c) corresponding NBPM spectrum.

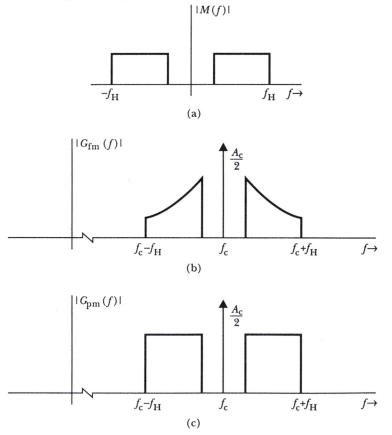

and

$$G_{\text{fm}}(f) \simeq \frac{A_c}{2}\left[\delta(f - f_c) + \delta(f + f_c) + k_f \frac{M(f - f_c)}{(f - f_c)} \right.$$
$$\left. - k_f \frac{M(f + f_c)}{(f + f_c)} \right] \qquad \text{(NBFM)} \qquad (4.13)$$

A typical message amplitude spectrum and the amplitude spectra of the corresponding FM and PM signals are shown in Figure 4.2. Note that the shape of the sidebands for PM is the same as the message spectrum, but the sidebands have a different shape for FM. The side frequencies become smaller for higher message frequencies due to the presence of the integral in the expression for a FM signal. Later, we will find that this fact makes the higher message frequencies more subject to the effects of noise. Therefore, practical FM systems often *preemphasize* the higher frequency message components at the transmitter followed by *deemphasis* at the receiver.

Equation 4.8 suggests that narrowband angle modulation can be generated by first DSB-SC modulating $\theta(t)$ on a phase-shifted version of the carrier and adding the result to the carrier. The block diagram of a system based on this observation for producing either NBFM or NBPM is shown in Figure 4.3. Note that it is necessary to integrate the message signal prior to DSB-SC modulation to obtain FM from this system. It is because of this integrator that the higher frequency message components have reduced amplitudes in the sidebands of a NBFM signal.

Narrowband tone modulation

We now consider the special case in which the message signal is a single sinusoidal tone given by

$$m(t) = A_{\text{m}} \cos\left(2\pi f_{\text{m}} t\right) \qquad \text{for FM}$$

FIGURE 4.3
Narrowband angle modulator block diagram.

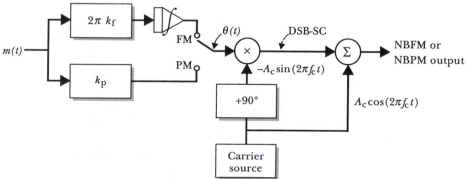

and

$$m(t) = A_m \sin (2\pi f_m t) \qquad \text{for PM}$$

Note that in either case the message is a single tone with a frequency of f_m. Different functional forms have been selected for the message for each modulation format simply to allow one development to serve both cases. For both modulation formats we obtain

$$\theta(t) = \beta \sin (2\pi f_m t) \qquad (4.14)$$

where

■ $$\beta = k_p A_m \qquad \text{for PM} \qquad (4.15)$$

and

■ $$\beta = \frac{k_f A_m}{f_m} = \frac{\Delta f}{f_m} \qquad \text{for FM} \qquad (4.16)$$

In either case, note that β is the peak deviation of the phase of the modulated signal from the phase of the unmodulated carrier. Recall from Equation 4.3 that k_f is the constant that relates the instantaneous frequency deviation to the message amplitude, and it has units of hertz per unit of $m(t)$. Thus, for the FM signal, $\Delta f = k_f A_m$ is the peak deviation of the instantaneous frequency from the carrier frequency.

When β is much less than 1 radian, we have narrowband angle modulation. Substituting Equation 4.14 into 4.8, we obtain the following expression for narrowband angle modulation with a single tone message signal:

$$g(t) = A_c \cos (2\pi f_c t) - A_c \beta \sin (2\pi f_m t) \sin (2\pi f_c t)$$

Using the identity for the product of sines, we obtain

$$g(t) = A_c \cos (2\pi f_c t) + \frac{A_c \beta}{2} \cos [2\pi (f_c + f_m)t] - \frac{A_c \beta}{2} \cos [2\pi (f_c - f_m)t]$$

$$(4.17)$$

Note that the narrowband angle modulated signal for a single tone message consists of a carrier component and a pair of side frequencies. This is almost identical to the situation for a tone-modulated AM signal, which is given by

$$g_{am}(t) = A_c [1 + \mu \cos (2\pi f_m t)] \cos (2\pi f_c t)$$

$$g_{am}(t) = A_c \cos (2\pi f_c t) + \frac{A_c \mu}{2} \cos [2\pi (f_c + f_m)t]$$

$$+ \frac{A_c \mu}{2} \cos [2\pi (f_c - f_m)t] \qquad (4.18)$$

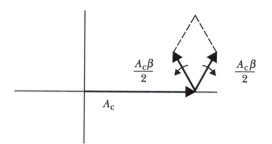

(a) Narrowband angle modulation

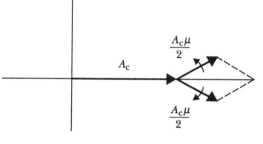

(b) AM

FIGURE 4.4
Phasor diagrams of tone modulated signals.

Comparison of Equations 4.17 and 4.18 shows that the chief difference between narrowband angle modulation and AM is the phasing of the side frequencies relative to the carrier. Figure 4.4 shows phasor diagrams for both tone-modulated angle modulation and AM. In the case of AM, note that the side frequencies add in phase with the carrier component producing amplitude fluctuations. In the case of angle modulation, the side frequency components add to produce a result at quadrature with the carrier resulting in fluctuations in the phase. When β becomes large, a single set of side frequencies results in amplitude fluctuations. Thus, when β is large, more than one set of side frequencies are necessary to describe a constant amplitude angle modulated signal, as we will see in the next section where wideband angle modulation is considered.

4.2
WIDEBAND FM

Wideband tone modulation

We now consider the spectra of angle modulated signals without the narrowband restriction. Again, due to the nonlinear nature of angle modulation, we must resort to special cases. We first consider modulation by a message consisting of a single tone. This is a continuation of the discussion of the last

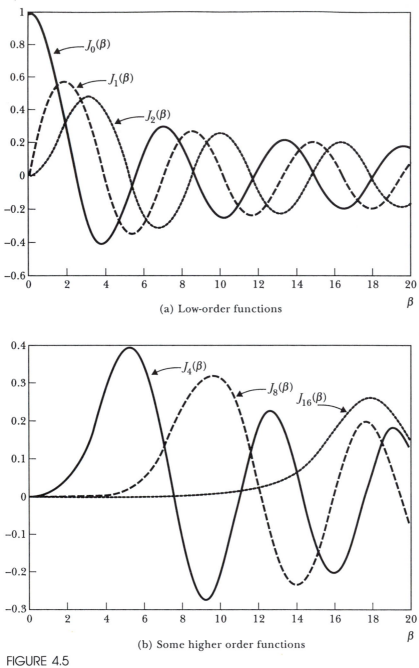

(a) Low-order functions

(b) Some higher order functions

FIGURE 4.5
Bessel functions.

section but without the restriction that β should be much less than 1 radian.

Substitution of Equation 4.14 into Equation 4.1 yields the following expression for the angle modulated signal with a single tone message:

$$g(t) = A_c \cos \left[2\pi f_c t + \beta \sin \left(2\pi f_m t \right) \right]$$

It is possible to show by the use of the Fourier series that this signal can also be written

$$g(t) = A_c \sum_{n=-\infty}^{\infty} \mathcal{J}_n(\beta) \cos \left[2\pi (f_c + n f_m) t \right] \tag{4.19}$$

where $\mathcal{J}_n(\beta)$ is an nth-order Bessel function of the first kind whose argument is β. This Bessel function is defined by

$$\mathcal{J}_n(\beta) = \frac{1}{2\pi} \int_{-\pi}^{\pi} \exp \left[j(\beta \sin x - nx) \right] dx \tag{4.20}$$

and has the following properties

1. $\lim_{\beta \to 0} \mathcal{J}_0(\beta) = 1$ $\qquad\qquad\qquad\qquad\qquad\qquad\qquad\qquad$ (4.21)

2. $\lim_{\beta \to 0} \mathcal{J}_1(\beta) = \beta/2$ $\qquad\qquad\qquad\qquad\qquad\qquad\qquad$ (4.22)

3. $\lim_{\beta \to 0} \mathcal{J}_n(\beta) = 0$ for all $n \geq 2$ $\qquad\qquad\qquad\qquad\qquad$ (4.23)

4. $\mathcal{J}_n(\beta) = (-1)^n \mathcal{J}_{-n}(\beta)$ $\qquad\qquad\qquad\qquad\qquad\qquad$ (4.24)

5. $\sum_{n=-\infty}^{\infty} [\mathcal{J}_n(\beta)]^2 = 1$ $\qquad\qquad\qquad\qquad\qquad\qquad\qquad$ (4.25)

Plots of the first several Bessel functions are shown in Figure 4.5, and Table 4.1 lists some useful selected values.

Study of Equation 4.19 shows that the tone-modulated signal contains an infinite number of side frequencies spaced at integer multiples of the message frequency above and below the carrier. Also the peak amplitude of the component at $f_c + n f_m$ is $A_c \mathcal{J}_n(\beta)$. This seems to indicate that angle modulation occupies an infinite bandwidth. However, study of the properties of the Bessel functions shows that the amplitudes of the side frequencies eventually become very small for large values of n so the side frequencies that are far removed from the carrier are not significant. In fact, it can be shown that over 98% of the power in a tone-modulated FM signal is contained within a bandwidth given by

$$\text{BW} = 2 f_m [\beta + 1] = 2(\Delta f + f_m) \tag{4.26}$$

TABLE 4.1
Selected Bessel function values

n	$\beta = 0$	$\beta = 0.1$	$\beta = 0.2$	$\beta = 0.5$	$\beta = 1$	$\beta = 2$	$\beta = 5$	$\beta = 8$	$\beta = 10$
0	1.0	0.998	0.990	0.938	0.765	0.224	−0.178	0.172	0.246
1		0.050	0.100	0.242	0.440	0.577	−0.328	0.235	0.043
2		0.001	0.005	0.031	0.115	0.353	0.047	−0.113	0.255
3					0.020	0.129	0.365	−0.291	0.058
4					0.002	0.034	0.391	−0.105	−0.220
5						0.007	0.261	0.186	−0.234
6						0.001	0.131	0.338	−0.014
7							0.053	0.321	0.217
8							0.018	0.223	0.318
9							0.006	0.126	0.292
10							0.001	0.061	0.207
11								0.026	0.123
12								0.010	0.063
13								0.003	0.029
14								0.001	0.012
15									0.004
16									0.001

Figure 4.6 shows the amplitude spectra of angle modulated signals for several values of β. The figure illustrates that the side frequencies outside the range of Equation 4.26 are indeed very small.

In contrast to amplitude modulation formats, it seems to be impossible to derive general estimates of the significant bandwidth of angle modulated signals for practical analog messages such as voice, music, or television. For example, in the case of DSB-SC, we could say that the bandwidth of the modulated signal is twice the highest message frequency, but the nonlinear nature of angle modulation precludes this kind of generalization for FM or PM. However, an empirical formula, known as *Carson's rule*, is often used in making an estimate of the bandwidth of an FM signal. Carson's rule gives the bandwidth as

$$\text{BW} \simeq 2f_H[D + 1] = 2(\Delta f + f_H) \qquad (4.27)$$

Where, as usual, f_H is the highest frequency present in the message signal. The parameter D, which is called the *deviation ratio*, is given by

$$D = \frac{\Delta f}{f_H} \qquad (4.28)$$

where Δf is the peak deviation of the instantaneous frequency of the FM signal from the carrier. One justification for Carson's rule is that it represents the

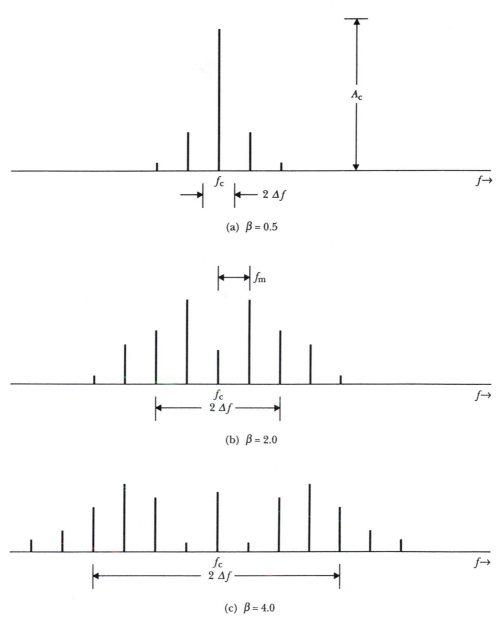

FIGURE 4.6
Amplitude spectra of tone modulated FM signals.

bandwidth given by Equation 4.26 for the worst case (highest frequency) tone modulation. If the actual message is replaced by a single tone at the highest message frequency with full deviation, the 98% power bandwidth of Equation 4.26 agrees with Carson's rule. In a critical situation, one might resort to laboratory measurements using a spectrum analyzer with an actual signal, or

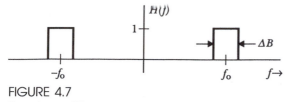

FIGURE 4.7
Bandpass filter transfer function.

to numerical techniques with a sample message to obtain a more confident estimate of bandwidth.

Spectrum of very wideband FM for a general message

Very wideband angle modulation results if the highest message frequency is reduced to a very small value while the peak frequency deviation is held constant. This gives a very large value for D (or, in the case of a single tone message, β). Thus, a very wideband FM signal is a sinusoid whose frequency is slowly swept back and forth by the low-frequency message. If this FM signal is passed through the ideal bandpass filter, whose transfer characteristic is shown in Figure 4.7, and if we consider the limiting case as the highest message frequency approaches zero, the output of the filter will be the same as the input when the instantaneous frequency falls in the passband of the filter, and the output will be zero when the instantaneous frequency is outside the passband. This is because the response time of the filter is negligible compared to the duration of time that the slowly varying signal is in the passband. Thus, the average power contained in the output signal of the filter depends on the fraction of the time that the instantaneous frequency is in the passband. This in turn depends on the fraction of the time that the message signal spends in the corresponding amplitude zone. The power spectral density of the FM signal at the center frequency of the filter f_0 can be estimated by dividing the output power of the filter by the filter bandwidth and taking the limiting case as the filter bandwidth ΔB approaches zero.

A refinement of the argument in the last paragraph leads to an expression for the power spectral density of a very wideband FM signal, given by

$$S_{\text{fm}}(f) = \frac{A_c^2}{4k_f} \left\{ p_m \left[\frac{f - f_c}{k_f} \right] + p_m \left[-\frac{f + f_c}{k_f} \right] \right\} \tag{4.29}$$

The function $p_m(x)$ depends on the relative amount of time that the message signal spends at the amplitude x. More precisely, for a random signal, $p_m(x)$ is the probability density function of the amplitude of the message signal. We will consider probability in Chapter 7 and you may want to return to this discussion again after reading that chapter. At this point, we consider an example for which we can find $p_m(x)$ by intuition and demonstrate the application of Equation 4.29.

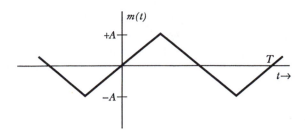

(a) Triangular message waveform

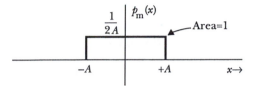

(b) Amplitude probability density function

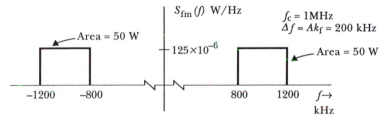

(c) Power spectral density of FM signal

FIGURE 4.8
Message waveform, amplitude probability density function, and PSD of Example 4.1.

EXAMPLE 4.1

The triangular message signal shown in Figure 4.8a is used to frequency modulate a $f_c = 1$ MHz, 10 V rms ($A_c = 14.14$ V) carrier with $k_f = 100$ kHz/V. Assume that the message amplitude A is 2 V and that the message frequency is low enough so very wideband FM results. Find the PSD of the FM signal and sketch to scale.

SOLUTION
The message signal spends equal amounts of time in the vicinity of each amplitude from $-A$ to A. From this we can conclude that $p_m(x)$ should be constant between $-A$ and A. Since the message amplitude is never outside

of this range, we expect $p_m(x)$ to be zero outside of this range. From a probability point of view we would expect that, if we select a random time, the amplitude of the message would be equally likely to take any value between $-A$ and A and have zero probability of having an amplitude outside of that range. Later in our discussion of probability, we will find that the area under a probability density function such as $p_m(x)$ must be unity. We can use this fact with the previous observations to determine that the value of $p_m(x)$ is $1/2A$ in the range from $-A$ to A. A graph of the resulting $p_m(x)$ is shown in Figure 4.8b.

The PSD of the FM signal is found by substituting $p_m(x)$ and the parameters given into Equation 4.29. A plot of the result is shown in Figure 4.8c. Note that the FM signal has an instantaneous frequency that sweeps slowly back and forth at a uniform rate from 800 to 1200 kHz. As we should expect, the power spectrum is constant over this interval and zero elsewhere. The power in the FM signal is simply the square of its rms amplitude or 100 W. Note that the area of the power spectral density agrees with this value.

Direct modulation

Many circuits exist for generating angle modulated signals. The *direct method* for producing FM is to construct an oscillator whose frequency depends on a control voltage. This type of oscillator is known as a *voltage-controlled oscillator* (VCO). these circuits often use variable capacitance diodes, which have a junction capacitance that depends on the reverse bias voltage. Many of the well-known oscillator circuits can be converted into VCOs by including a variable capacitance diode as one of the frequency determining elements in the oscillator. The control voltage, consisting of a dc component plus the message signal, is applied to reverse-bias the diode, changing the capacitance, which in

FIGURE 4.9
VCO circuit using a variable capacitance diode.

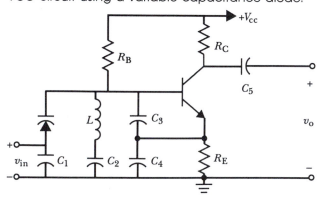

C_1: high impedance for message, low impedance at f_c

C_2 and C_5 : short circuit for f_c

$$f_c \cong \frac{1}{2\pi \sqrt{LC_{eq}}}$$

$$C_{eq} = C_{diode} + \frac{C_3\, C_4}{C_3 + C_4}$$

turn changes the frequency of the oscillator. A popular circuit is shown in Figure 4.9. As long as the peak frequency deviation is less than about 1% of the operating frequency, this approach often gives adequate linearity.

The block diagram of a popular approach for constructing a VCO that can be controlled over a wide range with excellent linearity is shown in Figure 4.10. As indicated in the figure, the circuit consists of three functional blocks: first, an amplifier whose gain can be switched between $+1$ and -1 by the state of a field effect transistor (FET) acting as an electronic switch; second,

FIGURE 4.10
A wide-range linear VCO.

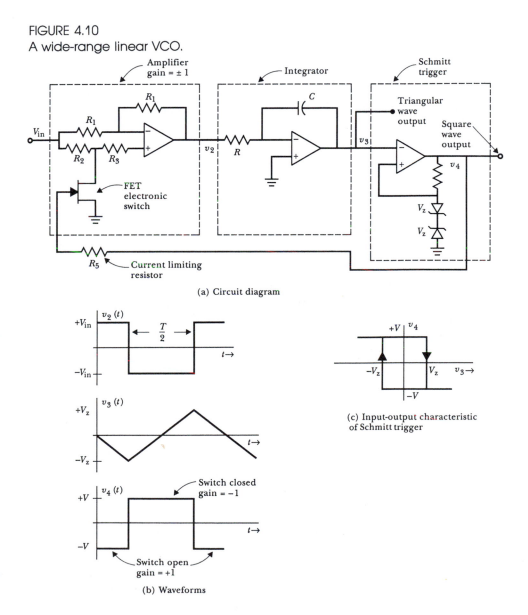

(a) Circuit diagram

(b) Waveforms

(c) Input-output characteristic of Schmitt trigger

an integrator circuit, which produces an output that is the (inverted) integral of the amplifier output; third, a Schmitt trigger whose output versus input characteristic displays hysteresis as indicated in part c of the figure. Typical waveforms are shown in Figure 4.10b.

Now we will explain the operation of the VCO of Figure 4.10. Suppose that the circuit is activated at $t = 0$ with the capacitor discharged and the output of the Schmitt trigger in its low output state. With the output of the Schmitt in the low state, the gate of the FET is negative, and the FET appears as an open circuit. In this condition, the gain of the amplifier is $+1$ so the input control voltage v_{in}, appears at the input to the integrator. Therefore, the output of the integrator ramps negative. When the output of the integrator just passes the negative switching point of the Schmitt, the output of the Schmitt switches to its positive state. This action, in turn, places the FET in its *on* state and the gain of the amplifier block becomes -1. Then the input to the integrator is the negative of the control voltage and the integrator ramps up until the upper switching point of the Schmitt is reached and the cycle repeats. As shown in Figure 4.10, a square wave is available at the Schmitt trigger output and a triangular waveform is available at the integrator output. If a sinusoidal output is required, it can be obtained by passing the triangle through a nonlinear waveshaping circuit. It should be mentioned that the performance of this circuit is poor with regard to phase noise. The output contains a fairly high level of angle modulation by circuit noise compared to, for example, the output of a typical crystal-controlled oscillator.

Indirect modulation

An alternative to the direct approach for producing wideband FM is to use the *indirect method*. In this method, a NBFM or NBPM signal is converted to wideband by the use of frequency multipliers. A frequency multiplier consists of a nonlinear device that produces harmonics of the input signal followed by a bandpass filter to select the desired harmonic multiple of the input. For example, consider the system shown in Figure 4.11, which passes an FM signal through a square-law device, followed by a bandpass filter to select the second harmonic of the input. The input signal is an angle modulated signal given by

$$v_{\text{in}} = A_c \cos \left[2\pi f_c t + \theta(t) \right]$$

FIGURE 4.11
Frequency doubler.

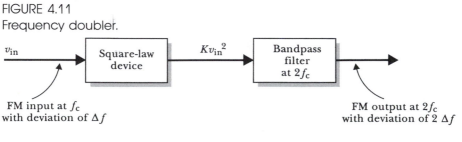

4 / FREQUENCY AND PHASE MODULATION

The output of the square-law device is given by

$$v_1(t) = K(v_{in})^2 = \frac{KA_c^2}{2} + \frac{KA_c^2}{2} \cos \left[4\pi f_c t + 2\theta(t) \right]$$

where we have made use of the identity for cosine squared. The bandpass filter selects only the second term on the right in the expression and rejects the dc component. Notice that the angle modulation term $\theta(t)$, as well as the carrier frequency, has been doubled. Thus, the modulation index of an FM signal is doubled in a frequency doubler.

If we use a nonlinearity having a linear term plus cubic and higher terms in addition to the square-law term, the output will contain the original angle modulated signal plus higher harmonics. Thus, it is possible, by using an appropiate nonlinearity and filter, to increase the frequency of an input signal by any integer multiple. In practice, it is found that the amplitudes of the harmonics fall off fairly rapidly as the multiplying factor goes up. It is often necessary to factor a large multiplier into a number of small factors and cascade the corresponding frequency multiplier blocks to obtain a workable circuit.

When the modulation index of an FM signal must be increased by a large factor, it may be necessary to include an up converter or down converter in the system to accommodate the required starting and ending carrier frequencies. In a frequency converter, a local oscillator signal at a fixed frequency is mixed with the FM signal in a nonlinear device to produce the sum and difference frequencies. A bandpass filter is used to select the desired component, either the sum or the difference. The modulation index of an angle modulated signal is not affected in a frequency converter because the argument of the local oscillator sinusoid is simply added to or subtracted from the argument of the input signal. Thus, the angle modulation term appears in the output with no change in peak deviation.

EXAMPLE 4.2

Figure 4.12 shows an indirect FM modulator with parameters similar to some of those used in commercial FM broadcast transmitters. The audio message signal with frequency content from 50 Hz to 15 kHz is first modulated onto a 100 kHz carrier using the narrowband modulator of Figure 4.3. This low starting carrier frequency may have been chosen to ease the implementation of the modulator circuits and because the high degree of frequency multiplication required raises the carrier frequency. Recall that, to produce an accurate FM signal, the modulation index must be much smaller than one for this narrowband modulator. Since the modulation index of an FM signal is inversely proportional to the message frequency, the highest value of β occurs with a full amplitude tone at the lowest message frequency. Allowing

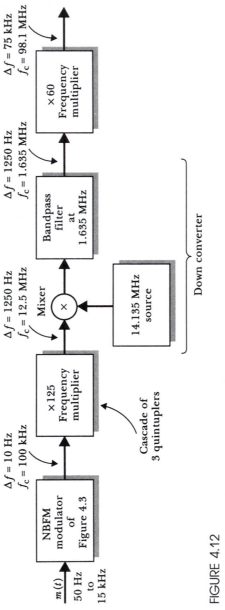

FIGURE 4.12
Indirect FM modulator of Example 4.2.

the modulation index for this worst case to be 0.2 results in a peak frequency deviation at the output of the narrowband modulator of $\Delta f = \beta f_m = 10$ Hz. To achieve the peak deviation required in a commercial transmitter of $\Delta f = 75$ kHz at the system output, the signal must undergo a frequency multiplication of 7500. This would result in a carrier frequency of 750 MHz, which is too high for the commercial FM band, so a down converter must be included at some convenient spot in the system. The required multiplication factor can be factored as

$$7500 = 2 \times 2 \times 3 \times 5 \times 5 \times 5 \times 5 = 125 \times 60$$

Therefore, the multiplication can be accomplished by cascading two doublers, one tripler, and four frequency quintuplers. A convenient place for the required down converter is immediately after the multiplication by 125 as shown in the figure.

Find the carrier frequency and peak frequency deviation at each point in the system.

SOLUTION

At the output of the NBFM modulator we have $f_c = 100$ kHz and $\Delta f = 10$ Hz. The $\times 125$ frequency multiplier chain increases both the carrier frequency and the peak frequency deviation so we have

$$f_c = 125 \times 100 \text{ kHz} = 12.5 \text{ MHz}$$

and

$$\Delta f = 125 \times 10 \text{ Hz} = 1250 \text{ Hz}$$

at the output of this first multiplier block.

The mixer produces sum and difference frequency components. One component has a carrier frequency of $12.5 + 14.135 = 26.635$ MHz and the other component has a carrier frequency of $14.135 - 12.5 = 1.635$ MHz. The bandpass filter passes only the component with a carrier frequency of 1.635 MHz. The peak frequency deviation is not affected in the mixer or in the bandpass filter so that at the output of the bandpass filter we have $f_c = 1.635$ MHz and $\Delta f = 1250$ Hz.

The final frequency multiplier chain affects both the carrier frequency and the frequency deviation. Thus, at the system output we have

$$f_c = 60 \times 1.635 = 98.1 \text{ MHz}$$

and

$$\Delta f = 1250 \times 60 = 75 \text{ kHz}$$

Demodulators

Demodulators for FM signals, also known as *frequency discriminators*, exist in wide variety. We discuss only a few approaches and refer the interested reader to the literature. An FM signal can be demodulated by differentiation with respect to time followed by envelope detection. Taking the derivative of a constant envelope FM signal converts it to a signal that has amplitude as well as an angle modulation. Recall that the FM signal defined in Equation

FIGURE 4.13
Typical discriminator with waveforms.

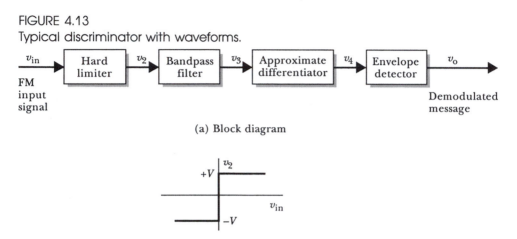

(a) Block diagram

(b) Input-output characteristic of hard limiter

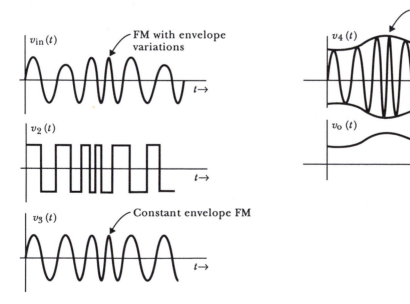

(c) Waveforms at several points

4.6 is

$$g_{\text{fm}}(t) = A_{\text{c}} \cos \left[2\pi f_{\text{c}} t + 2\pi k_{\text{f}} \int^t m(x)\, dx \right]$$

Taking the time derivative of this expression, we obtain

$$\frac{dg_{\text{fm}}(t)}{dt} = -A_{\text{c}} \left[2\pi f_{\text{c}} + 2\pi k_{\text{f}} m(t) \right] \sin \left[2\pi f_{\text{c}} t + 2\pi k_{\text{f}} \int^t m(x)\, dx \right]$$

Study of this expression shows that it does have both AM and FM characteristics. An envelope detector is commonly used for demodulating AM and is not sensitive to any FM that may be present on the input signal.

Usually, when demodulators based on this approach are used, the differentiator is preceded by a hard limiter followed by a bandpass filter. This combination is called a *bandpass limiter*. The hard limiter can consist of either an amplifier that is driven hard from saturation to cutoff, or a comparator. The hard limiter converts the received FM signal, which may contain incidental AM due to noise or filtering in the channel, into a constant amplitude squarewave while preserving the modulation of the zero crossings contained in the original FM signal. The bandpass filter removes the components near the harmonics of the carrier resulting in a constant amplitude sinusoidal FM signal to be processed by the differentiator and envelope detector. A block diagram of this type of discriminator is shown in Figure 4.13.

Recall from the properties of the Fourier transform that differentiation in the time domain corresponds to multiplication by $j2\pi f$ in the frequency domain. Thus, the transfer function of a differentiator is $j2\pi f$. Therefore, the gain of a differentiator is proportional to frequency. Many FM demodulators contain frequency selective elements that approximate the required linear gain characteristic over the frequency range of the FM signal to be demodulated. One possibility is to use a single resonant bandpass filter and tune it so that the carrier frequency of the signal to be demodulated is located on the slope of the bandpass characteristic as shown in Figure 4.14a. This discriminator works but suffers from a lack of good linearity unless the peak frequency deviation is only a very small fraction of the bandwidth of the circuit. A balanced circuit that plays off the nonlinearity of one tuned circuit against that of a second tuned circuit, as shown in Figure 4.14b, provides a significant improvement in linearity.

Another discriminator based on the approximation to the derivative

$$\frac{dg_{\text{fm}}(t)}{dt} \simeq \frac{g_{\text{fm}}(t) - g_{\text{fm}}(t - \Delta t)}{\Delta t}$$

is shown in Figure 4.15. Note that the approximation is valid only for small values of Δt.

The discriminators based on converting FM into AM, followed by envelope detection, need to be preceded by a bandpass limiter to eliminate AM

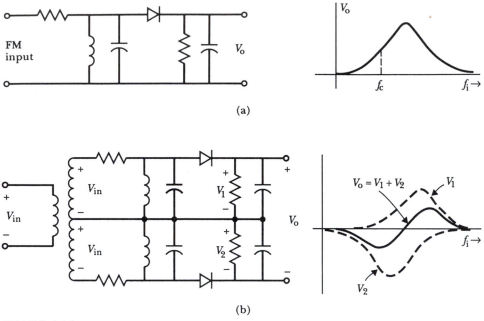

(a)

(b)

FIGURE 4.14
(a) Slope detector and (b) balanced discriminator.

due to noise and uncontrolled filtering in the channel or in earlier stages of the receiver. Then AM can be introduced in a controlled manner by a close approximation to the differentiator. However, as we will see later, it is not beneficial to the SNR of the demodulated signal to have a circuit that treats phase fluctuations that occur during periods of low signal amplitude on an equal basis with the phase during periods of high signal amplitude. The PLL, which is the topic of the next section, can be designed to partially ignore phase variations when the signal amplitude is low and achieve a better performance under marginal conditions.

FIGURE 4.15
Discriminator using a delay element.

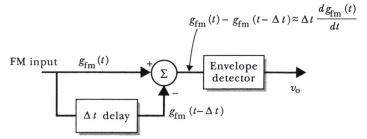

$$g_{\text{fm}}(t) - g_{\text{fm}}(t - \Delta t) \approx \Delta t\, \frac{d\,g_{\text{fm}}(t)}{dt}$$

4.3
PHASE-LOCKED LOOPS

Fundamental principles of operation

The *phase-locked loop* (PLL) is a feedback control system that is intended to cause the phase of a local voltage-controlled oscillator (VCO) to track the phase variations of an input signal. Figure 4.16 shows a message signal applied to the input of a VCO producing an FM signal, which is then applied to the input of a PLL. The PLL contains a phase detector that produces an output signal dependent on the phase difference between the input signal and the loop VCO output. The phase detector output is an error signal v_d that is passed through the loop filter to the VCO in such a manner that the error signal tends to cause the phase difference $\theta_i - \theta_o$ to be reduced. Thus, the phase detector, loop filter, and VCO form a feedback control system that generates a constant amplitude sinusoid whose phase follows the phase of the input signal.

If the VCO phase tracks the phase of the input signal exactly and if the loop VCO is identical to the modulator VCO, the message signal must appear at the input to the loop VCO. In this case, the loop would successfully demodulate the FM input signal. As we will see, the tracking is not exact, and the demodulated message is not precisely the same as at the modulator input. However, we will find that the overall effect is simply lowpass filtering of the message. By proper design of the loop, we can cause the cutoff frequency to be high enough so the demodulated message is practically the same as at the modulator input.

Another important application of PLLs is in carrier recovery in receivers for suppressed carrier modulation formats. In these applications, a carrier

FIGURE 4.16
Block diagram of FM modulator and PLL demodulator.

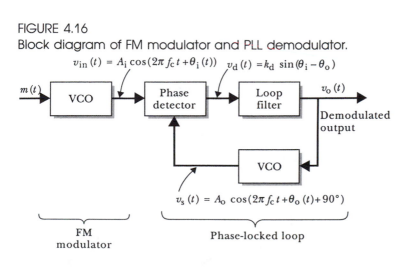

component containing phase variations due to noise or the message signal is available, and a version of the carrier without these phase variations is desired. A PLL can achieve the desired result if it is properly designed to track only the average phase of the noisy input signal and to smooth out the phase variations due to noise and the message modulation. The noisy carrier component is the input to the PLL, and the VCO produces the output signal. In this case, the PLL functions as a narrowband bandpass filter, which removes side frequencies due to noise and modulation. The PLL is sometimes called a *tracking filter* because the control action of the loop causes tracking of the carrier frequency. An example of such an application is indicated in Figure 3.8a.

Many types of phase detectors produce zero output for a nonzero phase difference between the input signals. As a result, the phase of the loop VCO tracks the phase of the input signal with a constant phase offset. In most of our discussion, we emphasize the use of the product phase detector, which has a constant phase offset of 90°. We have included this 90° in the expression for the loop VCO output shown in Figure 4.16. Therefore, when perfect phase tracking occurs we have $\theta_o = \theta_i$.

Both the phase detector and the loop VCO are nonlinear elements, so a PLL is a nonlinear system, which makes an exact analysis of its behavior difficult in the general case. However, when the phase difference between the input signal and the loop VCO is small, we will be able to find a linear model for the loop and perform an accurate analysis. We will qualitatively describe some of the nonlinear effects that occur when the phase difference is large, but we will not attempt an analysis of the nonlinear aspects of PLLs.

Phase detectors

One way to implement the phase detector is to use a multiplier circuit such as the double balanced mixer discussed in Section 3.1. The product device forms an output signal given by

$$v_d(t) = K v_{in}(t) v_s(t) \tag{4.30}$$

where K is the gain constant of the multiplier. The input to the PLL is given by

$$v_{in}(t) = A_i \cos \left[2\pi f_c t + \theta_i(t) \right] \tag{4.31}$$

and the loop VCO output signal is given by

$$v_s(t) = A_o \cos \left[2\pi f_c t + \theta_o(t) + 90° \right] \tag{4.32}$$

Substituting Equations 4.31 and 4.32 into Equation 4.30 results in

$$v_d(t) = K A_o A_i \cos \left[2\pi f_c t + \theta_i(t) \right] \cos \left[2\pi f_c t + \theta_o(t) + 90° \right]$$

Making use of trignometric identities results in

$$v_d(t) = \frac{1}{2} K A_o A_i \sin \left[\theta_i(t) - \theta_o(t) \right] + \frac{1}{2} K A_o A_i \sin \left[4\pi f_c t + \theta_i(t) + \theta_o(t) \right]$$

The second term on the right side of this expression is a double frequency term that usually should have no effect on the loop since it is of such a high frequency that it has little effect on the VCO. This is because, as we will see later, the VCO phase responds to the integral of its input and the integral of this high frequency term is often insignificant. Thus the effective part of the phase detector output is the lowpass (LP) part given by

$$\text{LP}(v_d) = k_d \sin (\theta_i - \theta_o) \qquad (4.33)$$

where $k_d = \frac{1}{2} K A_o A_i$ is the phase detector gain constant with units of volts per radian. Note that for the product phase detector, the gain constant k_d depends on the amplitude of the input signals. This can cause the characteristics of the loop to depend on the amplitude of the input signal. When we consider the noise characteristics of FM demodulators, we will see that this can be advantageous. In other applications, the dependence of loop parameters on signal amplitude may be objectionable. When the phase difference is small, Equation 4.33 can be approximated (dropping the LP notation) by

$$\blacksquare \qquad v_d \simeq k_d(\theta_i - \theta_o) \qquad (4.34)$$

Many alternative phase detector circuits are possible. Often the input signal to the PLL and the output of the loop VCO are either in the form of a square wave or are converted to a square wave in the phase detector. Two other phase detectors, which use logic elements with square-wave inputs, as well as their output characteristics, are shown in Figure 4.17. Figure 4.17b shows a phase detector based on an exclusive OR gate, which results in a triangular characteristic. A constant phase offset of $-90°$ and logic levels of $+V$ and $-V$ have been assumed in determining the characteristic of Figure 4.17b. The circuit of Figure 4.17c is based on an edge-triggered set-reset flip-flop and produces a zero error signal when there is a 180° phase difference between the input signals. For small phase differences (neglecting the fixed 90° or 180° phase offset), all these phase detectors produce an error signal proportional to the phase difference between the input signals as given by Equation 4.34. The value of the phase detector gain constant k_d is the slope of the phase detector characteristic at zero phase error. As we have seen, its value depends on the details of the phase detector circuit.

The VCO

The output signal of the loop VCO is given by

$$v_s(t) = A_o \cos \left[2\pi f_c t + 2\pi k_f \int_{-\infty}^{t} v_o(t)\, dt + 90° \right]$$

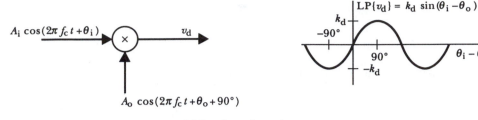

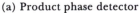

(a) Product phase detector

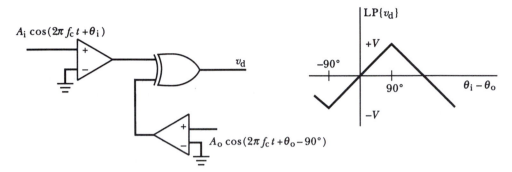

(b) Exclusive OR gate phase detector

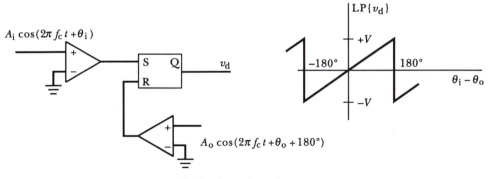

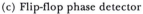

(c) Flip-flop phase detector

FIGURE 4.17
Phase detector circuits.

Note again that the VCO output signal $v_s(t)$ is a nonlinear function of the VCO input signal $v_o(t)$. However, if we consider the phase of the output signal rather than the actual output voltage, we have the following linear relationship between the output phase and the input to the VCO:

$$\theta_o(t) = 2\pi k_f \int_{-\infty}^{t} v_o(t)\, dt \qquad (4.35)$$

Recall that the constant k_f is the sensitivity constant of the VCO in hertz of frequency deviation per volt of input.

The linear model

To analyze the dynamic response of the PLL to transient input signals we use Laplace transform techniques. The Laplace transform of the phase detector output given in Equation 4.34 is

$$V_d(s) = k_d[\Theta_i(s) - \Theta_o(s)] \tag{4.36}$$

The Laplace transform of the relation between the VCO input signal and the VCO phase given in Equation 4.35 is

$$\Theta_o(s) = \frac{2\pi k_f}{s} V_o(s) \tag{4.37}$$

The loop filter is a linear element characterized by its transfer function $H(s)$. The Laplace transformed output of the loop filter is

$$V_o(s) = H(s)V_d(s) \tag{4.38}$$

Making use of the linear relationships of Equations 4.36 through 4.38, we can construct the linear model shown in Figure 4.18 for the system of Figure 4.16. This linear model is valid only if the phase error remains in the linear portion of the phase detector characteristic.

From Figure 4.18, we can write the following relation for the demodulated output signal

$$V_o(s) = k_d H(s)\left[\Theta_i(s) - \frac{2\pi k_f}{s} V_o(s)\right] \tag{4.39}$$

The phase of the input signal to the PLL demodulator is given by

$$\Theta_i(s) = \frac{2\pi k_f}{s} M(s) \tag{4.40}$$

where $M(s)$ is the Laplace transform of the message. Combining Equations 4.37, 4.39, and 4.40 yields the following transfer functions for the modulator

FIGURE 4.18
Transformed linear model of the system of Figure 4.16 valid only when $\theta_i(t) - \theta_o(t)$ is small at all times.

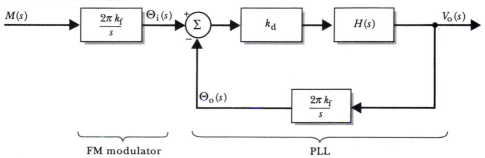

FM modulator   PLL

and PLL demodulator

$$\frac{V_o}{M}(s) = \frac{\Theta_o}{\Theta_i}(s) = \frac{2\pi k_d k_f H(s)}{s + 2\pi k_d k_f H(s)} \tag{4.41}$$

We will now discuss the linear behavior of the PLL for several alternative loop filters.

The first-order loop

When $H(s) = 1$ a *first-order loop* results, and the transfer functions of Equation 4.41 become

$$\frac{V_o}{M}(s) = \frac{\Theta_o}{\Theta_i}(s) = \frac{2\pi k_d k_f}{s + 2\pi k_d k_f} \tag{4.42}$$

This is exactly the same as the transfer function of a simple lowpass RC filter. If we substitute $s = j2\pi f$ into the transfer function for the first-order loop, we obtain the frequency response function

$$\frac{V_o}{M}(f) = \frac{1}{1 + j(f/f_b)}$$

where f_b is the break frequency of the lowpass filter given by $f_b = k_d k_f$. A plot of the magnitude of the transfer function is shown in Figure 4.19.

If an impulse $C\delta(t)$ is applied as the message signal, the phase at the input to the PLL makes a step change. The transforms of the input phase, the demodulated output, and the output phase are given by

$$\Theta_i(s) = \frac{2\pi k_f C}{s}$$

$$\Theta_o(s) = \frac{2\pi k_f C}{s} \frac{2\pi k_f k_d}{s + 2\pi k_f k_d}$$

FIGURE 4.19
Bode plot of the transfer function for a first-order loop. Note logarithmic frequency scale.

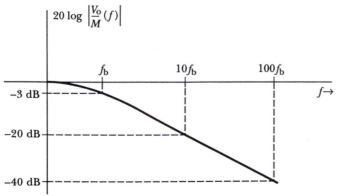

and

$$V_o(s) = C \frac{2\pi k_f k_d}{s + 2\pi k_f k_d}$$

The corresponding time functions are given by

$$\theta_i(t) = 2\pi k_f C u(t)$$

$$\theta_o(t) = 2\pi k_f C \left[1 - \exp\left(\frac{-t}{\tau}\right) \right] u(t)$$

and

$$v_o(t) = \frac{C}{\tau} \exp\left(\frac{-t}{\tau}\right) u(t)$$

where

$$\tau = \frac{1}{2\pi k_f k_d}$$

is the time constant of the system. These time responses are shown in Figure 4.20.

Notice that these results were obtained from the linear model and are valid only if the phase detector remains on the linear part of its characteristic. Immediately after the impulse is applied, the phase error is equal to the step in the input phase $2\pi k_f C$; therefore it is necessary to keep the step in phase less than about 10° for these results to be valid. This is necessary because of the sinusoidal nonlinearity of the product phase detector. The phase detectors of Figure 4.17b and c provide an extended linear range.

The parameter $f_b = k_f k_d$ is known as the *loop gain*, and it has units of hertz. The value of the loop gain completely characterizes the first-order loop in the linear range of operation.

We have assumed that the center frequency of the input signal and the loop VCO are the same. If we slowly vary the frequency of the input signal, the PLL will respond by changing the frequency of the loop VCO to track the

FIGURE 4.20
Response of the first-order loop to a step change in input phase.

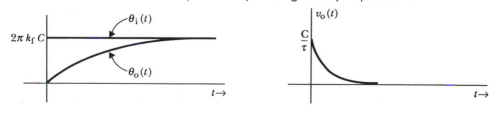

input frequency. In a first-order loop, as the input frequency slowly increases, a *static phase error* develops so as to produce an output voltage from the phase detector just large enough to cause the VCO frequency to match the incoming frequency. The maximum range of the input frequency that a loop is capable of tracking is known as the *hold-in range*. Since the maximum output of the product type phase detector is k_d, the maximum frequency deviation possible for the VCO is $k_d k_f$ Hz. Thus, the hold-in range of a first-order loop with a product phase detector is from the center frequency of the loop VCO minus the loop bandwidth to the center frequency plus the bandwidth.

The first-order loop is very satisfactory for demodulating FM signals as long as the loop bandwidth is wider than the message spectrum. However, PLLs are used in many other situations in communication systems in which a very small static phase error is necessary even though the input frequency is not the same as the center frequency of the VCO and at the same time the loop must have a narrow bandwidth. This combination is not possible with a first-order loop because the loop gain sets both the bandwidth and the static phase error for a given frequency offset. One system in which this conflict comes about is in carrier recovery systems such as those of Figures 3.8 and 3.9. If recovery of a carrier component in the presence of sidebands is wanted, it is necessary to have a narrow bandwidth so that the phase of the recovered carrier is not affected by the sidebands. In addition, it is important to have a small static phase error even with some offset of the carrier frequency. A loop filter that changes the loop into a second-order system can provide the solution to this conflict.

Second-order loops

Two popular loop filters used in second-order loops are shown in Figure 4.21. We will concentrate on the *active loop filter* since it gives better performance in most applications. The transfer function of the active loop filter is given by

$$H(s) = -\frac{s\tau_2 + 1}{s\tau_1} \tag{4.43}$$

FIGURE 4.21
Commonly used loop filter circuits.

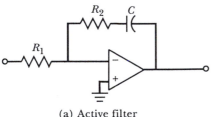

(a) Active filter

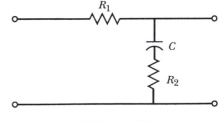

(b) Passive filter

where

$$\tau_1 = R_1 C \quad \text{and} \quad \tau_2 = R_2 C \tag{4.44}$$

Because of the negative sign in the transfer function of the active loop filter, a PLL locks at the other slope of the phase detector characteristic (i.e., at $180°$ on the detector characteristic of Figure 4.17a). For this case, the constant phase offset of the loop is $-90°$ rather than $+90°$ as it was for the first-order loop with $H(s) = 1$. If we change the phase offset in the expression for the VCO output to $-90°$ to account for this difference, the output of the phase detector is

$$v_d = -k_d \sin (\theta_i - \theta_o) \simeq -k_d(\theta_i - \theta_o)$$

for small phase errors.

Carrying this change through the preceding development gives the following expression for the loop transfer function:

$$\frac{V_o}{M}(s) = \frac{\Theta_o}{\Theta_i}(s) = \frac{-2\pi k_d k_f H(s)}{s - 2\pi k_d k_f H(s)} \tag{4.45}$$

Substitution of Equation 4.43 into 4.45 and some algebra produces

$$\frac{V_o}{M}(s) = \frac{\Theta_o}{\Theta_i}(s) = \frac{2\pi k_d k_f [s\tau_2 + 1]/\tau_1}{s^2 + s(2\pi k_d k_f \tau_2)/\tau_1 + 2\pi k_d k_f/\tau_1}$$

This can also be expressed in terms of the natural frequency $\omega_n = 2\pi f_n$ and the damping ratio ζ as

$$\frac{V_o}{M}(s) = \frac{\Theta_o}{\Theta_i}(s) = \frac{2\zeta\omega_n s + \omega_n^2}{s^2 + 2\zeta\omega_n s + \omega_n^2} \tag{4.46}$$

where

$$\zeta = \frac{\tau_2}{2}\left[\frac{2\pi k_d k_f}{\tau_1}\right]^{1/2} \tag{4.47}$$

and

$$\omega_n = \left[\frac{2\pi k_d k_f}{\tau_1}\right]^{1/2} \tag{4.48}$$

If the substitution $s = j2\pi f$ is made in Equation 4.46, the frequency response of the loop is found. The magnitude of this transfer function is plotted in Figure 4.22 versus the normalized frequency variable f/f_n for several values of the damping ratio. Notice that, as in the case of the first-order loop, the transfer function is that of a lowpass filter. In the second-order loop, some peaking of the transfer function may occur before roll-off. This peaking was

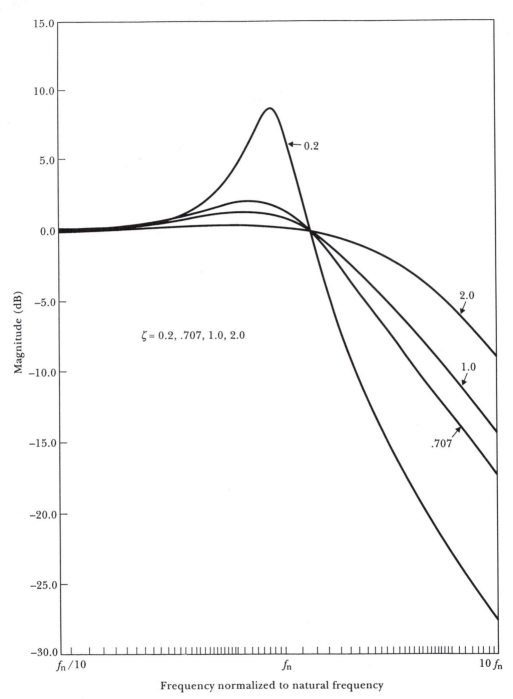

FIGURE 4.22
Transfer function of second-order loop.

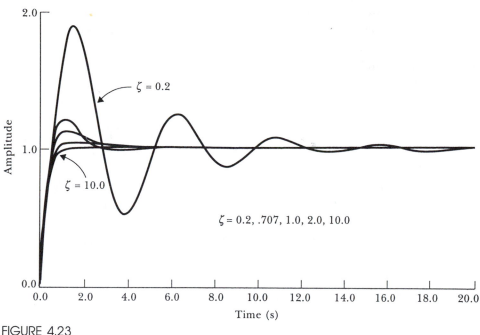

FIGURE 4.23
Second-order phase-locked loop unit step phase transient behavior for constant noise bandwidth of 1 Hz. (defined in Section 5.1.)

not possible in a first-order loop. The half-power bandwidth of the second-order loop transfer function is a function of both the natural frequency and the damping ratio. It is given by

$$f_{3\,\mathbf{dB}} = f_n[2\zeta^2 + 1 + [(2\zeta^2 + 1)^2 + 1]^{1/2}]^{1/2} \qquad (4.49)$$

If an impulse $C\delta(t)$ is applied as the message input in Figure 4.18, the input to the PLL will contain a step $2\pi k_f Cu(t)$ in its phase. The resulting response of the loop VCO phase can be found by use of the transfer function given in Equation 4.46. The resulting phase transient is shown in Figure 4.23 for several values of the damping ratio.

The PLL with the active loop filter has much better performance than the first-order loop with regard to static phase error when the incoming frequency is offset from the center frequency of the loop VCO. In the case of the first-order loop, a frequency offset causes a static phase error to develop that is just large enough to generate the phase detector output needed to cause the frequency of the loop VCO to match the incoming frequency. If the phase detector cannot produce a large enough output, the hold-in range has been exceeded, and the loop loses lock. When the second-order loop with the active filter experiences a frequency offset, a phase error is developed and the resulting phase detector output is integrated by the loop filter. In steady-state conditions, no phase error results because the voltage needed to cause the loop

VCO frequency to match the incoming frequency is supplied by the charge stored on the capacitor of the loop filter. In other words, the active filter does not require an input at dc to be able to produce the dc output needed to maintain the match in frequency. The active loop filter has infinite gain at dc. Note that the passive loop filter of Figure 4.21b does not have this characteristic (the gain of the passive filter is unity at dc) and a static phase error will occur due to a frequency offset. However, the passive filter does allow the use of a very high value for the loop gain $k_d k_f$. This can result in a very small static phase error without resulting in a wide bandwidth.

The hold-in range of the loop with the active loop filter would be unlimited if none of the loop elements had saturation limits. This is because a frequency offset generates only a temporary phase error that is integrated by the loop filter as described above. Eventually, if the frequency offset becomes large enough, either the amplifier saturates or the control range of the VCO is exceeded. Then the loop loses lock. Thus the hold-in range is determined by the saturation limits of the active filter or the VCO.

We have assumed that the phase error remains small in our analysis of the PLL and pointed out that the system becomes nonlinear when the phase difference is large. The linear analysis shows that the phase error is reduced by the action of the loop if the phase and frequency is closely aligned at the outset. PLLs are also capable of eventually reducing the phase error or "pulling into lock" when the initial phase and frequency difference is large. The range of frequencies for which the loop can eventually align the frequency of the loop VCO with the input frequency is called the *pull-in range* of the loop. Pull-in can require a very long time so this *acquisition* process is often aided by causing the center frequency of the loop VCO to sweep over the range of expected input frequencies until lock is achieved. A *lock detector* circuit is used to detect when the PLL is locked and to disable the sweep after lock is achieved. A sweep is particularly necessary with the active filter since offset voltages in the amplifier or the phase detector can cause the active filter output to integrate up to the saturation limit of the amplifier, which then prevents the proper control action of the loop during acquisition.

Design considerations

The design of a PLL depends heavily on its application. For an FM demodulator application in which the frequency offset is a small fraction of the message bandwidth, static phase error is of little concern. Here a first-order loop is satisfactory, provided that the loop gain is large enough to insure that the frequency response of the loop is flat beyond the highest message frequency.

In carrier recovery applications, a small static phase error is required, and the bandwidth must be narrow so that the phase modulation due to noise or the message signal is not passed to the loop VCO. These requirements indicate that a second-order loop with the active loop filter is needed. If the

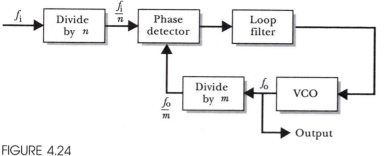

FIGURE 4.24
PLL frequency synthesizer.

incoming carrier frequency varies with time due to Doppler shifts caused, for example, by motion of the transmitter or receiver, the transient response of the loop must be considered to insure proper tracking. These considerations result in a determination of the required half-power bandwidth and damping factor for the loop. Usually, the damping factor is selected in the range between 0.5 and 1.0 because a smaller value for the damping factor results in excessive ringing in the transient response and pronounced peaking of the frequency response, and too large a value of damping ratio causes sluggish response. The bandwidth and damping factor can be used in Equation 4.49 to determine the natural frequency of the loop. Hardware considerations, such as the choice of a particular phase detector or VCO, may set values for k_d or k_f. The remaining parameters can then be determined from Equations 4.47 and 4.48.

PLL frequency synthesizers

Another application for the PLL is shown in Figure 4.24. This is a frequency synthesizer that can be used to generate a large number of output frequencies precisely related to the input frequency. The divide-by-n and divide-by-m blocks are digital counter circuits that produce an output cycle for each n or m input cycles. Thus, if the input frequency is f_i and the output frequency is f_o, the input frequencies to the phase detector are f_i/n and f_o/m. Under locked steady-state conditions, the PLL causes the input frequencies to the phase detector to be exactly equal or

$$\frac{f_i}{n} = \frac{f_o}{m}$$

Rearranging this, we find that the output frequency is given by

$$f_o = \frac{m}{n} f_i$$

One application for such a synthesizer is the local oscillator source in a radio receiver required to tune to a number of evenly spaced frequencies.

For example, consider an AM broadcast receiver that is required to tune from 550 kHz to 1600 kHz in 10 kHz steps with an intermediate frequency of 455 kHz. In this case, the local oscillator can be selected to tune from 1005 kHz to 2055 kHz in 10 kHz steps. This can be achieved precisely by using the system of Figure 4.24 if the input frequency is derived from an accurate, stable source. A convenient input reference could be a 1000 kHz crystal controlled oscillator. Then, if n is selected to be 200 and if m ranges from 201 to 411 in steps of 2, the output frequencies are the required local oscillator frequencies.

We have been able to give only an introduction to the important topic of PLLs in this section. The interested reader should consult the references for more information on the nonlinear characteristics of PLLs and the many possible circuit variations.

SUMMARY

1. Angle modulation results when the phase angle of a sinusoidal carrier is varied by a message signal. In PM, the phase is directly proportional to the message and in FM the phase is directly proportional to the running time integral of the message.

2. The instantaneous angular frequency of a sinusoid is the time derivative of its argument. For FM, the instantaneous frequency is linearly related to the message.

3. Narrowband angle modulation results when the peak phase deviation is much less than one radian. The amplitude spectrum of NBPM is exactly the same as that of AM: The message spectrum appears as upper and lower sidebands centered around a large discrete carrier component. The amplitude spectrum of NBFM has sidebands of the running time integral of the message centered about a discrete carrier component.

4. Integration of the message in FM reduces the relative amplitudes of the higher frequency message components making them more susceptible to noise. Practical FM systems often emphasize the high frequencies before modulation and de-emphasize them at the demodulator to reduce the effects of noise.

5. A large amount of the power in a narrowband angle modulated signal is contained in the carrier component and, like AM, the modulation efficiency is poor.

6. Narrowband angle modulation can be produced by DSB-SC modulating the message or its running integral onto a phase-shifted carrier and then adding the result to a large carrier component.

7. Wideband angle modulation by a single tone message signal results in a spectrum consisting of side frequency components spaced at integer mul-

tiples of the message frequency on both sides of the carrier. The peak amplitude of each member of the nth set of side frequencies is $A_c J_n(\beta)$. When n becomes large enough, the amplitude of the corresponding side frequencies is negligible.

8. Carson's rule is often used to obtain an empirical estimate of the bandwidth of angle modulated signals. Exact analysis of the spectrum in the general case is not available.

9. Very wideband FM results when the peak frequency deviation is much larger than the highest message frequency. In this case, the approximate power spectrum can be determined by considering the amplitude distribution of the message signal.

10. Wideband FM can be produced by the direct method in which the frequency of a VCO is controlled by the message signal. Numerous circuits to implement VCOs are possible.

11. Wideband FM can also be produced by the indirect method in which a narrowband FM signal is converted to wideband at the desired carrier frequency by the use of frequency multipliers and frequency converters. A frequency multiplier creates a harmonic of the input signal by distorting the sinusoidal input in a nonlinear circuit and then selecting the desired multiple with a bandpass filter. A frequency multiplier increases both the carrier frequency and the peak deviation, whereas a frequency converter changes only the carrier frequency.

12. Many FM demodulators or discriminators are based on the fact that FM can be converted to AM by differentiation of the signal. Typically, the received FM signal is passed through a bandpass limiter to eliminate any AM due to noise, then through a filter approximating the differentiator, and, finally, through an envelope detector, which recovers the message signal.

13. A phase-locked loop is a feedback control loop that causes the phase of a VCO to track the phase variations of an incoming signal. It is useful for demodulating FM signals and in many other places in communication systems.

14. A PLL is a nonlinear system, but an accurate linear model can be found when the phase error is small.

15. A first-order loop results when no loop filter is used. Its performance is characterized by its loop gain. The loop bandwidth and static phase error both depend on the loop gain and cannot be set independently.

16. Two types of loop filters are commonly used in second-order loops, one active and the other passive. The linear behavior of a second-order loop is characterized by the natural frequency and the damping ratio. The hold-in range of the loop with the active filter is limited only by the linear range of the amplifier and VCO. With either type of filter the frequency response of the loop and static phase error for a given frequency offset can be determined independently.

REFERENCES

F. M. Gardner. *Phaselock Techniques*, 2d ed. New York: Wiley, 1979.

J. Klapper and J. T. Frankle. *Phaselocked and Frequency Feedback Systems*. New York: Academic Press, 1972.

J. Smith. *Modern Communication Circuits*. New York: McGraw-Hill, 1986.

H. Taub and D. L. Schilling. *Principles of Communication Systems*, 2d ed. New York: McGraw-Hill, 1986.

PROBLEMS

1. The symmetrical square-wave message signal shown in Figure P4.1 is used to modulate a 10 kHz sinusoidal carrier. Write an equation for the modulated signal in terms of $m(t)$ and sketch the modulated waveform to scale if the modulation is (a) FM with a peak frequency deviation (from the unmodulated carrier) of 1 kHz, and (b) PM with a peak phase deviation of $90°$.

2. A message has the amplitude spectrum shown in Figure P4.2. The message is used to modulate a 1 MHz sinusoidal carrier having a peak value of 10 V. Sketch the amplitude spectrum of the modulated signal to scale if the modulation is (a) narrowband FM with $k_f = 1.0$ Hz/V, and (b) narrowband PM with $k_p = 0.05$ radian/V.

3. A NBFM signal is given by $10 \cos [2\pi 5000t + 0.05 \sin 2\pi 100t]$.
   (a) Sketch the amplitude spectrum to scale.
   (b) Find the power in the carrier component and in each of the side frequencies. What is the modulation efficiency of this signal?
   (c) Find the peak phase and frequency deviation.

4. Use the property of Bessel functions given in Equation 4.25 to find the total power contained in the angle modulated signal of Equation 4.19. Find the total power in the narrowband angle modulated signal of Equation 4.17 and note that there is a discrepancy compared to the previous result. Why?

5. A 1 kHz tone angle modulates a 100 kHz carrier with a peak frequency deviation of 5 kHz. The amplitude of the modulated signal is 5 V rms.
   (a) What is the total power in the modulated signal?
   (b) What is the power in the carrier and the first set of side frequencies?

FIGURE P4.1

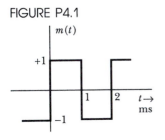

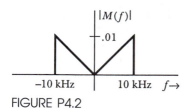

FIGURE P4.2

(c) How many side frequencies must be included with the carrier to contain at least 98% of the total power? Does this result agree with the bandwidth prediction of Equation 4.26?

6. A television message signal has a highest frequency of 4.5 MHz and is used to FM a 6.2 GHz carrier. If the bandwidth allowed for the modulated signal is 30 MHz, estimate the peak frequency deviation allowed. Is this narrowband or wideband FM?

7. (a) A 10 V rms, 1 MHz carrier is frequency modulated by the message waveform of Figure P4.3a with $k_f = 5$ kHz/V. Use Equation 4.29 to find the power spectrum of the modulated signal when the message frequency is very low and sketch the result to scale. Hint: The message waveform has only two amplitudes and spends equal amounts of time at each amplitude. Thus $p_m(x)$ consists of two impulses each of weight one-half.

   (b) Repeat for the message waveform of Figure P4.3b.

8. The capacitance of a typical variable capacitance diode is given by

$$C_{\text{diode}} = 16V^{-1/2} \text{ pF}$$

where V is the reverse bias voltage on the diode. If this diode is used in the VCO circuit of Figure 4.9 with $C_3 = C_4 = 100$ pF and $L = 5$ μH, calculate and plot the approximate oscillator frequency for V between 1 and 10 V. Is the variation of frequency with control voltage linear over this range?

9. Find the frequency of oscillation of the circuit of Figure 4.10 if v_{in} is 5 V, R is 1000 Ω, C is 0.01 μF, and V_z is 4 V.

10. Design the block diagram of a system to convert an angle modulated signal with a carrier frequency of 1.0 MHz and peak frequency deviation of 10 Hz into an angle modulated signal with a carrier frequency of exactly 50 MHz and a peak frequency deviation of approximately 10 kHz. To ease the eventual circuit design, avoid frequencies in excess of 200 MHz in your system.

FIGURE P4.3

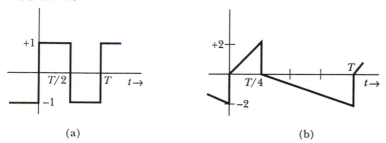

(a)   (b)

11. If the input to the FM discriminator shown in Figure 4.15 is $A \cos (2\pi f_i t)$, find and sketch to scale the steady-state output voltage of the circuit as a function of f_i if the delay Δt is 100 nanoseconds (ns) for f_i from zero to 20 MHz. Over what range of carrier frequencies is this circuit suitable for use as an FM demodulator if the peak deviation is 10 kHz? What carrier frequencies would not be suitable?

12. Assuming that the logic levels are $+V$ and $-V$, sketch the waveform at the output of the phase detector of Figure 4.17b and find its average value when (a) $\theta_i - \theta_o = 0°$, (b) $\theta_i - \theta_o = 90°$, and (c) $\theta_i - \theta_o = 180°$.

13. Repeat Problem 12 for the phase detector of Figure 4.17c if the flip-flop is triggered by the positive going edges of its input signals.

14. A PLL is constructed using the phase detector of Figure 4.17c with $V = 10$ V and a VCO with a center frequency of 10 MHz and a control sensitivity of 25 kHz/V. No loop filter is used (i.e., $H(s) = 1$).
    (a) Find the loop bandwidth.
    (b) Find the hold-in range of the loop.
    (c) If the loop is in the steady-state locked condition with a 10 MHz input signal before $t = 0$ and the input signal has a step increase in phase of 90° at $t = 0$, sketch the input voltage to the VCO to scale versus time. Neglect the high frequency (10 MHz) components of this voltage.
    (d) If the loop is locked to a 10 MHz signal that is angle modulated by a 100 kHz sinusoid, how large can the peak frequency deviation become before the linear range of operation of the loop is exceeded?

15. Design an active loop filter for the PLL of Problem 14 so that the loop half-power bandwidth is 100 Hz and the damping ratio is unity. Repeat part (d) of Problem 14 for this loop.

16. If the PLL of Problem 14 is used with an active loop filter, what is the static (steady-state) phase error when the loop is locked to an input sinusoid at 10.1 MHz? Repeat if the passive loop filter is used.

17. Consider the phase detector of Figure 4.17b with the two inputs $A \cos (2\pi f_c t)$ and $A \cos (6\pi f_c t + \theta_o)$. Find the average output voltage as a function of θ_o. Note that because this phase detector can produce an output when one frequency is three times the other, it is possible for a PLL using it to lock to an input frequency that is either three times or one-third the VCO center frequency.

5

NOISE IN AM
AND FM
SYSTEMS

Noise consists of random electrical signals added to the communication signal in the channel. These unwanted signals degrade the quality of the message at the receiver output. In this chapter, we discuss the characteristics of an idealized type of noise known as white noise, the sources of noise, the factors affecting the signal power received from a radio channel, and then we derive relationships for the signal-to-noise ratio (SNR) at the demodulator output for the modulation formats discussed in the preceding two chapters. The goal is to give the reader an appreciation for the factors that affect the output SNR in analog communication systems. In some cases, we are able to use this information to make accurate predictions of the performance of a system. This is demonstrated by a design example in the last section of the chapter.

White noise has a constant power spectral density (PSD) at all frequencies. In the first section of this chapter, we discuss the response of linear filters to white noise inputs, the concept of *noise bandwidth* of a filter, the effects of adding noise from various sources, and the characteristics of *bandpass noise*, obtained by passing white noise through a bandpass filter. White noise does not exist in the physical world because constant PSD from zero to infinite frequency implies infinite power. Nevertheless, it is a concept that is important because it approximates many real noise signals and thus provides a simple theoretical model for comparing the various modulation formats. Only those topics needed in the analysis of analog communication systems are presented. Other characteristics of noise are presented in Chapter 7 using probability concepts.

The second section of this chapter discusses the physical sources of noise in communication systems. One source is *thermal noise*, which is generated in resistors by the random thermal agitation of electrons. Thermal noise is also present at the terminals of a receiving antenna. It is the lower limit on the noise level of a radio communication system. Additional noise is added to the received signal by the first stages of the receiver. It is important to minimize this contribution by designing circuits with a good *noise figure*. In this chapter we develop the tools to enable the calculation of the total effective received noise power that is contributed by the receiver and thermal sources. Various other noise sources may also be present, including lightning, automobile ignitions, other transmitters, and so forth, but the performance limitations imposed by these noise sources are difficult to predict.

The third section of the chapter contains a discussion of the factors that affect the received signal power in radio communication. These include antenna characteristics and attenuation caused by distance and atmospheric absorption. We will see that the received power level can be calculated very accurately in the *free space* situation in which there are no absorbing or reflecting bodies between the transmitter and receiver.

The results of the first sections of the chapter enable us to calculate the signal power and the effective noise PSD at the input to the receiver for some radio communication systems of interest, such as satellite links. In these cases, factors such as antenna size, operating frequency, receiver noise figure, and range from transmitter to receiver must be included in the calculation of received signal power and the effective PSD of the system noise.

Sections four and five develop the theory for predicting the SNR at the demodulator output for the various AM and FM modulation formats. The performance characteristics of the modulation formats are compared and the chapter concludes with an example of a system design.

5.1
NOISE CHARACTERISTICS

White noise

A random noise $n(t)$ that has a constant PSD is called *white noise* because of its similarity to white light, which contains approximately equal power at every frequency. The PSD of a white-noise signal is

$$S_n(f) = \frac{N_0}{2} \text{ W/Hz} \tag{5.1}$$

Often, the actual noise signals in a communication system may not have a constant PSD. However, when the transmitted signal spans only a small range of the frequency domain, the actual noise PSD is often nearly constant over the frequency range of interest. Then a white-noise assumption can be used in theoretical considerations with accurate results.

If a signal $x(t)$ is passed through a filter, the PSD of the resulting output signal $y(t)$ was found in Section 2.5 to be

$$S_y(f) = |H(f)|^2 S_x(f) \tag{5.2}$$

where $S_y(f)$ is the PSD of the output signal, $H(f)$ is the transfer function of the filter, and $S_x(f)$ is the PSD of the input signal. If the input signal is white noise, this becomes

$$S_y(f) = |H(f)|^2 \frac{N_0}{2} \tag{5.3}$$

The total power contained in the output signal can be found by integrating its PSD over all frequencies. In equation form, we have

$$P_y = \frac{N_0}{2} \int_{-\infty}^{\infty} |H(f)|^2 \, df \tag{5.4}$$

EXAMPLE 5.1

Find the PSD and total power for the output signal of the lowpass filter shown in Figure 5.1a when the input signal $n(t)$ is white noise with a PSD of $N_0/2$.

FIGURE 5.1
A simple filter and its noise equivalent.

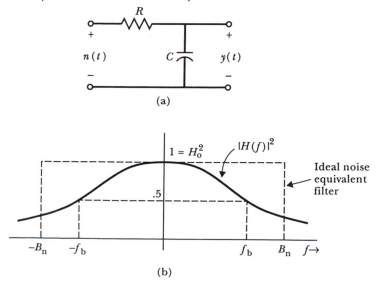

(a)

(b)

SOLUTION

The transfer function of the filter can be found by simple network analysis to be

$$H(f) = \frac{1}{1 + j(f/f_b)}$$

where $f_b = 1/2\pi RC$ is the 3 dB bandwidth of the filter. A plot of $|H(f)|^2$ is shown in Figure 5.1b. The PSD of the output signal is given by

$$S_y(f) = |H(f)|^2 \frac{N_0}{2} = \frac{N_0/2}{1 + (f/f_b)^2}$$

The total power in the output signal can now be found by integrating the PSD. In equation form:

$$P_y = \int_{-\infty}^{\infty} S_y(f)\, df = \int_{-\infty}^{\infty} \frac{N_0/2}{1 + (f/f_b)^2}\, df$$

$$= \frac{f_b N_0}{2} \tan^{-1} \frac{f}{f_b} \Big|_{-\infty}^{\infty} = \frac{\pi}{2} f_b N_0$$

Noise bandwidth

The concept of *noise bandwidth* is often used in connection with passing white noise through filters. The noise bandwidth of a filter is defined as the bandwidth of an equivalent ideal filter of the same type (i.e., lowpass or bandpass) that has the same nominal gain H_0 and passes the same noise power as the actual filter. The power gain of the equivalent filter for Example 5.1 is indicated by the dotted line in Figure 5.1b. Since the gain of the equivalent filter is a constant H_0 in the passband and zero elsewhere, the output power of the equivalent filter is given by

$$P_y = |H_0|^2 B_n N_0 \tag{5.5}$$

If the right-hand sides of Equations 5.4 and 5.5 are equated, the following expression for the noise bandwidth can be obtained:

$$B_n = \frac{1}{2|H_0|^2} \int_{-\infty}^{\infty} |H(f)|^2\, df \tag{5.6}$$

Since for real filters, the expression $|H(f)|$ is an even function of frequency, it can also be written as

$$B_n = \frac{1}{|H_0|^2} \int_0^{\infty} |H(f)|^2\, df \tag{5.7}$$

5 / NOISE IN AM AND FM SYSTEMS

Parseval's energy theorem (given in Equation 2.22) can be applied to Equation 5.6, resulting in the following alternate expression:

$$\blacksquare \qquad B_n = \frac{1}{2|H_0|^2} \int_0^\infty h^2(t)\, dt \qquad (5.8)$$

The lower limit has been set to zero in the last expression because the impulse response of the filter $h(t)$ is assumed to be zero for negative time.

EXAMPLE 5.2

Use Equation 5.8 to find the noise bandwidth of the filter of Figure 5.1a and then use Equation 5.5 to find the output power when the filter is driven with a white-noise signal. Compare this result with the power found in Example 5.1.

SOLUTION

The impulse response of the filter can be found to be

$$h(t) = \frac{1}{RC} \exp\left(\frac{-t}{RC}\right) u(t)$$

Since the filter is a lowpass filter with a dc gain of unity, we take the nominal gain H_0 as unity. Then using Equation 5.8 we obtain

$$B_n = \frac{1}{2(RC)^2} \int_0^\infty \exp\left(\frac{-2t}{RC}\right) dt = \frac{1}{4RC}$$

This can be expressed in terms of the 3 dB bandwidth, $f_b = 1/(2\pi RC)$, as

$$B_n = \frac{\pi}{2} f_b$$

Using Equation 5.5 we now find the total output power of the filter:

$$P_y = \frac{\pi}{2} f_b N_0$$

This agrees with the result found in Example 5.1

Note that the choice of H_0 is arbitrary, so the value of the noise bandwidth for a filter is meaningless unless the value of H_0 is stated or implied. (All choices for H_0 produce the same result for the output power from the filter provided the same value is consistently applied throughout a given calculation.)

Addition of noise signals

Noise from various sources often combines in communication systems. Consider the case in which two *ac* noise signals $n_1(t)$ and $n_2(t)$, whose powers are P_1 and

P_2, respectively, add together to form a total noise given by

$$n_s(t) = n_1(t) + n_2(t)$$

The power in the total noise signal is given by

$$P_n = \lim_{T \to \infty} \frac{1}{T} \int_{-T/2}^{T/2} [n_s(t)]^2 \, dt$$

$$= P_1 + P_2 + \lim_{T \to \infty} \frac{2}{T} \int_{-T/2}^{T/2} n_1(t) n_2(t) \, dt$$

The last term on the right-hand side of the last expression is often zero. When this is the case, the noise signals are said to be *uncorrelated*. Noise signals from physically independent sources are usually uncorrelated. For example, the thermal noises generated in two resistors are uncorrelated. For uncorrelated noise signals, the product is sometimes positive and sometimes negative, so the integral over a long period of time tends to have a small value. (Note that we have assumed *ac* noise signals having zero average value.) Thus, we have

$$P_s = P_1 + P_2$$

Now consider the case in which we have a message signal $m(t)$ available from two sources with uncorrelated noises, $n_1(t)$ and $n_2(t)$, added to the output of each source. In equation form, we have

$$r_1(t) = m(t) + n_1(t)$$

and

$$r_2(t) = m(t) + n_2(t)$$

We assume that the average power in $m(t)$ is P_m and that the powers in the two noise signals are equal, given by $P_1 = P_2 = P_n$. Now consider adding these two noisy versions of the message signal to form

$$r(t) = r_1(t) + r_2(t) = 2m(t) + n_1(t) + n_2(t)$$

The ratio of the signal power in $r_1(t)$ or $r_2(t)$ to the noise power is P_m/P_n. The message power in $r(t)$ is $4P_m$, because adding the two identical message components has doubled the amplitude of the message. The noise power in $r(t)$ is $2P_n$ if the noises are uncorrelated. Thus, the SNR of the combined signal $r(t)$ is $2P_m/P_n$ or twice that of either of the original signals. An intuitive explanation of why this happens is that the noise signals from uncorrelated sources sometimes have opposite signs so they add destructively. On the other hand, the message waveform is identical from both sources so it adds constructively. Clearly, the SNR can be raised even higher if more signal sources with independent noise components are available. In this situation, the messages are said to have added *coherently* whereas the noises have added *noncoherently*.

Increasing the SNR of a message signal by this means is possible whenever the same signal is available from several sources (or at several different times) and the noises are uncorrelated. One application for this is in seismic exploration for oil; echoes from a number of small explosions are added to enhance the SNR, which is often more convenient than using one large explosion. Another application is in recording a fetal electrocardiogram: The recurring heart signal of an unborn infant can be added coherently.

In demodulating DSB-SC signals, a message component is obtained from both the upper and lower sidebands, and these add together coherently. The demodulated noise component from the noise above the carrier is uncorrelated with the demodulated noise component from the noise below the carrier, and thus the SNR at the output of a DSB-SC demodulator is double the SNR at the demodulator input. We will demonstrate this result more fully later.

Bandpass noise

The white-noise signals at the receiver input are passed through a bandpass filter formed by the RF and IF filters of a typical radio receiver before they arrive at the demodulator. Thus, the noise at the input to the demodulator has a band-limited power spectrum. This type of noise is known as *bandpass noise*. If bandpass noise is observed on an oscilloscope, it will be seen to consist of a sinusoid with a slowly varying amplitude and phase angle. Only small changes in phase and amplitude occur during a carrier cycle. A typical waveform for a bandpass noise signal is shown in Figure 5.2. The frequency of the sinusoid approximately matches the center frequency of the PSD.

The narrower the spectrum becomes the more slowly the amplitude and phase of the bandpass noise change. An intuitive appreciation for the reason for this can be obtained by considering the application of white noise to the simple bandpass filter of Figure 5.3a, whose frequency response is shown in Figure 5.3b. If an impulse that, like white noise, has frequency content at all frequencies is applied to the input of the filter, the output consists of a damped sinusoid whose amplitude slowly decays after the impulse occurs. When white noise is applied, the noise signal continually feeds energy into the circuit and the output slowly fluctuates in amplitude and phase.

Since bandpass noise is a sinusoid with a slowly varying amplitude and phase, an equation for it can be written as

■
$$n(t) = a_\mathbf{n}(t) \cos{[2\pi f_c t + \theta_\mathbf{n}(t)]} \tag{5.9}$$

where $n(t)$ is the bandpass noise, $a_\mathbf{n}(t)$ is its slowly varying amplitude, and $\theta_\mathbf{n}(t)$ is its phase. The frequency f_c can be selected to be anywhere inside or slightly outside of the frequency range of the noise. (If it is picked too far away, the phase $\theta_\mathbf{n}$ is no longer slowly varying.) In our analysis of the effects of noise on the various modulation formats, the frequency f_c is selected to

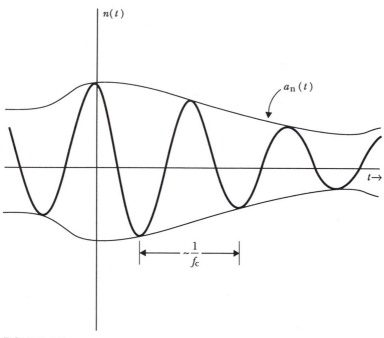

FIGURE 5.2
Typical bandpass noise waveform.

match the carrier frequency of the signal to be demodulated. Therefore, it falls within the noise spectrum and the phase slowly varies with time.

If the identity for the cosine of the sum of two angles is applied to Equation 5.9, an alternative expression for bandpass noise is found:

$$\blacksquare \qquad n(t) = n_i(t) \cos 2\pi f_c t - n_q(t) \sin 2\pi f_c t \qquad (5.10)$$

where $n_i(t)$ is called the *in-phase component*, and $n_q(t)$ is the *quadrature component* of the noise. These components are given by

$$\blacksquare \qquad n_i(t) = a_n(t) \cos \theta_n(t) \qquad (5.11)$$

FIGURE 5.3
A simple bandpass filter and its voltage transfer function.

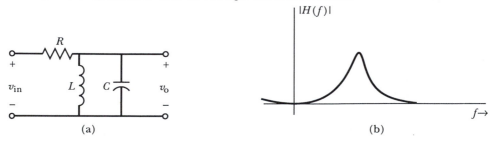

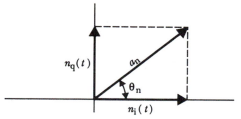

FIGURE 5.4
Phasor representation of bandpass noise.

and

$$n_q(t) = a_n(t) \sin \theta_n(t) \tag{5.12}$$

These relationships are indicated in the phasor diagram in Figure 5.4. It is often helpful in dealing with bandpass noise to visualize it as a phasor that is slowly varying in phase and amplitude.

The in-phase and quadrature components of a bandpass noise can be recovered by the DSB-SC demodulators shown in Figure 5.5. As we have seen, multiplication by a sinusoid forms sum and difference frequencies. It is the difference frequencies that are passed by the lowpass filters of Figure 5.5. Notice that if f_c falls inside the frequency range of the noise, a given difference frequency can be formed from noise components either above or below f_c. Thus, $n_i(t)$ and $n_q(t)$ can be formed from the sum of two noise components, one derived from noise above f_c and one derived from noise below f_c. If the noise originates from thermal sources, these two noise components are uncorrelated and their powers add. It is possible, if the noise is generated by a source that imposes some interdependence among the components at different frequencies (such as an intentional military jammer), that the noise components

FIGURE 5.5
Demodulators for obtaining the in-phase and quadrature components of bandpass noise.

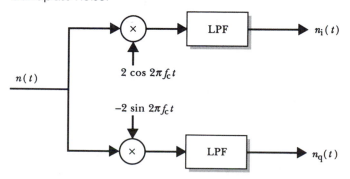

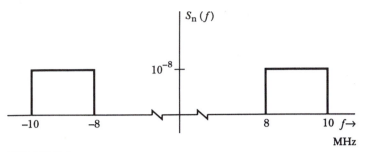

FIGURE 5.6
PSD of the bandpass noise of Example 5.3.

FIGURE 5.7
PSD of the in-phase and quadrature components of the bandpass noise of
Figure 5.6.

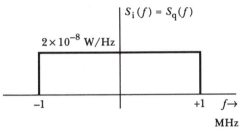

(a) f_c = 9 MHz

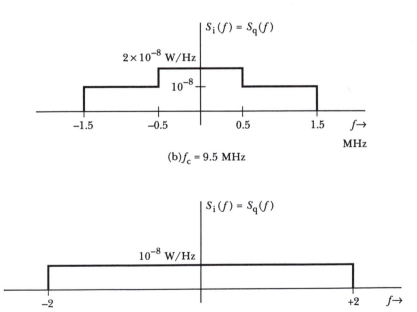

(b) f_c = 9.5 MHz

(c) f_c = 10 MHz

might add coherently. Our main interest at this time is thermal noise, so we assume that the noise signals above and below f_c produce results that add noncoherently in the demodulators of Figure 5.5. In this case, the PSD of $n_i(t)$ and $n_q(t)$ are the same and are given by

$$\blacksquare \qquad S_i(f) = S_q(f) = \mathrm{LP}\{S_n(f - f_c) + S_n(f + f_c)\} \qquad (5.13)$$

where LP indicates that the components in the vicinity of $2f_c$ inside the brackets should be discarded.

EXAMPLE 5.3

Find and plot the PSD of $n_i(t)$ and $n_q(t)$ for the bandpass noise whose PSD is shown in Figure 5.6 if the frequency f_c is chosen as (a) 9 MHz, (b) 9.5 MHz, and (c) 10 MHz. Also find the total power in $n_i(t)$ for each case.

SOLUTION
Equation 5.13 is used to find the PSD of $n_i(t)$ and $n_q(t)$. The term $S_n(f - f_c)$ is simply the PSD of $n(t)$ shifted to the right by f_c, and $S_n(f + f_c)$ is a left-shifted version. The parts of the spectrum inside the braces in the vicinity of $2f_c$ are discarded. The resulting PSD for each choice of f_c is shown in Figure 5.7. The power contained in $n_i(t)$ is the area under the PSD. It is the same for each choice of f_c and equals 40 mW.

In the next section, we discuss some of the physical sources of noise signals.

5.2
SOURCES OF SYSTEM NOISE

Thermal noise

A resistance at any temperature above absolute zero has a small noise voltage at its open-circuited terminals. The rms value of this *thermal noise* voltage is given by

$$\blacksquare \qquad v_n = [4kTRB_n]^{1/2} \text{ V rms} \qquad (5.14)$$

where $k = 1.38 \times 10^{-23}$ joules per kelvin (J/K) is Boltzmann's constant, T is the absolute temperature of the resistor in kelvins, R is the resistance in ohms, and B_n is the noise bandwidth (in Hz) of the system used to perform the measurement of the voltage. An intuitive explanation for this voltage is that thermal agitation of the electrons in the resistor causes them to move predominantly toward one terminal or the other. The terminal to which most of the electrons are closest becomes negative with respect to the other terminal.

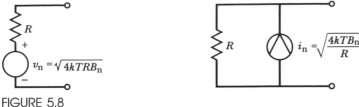

FIGURE 5.8
Noise equivalent circuits for a resistor.

Since the electrons are in constant motion, the voltage varies randomly with time. As indicated in Figure 5.8, this effect can be modeled either by a voltage source in series with the resistor or by the Norton equivalent, consisting of a current source in parallel with the resistor. The rms value of this parallel current source is

■
$$i_n = \left[\frac{4kTB_n}{R} \right]^{1/2} \text{ A rms}$$
(5.15)

The maximum power that can be extracted from the internal noise source is obtained when a load resistance of R ohms is connected to the terminals of the resistor. The value of this maximum *available power* is

$$P_n = kTB_n \text{ W}$$
(5.16)

Note that this is actual power delivered to the load R, in contrast to power normalized to a one-ohm load, which is usually used in theoretical considerations. If the load resistor is at the same temperature, it also has an internal noise source,·which results in a power flow in the opposite direction. Then the net power transfer is zero. If the resistors are at different temperatures, there is a net flow of power.

Notice that the available noise power from a resistor is directly proportional to bandwidth. This implies that the PSD of the noise is constant at all frequencies. From Equation 5.16 we conclude that thermal noise has a PSD given by

■
$$\mathcal{N}_o = kT \text{ W/Hz}$$
(5.17)

Actually, the thermal noise of a resistor is not white. The PSD is given more accurately by

$$S_n(f) = \frac{hf/2}{\exp{(hf/kT)} - 1} \text{ W/Hz}$$
(5.18)

where $h = 6.62 \times 10^{-34}$ is Planck's constant. However, it can be demonstrated that for frequencies below 100 GHz, with temperatures above about 10 K, this expression agrees very well with the white-noise approximation given in Equation 5.17. Thus, for the vast majority of communication systems (excluding optical), the white-noise approximation is very accurate.

Combining noise sources

Noise signals generated by resistors in a network are uncorrelated. Therefore, the power contributed by each noise source should be added as discussed in the previous section. Since power is proportional to the square of the effective voltage, noise voltages or currents should be combined by taking the square root of the sum of the squares of the contributions from the separate resistors.

EXAMPLE 5.4

Two resistors with values of 1500 Ω and 3000 Ω are in parallel. Find the open circuit noise voltage for the combination if the noise bandwidth of the measurement system is 10 kHz. Also, find a single equivalent resistor and its temperature if (a) both resistors are at 290 K and (b) the smaller resistor is at 100 K and the larger at 200 K.

SOLUTION
The parallel combination of the two resistors, including the noise current sources, is shown in Figure 5.9. The resistors can be combined and replaced by an equivalent resistor of 1000 Ω as shown in the figure. The noise current sources produce uncorrelated waveforms, so their values should be combined noncoherently. Thus, the rms value of the total current flowing through the equivalent resistor is given by

$$i_{eq} = [(i_1)^2 + (i_2)^2]^{1/2}$$

where i_1 and i_2 are the rms noise currents for the two resistors computed from Equation 5.15. When both resistors are at 290 K, we find that $i_{eq} = 4.00 \times 10^{-10}$ A, and the resulting open circuit voltage is $v_{eq} = 4.00 \times 10^{-7}$ V. Using Equation 5.14, we find that the 1000 Ω resistor would need to be at 290 K to produce this voltage. In general, when all the resistors in a network are at the same temperature, they can be combined into their equivalent values before noise calculations are carried out.

FIGURE 5.9
Parallel resistors and their equivalent of Example 5.4.

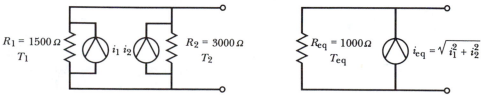

For part b, in which the resistors are at different temperatures, we find that the open circuit voltage of the combination is $v_{eq} = 2.7 \times 10^{-7}$ V. The equivalent 1000 Ω resistor would have to be at a temperature of 133 K to produce this open circuit voltage.

If a network contains frequency-dependent elements such as inductors or capacitors, the PSD due to each noise source can be found at the point of interest in the network. These contributions are then added to find the total PSD, which can then be integrated to find the associated power.

EXAMPLE 5.5

Find the rms value of the voltage across a $R = 50$ Ω resistor at 290 K if the only frequency limitation is caused by a $C = 2$ picofarad (pF) capacitor in parallel with the resistor.

SOLUTION

The noise model of the resistor and shunt capacitor is shown in Figure 5.10. From Equation 5.14, the rms voltage of the noise source is

$$v_n = [4kTRB_n]^{1/2} \text{ V rms}$$

The normalized power (one-ohm load) for this source is the square of this voltage. Since the noise is white, the (normalized) PSD is the power divided by the bandwidth. If we want the customary two-sided PSD, we must divide by the two-sided bandwidth $2B_n$. The result is

$$S_n(f) = \frac{N_0}{2} = 2kTR$$

Note that this is the normalized PSD associated with the noise voltage source and not the actual PSD available from the terminals of the noisy resistor. The actual PSD is given by Equation 5.17.

FIGURE 5.10
Circuit for Example 5.5.

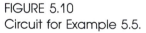
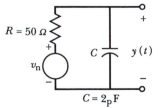

The voltage across the capacitor is a filtered version of the internal noise voltage of the resistor. The transfer function of the circuit is

$$H(f) = \frac{1}{1 + j(f/f_b)}$$

where $f_b = 1/2\pi RC = 1.59$ GHz. The (normalized) PSD of the voltage across the capacitor is given by

$$S_y(f) = |H(f)|^2 S_n(f)$$

The normalized power can now be obtained by integrating $S_y(f)$:

$$P_y = \int_{-\infty}^{\infty} S_y(f)\, df$$

$$P_y = (y_{\text{rms}})^2 = \int_{-\infty}^{\infty} \frac{2kTR}{1 + (f/f_b)^2}\, df$$

$$= 2kTRf_b \tan^{-1}\left(\frac{f}{f_b}\right)\Bigg|_{-\infty}^{\infty} = \frac{kT}{C} = 1.98 \times 10^{-9} \text{ W}$$

Taking the square root of this normalized power, we obtain the rms value of the voltage across the terminals of the RC circuit as

$$y_{\text{rms}} = 44.6 \ \mu\text{V}$$

Notice the somewhat surprising result that the voltage is independent of the size of the resistor. As the resistor becomes larger, its internal noise voltage becomes larger, but the cutoff frequency of the RC filter becomes lower; so more of the high frequency noise is filtered out. (We could have taken a short cut in this problem by using the noise bandwidth of the circuit found in Example 5.2 with Equation 5.5.)

In principle, the output noise voltage of complex networks can be found by following the procedure of Example 5.5. The PSD of the output due to all noise sources in the network acting individually is found. Of course, the transfer function for each source may be different. Then the power contributed to the output by each noise source can be found by integrating the corresponding PSD. Finally, the total power associated with the output voltage is found by adding the power contributed by all independent noise sources.

However, if all of the noise is due to resistors at the same temperature, a somewhat simpler method is possible. First, the impedance of the network looking into the terminals of interest is determined and separated into real and imaginary parts, which are denoted by

$$Z(f) = R(f) + jX(f) \tag{5.19}$$

The total noise voltage at the terminals due to the thermal noise of all of the resistors in the network can then be found from

$$\blacksquare \qquad (v_o)^2 = \int_0^{\infty} 4kTR(f)\,df \qquad\qquad (5.20)$$

EXAMPLE 5.6

Repeat Example 5.5 using Equation 5.20 to find the output voltage.

SOLUTION
The impedance of the parallel RC circuit is given by

$$Z(f) = \frac{R(1/j2\pi fC)}{R - j(1/2\pi fC)}$$

Multiplication of the numerator and denominator of this expression by the complex conjugate of the denominator, $R + j(1/2\pi fC)$, and some algebraic manipulation produces

$$Z(f) = \frac{R}{1 + (2\pi fRC)^2} - j\,\frac{2\pi fR^2C}{1 + (2\pi fRC)^2}$$

From this result, we find that the real part of the impedance is given by

$$R(f) = \frac{R}{1 + (2\pi fRC)^2}$$

Now the mean-square voltage can be found using Equation 5.20.

$$P_y = (y_{\text{rms}})^2 = \int_0^{\infty} \frac{4kTR}{1 + (2\pi RCf)^2}\,df$$

$$= \frac{2kT}{\pi C}\,\tan^{-1} 2\pi RCf \Big|_0^{\infty} = \frac{kT}{C}$$

This agrees with the result found in Example 5.5.

Antenna noise

Thermal noise is not limited to conventional resistors. For example, the theoretical model of an antenna includes a resistance known as the *radiation resistance*, which accounts for the energy that is radiated by the antenna when it is driven by a source. When an antenna is used for receiving, a thermal noise is associated with this radiation resistance. This noise is due to radio frequency thermal radiation received from whatever objects the antenna is facing. A dramatic demonstration of this effect occurs when a communication satellite moves between a receiving station and the sun. The noise from the sun is much greater than that from the cold black space that the receiving antenna normally "sees" behind the satellite. This increase in noise level is often large enough to disrupt communication totally. Satellites in geosynchronous orbit suffer this effect for a short time every day in the spring and the fall.

The noise received by an antenna is part of the noise that limits communication quality. It is often characterized by the *antenna temperature*, a fictitious temperature associated with the internal radiation resistance of the antenna. Antenna temperature depends on the objects the antenna is facing, the antenna design, and the frequency of operation. More information on antennas is included in the next section.

Shot noise

Another source of noise is caused by the granular nature of electrical charge. This type of noise is called *shot noise* because of the similarity of a small pulse of electrical current carried by a finite number of electrons to the cloud of shot in a shotgun blast. Shot noise occurs when current flows in such a manner that the electrons do not interact.

The classical example of a source of shot noise is a temperature-limited vacuum diode, which consists of a heated cathode separated from a plate by a vacuum. The heated cathode emits electrons that are attracted to the plate, which is held at a positive potential with respect to the cathode. The device operates in the temperature-limited region when the voltage difference is great enough so that electrons are pulled to the plate quickly enough to keep a *space charge* or cloud of electrons from forming between the cathode and plate. The current flowing in the device under these conditions shows small fluctuations due to the random emission of electrons. More electrons are emitted in some time intervals than in others of equal duration. The current through the diode consists of an average value or dc component I_{dc} plus a small fluctuating noise component. The noise component is nearly white (however, it eventually falls off at high frequencies due to device and circuit bandwidth limitations). The rms value of shot noise is given by

$$\blacksquare \qquad i_{sh} = [2qI_{dc}B_n]^{1/2} \text{ A rms} \qquad (5.21)$$

where $q = 1.59 \times 10^{-19}$ coulomb (C) is the charge on an electron, and as usual, B_n is the noise bandwidth of the measurement system.

Shot noise also occurs in semiconductor devices when electrons independently cross a barrier, as in a pn junction, but it does not occur when current flows in such a fashion that the electrons interact with one another. For example, if the voltage in the vacuum diode is lowered until an appreciable space charge builds up, the fluctuations in cathode emission cause the size of the electron cloud to fluctuate, but the flow of current to the plate is relatively smooth. The interaction of the electron flow in the space charge smooths the flow of current and the full effect of the shot current is not seen at the plate.

Similarly, shot current is not generated by the flow of current through a resistive medium in which there are electric field interactions among the electrons. Shot noise occurs when electrons independently cross a barrier.

Noise in amplifiers

Amplifiers, due to thermal and shot noise sources associated with their internal components, add noise to the signal being amplified. The *noise figure* of an amplifier is the ratio of the actual noise power available at the output terminals of the amplifier to the noise that would be available if the amplifier did not add noise. The input source is required to be at a standard temperature of $T_0 = 290$ K when noise figures are determined. In equation form, the noise figure is

$$F_r = \frac{P_{o,actual}}{P_{o,ideal}}\bigg|_{T=T_0=290 \text{ K}} \tag{5.22}$$

T is the temperature of the source of the signal to be amplified, $P_{o,actual}$ is the actual noise power available to a matched load at the output terminals, and $P_{o,ideal}$ is the noise power that would be available if the amplifier did not add any noise. Note that the term *available power* refers to the power that would be delivered to a matched load.

The noise figure is often expressed in decibels:

$$F_{dB} = 10 \log (F_r)$$

An ideal noise-free amplifier has a noise figure of $F_r = 1$ or $F_{dB} = 0$ dB.

An equivalent specification of the noise performance of an amplifier is its *noise temperature*, which is the amount that it is necessary to raise the temperature of the source to account for the noise added by the amplifier.

The actual noise power available at the output terminals of an amplifier consists of the amplified source noise plus the noise added by the amplifier. In equation form, we have

$$P_{o,actual} = P_{o,ideal} + P_{o,added} \tag{5.23}$$

The available output noise power due to the source can be written as the product of the available power gain G and the power available from the source. The *available power gain* is the ratio of the power available at the amplifier output for a matched load to the power available from the source under matched conditions. The noise power available from the source resistance is given in Equation 5.16. Thus, we have

$$P_{o,ideal} = GkTB_n \tag{5.24}$$

Substituting this into Equation 5.23, we obtain

$$P_{o,actual} = GkTB_n + P_{o,added} \tag{5.25}$$

Now, the noise added by the amplifier, $P_{o,added}$, can be accounted for by adding the effective noise temperature of the amplifier T_e to the source temperature. Thus, we have

$$P_{o,actual} = Gk[T + T_e]B_n \qquad (5.26)$$

Substituting Equations 5.24 and 5.26 into Equation 5.22, we obtain the relationship between noise figure and noise temperature of an amplifier:

$$F_r = 1 + \frac{T_e}{T_0} \qquad (5.27)$$

EXAMPLE 5.7

A certain signal source has an internal signal voltage of 100 μV, an internal resistance of 50 Ω, and a noise temperature of $T = 500$ K. This source is connected to an amplifier that has a noise figure of 3 dB when used with a 50 Ω source. The noise bandwidth of the amplifier is 2 MHz. Assuming that the signal is within the passband of the amplifier, find the SNR at the output.

SOLUTION
First, we convert the noise figure to a ratio as follows:

$$F_r = [10]^{F_{dB}/10} = 2$$

Now, the effective noise temperature of the amplifier can be found using Equation 5.27.

$$F_r = 2 = 1 + \frac{T_e}{T_0}$$

Since $T_0 = 290$, we find that $T_e = 290$. From Equation 5.26, we find the noise power available at the output of the amplifier to be

$$P_{o,actual} = Gk[T + T_e]B_n$$

We can substitute values for all of the variables except G to find

$$P_{o,actual} = G \times 2.18 \times 10^{-14} \text{ W}$$

Now, we find the signal power available at the output of the amplifier. The signal power available from the source is the power the source can deliver to a matched load. The source has an internal voltage of 100 μV. Thus, it produces 50 μV across a matched load of 50 Ω. The signal power available from the source is

$$\frac{(50 \ \mu V)^2}{50 \ \Omega} = 5 \times 10^{-11} \text{ W}$$

The signal power available at the amplifier output is the product of the available gain and the input power, or

$$P_{so} = G \times 5 \times 10^{-11} \text{ W}$$

The SNR in dB can now be found by dividing the signal power by the noise power found earlier and taking ten times the logarithm. This results in

$$\text{SNR} = 10 \log \left[\frac{P_{so}}{P_{o,\text{actual}}} \right] = 33.6 \text{ dB}$$

This result has been obtained assuming matched conditions both at the input and the output. However, both the noise power and the signal power are affected equally by a mismatch, so the SNR is not affected by a mismatch. As a last comment, we note that the noise figure is a function of the source impedance, so it is important to use the noise figure corresponding to the actual source impedance.

Noise performance of cascaded stages

As we have seen, a receiver consists of the cascade of several stages such as the RF amplifier, mixer, and IF amplifier. Each element adds some noise to the signal passing through it. It is useful to obtain the overall noise figure for cascaded systems in terms of the noise figures of the individual stages. Consider the two stages in cascade shown in Figure 5.11. F_{r1} is the noise figure of the first stage (as a ratio) and G_1 is its available gain. Similar notation is used for the second stage. The actual noise power available at the output of the second stage is given by

$$P_{o,2,\text{actual}} = G_2 P_{o,1,\text{actual}} + P_{o,2,\text{added}} \tag{5.28}$$

The actual power available from the first stage is

$$P_{o,1,\text{actual}} = F_{r1} P_{o,1,\text{ideal}} \tag{5.29}$$

Using Equation 5.24 to substitute for the ideal power, this becomes

$$P_{o,1,\text{actual}} = F_{r1} G_1 k T_0 B_n \tag{5.30}$$

FIGURE 5.11
Cascade of two amplifiers.

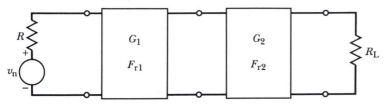

We have used the source temperature $T_0 = 290$ K since that is the required condition for determining the noise figure of a system. The power added by the second stage can be found as

$$P_{o,2,added} = (F_{r2} - 1)G_2 k T_0 B_n \qquad (5.31)$$

Substituting Equations 5.30 and 5.31 into Equation 5.28, we obtain

$$P_{o,2,actual} = G_1 G_2 F_{r1} k T_0 B_n + (F_{r2} - 1)G_2 k T_0 B_n \qquad (5.32)$$

The ideal output noise power can be obtained by setting $F_{r1} = F_{r2} = 1$ in the last expression, to find

$$P_{o,2,ideal} = G_1 G_2 k T_0 B_n \qquad (5.33)$$

Taking the ratio of the last two expressions gives the desired result for the overall noise figure of the cascade as

■ $$F_t = F_{r1} + \frac{F_{r2} - 1}{G_1} \qquad (5.34)$$

Notice that when the available gain G_1 of the first stage is large, the noise performance of the system is determined primarily by the first stage. When the first stage attenuates the signal (so that G_1 is less than one), the overall noise figure can become very large. Thus, it is important to avoid losses ahead of the first amplifier. The first amplifier should have a low noise figure and high gain in a high performance receiving system.

In the preceding development, we have used available power gains, which are the actual gains only if the system has matched impedances. When the system does not have matched impedances, Equation 5.34 still applies because both the amplified source noise and the added noise are reduced by the same amount due to the mismatch; so the ratio is unaffected. However, it is necessary to use available gains in Equation 5.34 even when dealing with an unmatched condition.

The amplifier noise model

The noise added by an amplifier can be modeled by a noise voltage source in series with the input terminals and a noise current source in parallel as indicated in Figure 5.12. The variables v_{na} and i_{na} denote the rms values of these noise sources. To simplify our discussion, we assume that the noise voltage and noise current are uncorrelated.

The three noise sources in Figure 5.12a can be combined by finding the *Thevenin equivalent* of the portion of the circuit to the left of the dotted line. The Thevenin impedance is equal to the source resistance, and the Thevenin voltage is given by

$$v_T(t) = v_{ns}(t) + v_{na}(t) + R i_{na}(t) \qquad (5.35)$$

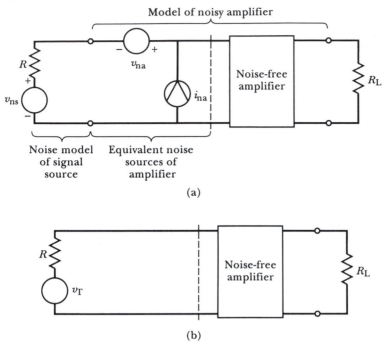

Model of noisy amplifier

Noise model
of signal
source

Equivalent noise
sources of
amplifier

(a)

(b)

FIGURE 5.12
(a) Noise model of a noisy amplifier driven by a noisy source and (b) equivalent circuit of (a).

where we have allowed $v_{ns}(t)$ to stand for the noise voltage waveform of the source as a function of time; but we still use v_{ns} for its rms value. Similar dual notation is used for the other quantities. If the noise sources are uncorrelated, the rms value of this Thevenin source is

$$v_{T} = [(v_{ns})^2 + (v_{na})^2 + R^2(i_{na})^2]^{1/2} \qquad (5.36)$$

As usual, we have combined the rms voltages from uncorrelated sources by taking the square root of the sum of the squares. The power available to the amplifier from the Thevenin source under matched conditions is

$$P_{a} = \frac{(v_{T})^2}{4R} \qquad (5.37)$$

The power available at the output terminals of the amplifier is

$$P_{o,actual} = GP_{a} \qquad (5.38)$$

where G is the available power gain of the amplifier. Substituting Equation 5.36 into 5.37 and the result into 5.38, we obtain

$$P_{o,actual} = \frac{G}{4R}[(v_{ns})^2 + (v_{na})^2 + (Ri_{na})^2] \qquad (5.39)$$

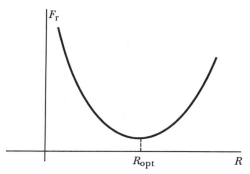

FIGURE 5.13
Noise figure versus source resistance.

If the amplifier were ideal, both noise sources i_{na} and v_{na} would be zero; then, we would have

$$P_{o,ideal} = \frac{G}{4R}(v_{ns})^2 \qquad (5.40)$$

If the source is at $T_0 = 290$ K, the mean-square source noise is

$$(v_{ns})^2 = 4kT_0RB_n \qquad (5.41)$$

Substituting Equation 5.41 into 5.40 and taking the ratio of Equations 5.39 and 5.40, we find an expression for the noise figure of the amplifier:

■
$$F_r = 1 + \frac{(v_{na})^2}{4kT_0RB_n} + \frac{R(i_{na})^2}{4kT_0B_n} \qquad (5.42)$$

The noise figure given in the last expression is a function of the source resistance, illustrated in the plot of noise figure of a typical amplifier versus source resistance shown in Figure 5.13. Notice that an optimum source resistance exists which results in the minimum noise figure. This is because, for very small source resistances, the second term on the right side of Equation 5.42 becomes large whereas for large resistances the last term becomes large. If the noise sources are uncorrelated, the *optimum source resistance* is given by the rather simple formula

■
$$R_{opt} = \frac{v_{na}}{i_{na}} \qquad (5.43)$$

Impedance-shifting circuits or transformers can be included between the source and the amplifier so that the amplifier "sees" the optimum source impedance leading to the lowest noise figure. The best noise figure does not necessarily occur when the source is matched for maximum power transfer to the receiver. Amateur radio enthusiasts are aware of this fact and advise tuning the matching network in a receiver to obtain the clearest reception from a weak signal rather than the loudest reception. However, matching for optimum noise figure may not always be the most desirable procedure. For example, we might want to avoid reflections on a transmission line leading to the receiver.

When the amplifier itself determines the noise bandwidth, the result is referred to as a *broadband noise figure*. If the measuring instrument responds only to a narrow band of frequencies, the result is called a *spot noise figure*. A spot noise figure can be a function of the center frequency of the narrowband filter, due to the fact that the internal noise of the amplifier may not be white.

When using noise figure or equivalent noise temperature, it is important to be sure that the parameter has been determined for the same conditions that exist in the circuit of interest. They are a function of the source impedance, and they may also be a function of frequency if they are spot measurements.

5.3
RECEIVED SIGNAL POWER: THE LINK BUDGET

Received power under free-space conditions

In a radio communication system, a transmitting antenna converts electrical power from the transmitter to an electromagnetic wave that propagates through space to the receiving antenna. The wave spreads as it propagates, so the power density available to the receiving antenna declines with communication distance. In point-to-point communication, the transmitting antenna is designed to radiate as much power as possible in the direction of the receiver and very little in other directions. The object of this section is to give an overview of the characteristics of antennas and propagation so that the received power can be calculated.

An *isotropic antenna* radiates power equally in all directions. At a distance d from the antenna, the radiated power is spread over a spherical surface whose area is $4\pi d^2$. Thus, the power density resulting from an isotropic transmitting antenna is given by

$$\frac{P_t}{4\pi d^2} \text{ W/m}^2$$

where P_t is the total radiated power.

The *gain* of a directional antenna is the ratio of the power density in the preferred direction of radiation to the power density that would be present at the same distance if the antenna were isotropic. Thus, the power density in the preferred direction due to a transmitting antenna with gain G_t is

$$\frac{G_t P_t}{4\pi d^2} \text{ W/m}^2$$

Antenna gain is often expressed in decibel form as

$$G_{t\text{ dB}} = 10 \log G_t$$

The *effective aperture* of a receiving antenna is the area from which the receiving antenna absorbs the power of the propagating electromagnetic wave. Thus, the signal power available at the terminals of the receiving antenna is given by

$$P_r = \frac{G_t P_t}{4\pi d^2} A_r \text{ W} \tag{5.44}$$

where A_r is the effective aperture of the receiving antenna.

The effective aperture of an antenna when it is used for receiving is related to the gain of the antenna, and when it is used for transmitting by

$$A = G \frac{\lambda^2}{4\pi} \tag{5.45}$$

where λ is the wavelength of the signal. The wavelength is given in terms of the center frequency f_c of the (assumed narrowband) signal and the velocity of light $c = 3 \times 10^8$ m/s as

$$\lambda = \frac{c}{f_c} \tag{5.46}$$

If Equation 5.45 is used to substitute for the aperture of the receiving antenna in Equation 5.44, the following expression for received power results:

$$P_r = P_t G_t G_r \left[\frac{\lambda}{4\pi d} \right]^2 \tag{5.47}$$

Communication engineers often convert this equation to decibel form by dividing both sides by a reference power level (typically one watt or one milliwatt) and taking the logarithm multiplied by 10 to obtain

$$P_{r\text{ dB}} = P_{t\text{ dB}} + G_{t\text{ dB}} + G_{r\text{ dB}} - 20 \log \left[\frac{4\pi d}{\lambda} \right] \tag{5.48}$$

where $P_{r\text{ dB}} = 10 \log (P_r/P_{\text{reference}})$ is the received power in decibels relative to the reference power. If the reference power is one watt, the units of $P_{r\text{ dB}}$ are dBW. Alternatively, the reference power can be one milliwatt (mW), in which case the units are dBm. For example, a power of one-tenth of a watt may be given as either -10 dBW or as $+20$ dBm.

The last term on the right side of Equation 5.48 is called the *free space loss*. The term *free space* implies that there are no absorbing or reflecting bodies between the antennas. Equations 5.44, 5.47, and 5.48 give the received power under these ideal free-space conditions. In practical applications, losses must be included to account for many other factors such as *pointing loss* due to misalignment of the antennas, *atmospheric attenuation*, power loss in cables leading to the antenna, and so forth.

Antenna characteristics

It is useful to classify antennas in terms of their maximum dimension in wavelengths. First, we consider antennas that are small compared with a wavelength (maximum dimension of $\lambda/10$ or less). Examples of such antennas are short dipoles and small loops (often they are multiple turns on a ferrite core). For example, such small antennas are used in AM broadcast receivers. These antennas have low values of gain that are close to that of an isotropic antenna. They also have poor radiation efficiency. Much of the power delivered to the terminals of such small antennas is converted to heat in the ohmic resistance of the antennas instead of being radiated into space. For this reason, small antennas are not well suited for transmitting. On the other hand, antennas with a minimum dimension of a half wavelength or more can be easily designed to radiate nearly all of the power delivered to their terminals.

Examples of antennas in the size range from one-half to two wavelengths can be seen in receiving antennas for television or FM broadcast. Gains run from 2 dB to perhaps 20 dB.

Parabolic dish and exponential horn antennas, such as those in terrestrial microwave or satellite communications, are examples of large antennas whose maximum dimension is ten wavelengths or more. The effective aperture of large dish and horn antennas is typically in the range of 50 to 70% of the physical area, as seen from the preferred direction.

Received noise and antenna temperature

As we pointed out in the previous section, a source such as an antenna delivers noise power to its terminals as well as signal power. The level of this noise is dependent on many factors such as the frequency of operation, the direction the antenna is pointed, the season of the year, the time of day, and whether the antenna is in an urban or rural environment. Noise can be due to electrical storms, black body radiation from nearby objects, manmade interference, or galactic radiation from sources in space. The level of such noise can be characterized by the temperature necessary for the internal resistance of the antenna to develop an equivalent noise level from thermal sources. Due to the many sources of noise that may be present, the effective *antenna temperature* T_a can be rather difficult to predict. A plot of typical antenna temperatures is shown versus frequency in Figure 5.14. Antennas that are pointed upward experience the lowest temperatures, in the range from 1 to 10 GHz, and so this range of frequencies is attractive for space communication. At about 10 GHz atmospheric absorption begins to become significant, and black body radiation from this absorbing medium raises the antenna temperature. Below 100 MHz, electrical noise due to electrical storms and man-made sources causes the effective antenna temperature to become very high.

The noise power available to the receiver from the antenna, denoted by P_a, is given by substituting the antenna temperature into Equation 5.16. This

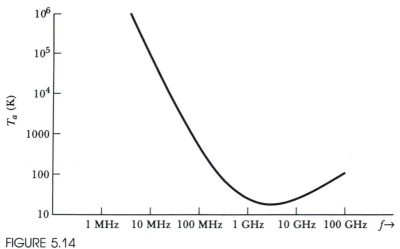

FIGURE 5.14
Typical equivalent antenna temperatures versus frequency.

results in

$$P_a = kT_aB_n$$

As we discussed in the previous section, the effective noise temperature of the receiver can be added to the source temperature to determine the noise added by the receiver. Thus, the total system noise is given by

■ $$P_n = k[T_a + T_e]B_n = kT_{syst}B_n \qquad (5.49)$$

where the *system temperature* T_{syst} is given by

■ $$T_{syst} = T_a + T_e \qquad (5.50)$$

EXAMPLE 5.8

Find the received power and SNR for the radio communication system with the parameters given below. These values are similar to those of a "backyard" receiver for the reception of television signals from satellites in geosynchronous orbit.

$$P_t = 8 \text{ W}$$

$$G_{t\,dB} = 26 \text{ dB}$$

The receiving antenna is a 10-ft diameter parabolic dish with an effective aperture equal to 50% of the physical area and an equivalent noise temperature of $T_a = 50$ K.

$$f_c = 4 \text{ GHz}$$

The receiver noise temperature is $T_e = 100$ K.

$$B_n = 30 \text{ MHz}$$

$$d = 25,000 \text{ miles}$$

SOLUTION

The gain of the receiving antenna can be calculated from Equation 5.45 using an effective aperture that is 50% of the area of a 10-ft (3.05-m) diameter circle. The result is $G_r = 8160$, which can be expressed in decibels by taking 10 times the logarithm of the gain to obtain $G_{r\,dB} = 39.1$ dB. The transmitted power is $10 \log 8 = 9.0$ dBW. Now, the received power can be found using Equation 5.48. In tabular form, we have the following:

$P_{t\,dB}$	9.0 dBW
$G_{t\,dB}$	26.0 dB
$G_{r\,dB}$	39.1 dB
free space loss $= -20 \log (4\pi d/\lambda)$	-196.6 dB
$P_{r\,dB}$	-122.5 dBW

The SNR can now be found by dividing the received power by the effective noise power given by Equation 5.49. Alternatively, we can work with decibels in tabular fashion as follows:

$P_{r\,dB}$	-122.5 dBW
$-10 \log k$	228.6 dBJ/K
$-10 \log T_{\text{syst}}$	-21.8 dBK
$-10 \log B_n$	-74.8 dBHz
SNR	9.5 dB

This SNR is low, but we will find in our analysis of FM demodulators that the SNR at the output of the demodulator can be considerably higher than at the input. Clear television signals can be obtained by demodulating such a low SNR input signal.

FIGURE 5.15
Simplified model of radio communication system.

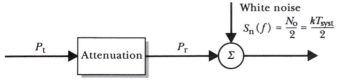

At this point, we have seen how to determine the received signal power and the PSD of the noise added to the signal in a free-space radio communication system. The received power can be calculated as shown in Example 5.8, and the PSD of the added noise is

$$\blacksquare \qquad S_n(f) = \frac{N_0}{2} = \frac{kT_{\text{syst}}}{2} \text{ W/Hz} \qquad (5.51)$$

Thus, the effects of transmission of the signal are attenuation and the addition of white noise, as shown in Fig. 5.15. In the next two sections we will use this simple model for the channel in an analysis of the SNR at the demodulator output for various modulation formats.

Non-line-of-sight propagation

We should point out, before leaving the topic of link budget calculations, that not all communication systems have channel characteristics that can be modeled by simple attenuation and added white noise. For example, it is often the case that a signal arrives at the receiver by two paths that may have significantly different time delays. The signal may travel directly from the transmitter to the receiver, and it may also be reflected to the receiver from a building or other object. The bending of radio waves by temperature variations in the atmosphere is another source of multipath reception. Examples of communication channels that are not adequately modeled by Figure 5.15 are prevalent in military communications in which a noise signal due to a jammer may cover only part of the frequency band so that it is not white, or in which it is pulsed so that the noise power is not steady. Even though the model of Figure 5.15 does not apply in every case, it is important enough to warrant detailed consideration. It is the channel model we will use in our analyses of the noise performances of the various modulation formats.

5.4
OUTPUT SNR FOR BASEBAND AND AM SYSTEMS

In this section, we derive relationships for the output SNR for baseband and various forms of AM. In the next section, we will extend the analysis to FM and PM. In each case we assume that the channel simply attenuates the transmitted signal and adds white noise. Furthermore, the message is assumed to extend from approximately dc to a highest frequency f_H, and the filters in each system are chosen to obtain a flat overall frequency response from dc to f_H.

Baseband

The model for analysis of a baseband communication system is shown in Figure 5.16a. The message signal $m(t)$ is applied to the input of the channel, which

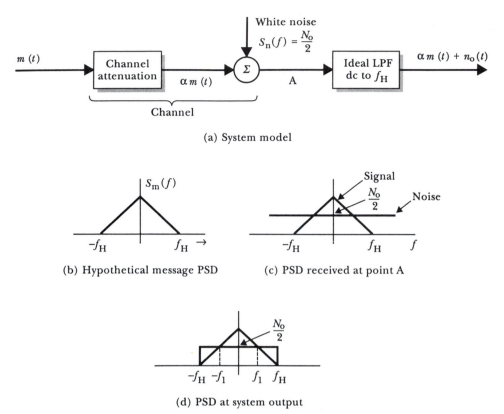

FIGURE 5.16
Model for analysis of baseband communication.

attenuates the signal by multiplying by α and adds white noise whose PSD is $N_0/2$ at all frequencies. A hypothetical PSD for the message is shown in part b of the figure. Part c shows the PSD of both the signal and noise components at the output of the channel. The receiver simply consists of an ideal lowpass filter that passes the message with constant gain from dc to f_H. This filter eliminates noise components above f_H. For simplicity, we assume that the gain of the filter is unity. (The resulting SNR at the output is unaffected by the value of the gain since both the noise and signal power depend on the gain.) The PSD resulting for the signal and noise components at the output of the filter is shown in part d of the figure.

The signal power at the receiver output is $\alpha^2 P_m$. This is also the received signal power P_r at the channel output since the gain of the lowpass filter is assumed to be unity. The noise power at the system output is the area under the noise PSD shown in part d of Figure 5.16. This power is $N_0 f_H$; thus, the output SNR is given by

$$\blacksquare \qquad \text{SNR}_o = \frac{\text{signal power out}}{\text{noise power out}} = \frac{P_r}{N_0 f_H} \qquad \text{(baseband)} \qquad (5.52)$$

Notice that the SNR could be improved by slightly reducing the cutoff frequency of the receiver filter because this would cause a greater percentage reduction in noise power at the output than it would decrease the signal power. However, reducing the cutoff frequency of the receiver filter would also cause some distortion of the signal, which may be objectionable. Filters have been developed for producing the "best" estimate of a signal imbedded in noise, but the response of such a filter depends on the exact PSD of the signal and noise. We will not consider this type of filtering in our analysis of the effect of noise on modulated systems. Instead, we require the overall frequency response of the systems to be flat over the frequency range of the message.

DSB-SC

The model of a system using DSB-SC modulation is shown in Figure 5.17. The message signal is DSB-SC modulated onto the carrier by a product device. The modulated signal is passed through the channel, which attenuates it and adds white noise. The receiver first passes the incoming signal and noise through a bandpass filter to eliminate the out-of-band noise and then to the demodulator which multiplies by the recovered carrier and finally to a lowpass filter to eliminate the components at the second harmonic of the carrier.

Some of the details of a typical communication system have been omitted from the model shown in Figure 5.17 for simplicity. In an actual receiver, there is usually one or more frequency converters. The predetection filter corresponds (approximately) to the IF filter as discussed in Section 3.6. In a complete receiver, as in the simplified model of Figure 5.17, the modulated signal arrives at the demodulator imbedded in bandpass noise. Thus, the simplifications we have made by not including all of the details of a practical receiver do not affect the resulting expression for the output SNR.

The received signal component at the output of the channel is given by

$$g_{sc}(t) = \alpha A_c m(t) \cos{(2\pi f_c t)} \tag{5.53}$$

The average power P_r in the received signal is given by

$$P_r = \lim_{T \to \infty} \frac{1}{T} \int_{-T/2}^{T/2} [g_{sc}(t)]^2 \, dt \tag{5.54}$$

If f_c is much higher than the highest message frequency, this equation can be reduced to

$$P_r = \frac{1}{2} \alpha^2 A_c^2 P_m \tag{5.55}$$

where P_m is the average power in the message signal $m(t)$.

Assuming that the gain of the filters is unity in the passband, the output signal from the demodulator is given by

$$\text{LP}\{[\alpha A_c m(t) \cos{(2\pi f_c t)}][2 \cos{(2\pi f_c t)}]\} = \alpha A_c m(t) \tag{5.56}$$

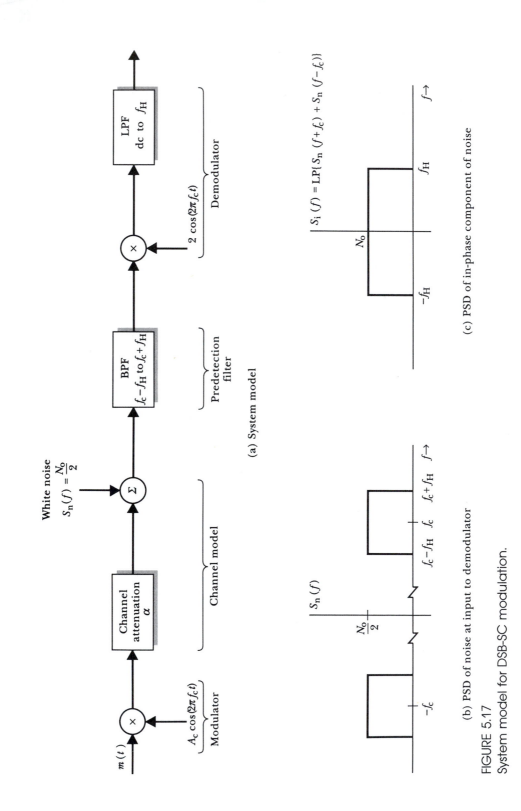

(a) System model

(b) PSD of noise at input to demodulator

(c) PSD of in-phase component of noise

FIGURE 5.17
System model for DSB-SC modulation.

and the resulting output signal power is

$$P_{so} = \alpha^2 A_c^2 P_m = 2P_r \qquad (5.57)$$

The noise at the output of the predetection filter is bandpass noise. The PSD of this noise is denoted as $S_n(f)$ and is shown in Figure 5.17b. We showed in the discussion leading to Equation 5.10, that such bandpass noise can be expressed as

$$n(t) = n_i(t) \cos 2\pi f_c t - n_q(t) \sin 2\pi f_c t \qquad (5.58)$$

The noise resulting at the output of the demodulator is given by

$$LP\{n(t)2 \cos 2\pi f_c t\} = n_i(t) \qquad (5.59)$$

Thus, only the in-phase component of the noise contributes to noise at the output of the demodulator. The PSD of the in-phase component of the noise is given in Equation 5.13 as

$$S_i(f) = LP\{S_n(f - f_c) + S_n(f + f_c)\} \qquad (5.60)$$

The resulting PSD is shown in Figure 5.17c. The noise power at the output of the demodulator is the area under $S_i(f)$, given by

$$P_{no} = 2N_0 f_H \qquad (5.61)$$

The output SNR of the system is therefore given by

$$\blacksquare \qquad SNR_o = \frac{P_{so}}{P_{no}} = \frac{2P_r}{2N_0 f_H} = \frac{P_r}{N_0 f_H} \qquad \text{(DSB-SC)} \qquad (5.62)$$

Note that the output SNR of the DSB-SC system is exactly the same as for the baseband system. DSB-SC modulation does nothing to combat noise in the channel. Later, we will show that the use of FM can improve the output SNR.

The noise power at the input to the demodulator is the area under $S_n(f)$, which is $2N_0 f_H$. The signal power at the input to the demodulator is P_r, so the SNR at the input to the demodulator is

$$SNR_i = \frac{P_r}{2N_0 f_H} \qquad (5.63)$$

Note that this is half of the output SNR, showing that the SNR is doubled in the demodulation process. One way of explaining this fact is to notice that the demodulator does not respond to the quadrature component of the noise $n_q(t)$. Thus, the demodulator responds to all of the signal but only to half of the noise, and thus the SNR is doubled.

An alternative explanation was brought out in our discussion of adding uncorrelated and correlated signals in Section 5.1. We saw that the signal components from the upper and lower sidebands add coherently, whereas the noise components above the carrier add noncoherently to the noise components

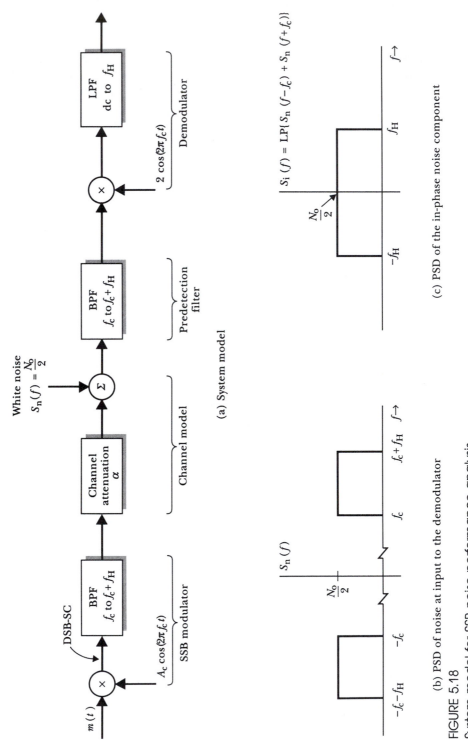

(a) System model

(b) PSD of noise at input to the demodulator

(c) PSD of the in-phase noise component

FIGURE 5.18

System model for SSB noise performance analysis.

below the carrier in the demodulator. Our analysis in Section 5.1. showed that this would improve the SNR by a factor of two.

SSB

Figure 5.18 shows a model for the analysis of the effect of noise on a SSB communication system. We assume that the upper sideband version of SSB is in use, but the analysis and results are almost identical for the lower sideband version. The modulator indicated in the figure produces SSB by the filtering method as discussed in Section 3.3. The channel attenuates the signal and adds white noise. The receiver first removes the out-of-band noise in the predetection filter. The filtered signal and noise are then multiplied by the local carrier, and the product is lowpass filtered to eliminate the components near twice the carrier frequency. This process demodulates the SSB signal.

The signal component at the output of the predetection filter is the SSB signal given by Equation 3.16. Including a constant αA_c to account for the received amplitude, the received signal is given by

$$g_{\mathrm{ssb}+}(t) = \alpha A_c m(t) \cos (2\pi f_c t) - \alpha A_c \hat{m}(t) \sin (2\pi f_c t) \tag{5.64}$$

Recall from Section 3.3 that $\hat{m}(t)$ is the Hilbert transform of $m(t)$, obtained by phase shifting all of the frequency components of $m(t)$ by $-90°$. The power in the received signal at the output of the channel is given by

$$P_r = \lim_{T \to \infty} \frac{1}{T} \int_{-T/2}^{T/2} [g_{\mathrm{ssb}+}(t)]^2 \, dt \tag{5.65}$$

Substituting Equation 5.64 into 5.65, expanding the integrand, neglecting terms including a double carrier-frequency sinusoid, under the assumption that the message and its Hilbert transform are slowly varying compared with the carrier, and using the fact that the power in the Hilbert transform of $m(t)$ is the same as in $m(t)$, we obtain

$$P_r = \alpha^2 A_c^2 P_m \tag{5.66}$$

The signal at the output of the demodulator is given by

$$\mathrm{LP}\{g_{\mathrm{ssb}+}(t)2 \cos (2\pi f_c t)\} = \alpha A_c m(t) \tag{5.67}$$

The signal power at the demodulator output is

$$P_{so} = \alpha^2 A_c^2 P_m \tag{5.68}$$

The noise at the input to the demodulator in Figure 5.18 is bandpass noise given by

$$n(t) = n_i(t) \cos (2\pi f_c t) - n_q(t) \sin (2\pi f_c t) \tag{5.69}$$

The PSD of this bandpass noise is shown in Figure 5.18b. The noise component at the output of the demodulator is simply $n_i(t)$. The PSD of the noise at the demodulator output is given by Equation 5.60 and is shown in Figure 5.18c. The noise power at both the input and output of the demodulator can be found by computing the area under the corresponding PSD. Both powers turn out to have the same value, given by

$$P_{no} = P_{ni} = N_0 f_H \tag{5.70}$$

Notice, by comparing Equations 5.66 and 5.68, that the signal power into and out of the demodulator are the same. Therefore, the SNR is the same at the demodulator input and output. They are both given by

$$SNR_o = SNR_i = \frac{P_r}{N_0 f_H} \quad \text{(SSB)} \tag{5.71}$$

Comparison of this result with Equations 5.52 and 5.62 for baseband and DSB-SC systems shows that all three systems have the same performance for additive white noise. SSB simply shifts the message spectrum up in frequency at the modulator, noise is added in the channel, then the receiver eliminates the out-of-band noise and converts the signal and in-band noise back to baseband. It is not surprising that noise performance is unaffected by the frequency translation since we have assumed that the noise is white. (If the noise were not white, then better performance could be obtained by shifting into a region of low noise PSD.)

AM

The system model shown in Figure 5.17 also serves for the analysis of AM. If a large dc component is added to the message ahead of the modulator, the transmitted signal is AM. Since this dc component is added simply to cause a carrier component to appear in the transmitted signal, it is discarded after demodulation. The portion of the received power that contributes to the useful part of the demodulated message is the power in the sidebands. Therefore, Equation 5.62 can be used to predict the performance of an AM system that uses a coherent demodulator if the received power in the sidebands is used instead of the total received power. The fraction of the total power of an AM signal that is contained in the sidebands is given by Equation 3.10, which is

$$\eta = \frac{P_m}{P_m + 1} \tag{5.72}$$

Therefore, the output SNR for an AM system is the product of the SNR for a DSB-SC system and η. The result is

$$SNR_o = \frac{P_r}{N_0 f_H} \frac{P_m}{P_m + 1} \tag{5.73}$$

Thus, AM has a much poorer noise performance than baseband, DSB-SC, or SSB communication systems due to the fact that η is typically much less than one. This is particularly true for audio messages with very low rms-to-peak-value ratios if the modulation level is kept below 100% to insure that the envelope is identical to the message waveform so that peak detectors can be used.

The result in Equation 5.73 is valid when AM is demodulated by a coherent demodulator (i.e., multiplication by the original carrier followed by lowpass filtering). However, one of the primary advantages of AM is the fact that it can be demodulated by the much simpler envelope detector. Therefore, we now consider the noise performance of AM when it is demodulated with an envelope detector. The signal plus noise at the input to the demodulator of Figure 5.17 when AM is transmitted is given by the sum of an AM signal given by Equation 3.8 and a bandpass noise given by Equation 5.69. The resulting expression for the received signal, including the factor α to account for channel attenuation, is

$$r(t) = \alpha A_c \cos (2\pi f_c t) + \alpha A_c m(t) \cos (2\pi f_c t)$$
$$+ n_i(t) \cos (2\pi f_c t) - n_q(t) \sin (2\pi f_c t) \qquad (5.74)$$

The phasor diagram shown in Figure 5.19a illustrates each of the terms in Equation 5.74 and shows that this expression can be put into the form

$$r(t) = a(t) \cos (2\pi f_c t + \theta(t)) \qquad (5.75)$$

The envelope $a(t)$ is given by

$$a(t) = [\{\alpha A_c + \alpha A_c m(t) + n_i(t)\}^2 + \{n_q(t)\}^2]^{1/2} \qquad (5.76)$$

The phase angle $\theta(t)$ is given by

$$\theta(t) = \tan^{-1} \left[\frac{n_q(t)}{\alpha A_c + \alpha A_c m(t) + n_i(t)} \right] \qquad (5.77)$$

The output of an ideal envelope detector is $a(t)$. The noise components $n_i(t)$ and $n_q(t)$ and the message are all randomly varying signals whose amplitudes are unpredictable, but when the SNR is very high, we can expect that the term $\alpha A_c + \alpha A_c m(t) + n_i(t)$ is much larger than $n_q(t)$ almost all of the time. As a result, when the SNR is high, the $n_q(t)$ term in Equation 5.76 can be dropped, and the output of the envelope detector can be approximated by

$$a(t) \simeq \alpha A_c + \alpha A_c m(t) + n_i(t) \qquad (5.78)$$

The output of the coherent demodulator shown in Figure 5.17 when the input is an AM signal plus narrowband noise is given by

$$\text{LP}\{[\alpha A_c \cos (2\pi f_c t) + \alpha A_c m(t) \cos (2\pi f_c t)$$
$$+ n_i(t) \cos (2\pi f_c t) - n_q(t) \sin (2\pi f_c t)][2 \cos (2\pi f_c t)]\} \qquad (5.79)$$

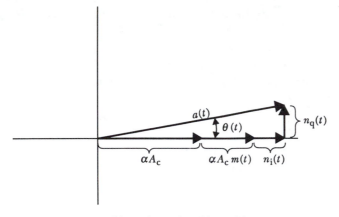

(a) High SNR $a(t) \simeq \alpha A_c + \alpha A_c\, m(t) + n_i(t)$

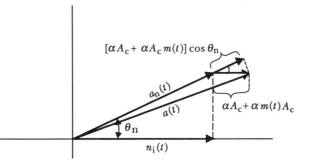

(b) low SNR $a(t) \simeq a_n(t) + [\alpha A_c + \alpha A_c\, m(t)] \cos \theta_n$

FIGURE 5.19
Phasor diagram of signal and noise components at the input to an envelope detector for AM.

Equation 5.79 can be reduced to exactly the same result as Equation 5.78. Thus, the noise performance of the envelope detector is the same as that of the coherent demodulator when the SNR is high.

When the SNR is very low, the performance of the envelope detector is much worse than that of the coherent detector. To demonstrate this, we first rewrite Equation 5.74 using the expression for bandpass noise given in Equation 5.9. The result is

$$r(t) = [\alpha A_c + \alpha A_c m(t)] \cos (2\pi f_c t) + a_n(t) \cos (2\pi f_c t + \theta_n(t))$$

When the SNR is very low, $a_n(t)$ is much greater than $\alpha A_c + \alpha A_c m(t)$ almost all of the time. As indicated in Figure 5.19b, the output of the envelope detector is then approximately

$$a(t) \simeq a_n(t) + [\alpha A_c + \alpha A_c m(t)] \cos \theta_n(t)$$

Notice that the message signal appears in the envelope detector output multiplied by the cosine of the phase of the noise. This product is probably little more intelligible or useful than the noise itself, and thus the envelope detector does not contain a useful signal component. We conclude that the SNR at the output of an envelope detector falls very rapidly when the SNR at the input becomes small. When the output SNR of a demodulator falls very rapidly as the input SNR decreases, the demodulator is said to exhibit a *threshold effect*. We will see another example of this threshold effect when we consider demodulation of FM signals in the presence of noise.

5.5
OUTPUT SNR FOR FM AND PM SYSTEMS
Approximate analysis under high SNR conditions

The system model for the noise performance of FM and PM is shown in Figure 5.20a. The message signal is modulated onto the carrier using either FM or PM. The modulated signal is then passed through the channel where it is attenuated and white noise is added. At the receiver, the received signal and noise are passed through a predetection bandpass filter to eliminate the out-of-band noise. The bandwidth of the filter is selected using Carson's rule. This bandwidth is just wide enough to pass nearly all of the signal but not the noise that falls outside the frequency range of the modulated signal. The output of the predetection filter is then demodulated using a phase detector or a frequency discriminator, depending on the type of modulation under consideration.

The input to the demodulator is the sum of the attenuated angle modulated signal given by Equation 4.1 and narrowband noise given by Equation 5.9. This sum is given by

$$r(t) = \alpha A_c \cos\left[2\pi f_c t + \theta(t)\right] + a_n(t) \cos\left[2\pi f_c t + \theta_n(t)\right] \qquad (5.80)$$

As usual, the factor α has been included to account for channel attenuation. The phase angle of the modulated signal $\theta(t)$ is given by Equation 4.4 for FM and by 4.5 for PM. These equations are repeated here for convenience.

$$\theta(t) = 2\pi k_f \int^t m(x)\, dx \qquad \text{for FM} \qquad (5.81)$$

$$\theta(t) = k_p m(t) \qquad \text{for PM} \qquad (5.82)$$

As illustrated in the phasor diagram in Figure 5.21, the received signal can also be expressed as

$$r(t) = a(t) \cos\left[2\pi f_c t + \theta(t) + \psi(t)\right]$$

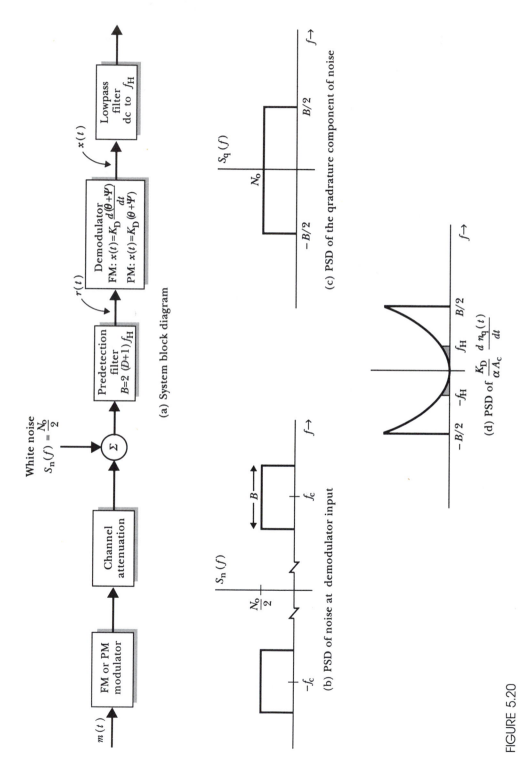

FIGURE 5.20
System model for noise analysis of angle modulation.

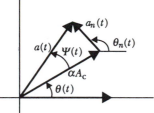

FIGURE 5.21
Phasor diagram of the signal and noise components at the input to the demodulator.

The angle $\psi(t)$ represents the effect of the noise on the demodulator output because the ideal phase detector and frequency discriminator respond only to the phase angle of their input signals. From Figure 5.21 we can determine that this angle is given by

$$\psi(t) = \tan^{-1} \left[\frac{a_n(t) \sin [\theta_n(t) - \theta(t)]}{\alpha A_c + a_n(t) \cos [\theta_n(t) - \theta(t)]} \right] \qquad (5.83)$$

Inspection of this expression for $\psi(t)$ shows that it is a nonlinear combination of the modulated signal and the noise. Thus, demodulators for angle modulated signals are nonlinear devices. Therefore, superposition does not hold for these demodulators. In contrast, the behavior of coherent demodulators for AM signals is linear, and we found the demodulator output due to signal and noise acting separately. An exact analysis of angle modulated signals is much more difficult because of this nonlinearity.

To proceed, we need to make several assumptions to simplify the expression for $\psi(t)$. Of course, the results are only valid when the assumed conditions are met. First, we assume that the SNR at the input to the demodulator is high, so $a_n(t)$ is much less than αA_c almost all of the time. Second, in finding the noise level at the demodulator output, we assume that the message signal is zero. Because the phase angle $\psi(t)$ depends on both the message and the noise in a nonlinear way, the noise component of the demodulator output is affected by the presence of the message. Due to this nonlinearity, an exact analysis is difficult. However, experimental results show that the output noise is not drastically different when the message is present from when the message is zero. As a result of these assumptions, we will find the ratio of the signal power at the demodulator output to the noise power when the message is zero, as it is between words in speech or between pieces of music. In addition, the result is only valid under large SNR conditions.

Using the assumption that the message is zero, we can substitute $\theta(t) = 0$ into Equation 5.83. Then we can use Equation 5.12, which is

$$n_q(t) = a_n(t) \sin \theta_n(t)$$

to substitute for the numerator in the argument of the $\tan^{-1}$ function, resulting in

$$\psi(t) = \tan^{-1}\left[\frac{n_q(t)}{\alpha A_c + a_n(t) \cos \theta_n(t)}\right]$$

This is an exact expression for $\psi(t)$ when the message is quiescent. Making the additional assumption that the SNR is high enough for αA_c to be much greater than $a_n(t)$ almost all of the time and using the fact that the arctangent is nearly equal to its argument for small values, we have

$$\psi(t) \simeq \frac{n_q(t)}{\alpha A_c} \tag{5.84}$$

We find this expression for $\psi(t)$ much easier than Equation 5.83 to use in our analysis because it results in a linear relationship between the noise signal and the demodulator output.

The demodulator output in Figure 5.20 is given by

$$x(t) = K_D \frac{d}{dt}[\theta(t) + \psi(t)] \qquad \text{for FM} \tag{5.85}$$

and

$$x(t) = K_D[\theta(t) + \psi(t)] \qquad \text{for PM} \tag{5.86}$$

where K_D is the gain constant of the demodulator. Substitution, using Equations 5.81, 5.82, and 5.84 results in

$$x(t) = 2\pi K_D k_f m(t) + \frac{K_D}{\alpha A_c} \frac{dn_q(t)}{dt} \qquad \text{for FM} \tag{5.87}$$

and

$$x(t) = K_D k_p m(t) + \frac{K_D}{\alpha A_c} n_q(t) \qquad \text{for PM} \tag{5.88}$$

In both cases, the first term represents the demodulated message and the second term is noise. (Actually, the noise terms in Equations 5.87 and 5.88 are valid only when the message is zero and the SNR is high.) The demodulated message signal passes entirely through the final lowpass filter of Figure 5.20, so the output message power is given by

$$P_{so} = (2\pi K_D k_f)^2 P_m \qquad \text{for FM} \tag{5.89}$$

and

$$P_{so} = (K_D k_p)^2 P_m \qquad \text{for PM} \tag{5.90}$$

To evaluate the noise power at the demodulator output we must find the PSD of the noise signal at the output of the demodulator in Figure 5.20a.

The noise at the input to the demodulator has the PSD shown in Figure 5.20b since it is simply white noise filtered by the predetection filter (which has an assumed gain of unity). The PSD of the quadrature component $n_q(t)$ is given by Equation 5.13. The resulting PSD is shown in Figure 5.20c. As indicated in Table 2.1, taking the time derivative of a signal is equivalent to passing it through a linear system with a transfer function given by

$$H(f) = j2\pi f \tag{5.91}$$

The PSD of the output of a linear system is given by Equation 2.38 as

$$S_{out}(f) = |H(f)|^2 S_{in}(f) \tag{5.92}$$

Therefore, the PSD of the derivative of $n_q(t)$ is

$$(2\pi f)^2 S_q(f) \tag{5.93}$$

The noise term in Equation 5.87 is $K_D/\alpha A_c$ times the derivative of $n_q(t)$, so the PSD of the output noise of the FM demodulator is

$$\left(\frac{K_D}{\alpha A_c}\right)^2 (2\pi f)^2 S_q(f) \tag{5.94}$$

This is plotted in Figure 5.20d. Note that the noise signal at the output of the frequency discriminator has little power at frequencies near dc but large power at the higher frequencies. The noise power, contained in the derivative of $n_q(t)$, which passes through the final lowpass filter is the shaded area in Figure 5.20d. This is given by

$$P_{no} = \left[\frac{2\pi K_D}{\alpha A_c}\right]^2 N_0 \int_{-f_H}^{f_H} f^2 \, df$$

Integrating and evaluating, we obtain

$$P_{no} = \left[\frac{2\pi K_D}{\alpha A_c}\right]^2 N_0 \frac{2}{3} f_H^3 \qquad \text{for FM} \tag{5.95}$$

Thus, we have found the noise power at the system output for the FM case.

Now we find the output noise power for the PM case. Referring to Equation 5.88, we see that the noise component at the demodulator output for the PM case is $(K_D/\alpha A_c)$ times $n_q(t)$. Therefore, the PSD of the output noise is

$$\left(\frac{K_D}{\alpha A_c}\right)^2 S_q(f) \tag{5.96}$$

The output noise power is the integral

$$P_{no} = \left(\frac{K_D}{\alpha A_c}\right)^2 \int_{-f_H}^{f_H} S_q(f) \, df \tag{5.97}$$

Since $S_q(f)$ is equal to $\mathcal{N}_0$ in the range of integration, we obtain the noise power at the system output for the PM case as

$$P_{no} = \left(\frac{K_D}{\alpha A_c}\right)^2 2\mathcal{N}_0 f_H \qquad \text{for PM} \qquad (5.98)$$

Now we can find the output SNR for either modulation format. Taking the ratio of Equations 5.89 and 5.95 we find

$$\text{SNR}_o = \frac{P_{so}}{P_{no}} = \frac{3\alpha^2 A_c^2 k_f^2 P_m}{2\mathcal{N}_0 f_H^3} \qquad \text{for FM} \qquad (5.99)$$

At this point it is convenient to assume that the peak extremes of the message signal $m(t)$ are ± 1. This is no loss of generality since the degree of modulation can still be adjusted to any value by selection of k_f or k_p. As a reminder of the fact that we have normalized the message, we will use P_{m_n} instead of P_m for the message power. Since the message has been normalized and k_f is the constant of proportionality between instantaneous frequency deviation and the message, the peak frequency deviation is given by $\Delta f = k_f$. The deviation ratio given by Equation 4.28 is $D = \Delta f / f_H = k_f / f_H$. Also we can find from Equation 5.80 that the average received signal power is

$$P_r = \frac{\alpha^2 A_c^2}{2}$$

When these substitutions are made in Equation 5.99, we obtain

$$\blacksquare \qquad \text{SNR}_o = \frac{P_r}{\mathcal{N}_0 f_H} 3 D^2 P_{m_n} \qquad \text{for FM} \qquad (5.100)$$

where P_r is the received power of the FM signal at the output of the channel, $\mathcal{N}_0/2$ is the effective noise PSD at the receiver input, f_H is the highest message frequency, $D = \Delta f / f_H$ is the deviation ratio, and P_{m_n} is the power in the normalized message. This is an important result, and we will spend some time discussing its implications; but first we will derive the corresponding result for PM.

Taking the ratio of Equations 5.90 and 5.98 results in

$$\text{SNR}_o = \frac{\alpha^2 A_c^2 k_p^2 P_m}{2\mathcal{N}_0 f_H} \qquad \text{for PM} \qquad (5.101)$$

Again, we find it convenient to assume that the message signal has been normalized. In this case, the peak phase deviation of the PM signal from the unmodulated carrier is $\Delta\theta = k_p$. Incorporating these changes into Equation 5.101 results in

$$\blacksquare \qquad \text{SNR}_o = \frac{P_r}{\mathcal{N}_0 f_H} (\Delta\theta)^2 P_{m_n} \qquad \text{for PM} \qquad (5.102)$$

Comparison of SNR performance with DSB-SC and SSB

Referring to Equations 5.52, 5.62, and 5.71, we see that the output SNR for baseband, DSB-SC, and SSB is given by

$$\blacksquare \qquad \mathrm{SNR}_o = \frac{P_r}{\mathcal{N}_0 f_H} \qquad (5.103)$$

Comparing Equations 5.100 and 5.102 with Equation 5.103 shows that the output SNR for FM has the additional factor $3D^2 P_{m_n}$, and PM has the factor $(\Delta\theta)^2 P_{m_n}$. These results seem to say that we can obtain as good an output SNR as we want simply by making the deviation ratio D larger for FM and the peak phase deviation $\Delta\theta$ larger for PM. This is true, as long as the assumptions we made in deriving these results are valid. Recall that we found it necessary to assume that the SNR at the input to the demodulator is large in order to be able to reduce the expression for $\psi(t)$ to the simple result given in Equation 5.84. As the deviation of an angle modulated signal is increased, the bandwidth of the predetection filter must be increased to accommodate the wider signal bandwidth, and so more noise reaches the demodulator. As a result, the input SNR falls as D or $\Delta\theta$ is increased until the high SNR assumption is no longer valid. At that point, we find that the output SNR begins to fall with increasing deviation rather than increasing as predicted by Equations 5.100 and 5.102. When the output SNR falls by some predetermined amount (1 dB is typical) below the value predicted by those equations, we say that the system has reached threshold. When the system is operating at or below threshold, the output SNR drops very rapidly with additional decrease in SNR at the demodulator input.

An intuitive appreciation for why threshold occurs can be obtained from Figure 5.22 which shows the sum of two phasors, a larger one and a smaller one. Notice that the phase of the sum of the two phasors is controlled primarily by the phase of the larger phasor. The smaller phasor makes only a small change in phase of the sum. (When the smaller phasor is in phase with the larger phasor, it has no effect on the phase of the sum and when it is at

FIGURE 5.22
When a large phasor and a small phasor are added, the phase angle of the sum is almost the same as the phase angle of the larger phasor.

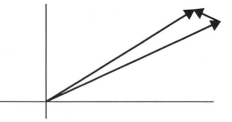

quadrature to the larger phasor it has its maximum, but still small, effect.) For high SNR at the input to the demodulator, the phase is primarily determined by the angle modulated signal with only a small perturbation by the noise. On the other hand, when the system is far below threshold, the phase seen by the demodulator is primarily due to noise and the demodulator output contains a distorted signal component. The fact that the strongest signal at the input to the demodulator in an angle modulated system dominates the output is called *capture effect*: The strongest signal (or noise) "captures" the receiver. Capture can also occur when two interfering angle modulated signals are present. If one signal is much stronger than the other, its message will appear at the demodulator output in preference to the other message. Threshold occurs when the signal is about to lose its capture effect to the noise.

In comparing Equation 5.100 with 5.103, we found that the output SNR of an FM system contains the additional factor $3D^2 P_{m_n}$, compared with a DSB-SC system. For wideband FM, D is larger than one, which tends to make the output SNR larger than for DSB-SC. However, the average power in the normalized message signal P_{m_n} may be small, particularly for high fidelity audio messages, because the peak value of such messages is much larger than their rms value since loud sounds are relatively rare. Thus, the relative performance of FM and DSB-SC depends on the peak-to-rms ratio of the message signal. For a sinusoidal test message, we can show that P_{m_n} is one half; so FM has an advantage over DSB-SC even when the deviation ratio is quite low. Television signals have P_{m_n} on the same order as a sinusoid; so they give FM a greater advantage than audio messages.

When we compare Equations 5.100 and 5.103, the received power P_r is the average received power. In some systems, it may be a *peak power limitation* of the transmitter circuits that sets the limit on the received power. For an angle modulated signal, the peak power (averaged over a carrier cycle) is the same as the average power and is independent of the message signal. Referring to Equation 4.1 for an angle modulated signal, we can show that this power is given by

$$\blacksquare \qquad P_{avg} = P_{peak} = \frac{A_c^2}{2} \qquad \text{for FM} \qquad (5.104)$$

Referring to Equation 3.1 for a DSB-SC signal and assuming that the message signal $m(t)$ has been normalized, we can show that

$$P_{peak} = \frac{A_c^2}{2} \qquad \qquad \text{for DSB-SC} \qquad (5.105)$$

$$\blacksquare \qquad P_{avg} = \frac{P_{m_n} A_c^2}{2} = P_{m_n} P_{peak} \qquad \text{for DSB-SC} \qquad (5.106)$$

Thus, if the limitation is on peak power, the average power allowed for the DSB-SC signal is P_{m_n} times the average power allowed for the FM signal. As we have noted, P_{mn} can be quite small for audio signals. Thus, when there is a peak power limitation, FM has an advantage because it is allowed a higher average power than DSB-SC. Similarly, FM has a peak power advantage compared to SSB and baseband communication systems.

Threshold effects and extended threshold demodulation with PLLs

The character of the noise at the output of a frequency discriminator changes as the SNR is reduced below threshold. At threshold, high amplitude *spikes* begin to appear in the demodulator output. An explanation of this phenomenon follows.

Consider an unmodulated carrier-plus-noise. The unmodulated carrier can be represented by the phasor at zero phase in Figure 5.23a and b. The noise phasor has a randomly varying phase and amplitude, so the composite signal also varies in phase and amplitude, and the tip of the composite phasor traces out a randomly varying path as shown in the figure. When the SNR

FIGURE 5.23
Input and output of a frequency discriminator.

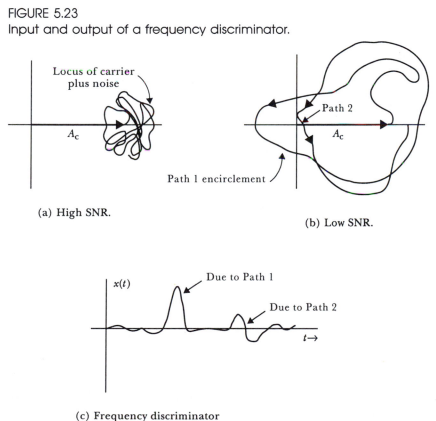

(a) High SNR.

(b) Low SNR.

(c) Frequency discriminator output for (b).

is high as in Figure 5.23a, the variations in phase are relatively small. However, if the noise becomes large enough, the tip of the composite phasor occasionally encircles the origin, and the phase of the composite phasor rapidly changes by 2π. The output of the frequency discriminator depends on the rate of change of the phase, so it contains a spike every time an encirclement occurs, as indicated by the waveform sketch in Figure 5.23c for path 1. Path 2 represents a close approach to the origin but no encirclement. Note that the output pulse due to path 2 has a net area of zero since the net phase change on path 2 is zero. On the other hand, path 1 has a net phase increase of 2π. Therefore, its spike output has a net area equal to $2\pi K_D$. (This is due to the fact that the discriminator output is K_D times the derivative of the input phase angle.) The contribution of these noise signals to the system output depends on their low frequency content because of the final lowpass filter of Figure 5.20a. The value of the Fourier transform at dc is equal to the area under the signal. (This can be verified by substituting $f = 0$ in the definition given in Equation 2.18.) Thus, the contribution to the output noise is much greater for path 1 than for path 2 of Figure 5.23. When encirclements of the origin begin to occur fairly often, the SNR at the system output deteriorates very rapidly. The encirclements occur when the probability of having the noise amplitude greater than the carrier amplitude becomes significant. As we will see when we discuss probability in Chapter 7, this happens abruptly for a thermal noise signal (gaussian amplitude distribution) when the rms value of the noise becomes greater than about one-third of the carrier amplitude.

The deterioration of SNR at threshold is due to origin encirclements that produce rapid phase changes. These encirclements occur when the tip of the phasor encircles the origin. The composite amplitude is likely to be very small at this time. Thus, it would seem to be better to design an FM demodulator that ignores phase changes when the input amplitude is small. A PLL demodulator can be designed so that it tends to ignore rapid phase changes when the input amplitude is low. Recall from Section 4.3 that the loop gain of a first-order PLL is proportional to the phase detector gain constant k_d, which in turn is proportional to the amplitude of the input signal for a product phase detector. Thus, it is possible to design a PLL so that the loop gain varies with input signal amplitude. This characteristic is helpful for an FM demodulator operating close to threshold because the input amplitude is likely to be low when the encirclements occur. Fast changes in phase are not tracked by a PLL with a sufficiently low value of loop gain. Thus, the PLL can have a high loop gain and track the input phase when the composite input amplitude is large but have a low loop gain when the input amplitude is low, and encirclements are likely. It is possible to extend the onset of threshold by about 3 dB with a properly designed PLL as compared to a bandpass limiter and frequency discriminator. The interested reader is referred to Taub and Schilling for more information on this topic.

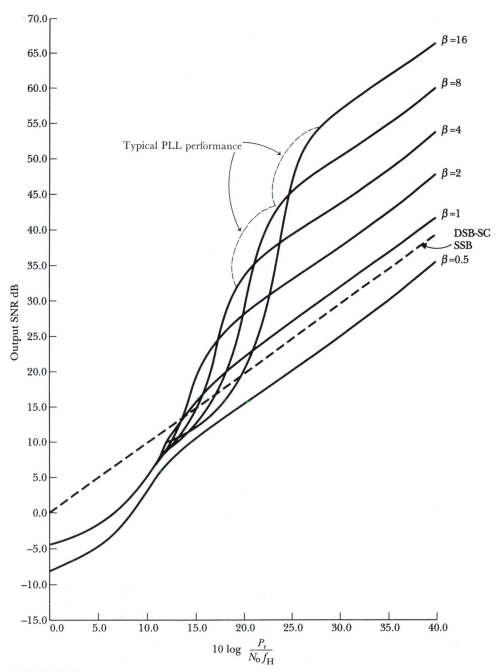

FIGURE 5.24
Output SNR for an FM system with a sinusoidal message.

Taub and Schilling also give an analysis of the SNR of a discriminator in the threshold region and arrive at the following result for the output SNR for a sinusoidal message signal of frequency f_H.

$$\text{SNR}_o = \frac{(P_r/N_0 f_H)(3/2)\beta^2}{1 + (12\beta/\pi)(P_r/N_0 f_H)\exp\left\{-[P_r/2(\beta+1)N_0 f_H]\right\}} \quad (5.107)$$

Note that the numerator of this expression is the same as the result we have found in Equation 5.100, specialized to the case in which the message is a sinusoid. (D has been replaced by β and the normalized message power is $P_{m_n} = 1/2$ for the sinusoidal message.) The denominator of Equation 5.107 accounts for the reduction of SNR due to the spikes that occur below threshold. Equation 5.107 is plotted in Figure 5.24 for several values of β. The output SNR for baseband, DSB-SC, and SSB systems is also plotted for comparison. The output SNR achievable by typical PLLs is also shown in the figure. Note the threshold extension for the PLL demodulators.

The SNR at the input to the demodulator in Figure 5.20 for the case of sinusoidal modulation is given by

$$\text{SNR}_i = \frac{P_r}{2(\beta+1)N_0 f_H} \quad (5.108)$$

This expression can be used with Equation 5.107 to find an expression for the output SNR of a frequency discriminator in terms of the input SNR to the discriminator for the case of a sinusoidal message. The result is given by

$$\text{SNR}_o = \frac{3\beta^2(\beta+1)\text{SNR}_i}{1 + [24\beta(\beta+1)/\pi]\text{SNR}_i \exp(-\text{SNR}_i)} \quad (5.109)$$

This is plotted in Figure 5.25. Note that threshold occurs when the SNR at the input to the frequency discriminator falls below about 10 dB.

Preemphasis and deemphasis

Another technique for increasing the output SNR of an FM communication system is illustrated in the block diagram of Figure 5.26. As shown in the figure, a lowpass filter has been added to the output of the system of Figure 5.20. Since the output noise has a PSD that increases with frequency, the output noise can be decreased significantly by adding a lowpass filter at the system output. Of course, the lowpass filter also attenuates the high frequency components of the signal. Therefore, a filter has also been added at the input of the system to increase the high frequency content of the signal. These filters are known as *preemphasis* filters at the input and *deemphasis* filters at the demodulator output. Since the preemphasis filter may affect the amplitude of the signal passing through it, a gain adjust block with amplitude gain K has been included at its output so that the peak amplitude into the FM transmission system is

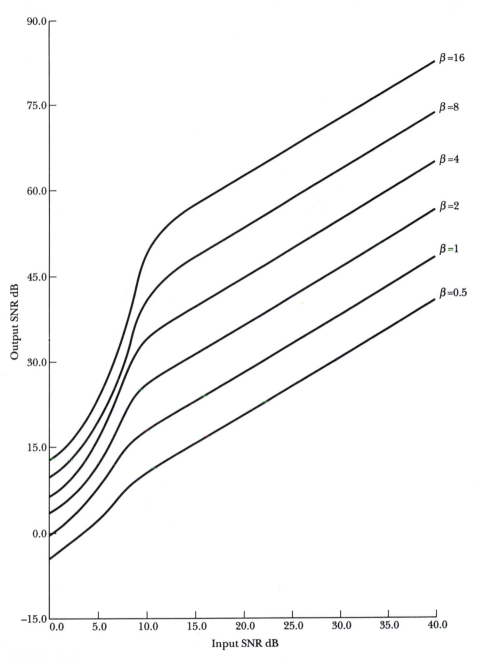

FIGURE 5.25
Output SNR versus input SNR for a frequency discriminator with a sinusoidal message.

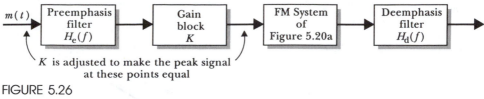

$m(t)$ → [Preemphasis filter $H_e(f)$] → [Gain block K] → [FM System of Figure 5.20a] → [Deemphasis filter $H_d(f)$] →

K is adjusted to make the peak signal at these points equal

FIGURE 5.26
FM system with preemphasis and deemphasis.

unchanged. This block is necessary to keep the peak frequency deviation and, hence, the transmitted bandwidth the same as before the filters are added.

The transfer functions of the preemphasis filter and the deemphasis filter are denoted by $H_e(f)$ and $H_d(f)$, respectively. To have a constant overall frequency response, we require

$$H_e(f)H_d(f) = 1 \qquad (5.110)$$

for all frequencies in the message signal. The circuit diagrams and transfer functions of typical filters used for preemphasis and deemphasis in commercial FM broadcasting are shown in Figure 5.27.

Let us now consider the effects of preemphasis and deemphasis on the output SNR of the system. In our discussion, we assume that the gain of the filters is unity (0 dB) at low frequencies as shown for the filters of Figure 5.27. Assuming that the bandwidth of the predetection filter is unchanged, the noise power at the demodulator output is unchanged when the filters are added to the system. Since the deemphasis filter attenuates the high frequency components of the noise, the noise power at the output of the deemphasis filter is less than at the demodulator output. If the filters are selected so that Equation 5.110 is satisfied, the signal at the output of the deemphasis filter is K times the signal present at the output of the system before the filters and gain adjustment block were added. The output signal power is therefore K^2 times larger in the system with emphasis. Thus, the change in the output SNR depends on two factors: the amount that the deemphasis filter reduces the output noise power and the value of K^2. Recall that the gain adjustment block with gain K is needed to adjust the peak amplitude of the emphasized signal at the FM modulator input to be equal to the peak value before adding the preemphasis filter. Since the preemphasis filter increases the high frequency components of the signal, we can expect that it also increases the peak amplitude of the signal and K must be less than unity. However, experience has shown that typical voice and music signals have little power in the high frequency components; thus, K should be only slightly less than one for these signals. Therefore, the output signal power can remain almost the same when emphasis is added to the system whereas the noise power is substantially reduced, resulting in an improved SNR at the output. Note that the characteristics of the message signal have a strong effect on the amount of improvement possible.

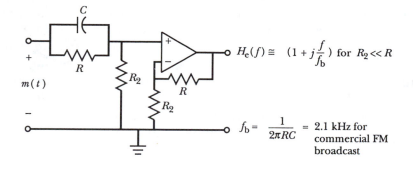

$$H_e(f) \cong \left(1 + j\frac{f}{f_b}\right) \text{ for } R_2 \ll R$$

$$f_b = \frac{1}{2\pi RC} = 2.1 \text{ kHz for commercial FM broadcast}$$

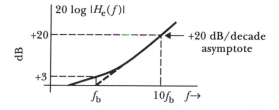

(a) Preemphasis filter

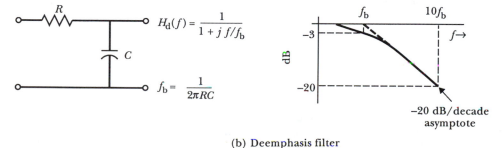

$$H_d(f) = \frac{1}{1 + j\, f/f_b}$$

$$f_b = \frac{1}{2\pi RC}$$

(b) Deemphasis filter

FIGURE 5.27
Preemphasis and deemphasis filters and Bode plots of their transfer functions.

5.6
SYSTEM DESIGN EXAMPLE

At this time, we consider a (hopefully) realistic system design problem to illustrate how some of the material presented so far can be applied. However, in any effort to design any but the most trivial systems, a good deal of knowledge of topics outside the realm of this book must be applied. The range of problems that can occur is large and the practical state of the art of such things as electronic circuit design is constantly changing. Nevertheless, the concerns of the communication system designer are similar from problem to problem.

As one practices the art of engineering design on a number of projects, a good deal of the necessary knowledge and skill becomes second nature.

The example we consider is the design of a radio system to communicate a compressed voice signal from the moon to an earth receiving station. The term *compressed* refers to the process of limiting the peak values of the signal to reduce the ratio of peak to rms values of the signal. This is done to improve the received SNR, and therefore intelligibility, even though it causes some distortion. The signal to be communicated and the system to be designed have the following characteristics:

The highest message frequency is $f_H = 5$ kHz.

The ratio of rms to peak value of the message is 0.1.

$SNR_o = 40$ dB is required at the receiver output.

A 10 m diameter dish antenna is available at the receiving site if it turns out to be suitable.

"Reliable" reception should be possible whenever the moon is at least 10° above the horizon at the receiving site.

The message signal is 100 mV rms from a 50 Ω source.

The transmitter should contain its own power source and antenna. It should occupy the minimum volume and have the smallest mass possible consistent with other requirements. It should be capable of continuous operation for three years.

The modulation type, carrier frequency, and transmitted bandwidth are to be determined during the design process.

Perhaps the first thing to consider is the choice of carrier frequency. Equations 5.44 and 5.45 show that, for antennas with fixed aperture, the received signal power increases with an increase in the carrier frequency. This is because the gain of the transmitting antenna increases with frequency (Equation 5.45) and the received power is proportional to the transmitting antenna gain (Equation 5.44).

On the other hand, if we increase the frequency above about 10 GHz, then *rain fades* caused by absorption of the signal by water become a problem. The degree to which this is a problem depends on the receiver location and on exactly what is meant by the vague requirement for "reliable" reception. (Often in practice, engineers must deal with such vague specifications. Sometimes a more refined specification can be elicited from the customer, and at other times, knowledge of the ultimate mission objective provides a guide.)

Another consideration in selecting the carrier frequency is that generation of the transmitter signal power becomes more difficult and consumes more power from the power supply as the frequency goes up. Atmospheric absorption begins to increase significantly above about 30 GHz even in the absence of precipitation, though there are "windows" at higher frequencies. Interference with and by other systems may also be a factor.

After considering of all of these factors, we come to a "first cut" on our choice of carrier frequency. We should realize that our first choice may create unresolvable problems later in the design, and we may need to come back to this step. For purposes of illustration suppose that, at this point, a carrier frequency of 3 GHz seems to be most appropriate. Therefore, the wavelength of the transmitted signal is given by

$$\lambda = \frac{c}{f_c} = \frac{3 \times 10^8}{f_c} = 0.1 \text{ m}$$

where $c = 3 \times 10^8$ m/s is the speed of light.

At this point, we can begin to estimate the eventual system temperature. Recall that the system temperature depends on the noise received by the receiving antenna and the noise added by the receiver. The receiving system temperature is the sum of the antenna temperature and the effective noise temperature of the receiver (Equation 5.50). Antenna temperature depends on the carrier frequency, the antenna characteristics, and where the antenna is pointed. Our choice of 3 GHz for the carrier frequency is about optimum for minimizing temperature for an antenna pointed upward through the earth's atmosphere. (One additional consideration here is the fact that the sun is nearly behind the moon during a new moon. What effect is this going to have on antenna temperature and, again, what does "reliable" communication mean?) Suppose that after some study of the variables we decide that a "worst case" antenna temperature of 50 K seems reasonable.

The effective noise temperature of the receiver is a function of how much money we want to spend. Since low cost is always a goal, we decide at this point in the design to select a low-noise amplifier (LNA) having a noise temperature of 50 K. If the available gain of the LNA is very high, it sets the noise temperature of the receiving system. This results in a system noise temperature of 100 K, the sum of the antenna temperature and the effective noise temperature of the receiver. (Note that a more expensive LNA with a much lower noise temperature would lower the system temperature to about 60 K, and this would reduce the transmitter power to 60% of the value required for a 100 K system temperature. As we will see, this would probably be more economical overall, considering the cost of putting additional power supply mass on the moon.) The effective noise PSD available at the receiving antenna can now be calculated using Equation 5.51. The result is

$$\mathcal{N}_0 = k T_{\text{syst}} = 1.38 \times 10^{-21} \text{ W/Hz} \tag{5.111}$$

Now we turn to the question of what type of modulation we should use. We confine our consideration to the analog formats that we have studied. (A complete design should also consider digital methods, and we will return to this example in Section 8.7 after we have studied digital modulation.) Since we have seen from Equation 5.100 that output SNR can be improved at the expense of increased bandwidth by using FM, we certainly want to consider the use of FM in the system design.

Sometimes, available bandwidth is limited by regulations or technical constraints. As a rule of thumb, we can say that it becomes difficult to design antennas and other system components if the bandwidth is larger than about 10% of the carrier frequency. Since the message bandwidth is 5 kHz and we are contemplating a carrier frequency of 3 GHz, we can have a large amount of bandwidth expansion before this constraint comes to bear. For purposes of this example, we ignore regulatory constraints, but that might not be a good idea in engineering practice. Thus, bandwidth is not a serious limitation in this design.

Now we can find the optimum bandwidth for our system or, equivalently, the deviation ratio D if we use FM. From Figure 5.24 we see that the maximum output SNR is obtained for fixed P_r, $\mathcal{N}_0$, and f_H when the deviation ratio is selected so that the system is operating just above threshold. From Figure 5.25, we see that threshold occurs when the SNR at the demodulator input is about 10 dB, or somewhat less if an extended threshold demodulator such as a PLL is used. Note that the output SNR drops very rapidly below threshold, so we need to be sure that enough margin is designed into the final system to guarantee that operation is above threshold. For now, we design for a SNR at the demodulator input of 10 dB. Reference to Figure 5.20a shows that the SNR at the demodulator input is given by

$$\text{SNR}_i = \frac{P_r}{\mathcal{N}_0 B} \tag{5.112}$$

Using Carson's rule given in Equation 4.27 for the predetection bandwidth, this becomes

$$\text{SNR}_i = \frac{P_r}{\mathcal{N}_0 f_H 2(D+1)} \tag{5.113}$$

Now, the requirement for an input SNR of 10 dB (which we note is also 10 as a ratio, and it is the ratio we should use in the previous equation) results in

$$\frac{P_r}{\mathcal{N}_0 f_H} = 20(D+1) \tag{5.114}$$

Using this to substitute in Equation 5.100 for the SNR at the receiver output, we find

$$\text{SNR}_o = 60D^2(D+1)P_{m_n} \tag{5.115}$$

where P_{m_n} is the power in a normalized version of the signal. One of the specifications for the system is that the ratio of rms to peak value is 0.1. Thus, if the message is normalized so that its peak value is unity, the rms value is 0.1, and the power it delivers to a one-ohm resistor is

$$P_{m_n} = 0.01$$

Substituting this into Equation 5.115 and solving for D, we find

$$D = 25.2$$

This can be used in Carson's rule of Equation 4.27 to determine the bandwidth of the transmitted signal:

$$B = 2f_H(D + 1) = 262 \text{ kHz}$$

Now we can determine the available power needed at the receiving antenna to fill the specification for the SNR at the input to the demodulator. Substituting previously found values for B and N_0 into the expression

$$SNR_i = 10 = \frac{P_r}{N_0 B}$$

we find that

$$P_r = 3.62 \times 10^{-15} \text{ W} \qquad \text{(for FM)}$$

Taking 10 times the logarithm of this value, we find the received power in dB relative to one watt.

$$P_r = -144.4 \text{ dBW} \qquad \text{(for FM)}$$

At this point, it is interesting to calculate the received power required for a DSB-SC or a SSB system. For these systems, the output SNR is given by

$$SNR_o = \frac{P_r}{N_0 f_H}$$

Assuming that the same antenna and LNA are used, N_0 is the same as in the FM system. Substituting values, we find that

$$P_r = 6.90 \times 10^{-14} \text{ W} \qquad \text{(for DSB-SC and SSB)}$$

The received and, ultimately, the transmitted power requirement for DSB-SC is 19.1 times higher than for the FM system. The situation is even worse if we consider peak power, which is 100 times higher yet for the DSB-SC approach than for FM, due to the ratio of rms to peak message amplitudes given by Equations 5.104 and 5.106. At this point, we can confidently conclude that an FM modulation format is the best choice of the formats we have considered to this point.

Now we work toward a determination of the required transmitter power and the antennas to be used. Since a 10 m parabolic dish is available at the receiving site we can, at least initially, assume that it will be the receiving antenna. We assume that it is suitable for use at the carrier frequency we have chosen and that the effective aperture is 50% of the physical area of the dish. This is a slightly conservative guess at the effective aperture but it should be sufficiently accurate for initial design calculations. Based on these estimates, the effective aperture of the receiving antenna is

$$A_r = 0.50(\pi 5^2) = 39.3 \text{ m}^2$$

The gain of the receiving antenna can then be calculated using Equation 5.45:

$$G_r = \frac{4\pi A_r}{\lambda^2} = 49,350$$

In decibels this is

$$G_{r\,dB} = 46.9 \text{ dB}$$

The approximate maximum distance from the moon to the earth is $d = 252,000$ miles or 4.06×10^8 m. The free space loss can now be calculated:

$$20 \log \left[\frac{4\pi d}{\lambda} \right] = 214.1 \text{ dB}$$

Substituting the previously found values into Equation 5.48, we find

$$P_{t\,dB} = P_{r\,dB} - G_{t\,dB} - G_{r\,dB} + 20 \log \left[\frac{4\pi d}{\lambda} \right] \tag{5.116}$$

$$P_{t\,dB} = 22.8 - G_{t\,dB}$$

This must be satisfied by our choice of transmitter power and transmitting antenna gain. We can choose a small low-gain antenna and high transmitter power or a large antenna and low power.

Another consideration in the choice of both antennas is the matter of keeping them pointed at one another. The gains we have calculated up until now have assumed perfect pointing. As either antenna is rotated away from the other, the received power falls off. The 3-dB or half-power beamwidth of an antenna is the angular rotation of the antenna from the point on one side of exact alignment at which the power is 3 dB below the value for exact alignment, to the same point on the other side. The receiving antenna must have a tracking system to keep it pointed at the transmitter as the moon moves relative to the earth. For a high-gain antenna, the beamwidth θ is related to the gain G by the approximate formula

$$\theta^2 = \frac{27,000}{G}$$

where θ is in degrees. Thus the beamwidth of the receiving antenna is about $0.55°$, and the pointing error should be somewhat less than this to avoid significant loss in received power.

In contrast with the motion of the moon as seen from the earth, the earth appears to remain stationary in the lunar sky because the moon rotates only once for each of its revolutions around the earth. Actually the earth moves by about $\pm 4°$ around its average position. Therefore, we should consider using a transmitting antenna with a wide enough beamwidth so that tracking is not necessary by the transmitter. Perhaps an antenna with a beamwidth of 8° would be a good choice because it would only require doubling the transmitter power to account for the worst case pointing loss; however, this is a tradeoff that should be studied carefully before the design is completed. This choice of transmitter antenna beamwidth leads to a gain of 422 or 26.3 dB.

Using this value in Equation 5.116 leads to a required transmitter power of -3.5 dBW. Since we chose to allow 3 dB of pointing loss in the worst case, this means the transmitter power would be -0.5 dBW or 0.89 W.

At this point, we have completed a cursory design of the desired system. It may (probably does) have some very serious errors or oversights. Certainly we have made some progress, the choice of FM over DSB-SC or SSB seems clear cut, for example. The point here is not to complete the design, since that would take hundreds or even thousands of pages to present in detail, but to illustrate some of the typical concerns and the use of the theory in design situations.

SUMMARY

1. White noise has a constant PSD at all frequencies. Although it does not exist in nature, it is an important theoretical concept for the analysis of communication systems.

2. The PSD of a noise signal at the output of a linear system can be determined from the PSD of the input noise and the transfer function of the system.

3. The total power in a noise signal can be determined by integrating its PSD over frequency.

4. The noise bandwidth of a filter is the bandwidth of an ideal brickwall filter with the same nominal gain as the actual filter, provided that the output power of both filters are equal when the inputs are driven with white noise. Noise bandwidth is useful for calculating the output noise power when the input noise is white. It cannot be used if the input noise is not white. Noise bandwidth can be calculated either from the transfer function or the impulse response of the filter.

5. If the time average of the product of two noise signals is zero, the signals are said to be uncorrelated. If uncorrelated noise signals are added, the resulting power is the sum of the power in the individual signals. Equivalently, the rms values of uncorrelated signals are combined by taking the square root of the sum of the squares.

6. When a desired signal is available from several sources with uncorrelated noises, summing the signals results in coherent addition of the desired signal and noncoherent addition of the noises. This produces an improved SNR at the summer output. This effect is responsible for doubling of the SNR in DSB-SC demodulation.

7. Bandpass noise results from passing white noise through a bandpass filter. Bandpass noise waveforms are sinusoidal with frequencies approximately matching the center frequency of their PSD. The envelop and phase of bandpass noise vary slowly with time. Bandpass noise can be resolved into in-phase and quadrature components. The PSD of the in-phase and quadrature components can be determined from the PSD of the bandpass noise and the (assumed) carrier frequency.

8. Thermal noise is caused by the random motion of electrons in resistors and the reception of black body radiation by receiving antennas. It is approximately white for most communication systems of interest, excluding optical systems. Thermal noise is always present and sets the lower limit of the noise level in communication systems.

9. Thermal noise can be modeled by a noise voltage source in series with each resistor in a circuit. Noise from various sources is combined by adding the PSD contributions from all sources and then integrating the total PSD to find total noise power. In the special case in which all the resistors in a passive network are at the same temperature, the total noise can be computed from the resistive part of the network impedance as seen from the terminals of interest.

10. Shot noise results from the granular nature of electric charge when a current flows through a barrier as in a temperature limited vacuum diode or a *pn* junction.

11. The noise figure of an amplifier is the ratio of the actual noise output power to the noise output power that is due to thermal noise from the source, provided that the source impedance is at the standard temperature of 290 K. Noise figure is a function of the source impedance. Each amplifier has an optimum source impedance for minimum noise figure.

12. The effective noise temperature of an amplifier can be computed from the noise figure. It is the amount that is necessary to add to the source temperature in noise calculations to account for the noise added by the amplifier.

13. The noise figure of a cascaded system can be calculated from the noise figures and available gains of the individual stages. If the first stage has a high gain, it has the dominant effect on the overall noise figure.

14. The signal power available at the terminals of a receiving antenna under free-space propagation conditions can be calculated from the antenna characteristics, the transmitted power, the distance between the antennas, and the carrier frequency.

15. The effective noise power available at the receiving antenna terminals is a function of the noise power received from external sources and the noise added by the receiver. It is characterized by the system temperature, which is the sum of the antenna temperature and the effective noise temperature of the receiver.

16. The output SNR of DSB-SC and SSB systems using coherent demodulators is the same as for a baseband system in the presence of white noise. These modulation formats do nothing to improve noise performance unless the noise is not white and the carrier frequency can be selected to place the modulated signal in a range of frequencies with low noise.

17. The noise performance of AM is poorer than other modulation formats. It has been used historically to allow the use of envelope detectors for demodulation. The noise performance of an envelope detector is the same as that of a coherent demodulator, provided that the input SNR is high. At low SNR, the performance of an envelope detector displays a threshold effect, and its performance becomes very poor.

18. FM provides improved noise performance compared with that of linear modulation formats. The improvement is gained at the expense of bandwidth and depends on the SNR at the input to the demodulator being above threshold, which occurs at an input SNR of about 10 dB. Below threshold, the output SNR of an FM system becomes very poor.

19. Wideband FM systems display a capture effect in which the strongest signal present at the input to the demodulator suppresses the effect of any other signal present.

20. Near threshold, the output of a frequency discriminator begins to contain noise spikes when the noise overcomes the signal, which causes encirclements of the origin resulting in rapid phase changes of 360°.

21. PLLs can be designed to extend the onset of threshold by about 3 dB. It is important to avoid using a limiter when threshold extension is desired.

22. Preemphasis and deemphasis is the process of increasing the high-frequency content of a signal before transmission through an FM system and then lowpass filtering at the receiver to restore the original amplitude to each frequency component. It can improve the output SNR of a system. The amount of improvement depends on characteristics of the message signal. It works well for voice and music signals because preemphasis increases the signal amplitude only a small amount, due to the fact that the high-frequency content of these messages is small.

23. The theory of communication must be combined with knowledge of antennas, propagation, and electronics in the design of a radio communication system. System requirements are often vague or stated in such a manner that the application of theoretical results is difficult. Practical knowledge and good judgement gained through experience is an essential ingredient of the design process. Nevertheless, an excellent understanding of communication theory is indispensable to good design.

REFERENCES

K. Feher. *Digital Communications*. Englewood Cliffs, New Jersey: Prentice-Hall, 1981.

H. L. Krauss, C. W. Bostian, and F. H. Raab. *Solid State Radio Engineering*. New York: Wiley, 1980.

C. D. Motchenbacher and F. C. Fitchen. *Low-Noise Electronic Design*. New York: Wiley, 1973.

J. Smith. *Modern Communication Circuits*. New York: McGraw-Hill, 1986.

H. Taub and D. L. Schilling. *Principles of Communication Systems*, 2d ed. New York: McGraw-Hill, 1986.

PROBLEMS

1. Find the total power in the noise signals whose PSDs are shown in Figure P5.1.
2. Find the noise bandwidths of the filters with the transfer functions shown in Figure P5.2. Use the nominal gain H_o indicated for each filter.

FIGURE P5.1

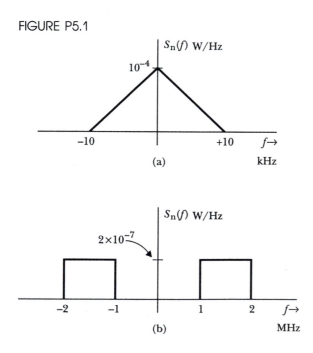

(a)

(b)

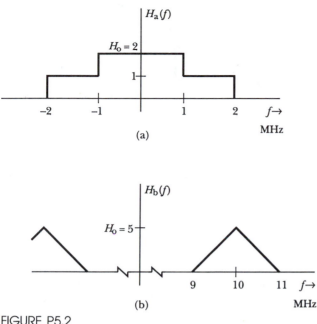

(a)

(b)

FIGURE P5.2

3. The filters whose impulse responses are shown in Figure P5.3 are lowpass filters. Find the noise bandwidth for each filter. Take H_o as the dc gain of the filter in each case.

4. A message signal with power P_m is transmitted by two paths to a receiver. Uncorrelated noises $n_1(t)$ and $n_2(t)$ are added to the signal in the respective paths. The power in the noise signals is P_{n1} and P_{n2}. The receiver produces an output by taking a weighted sum of the noisy received signals. The system is shown in Figure P5.4. Find the output SNR of the receiver. Find the value of the weighting factor k that optimizes the output SNR. What is the optimum value of k when $P_{n1} = P_{n2}$? What is the optimum value of k when P_{n1} is much larger than P_{n2}? Do these results seem reasonable?

FIGURE P5.3

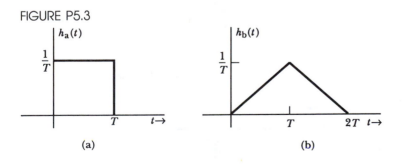

(a)

(b)

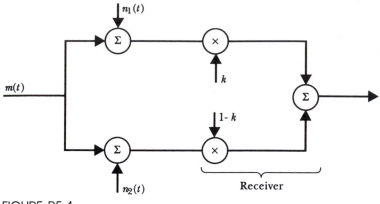

FIGURE P5.4

5. In the system shown in Figure P5.5, signal power is split between two transmission channels by selection of the parameter, α, which may range from zero to one. If the power in $m(t)$ is P_m, show that the sum of the powers entering the two channels is also P_m. Uncorrelated noise signals are added in each channel. Assume that the noise signals have the same power denoted by P_n. The receiver produces an output by taking a weighted sum of the channel outputs. Find the output SNR and determine what value of k is optimum as a function of α. Compare the output SNR for optimum k with the case in which only one channel is used (i.e., $\alpha = k = 0$). Is there any advantage in splitting the message power between two channels?

6. Bandpass noise has been produced by passing white thermal noise through a bandpass filter. The PSD of the noise is shown in Figure P5.6. Determine and sketch to scale the PSD of the in-phase and quadrature components of the noise if the carrier frequency is (a) 5.0 MHz, (b) 5.5 MHz and (c) 6.0 MHz.

FIGURE P5.5

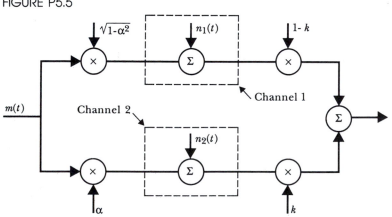

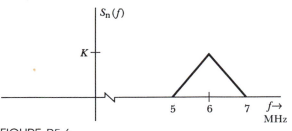

FIGURE P5.6

7. Two resistors R_1 and R_2, at temperatures T_1 and T_2, respectively, are connected in series. Find a single resistor and the appropriate temperature that are equivalent to the series combination.

8. One 50 Ω resistor is at 1000 K, and a second 50 Ω resistor is at absolute zero. If these resistors and a 0.5 pF capacitor are connected in parallel, find the net flow of power from the hot resistor to the cold one due to the thermal noise voltage.

9. Write an expression in the form of an integral for the mean-square noise voltage at the terminals of the network shown in Figure P5.7.

10. A 1 mA current flows through the series combination of a 50 Ω resistor and a device that creates shot noise. Find the rms voltage across the resistor due to the shot noise in a 1 MHz bandwidth. Compare this to the thermal noise source associated with the resistor if the resistor is at 290 K.

11. A certain amplifier has a noise figure of 6 dB. How much does this amplifier degrade the output SNR from that which would be produced by an ideal amplifier if the effective noise temperature of the source is (a) 30 K, (b) 3000 K, and (c) 290 K? Express the answers in decibels.

12. Consider the system shown in Figure P5.8. With the switch in position A and the attenuator set at 10 dB, the voltmeter reads 50 mV. With the switch in position B and the attenuator set at 12 dB, the voltmeter again reads 50 mV. Find the effective noise temperature and the noise figure of the amplifier. For what source impedance is this measurement valid? If the bandwidth of the attenuator and voltmeter is much broader than that of the amplifier, this would be a *broadband measurement*. If the measuring system has a very narrow bandwidth, then we have a *spot measurement* whose value depends on the center frequency of the passband.

FIGURE P5.7

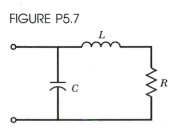

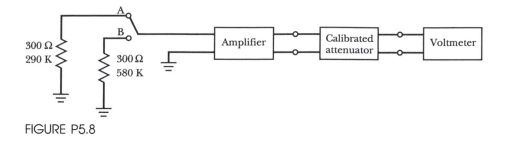

FIGURE P5.8

13. Consider the system shown in Figure P5.9. By how many decibels does the SNR at the load decrease when the switch is closed? This is a crude model for what happens when the cable connecting the antenna feed to the LNA in a space communication system becomes lossy. Note that the loss resistance is "hot" compared with the source, making its effect much greater than simple attenuation of the signal.

14. In the system shown in Figure P5.10, the noise added by the amplifier can be modeled by the uncorrelated current and voltage source shown. (a) What turns ratio results in the maximum signal power delivered to the load? What is the noise figure of the system in this case? (b) What turns ratio results in the maximum SNR at the load? What is the noise figure of the system in this case?

15. A certain television station operating on channel 6 ($f_c = 83.25$ MHz) has an effective radiated power of 5000 W. (This is the product of the transmitted power and the transmitting antenna gain.) A receiving antenna with a gain of 10 dB and a noise temperature of 290 K is used with a receiver having a noise figure of 3 dB. Find the maximum free space distance that results in a SNR of at least 30 dB in a 6 MHz noise bandwidth.

16. A certain message signal has a ratio of peak value to rms value of 100. Using DSB-SC with a coherent demodulator in a communication system, the output SNR is 40 dB. It is proposed to change to AM so that an envelope detector can be used. By what factor must the average transmitted power be increased to keep the output SNR at 40 dB? The degree of modulation of the AM signal should be as high as possible consistent with the use of the envelope detector.

17. The receiver intended for reception of DSB-SC shown in Figure 5.17 is used to receive a SSB signal. Assuming that white noise is the only source of interference, by how many decibels is the output SNR degraded by the wrong receiver? Assume, as shown in the figure, that f_c falls in the center of the predetection filter.

FIGURE P5.9

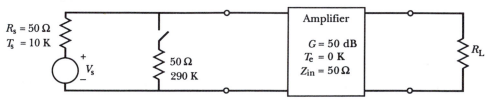

$R_s = 50\ \Omega$
$T_s = 10$ K

V_s

$50\ \Omega$
290 K

Amplifier
$G = 50$ dB
$T_e = 0$ K
$Z_{in} = 50\ \Omega$

R_L

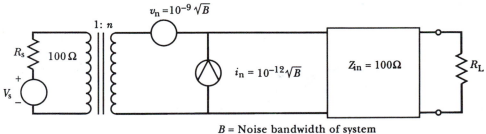

$$v_n = 10^{-9} \sqrt{B}$$

1: n

R_s 100 Ω

V_s

$i_n = 10^{-12}\sqrt{B}$

$Z_{in} = 100\,\Omega$

R_L

B = Noise bandwidth of system

FIGURE P5.10

18. Suppose that the passband of the predetection filter in Figure 5.17 is increased so that it extends from $f_c - 2f_H$ to $f_c + 2f_H$. What is the effect on the output SNR when receiving a DSB-SC signal in the presence of white noise?

19. Find an expression for the PSD of the noise at the output of the FM system shown in Figure 5.26 when the filters shown in Figure 5.27 are used. Assuming that the preemphasis does not increase the peak value of the signal, because the signal has very low components at high frequencies, K is unity for the same peak frequency deviation. Under this condition, the signal power at the output does not change when preemphasis and deemphasis are added to the system. Therefore, the SNR improvement depends only on how much the deemphasis filter reduces the noise. Evaluate this improvement for $f_H = 15$ kHz and $f_b = 2.1$ kHz.

6

PULSE
MODULATION
TECHNIQUES

We have discussed various forms of continuous wave (CW) modulation, in which the carrier is a sinusoidal waveform. In this chapter, we will continue our discussion with a form of modulation in which the carrier is a pulse train. In pulse modulation techniques, sample values of the message signal are taken periodically and used to vary the amplitude, width, or some other parameter of every pulse in the pulse train. This is in contrast to CW modulation techniques in which the value of the amplitude, frequency, or phase of the sinusoidal carrier is continuously varied by the message signal.

In the first section of this chapter, we investigate several methods for periodic sampling of a message signal, the rate at which samples must be taken to preserve the message content, and how the message can be recovered from the samples.

One of the advantages of pulse train modulation by message samples is that it is possible to interleave pulses that are modulated by samples of several different messages. The pulse train of interleaved message samples is then transmitted over the channel. At the receiving end, the pulses of the various messages are separated and routed to their appropriate destinations. Thus, several messages can be transmitted over one channel at one time. This technique is called *time division multiplexing* (TDM). This is in contrast with the CW modulation formats, in which we saw that FDM can be used to transmit messages simultaneously by choosing carrier frequencies far enough apart so that the signals do not overlap in the frequency domain. We discuss TDM in the second section of the chapter.

In the third section of this chapter, we consider the bandwidth of *pulse amplitude modulated* (PAM) signals. Constricting the bandwidth of a pulse train causes the pulses to be smeared so that they overlap. This can cause the modulation of one pulse to appear in the time slot of the following pulse so that crosstalk occurs when the pulses are separated at the receiving end. This is also called *intersymbol interference* (ISI). We show that if the spectrum of the pulse train is shaped in a certain way, ISI can be avoided.

In the fourth section, we consider a PAM TDM system including the effects of noise. We show that the noise performance and bandwidth requirements are comparable to those of a SSB FDM system.

The last section discusses a digital technique known as *pulse code modulation* (PCM). In PCM, the message is sampled by measuring its amplitude periodically and then converting each sample value into a binary code word consisting of a string of 1s and 0s. This process is often called analog-to-digital (A/D) conversion. It is impossible to exactly represent, by a finite code word, a sample value that can take on a continuum of values. Thus, A/D conversion introduces errors known as *quantizing errors* into the value of every sample. We will consider the limitation on SNR imposed by quantizing error. The noise performance of PCM will be discussed and compared with analog modulation techniques.

The code words generated in a PCM system can be used to modulate a carrier in many different ways. If the carrier is a pulse train, a positive pulse can be transmitted for each 1 and a negative pulse for each 0. Another way is to transmit a pulse for a 1 and no pulse for a 0. A sinusoidal carrier can also be used, in which case the code bits determine the amplitude, frequency, or phase of the modulated signal. Digital methods of communication have many advantages over analog methods, and a good deal of the remainder of this book will be devoted to their study.

6.1
SAMPLING THEORY

Instantaneous sampling

Now we consider a message signal $m(t)$ that has frequency content from approximately dc to an upper frequency limit f_H and determine the minimum sampling rate required as well as methods for reconstructing the entire message from its samples. There are several sampling methods. We first consider the case in which the amplitude of the message is measured periodically. A convenient theoretical model for the sampling process in this case is to represent the sampled signal as the product of the message signal and a periodic train of impulses given by

$$\blacksquare \qquad m_s(t) = m(t)s(t) = m(t) \sum_{n=-\infty}^{\infty} T_s \delta(t - nT_s) \qquad (6.1)$$

where $m_s(t)$ is the sampled version of the signal and T_s is the sampling interval (i.e., the time between samples). A typical message, the sampling waveform $s(t)$, and the sampled version of the message are shown in Figure 6.1. Note that the only amplitude values of the message retained in the sampled version are the values at $t = nT_s$. The weights of the sampling impulses have been chosen as T_s for theoretical convenience in the following discussion.

The sampling signal $s(t)$ is periodic and so can be represented by a Fourier series:

$$s(t) = \sum_{k=-\infty}^{\infty} \alpha_k \exp (jk2\pi f_s t) \qquad (6.2)$$

where $f_s = 1/T_s$ is the sampling rate and the α_k are the Fourier coefficients given by Equation 2.16. Thus, we have

$$\alpha_k = \frac{1}{T_s} \int_{-T_s/2}^{T_s/2} s(t) \exp (-jk2\pi f_s t) \, dt \qquad (6.3)$$

Substituting the fact that, in the range of the integral,

$$s(t) = T_s \delta(t) \qquad (6.4)$$

FIGURE 6.1
Sampling by an impulse train.

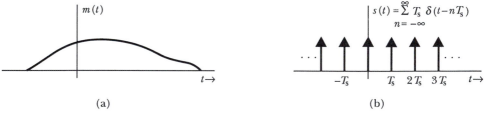

(a)

(b)

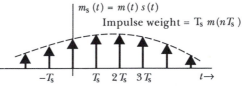

(c)

results in

$$\alpha_k = \frac{1}{T_s} \int_{-T_s/2}^{T_s/2} T_s \delta(t) \exp\left(-jk2\pi f_s t\right) dt \tag{6.5}$$

Using the integral property of the impulse function results in

$$\alpha_k = 1 \tag{6.6}$$

Substituting this into Equation 6.2, we obtain

$$s(t) = \sum_{k=-\infty}^{\infty} \exp\left(jk2\pi f_s t\right) \tag{6.7}$$

which we can substitute for $s(t)$ in Equation 6.1 to obtain the following expression for the sampled signal:

$$m_s(t) = m(t) \sum_{k=-\infty}^{\infty} \exp\left(jk2\pi f_s t\right) \tag{6.8}$$

Assuming that the message can be Fourier transformed, we can find the Fourier transform of the sampled signal by using the modulation property from Table 2.1. The result is given by

$$\blacksquare \qquad M_s(f) = \sum_{k=-\infty}^{\infty} M(f - kf_s) \tag{6.9}$$

A hypothetical message amplitude spectrum is shown in Figure 6.2a, and the amplitude spectrum of the sampled signal is shown in Figure 6.2b and c for two choices of the sampling frequency. Notice that the spectrum of the sampled signal consists of the original message spectrum plus versions of it that are centered around each integer multiple of the sampling frequency. When f_s is greater than $2f_H$, as in Figure 6.2b, the original spectrum appears separated from the other terms. In this case, the original message signal can be recovered by passing the sampled version through a lowpass filter to eliminate everything but the original spectrum. The transfer function of a suitable filter is indicated in Figure 6.2b, and a block diagram of the sampling and reconstruction process is shown in Figure 6.3.

Also notice that if we try to recover the message from the sampled version when the sampling frequency is less than twice the highest message frequency, as in Figure 6.2c, the filter output will contain some of the message spectrum terms centered around the sampling frequency. These components are known as *alias* terms and contribute distortion to the recovered message.

Thus, we have established the very important fact that it is possible to exactly reproduce a message signal from knowledge of its amplitudes at periodic sampling instants, provided that the sampling frequency is at least twice the highest message frequency. In principle, this can be done by passing an impulse train, modulated by the sample values, through a lowpass filter. In practice, impulses are not used, of course, because they have infinite

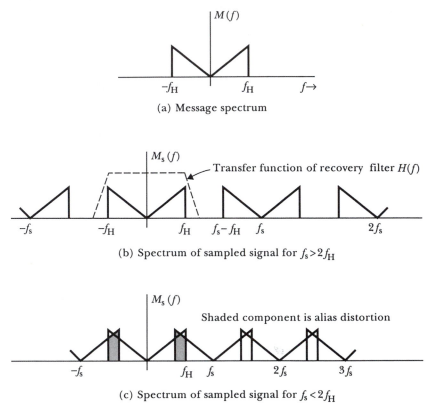

(a) Message spectrum

(b) Spectrum of sampled signal for $f_s > 2f_H$

(c) Spectrum of sampled signal for $f_s < 2f_H$

FIGURE 6.2
Spectra of a typical message and its sampled version.

amplitude in the time domain and content out to infinite frequency in the frequency domain. Instead, pulses with finite amplitude and nonzero width are used. We consider these types of sampling in the rest of this section.

Even though weighted impulses are not practical, instantaneous sampling does find many applications. For example, signals are often sampled, and the

FIGURE 6.3
Sampling by multiplication with an impulse train and message recovery by low-pass filtering.

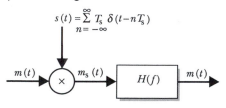

samples are converted to binary words that represent the sample values. These words can be used in transmission of the signal through a PCM system. The familiar compact disk for audio recording is a good example of this. In signal processing, the digital words are processed by a digital computer. Thus, even though the samples are represented by digital words rather than weighted impulses, the concepts of instantaneous sampling apply.

Flat-topped sample pulses

Now we consider the reconstruction of the message signal from a train of square pulses whose amplitudes have been modulated with the sample values of the message. To do this, we consider the system shown in Figure 6.4a, which is useful for establishing the desired theoretical result.

FIGURE 6.4
Flat-topped pulse sampling and message recovery.

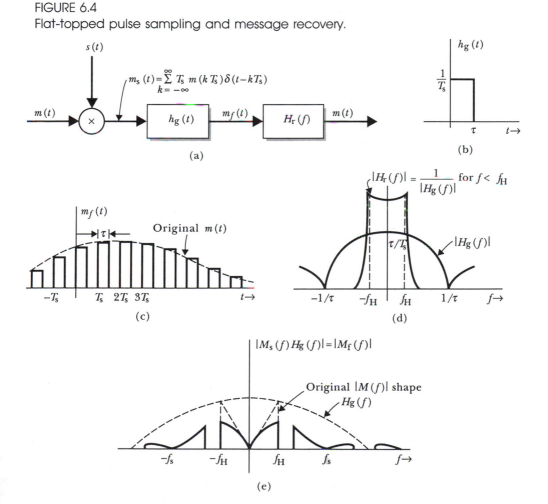

In the system of Figure 6.4a, the message signal is first multiplied by the impulse train sampling signal $s(t)$ as in the preceding analysis. The sampled signal is then passed through a filter with an impulse response $h_g(t)$, shown in Figure 6.4b. The output of the filter contains a *flat-topped pulse* for each impulse at its input. The output waveform from the filter is shown in Figure 6.4c. Notice that the amplitude of each pulse in the filter output is equal to the amplitude of the original message at the corresponding sample point.

We now determine the requirements for insuring recovery of the original message waveform from this flat-topped pulse train by considering the frequency domain representation. The impulse response of the filter that generates the flat-topped pulses is given by

$$h_g(t) = \frac{1}{T_s} \qquad \text{for } 0 < t < \tau$$

$$= 0 \qquad \text{otherwise}$$

The transfer function of the filter is the Fourier transform of its impulse response, given by

$$H_g(f) = \left(\frac{\tau}{T_s}\right) \text{sinc } (f\tau) \exp{(-j\pi f\tau)} \tag{6.10}$$

The exponential term represents the time delay of the center of the pulse by $\tau/2$. It does not affect the magnitude of the transfer function plotted in Figure 6.4d. The spectrum of the flat-topped pulse train at the output of the generation filter can be obtained by multiplying the spectrum of the impulse-sampled signal (given in Equation 6.9 and shown in Figure 6.2b) by the transfer function $H_g(f)$. The result is

$$\blacksquare \qquad M_f(f) = \left(\frac{\tau}{T_s}\right) \text{sinc } (f\tau) \exp{(-j\pi f\tau)} \sum_{k=-\infty}^{\infty} M(f - kf_s) \tag{6.11}$$

The amplitude spectrum of this result is shown in Figure 6.4e. Notice that the spectrum of the flat-topped sampled signal contains the original message spectrum, but the higher frequency components have been attenuated by the transfer function of the filter $H_g(f)$. Components of the message are also centered around multiples of the sampling frequency. As before, these do not extend below the highest message frequency as long as the sampling frequency is higher than twice the highest message frequency.

Examination of the spectrum of the flat-topped signal shown in Figure 6.4e indicates that the original message can be recovered by passing the flat-topped pulses through a reconstruction filter that eliminates the frequency components above f_H and restores the higher message signal components to their original amplitude. In the frequency range of the message, the product of the transfer functions of the filters must be unity if the original message is to be recovered. Thus, the magnitude of the transfer function of the recovery filter is the reciprocal of the magnitude of the transfer function of the pulse generating filter. Thus we have

$$\blacksquare \qquad |H_r(f)| = \frac{1}{|H_g(f)|} = \frac{T_s}{\tau|\text{sinc}(f\tau)|} \qquad \text{for } 0 < f < f_H$$
$$= 0 \qquad\qquad \text{for } f > f_s - f_H \qquad (6.12)$$

In the frequency range from f_H to $f_s - f_H$, the magnitude of the recovery filter transfer function can take on any values that make the implementation of the filter easiest. The magnitude of the transfer function of the recovery filter is shown in Figure 6.4d. The phase of the recovery filter should be linear with frequency to avoid waveform distortion, as discussed in Section 2.5. If there is a net linear phase term for the overall transfer function of the filters in Figure 6.4a, the output is simply a delayed version of the input message.

Often, signals that have been converted to digital PCM for storage (the audio compact disk is a good example) or transmission are reconstructed at the receiving end by applying the code words generated from each message sample to a *digital-to-analog converter* (DAC). The DAC produces an output voltage corresponding to the input value for each code word. The resulting DAC output is a flat-topped sampled version of the original signal. The pulse width τ is equal to the interval between samples. A typical message waveform and the corresponding DAC output are shown in Figure 6.5. As we have seen in the preceding analysis, the higher frequency components in the message are reduced in amplitude by this process so it is necessary to pass the DAC output through a compensating filter with the reciprocal sinc transfer function given by Equation 6.12 to obtain a flat overall transfer function. The reduction of the high-frequency amplitudes by flat-topped sampling is sometimes called the *aperture effect*. Study of Equation 6.10 will show that aperture effect becomes less pronounced as the pulse width or aperture τ becomes smaller.

Natural sampling

Another variation of sampling is illustrated by the system shown in Figure 6.6a. Here the message signal is sampled by a switch that alternates between the

FIGURE 6.5
A typical message waveform and its reconstruction by a DAC.

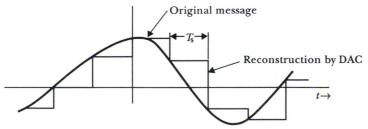

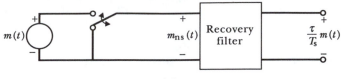

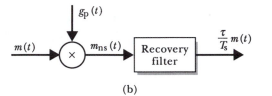

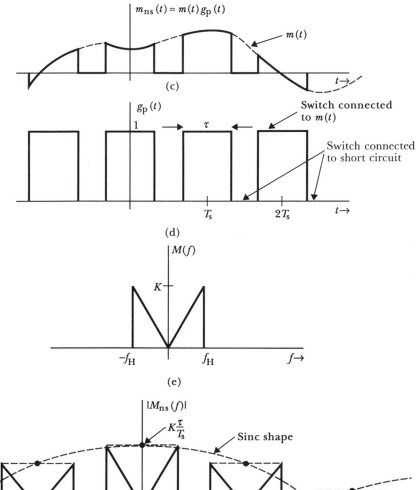

FIGURE 6.6
Natural sampling.

message source and a zero input provided by the short circuit. The sampled waveform is shown in Figure 6.6c and consists of sections of the original message separated by intervals during which the sampled signal $m_{ns}(t)$ is zero. This type of sampling is sometimes called *natural sampling*. It is different from the flat-topped version of sampling in which the amplitude during the sample pulses is constant and equal to the message amplitude at the beginning of the pulse, as shown in Figure 6.4c.

The action of the switch can be modeled by the multiplier in the block diagram of Figure 6.6b. The message is multiplied by the pulse train $g_p(t)$, which is 1 when the switch is connected to the message and 0 when the switch is connected to the short circuit. The sampling waveform $g_p(t)$ is shown in Figure 6.6d. It is the same as the waveform analyzed earlier in Example 2.2 with $A = 1$. In that example, we found the Fourier series for $g_p(t)$ given by

$$g_p(t) = \sum_{n=-\infty}^{\infty} \frac{\tau}{T_s} \operatorname{sinc}\left[\frac{n\tau}{T_s}\right] \exp\left(jn2\pi f_s t\right) \tag{6.13}$$

(We have changed the notation of Example 2.2 slightly to fit the present discussion.) Now if we substitute for $g_p(t)$ in the expression

$$m_{ns}(t) = m(t)g_p(t) \tag{6.14}$$

and use the modulation property of the Fourier transform, we find that the transform of the naturally sampled signal is

$$\blacksquare \qquad M_{ns}(f) = \sum_{n=-\infty}^{\infty} \frac{\tau}{T_s} \operatorname{sinc}\left[\frac{n\tau}{T_s}\right] M(f - nf_s) \tag{6.15}$$

Thus, the naturally sampled signal consists of the original message spectrum and weighted versions of it centered around integer multiples of the sampling frequency. The amplitude spectra of a hypothetical message and the corresponding naturally sampled signal are shown in Figure 6.6e and f. Note that this is similar to the spectrum of an impulse-sampled signal, shown in Figure 6.2b. The difference is that the versions of the message spectrum are weighted by the sinc function in the case of natural sampling but have equal amplitude all the way out to infinite frequency in the case of impulse sampling. As before, if the sampling frequency is greater than twice the highest message frequency, no aliasing occurs. Then the original message can be recovered by lowpass filtering. The naturally sampled signal contains the original unaltered message spectrum centered at dc. Thus, the message waveform can be recovered by a lowpass filter having constant gain out to f_H and zero gain beyond $f_s - f_H$. This is in contrast to the recovery filter for flat-topped sampling, for which the gain must be greater for the higher message frequencies to compensate for the aperture effect. Comparison of Equations 6.9, 6.11, and 6.15 shows the differences among the spectra of the sampled signals for impulse sampling, flat-topped sampling, and natural sampling, respectively.

Notice that the dc-centered version of the message spectrum in the naturally sampled signal spectrum, given by the $n = 0$ term of Equation 6.15, is weighted by the factor τ/T_s. Therefore, if the recovery filter has a passband gain of unity, the output signal will be $(\tau/T_s)m(t)$. Note that the filter output decreases in amplitude as the switch spends more time in the short-circuit position. If noise is added to the sampled signal before the recovery process, this can lead to poor output SNR when very narrow samples are used. The same effect also occurs with flat-topped sampling.

Bandpass signals

The requirement that the sampling frequency should be greater than twice the highest message frequency applies to baseband signals, which have a lowest frequency approaching dc. However, when bandpass signals are to be sampled, lower sampling rates can sometimes be used. Consider a bandpass signal, such as the example shown in Figure 6.7, with a highest frequency denoted by f_H, a lowest frequency denoted by f_L, and a bandwidth $B = f_H - f_L$. For this signal, it can be shown (Taub and Schilling, Chapter 5) that the sampling frequency must satisfy the constraints given by

$$2B\left[\frac{k}{N}\right] < f_s < 2B\left[\frac{k-1}{N-1}\right] \tag{6.16}$$

where

$$k = \frac{f_H}{B} \tag{6.17}$$

and N takes on all integer values between 1 and k.

EXAMPLE 6.1

Assume that a signal has the amplitude spectrum shown in Figure 6.7 with $f_H = 4$ kHz and $f_L = 3$ kHz. Find the sampling frequencies that do not result in aliasing. Sketch the spectrum of the instantaneously sampled signal (similar to Figure 6.2b) for one of the allowed sampling rates to verify that aliasing does not occur.

FIGURE 6.7
Typical bandpass signal spectrum.

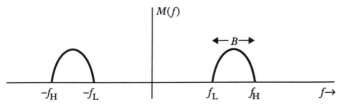

SOLUTION

We find that $B = f_H - f_L = 1$ kHz. Substituting into Equation 6.17, we obtain

$$k = \frac{f_H}{B} = 4$$

Thus, $\mathcal{N}$ ranges from one to four. Substituting into Equation 6.16, we find that for $\mathcal{N} = 1$

$$8 \text{ kHz} < f_s < \infty$$

This corresponds to the result we found by considering baseband signals. For $\mathcal{N} = 2$ we find

$$4 \text{ kHz} < f_s < 6 \text{ kHz}$$

For $\mathcal{N} = 3$:

$$2.667 \text{ kHz} < f_s < 3 \text{ kHz}$$

For $\mathcal{N} = 4$:

$$2 \text{ kHz} < f_s < 2 \text{ kHz}$$

The spectrum of the sampled signal consists of the original spectrum plus versions of it translated to all integer multiples of the sampling frequency. The spectrum of the sampled signal for a sampling rate of 5 kHz is shown in Figure 6.8. Note that recovery of the original message signal in this case requires the use of a bandpass filter, as indicated in the figure, rather than a lowpass filter. If the sampling rate approaches any of the limits given above (except ∞), the recovery filter must make a rapid transition from passband to total rejection. This can make the filter difficult to implement; therefore, the sampling rate should not be at the theoretical limits for practical systems.

FIGURE 6.8
Result of sampling a bandpass signal with $f_L = 3$ kHz and $f_H = 4$ kHz at a sampling rate of $f_s = 5$ kHz.

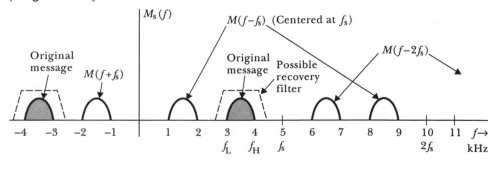

6.2
TIME DIVISION MULTIPLEXING

After a number of message signals have been sampled, it is possible to transmit them over a single channel by interleaving the samples in time. The diagram for a simple system that does this for two message signals is shown in Figure 6.9a. In the figure, the switch at the transmitting end alternates rapidly between the two message sources, so the signal sent through the channel consists of alternating sections of the two messages, as shown in Figure 6.9b. At the receiving end, another switch, synchronized with the one at the sending end (allowing for any delay in transmission), separates the two signals. Thus, the inputs to the recovery filters are naturally sampled versions of the message signals as shown by the waveforms in Figure 6.9c and d.

It is interesting to consider the amplitude spectra of the waveforms at various points in the system of Figure 6.9a. Example messsage spectra are shown in Figure 6.10a. It has been assumed for clarity of explanation that message 2 is a sine wave. The spectrum of a naturally sampled signal was given in Equation 6.15, which can be used with the fact that $\tau = T_s/2$ in the present case to obtain the sketches of the naturally sampled versions of the message waveforms at the inputs to the recovery filters. The results are shown in Figure 6.10c and d. The time division multiplexed (TDM) signal on the channel is the sum of these spectra, as shown in Figure 6.10b. Notice that the message spectra overlap. It is in the time domain that the messages are distinct and can be separated by the switch. In general, it is not possible to identify the two message signals by observing the amplitude spectrum on the channel. It is possible in the case illustrated because we selected a sinusoidal test signal, which has a distinctive spectrum, for message 2.

Practical systems based on the principles of the system in Figure 6.9a have been designed. For example, the Bell System 101 ESS (Electronic Switching System) provides for two-way interconnection of pairs of users in a private branch exchange (PBX) by the use of TDM analog switching. The system is limited to a maximum of about 25 users due to practical bandwidth limitations.

One of the serious limitations of the system of Figure 6.9a is illustrated by the amplitude spectrum of the TDM signal shown in Figure 6.10b. The spectrum of the signal eventually decays to zero at high frequencies, but it does so at the slow rate determined by the sinc function envelope. When more signals are multiplexed, the bandwidth becomes wider and the channel bandwidth restrictions are quickly exceeded. When the multiplexed signal is band limited, the pulses spread in time, and unless care is taken in the selection of the filtering, crosstalk or ISI occurs. We treat this subject in the next section.

Another reason for band limiting the multiplexed signal before recovery of the message signals is to improve the SNR. If white noise is added to the signal in the channel, the SNR at the receiving end will theoretically be zero because the total power in white noise is infinite. In this case, however, band

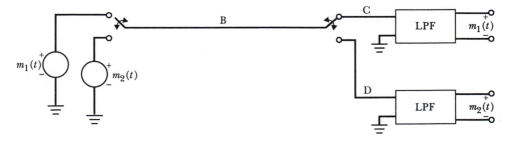

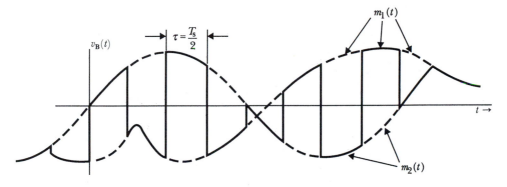

(a)

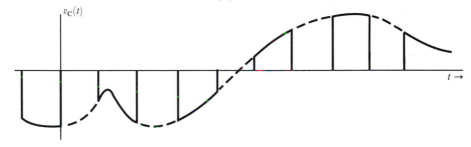

(b)

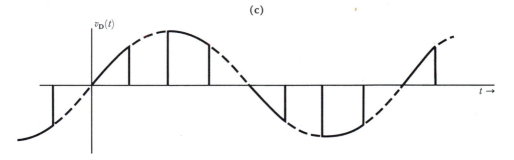

(c)

(d)

FIGURE 6.9
Simple TDM system.

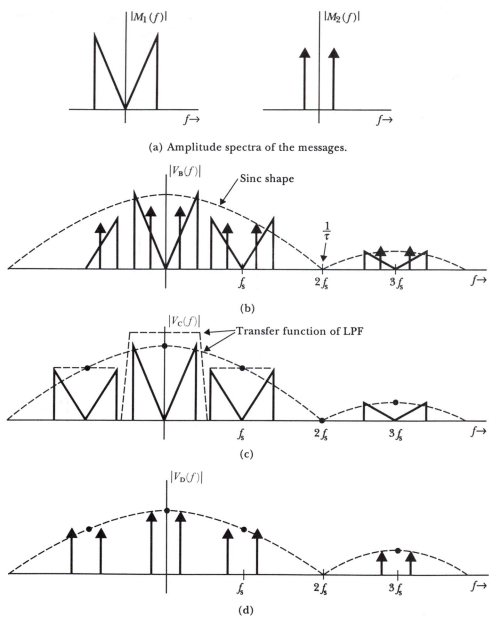

(a) Amplitude spectra of the messages.

(b)

(c)

(d)

FIGURE 6.10
Amplitude spectra of the waveforms of the TDM system of Figure 6.9. Parts b, c, and d of this figure correspond to b, c, and d of Figure 6.9, respectively.

limiting the noise reduces the noise power to a finite value, which gives a useful SNR. Even if the noise is not white, a significant improvement in SNR can often be gained by filtering the received TDM signal before recovering the messages.

When message signals with widely varying bandwidths are to be multiplexed in this fashion, it can be advantageous to interleave a number of samples of the higher bandwidth messages between samples of the lower bandwidth messages. In this manner, a sampling rate appropriate for each message can be obtained. (See problem 10 at the end of the chapter for an example.)

6.3
THE NYQUIST CRITERION FOR ZERO INTERSYMBOL INTERFERENCE

Intersymbol interference in PAM systems

The block diagram of the system we will use for discussing *intersymbol interference* (ISI) is shown in Figure 6.11a. An impulse train weighted by the message values to be transmitted is indicated at the input to the system. The message values can be time interleaved samples from a number of analog sources, or they can be from binary data, in which case the weights would take on only two values. A typical weighted impulse train is shown in Figure 6.11b.

The impulse train is applied to the input of a pulse-shaping filter, which determines the shape of the pulses placed on the channel. If flat-topped square pulses are desired, then the filter with the impulse response illustrated in Figure 6.11c is used. In this case, each impulse sends a flat-topped pulse into the channel. The amplitude of each pulse is modulated by its corresponding message value. Thus, the signal entering the channel is an analog or digital *pulse-amplitude modulated* (PAM) signal. In practice, of course, the system for generating the transmitted signal would not employ impulses, due to their infinite amplitude and bandwidth. Nevertheless, the block diagram of Figure 6.11a does produce the same input signals as a practical system. Furthermore, it is amenable to analysis.

The channel filters the transmitted signal and adds noise. To limit the noise power and improve the SNR at the input to the message recovery system, the received signal is further filtered. Thus, the weighted impulse train passes through a cascade of filters as indicated in Figure 6.11a. The equivalent overall transfer function of these filters is given by

$$H_e(f) = H_p(f)H_c(f)H_n(f) \qquad (6.18)$$

The impulse response of this equivalent filter, denoted by $h_e(t)$, is the inverse Fourier transform of the transfer function. Each impulse at the system input results in a weighted and delayed version of $h_e(t)$ at the system output. Thus, the signal component at the output of the noise reduction filter is given by

$$s_o(t) = \sum_{k=-\infty}^{\infty} m_k h_e(t - kT) \qquad (6.19)$$

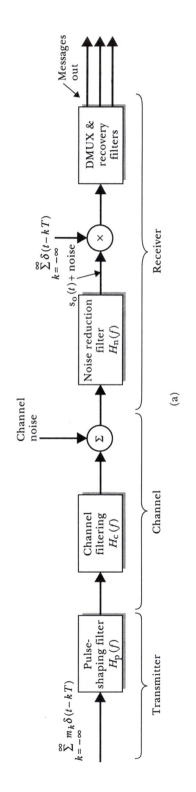

(a)

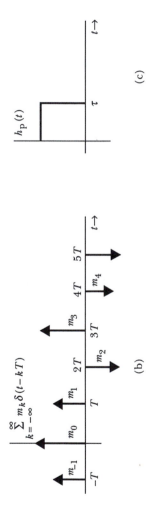

(b)

$h_p(t)$

τ $t \rightarrow$

(c)

FIGURE 6.11
PAM system.

where m_k is the kth message value, and T is the duration of the time slot for each message value.

We are interested in finding an equivalent overall filter $H_e(f)$ such that the bandwidth of the received signal is limited to a low value so that the system is bandwidth efficient. At the same time, we require that no ISI occurs.

For convenience in our discussion, we are not concerned if the impulse responses of our filters begin before $t = 0$ such that the filters are theoretically unrealizable. Later, when we need to design the filters, a sufficient time delay can be included in the impulse responses to insure that they are realizable. This simply contributes an additional linear phase term to their transfer functions.

At this point, we might be tempted to try to find a time domain pulse that is zero outside some (hopefully small) time slot to guarantee no ISI and has a Fourier transform equal to zero for frequencies above some (hopefully low) limit. Unfortunately, no such pulse exists (Wozencraft and Jacobs, 1965). However, there is another way to avoid ISI with a bandlimited pulse, which is to select a pulse shape that is zero at all integer multiples of the pulse spacing. One such possibility is the sinc pulse, given by

$$h_e(t) = \text{sinc}\left(\frac{t}{T}\right) \tag{6.20}$$

If this is substituted into Equation 6.19, we have the following expression for the signal component at the output of the noise limiting filter of Figure 6.11a:

$$s_o(t) = \sum_{k=-\infty}^{\infty} m_k \, \text{sinc}\left(\frac{t - kT}{T}\right) \tag{6.21}$$

Some of the terms of Equation 6.21 are plotted in Figure 6.12 for typical values of the message m_k. Notice that at the instant $t = 0$, all of the terms except

FIGURE 6.12
Pulses at input to sampler when $h_e(t) = \text{sinc}(t/T)$.

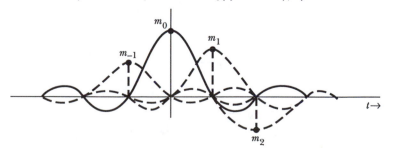

$k = 0$ vanish, and the value of $s_o(0)$ is m_0. Similarly, $s_o(kT) = m_k$. Thus, the message values can be recovered by taking instantaneous samples of the received signal at the output of the noise-reduction filter in the system of Figure 6.11a. These samples can then be demultiplexed (DMUX) and used to reconstruct the original message waveforms.

The overall frequency response $H_e(f)$ that corresponds to the selection of the sinc pulses of Equation 6.20 is that of an ideal brickwall filter, shown in Figure 6.13. The use of the sinc pulse has several drawbacks. First, as the overall frequency response of Figure 6.13 shows, the transfer function is required to fall from a constant gain in the passband abruptly to zero at $f = 1/(2T)$, which is difficult to implement. Second, even if the filters could be implemented, the sinc pulses die out slowly; therefore, if the received signal is not sampled at exactly the correct instants, ISI from many other pulses results. This is evident in Figure 6.12.

The Nyquist criterion

To achieve more useful pulse shapes, Nyquist was led to search for other transfer functions that would result in uniformly spaced zeroes in the time domain. He was able to show that if the overall frequency response is pure real (i.e., zero phase shift at all frequencies) and has the property

$$\blacksquare \quad H_e\left(\frac{1}{2T} - \Delta f\right) + H_e\left(\frac{1}{2T} + \Delta f\right) = C \qquad \text{for } 0 < \Delta f < \frac{1}{2T} \qquad (6.22)$$

where C is an arbitrary constant, then $h_e(t)$ is zero at integer multiples (except zero) of T. Therefore, ISI will not be present in the samples taken at the system output. This is a symmetry requirement on the overall frequency response of the system as illustrated in Figure 6.14. The transfer function of this Nyquist filter is zero for frequencies above $1/T$.

FIGURE 6.13
$H_e(f)$ when $h_e(t) = \text{sinc}(t/T)$.

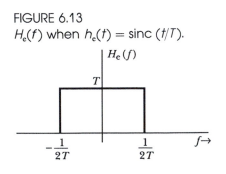

FIGURE 6.14
Nyquist filter for zero ISI with
$H_e[1/(2T) - \Delta f] + H_e[1/(2T) + \Delta f] = C$.

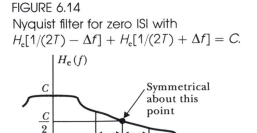

The raised-cosine filter

One well-known Nyquist filter is the raised-cosine transfer function, given by

$$
\begin{aligned}
\bullet \ H_e(f) &= T && \text{for } 0 \le f \le \frac{1-r}{2T} \\[2mm]
&= \frac{T}{2}\left[1 - \sin\left(\pi T\,\frac{[f - 1/(2T)]}{r}\right)\right] && \text{for } \frac{1-r}{2T} \le f \le \frac{1+r}{2T} \\[2mm]
&= 0 && \text{for } \frac{1+r}{2T} \le f \le \infty
\end{aligned}
\tag{6.23}
$$

where r ranges from zero to one and is called the *roll-off factor*. The corresponding impulse response is given by

$$
h_e(t) = \operatorname{sinc}\left(\frac{t}{T}\right)\frac{\cos(\pi rt/T)}{1 - (2rt/T)^2}
\tag{6.24}
$$

Note that when the roll-off factor r is zero, $h_e(t)$ is the same as the sinc pulse of Equation 6.20. The transfer function and impulse response of the raised-cosine filter are shown in Figure 6.15 for several values of r. Note that as the roll-off factor becomes larger, the impulse response decays faster outside the range $-T < t < T$. This results in improved (compared with the sinc pulse) ISI performance when the sampling is not done at precisely the correct times. Also, the transfer function makes a much less abrupt change in gain for larger values of r. This makes the filters easier to implement. These improvements are at the expense of increased bandwidth, due to the fact that the absolute bandwidth of $H_e(f)$ becomes larger with higher values of the roll-off factor.

Division of the overall transfer function

Now we briefly consider how the overall transfer function $H_e(f)$ should be apportioned between the various blocks of Figure 6.11a. The signal at the input to the noise reduction filter can be written as

$$
s_r(t) = \sum_{k=-\infty}^{\infty} m_k h_r(t - kT)
\tag{6.25}
$$

where $h_r(t)$ is the impulse response of the cascade of the pulse-forming filter and the channel filter. This can be found by taking the inverse Fourier transform of

$$
H_r(f) = H_p(f)H_c(f)
\tag{6.26}
$$

Thus, each message value produces a weighted pulse of the form $m_k h_r(t - kT)$ at the input to the noise reduction filter. We want to choose the transfer function of the noise reduction filter in such a manner that the SNR of the samples at the system output is maximized. We will show in Chapter 8 that this

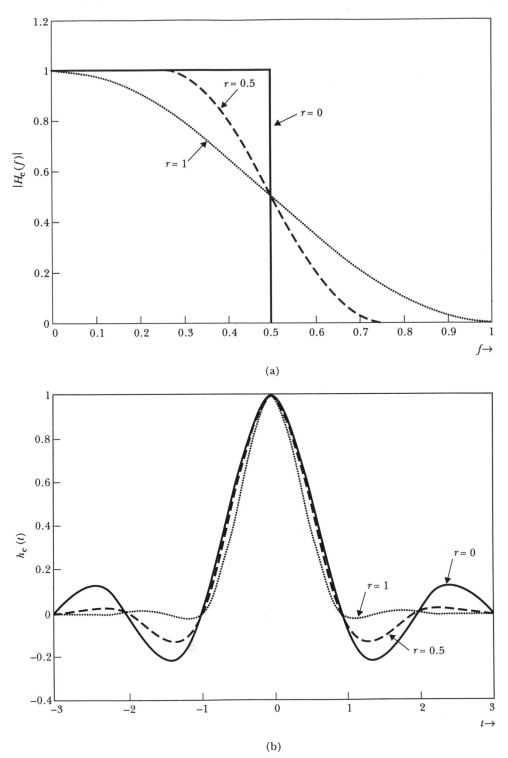

(a)

(b)

FIGURE 6.15
Transfer functions of (a) raised-cosine filters and (b) impulse responses of raised-cosine filters. (Values shown are for $T = 1$.)

can be accomplished for white noise by choosing the impulse response of the noise reduction filter to be a time-reversed version of the received pulse shape given by

$$h_n(t) = Kh_r(-t) \qquad (6.27)$$

where K is an arbitrary positive constant. A filter selected in this manner is known as a *matched filter*. Application of the scaling property of the Fourier transform to Equation 6.27 results in

$$H_n(f) = KH_r(-f) \qquad (6.28)$$

If we consider only filters with zero phase shift, the transfer functions are even functions of frequency. Thus, we have

$$H_n(f) = KH_r(f) \qquad (6.29)$$

Using Equations 6.26 and 6.29 to substitute into 6.18, we obtain

$$H_e(f) = \frac{1}{K}[H_n(f)]^2 \qquad (6.30)$$

This results in

$$H_n(f) = KH_p(f)H_c(f) = [KH_e(f)]^{1/2} \qquad (6.31)$$

Thus, to avoid ISI, we need to select the overall transfer function to be a Nyquist filter such as the raised-cosine filter. Assuming that we elect to use a raised-cosine filter, the roll-off factor is selected to obtain the best compromise between bandwidth consumed and the difficulty of implementing the filters or pulse shapes. Then, if the channel transfer function $H_c(f)$ is known, the transform of the transmitted pulse and the required noise reduction filter can be determined from Equation 6.31.

Equalizers

In practice, the channel response may not be known in advance or it may vary from time to time. Then either a manually adaptable or self-adaptable filter can be used ahead of the noise reduction filter. These filters are often called *equalizers* because they are used to make the gain of the channel-plus-equalizer constant or equal at all frequencies. A known *training sequence* is often used to enable the equalizer to adjust to the channel before messages are transmitted. In other cases, it is possible to continually adjust the equalizer during message transmission. When the equalizer control tries to force zero ISI it is called a *zero-forcing equalizer*. Another possibility is for the control to produce the least-mean-squared error between the actual message values and what is received at the system output. A great deal of work has been done on this problem so the interested reader can find much additional information in the literature.

In the next section, we consider the bandwidth requirements and noise performance of an analog TDM PAM system using ideal brickwall filters.

6.4
PERFORMANCE OF A TDM PAM SYSTEM

Bandwidth requirements

In this section, we consider the bandwidth requirements and noise performance of the system shown in the block diagram of Figure 6.16a. In this system, N message signals are sampled, time division multiplexed, and transmitted through an additive white noise channel using PAM. The message signals are assumed to have equal power, denoted by P_m, and each has a highest frequency denoted by f_H. The message signals enter the first block in the system where they are sampled at the minimum sampling rate allowed without aliasing. This rate is

$$f_s = 2f_H = \frac{1}{NT} \qquad \text{samples per second} \qquad (6.32)$$

We assume that the sampling interval is NT seconds for each signal, and the samples are interleaved in time. Thus, the samples taken at $0, NT, 2NT, 3NT,$ and so on are from the signal $m_0(t)$, whereas the samples taken at $T, (N+1)T,$ $(2N+1)T, (3N+1)T,$ and so on are from the message $m_1(t)$, and so forth for all N messages. Typical messages and their interleaved samples are shown in Figure 6.17 for the case of $N = 3$ messages.

The overall transfer function $H_e(f)$ up to the output of the noise-reduction filter has been chosen to be an ideal lowpass filter, as shown in Figure 6.16b. This corresponds to choosing the impulse response to be the sinc pulse of Equation 6.20. (It is also equivalent to a raised-cosine pulse with a roll-off factor r of zero.) The bandwidth of the overall transfer function extends from dc to

$$B = \frac{1}{2T} \qquad (6.33)$$

As we have shown in the preceding section, this choice for the overall transfer function results in no ISI at the output, provided that the output samples are taken at multiples of T.

Using Equation 6.32 to substitute for T in Equation 6.33, we find that

$$B = Nf_H \qquad (6.34)$$

Thus, the bandwidth required by the TDM system is the same as that of an FDM system using SSB, as discussed in Chapter 3. In practical versions of these systems, a larger bandwidth than indicated by Equation 6.34 is required. In the case of TDM, a more gentle roll-off, afforded by choosing a roll-off

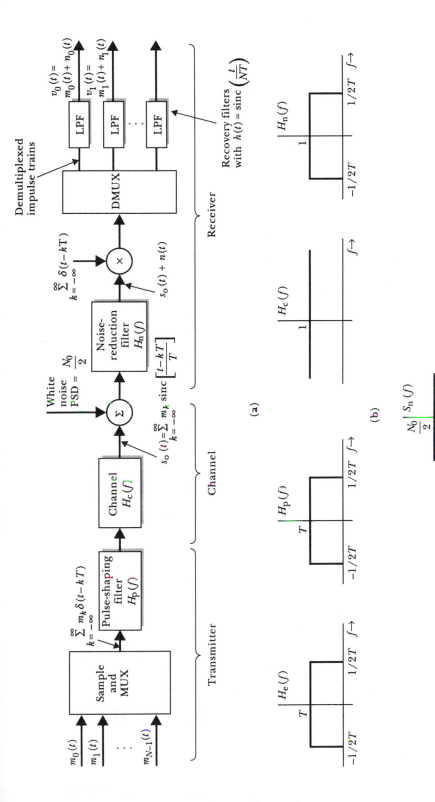

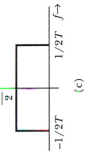

FIGURE 6.16
TDM PAM system model.

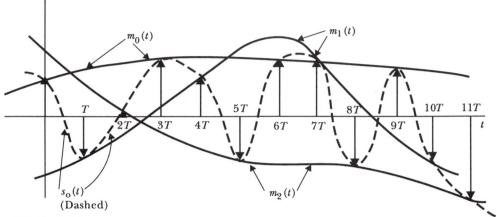

FIGURE 6.17

Three message waveforms, their samples, and the resulting TDM signal at the output of the channel filter. Note that $m_0(t)$ is sampled at 0, 3T, 6T, etc., whereas $m_1(t)$ is sampled at T, 4T, 7T, etc.

factor r greater than zero, makes the filters easier to implement. With SSB and FDM, some guardband is allowed between channels so that crosstalk does not occur.

Noise performance

Now we consider the noise performance of the system of Figure 6.16a. Our goal is to develop an expression for the SNR at the system outputs.

The transfer functions of the pulse-shaping filter, the channel filter, and the noise-reduction filter are shown in Figure 6.16b. The channel is assumed to have unity gain over the entire signaling bandwidth. The pulse-shaping filter is an ideal lowpass filter. The noise-reduction filter transfer function has been chosen by use of Equation 6.31 (with $K = 1/T$) to optimize the noise performance of the system. Note that the noise-reduction filter is also an ideal lowpass filter.

The signal component at the output of the noise-reduction filter is a band-limited signal $s_0(t)$, given by Equation 6.21. Notice that $s_0(t)$ takes on values equal to the message samples at the sampling instants. The signal $s_0(t)$ is shown for the example waveforms in Figure 6.17. Note that because the gain of the noise-reduction filter is unity, $s_0(t)$ appears at the output of the channel (i.e., the signal is unchanged in passing through the noise reduction filter).

We denote the received signal power at the input to the noise-reduction filter per message signal as P_r. Since there are N message signals, the total received power is NP_r. Note again from Figure 6.17 that the received signal $s_0(t)$ at the output of the channel takes on amplitudes equal to the message signals at the sampling instants. Thus, we expect that the total received power contained in $s_o(t)$ is equal to the power in a message signal, or

$$N P_r = P_m \tag{6.35}$$

This intuitive result can also be obtained by a more rigorous mathematical approach using the orthogonality property of sinc functions. However, the derivation is somewhat tedious, and we will not take the space to present it.

The noise PSD at the output of the noise-reduction filter is given by

$$S_n(f) = \frac{1}{2} N_0 |H_n(f)|^2 \tag{6.36}$$

and is shown in Figure 6.16c. The total noise power at the output of the noise reduction filter can be found by integrating this PSD. The result is

$$P_{no} = \frac{N_0}{2T} \tag{6.37}$$

The sum of the noise and signal at the output of the noise-reduction filter is sampled by multiplication with a unit impulse train and the resulting weighted impulses are de-interleaved to separate the messages. The de-interleaved impulse train at the input to the ith recovery filter is given by

$$\sum_{k=-\infty}^{\infty} (m_{kN+i} + n_{kN+i}) \delta[t - (kN + i)T] \tag{6.38}$$

where n_{kN+i} is sample number $kN + i$ of the noise $n(t)$ at the output of the noise-reduction filter. Note that the message samples in Equation 6.38 are the separated samples of the ith message.

These separated impulse trains are then passed through ideal lowpass filters with a passband from dc to f_H to recover the individual message waveforms. The impulse response of these lowpass filters is given by

$$h_f(t) = \operatorname{sinc}\left(\frac{t}{NT}\right) \tag{6.39}$$

The resulting output signals from the recovery filters contain a weighted and delayed sinc pulse for each impulse at their inputs and are given by

$$
\begin{aligned}
v_i(t) &= \sum_{k=-\infty}^{\infty} (m_{kN+i} + n_{kN+i}) \operatorname{sinc}\left[\frac{t - kNT}{NT}\right] \\
&= m_i(t) + n_i(t)
\end{aligned}
\tag{6.40}
$$

where $m_i(t)$ is the ith message signal, and $n_i(t)$ is the noise waveform at the output of the ith filter resulting from lowpass filtering of the de-interleaved noise samples. An example noise waveform $n(t)$ at the output of the noise-reduction filter, its impulse-sampled version, and the resulting noise waveform

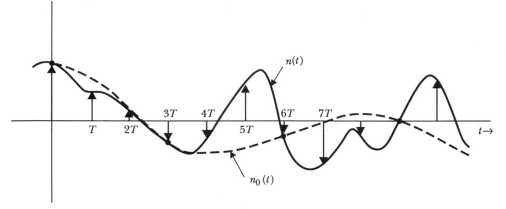

FIGURE 6.18

A typical noise waveform $n(t)$ at the output of the noise-reduction filter, its samples, and the noise waveform $n_0(t)$ resulting from lowpass filtering of the samples taken at 0, 3T, 6T, etc.

$n_0(t)$ at the top $(i = 0)$ system output are shown in Figure 6.18 for the case of $N = 3$. Note from the figure that the noise amplitude at the outputs of the recovery filters and at the output of the noise reduction filter are the same at the sampling instants. Thus, we can see intuitively that the noise power at the output of each message recovery filter is the same as the noise power at the output of the noise-reduction filter.

Since the messages appear at the outputs of the corresponding recovery filters, the signal power out of each channel is the same as the original message power P_m. Thus, the output SNR is

$$\mathrm{SNR_o} = \frac{P_\mathrm{m}}{P_\mathrm{no}} \tag{6.41}$$

Using Equation 6.35 to substitute for P_m and Equation 6.37 to substitute for P_no, we have

$$\mathrm{SNR_o} = \frac{2TNP_\mathrm{r}}{N_0} \tag{6.42}$$

Using Equation 6.32 to solve for NT and substituting the result into the last equation results in

$$\mathrm{SNR_o} = \frac{P_\mathrm{r}}{N_0 f_\mathrm{H}} \tag{6.43}$$

This is exactly the result we found for the noise performance of SSB, DSB-SC and baseband communication of a single message in our analysis of Chapter 5. Thus, we have found that the TDM PAM system performs equally with FDM SSB with respect to both bandwidth requirements and noise performance.

To avoid crosstalk with the TDM PAM system, it is necessary to control the shape of the overall system transfer function so that the Nyquist criterion of Equation 6.22 is met. With FDM, it is not necessary to control the overall frequency response as carefully since a slow variation of gain with frequency causes only a small amount of linear frequency distortion over the band of a single channel. Some gain variation from channel to channel may result, but this does not usually have serious consequences unless it is very great.

However, nonlinear distortion can be more serious for FDM than for TDM systems. If flat-topped pulses are passed through a memoryless non-linearity, the amplitude of each pulse is changed nonlinearly. This causes each recovered signal to be distorted, but crosstalk does not occur. On the other hand, high-order nonlinear distortion can cause the various FDM channels to mix and produce products that fall in the communication band, resulting in crosstalk.

Pulse width and pulse position modulation

In addition to PAM, other forms of analog pulse modulation exist. Pulse-width modulation (PWM) and pulse-position modulation (PPM) are two possibilities. Example waveforms of these are shown for a typical message waveform in Figure 6.19. As indicated in the figure, the message is sampled periodically, and the sample values are used to modulate either the width or the time position of the pulses in the pulse train. In PWM, the pulse has a given width when the message is zero. The width of each pulse is varied by the corresponding message sample value. In the waveform shown, a positive sample results in a wider pulse, and a negative sample value results in a narrower pulse. In PPM, the leading edge of each pulse is varied around a given instant by the message sample. In the waveform shown in Figure 6.19c, positive message samples result in earlier pulses and negative message samples result in later pulses.

PPM and PWM can be used as a means of expending bandwidth to gain a higher output SNR than is afforded by PAM. We will not give a detailed discussion of this because PCM, which is discussed in the next section, usually provides a better alternative.

6.5
DIGITAL PULSE MODULATION SCHEMES

In *pulse code modulation* (PCM) schemes for transmitting analog messages, the message is sampled, and the samples are quantized to a discrete set of amplitude levels. Then each level is converted to a code word that is transmitted through the channel. At the receiving end, the code words are used to reconstruct an approximation of the original message. As we pointed out in Chapter 1, there

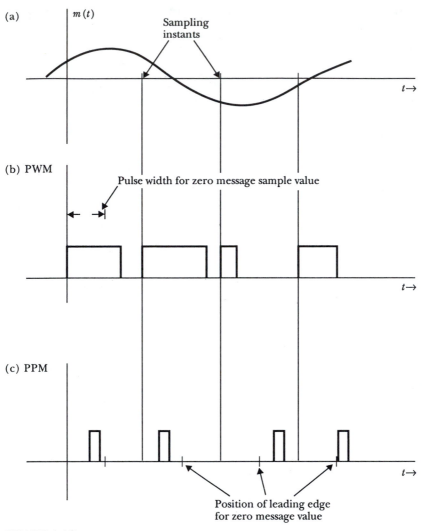

FIGURE 6.19
Typical message and corresponding PWM and PPM waveforms.

are a number of advantages to digital methods. In this section, we consider the degradation of signal quality due to quantization, compare the bandwidth requirements and noise performance to those of the analog methods we have studied, and briefly consider an alternative digital technique known as *delta modulation*.

Quantization

Quantization of an analog message is required by digital methods because it is possible to represent only a finite number of amplitude levels with a fixed-

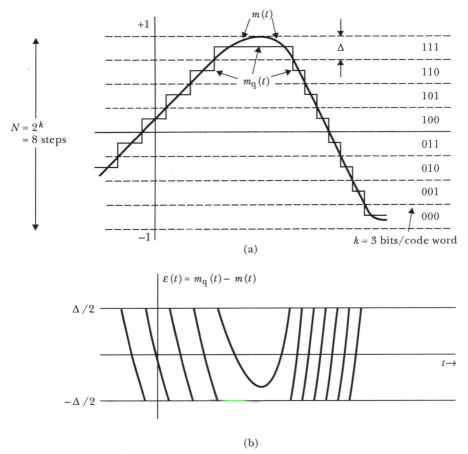

FIGURE 6.20
(a) A message and its quantized version and (b) corresponding quantization error $\epsilon(t)$.

length code word. The quantized version $m_q(t)$ of a message signal $m(t)$ is an approximation that replaces the actual message amplitudes with discrete steps. This is illustrated in Figure 6.20a. We assume that the message amplitude is confined to the range from -1 to $+1$. As illustrated, this range is divided into N zones, each of width Δ. The quantized signal takes on a value in the middle of the zone that contains the message amplitude at each instant of time. Thus, the quantized signal changes its amplitude by a step of height Δ when the message signal crosses into a new zone.

The input-output characteristic of the quantization process is shown in Figure 6.21. Notice in the figure that quantization is a nonlinear operation on the input signal. As a result, we can expect the quantized output to contain harmonics, as well as sums and differences, of the input frequencies.

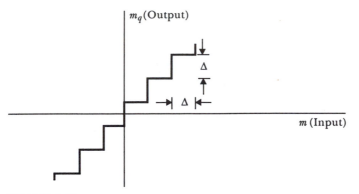

m_q(Output)

Δ

Δ

Δ

m (Input)

FIGURE 6.21
Input-output characteristic of quantizer.

The symbols in the code words are usually 1 and 0. Then the number of amplitude zones $\mathcal{N}$ is related to the number of bits k in each code word by

$$\mathcal{N} = 2^k \tag{6.44}$$

The width of each quantization zone is the total range divided by the number of zones. Since we assume that the message is confined to the amplitude range from -1 to $+1$, this is given by

$$\Delta = \frac{2}{\mathcal{N}} = 2^{1-k} \tag{6.45}$$

A possible binary code word assignment is shown for each zone in Figure 6.20a. This code is obtained simply by counting in binary from zero at the bottom zone to the top zone. Other binary codes are sometimes used in which, for example, the first bit indicates the sign of the value assigned to each zone, and the remaining bits indicate the magnitude.

The difference between the original message and the quantized version is the distortion added to the signal in the nonlinear quantization process. This distortion signal $\epsilon(t)$ is shown in Figure 6.20b for the waveforms in part (a) of the figure. The quantization error has much the same effect as additive noise, and its important parameter is its power.

Note that the quantization error consists of segments that closely resemble a period of the periodic sawtooth signal of Figure 6.22. When the number of quantization zones is large, this resemblance becomes greater because sections of the message waveform appear to be straighter as they become shorter. The average power in the periodic sawtooth waveform of Figure 6.22 can be calculated as

$$P_v = \frac{1}{T} \int_{-T/2}^{T/2} [v(t)]^2 \, dt = \frac{1}{T} \int_{-T/2}^{T/2} \left[\frac{\Delta t}{T}\right]^2 \, dt = \frac{\Delta^2}{12} \tag{6.46}$$

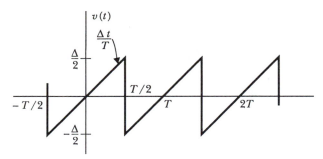

FIGURE 6.22
Sawtooth waveform.

Note that the power in the sawtooth does not depend on its period. Since the quantization error consists of segments very similar to the sawtooth waveform but with a variable period, we conclude that the power in the quantization noise $\epsilon(t)$ is also given closely by

$$P_q = \frac{\Delta^2}{12} \qquad (6.47)$$

Although the total power in the quantization noise is very nearly the same as for the periodic sawtooth waveform, the distribution in the frequency domain is quite different. The quantization noise $\epsilon(t)$ results from frequency multiplication and mixing effects in the nonlinear quantization process. The PSD of the quantization noise depends in some way on the message signal and the width of the quantization zones. In general, we can expect that the quantization noise will extend in the frequency domain well above the highest frequency of the message due to the frequency multiplication effects of the nonlinear quantizer. A hypothetical message PSD and the PSD of the quantization noise are shown in Figure 6.23. The area under the PSD of the quantization noise is equal to its total power, which is $\Delta^2/12$.

A convenient theoretical model of the quantization, sampling, and message recovery process is shown in Figure 6.24. The message signal is first quantized to discrete amplitude levels and then sampled by multiplication by an impulse train. Finally, the original message is recovered from the quantized samples by the lowpass filter. Due to the quantization noise $\epsilon(t)$, added by the quantizer, the output contains a noise signal $n_q(t)$. Only the portion of the power in $\epsilon(t)$ that contributes to output noise power in $n_q(t)$ is of significance in a PCM system.

As we have seen in our discussion of sampling, the impulse train contains equal components at all integer multiples of the sampling frequency. If these components of the sampling waveform are multiplied by a signal waveform,

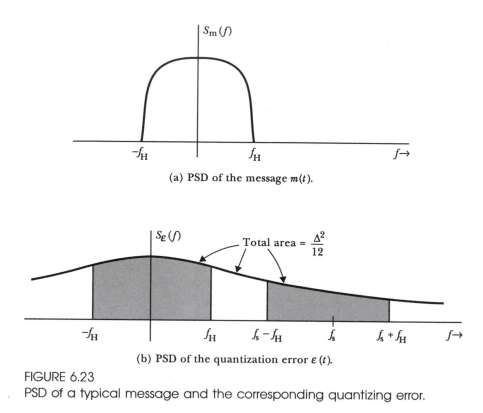

(a) PSD of the message $m(t)$.

(b) PSD of the quantization error $\varepsilon(t)$.

FIGURE 6.23
PSD of a typical message and the corresponding quantizing error.

the signal spectrum is translated to each integer multiple of the sampling frequency. Thus, each frequency component of the quantization noise $\epsilon(t)$ that lies closer than f_H to one of the multiples of the sampling frequency produces a difference-frequency component in the frequency range of the message (i.e., dc to f_H). The portion of the quantization noise that contributes to noise at the output of the recovery filter in the system of figure 6.24 is shown as the shaded regions of Figure 6.23b. It is clear from this figure that if the sampling

FIGURE 6.24
Quantization, sampling, and message recovery process.

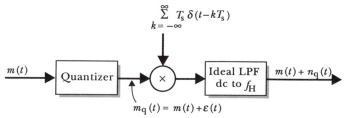

frequency is chosen to be twice the highest message frequency, then all of the quantization noise contributes to the noise at the output of the recovery filter, and the output noise power is also $\Delta^2/12$. On the other hand, if the sampling rate is much higher, not all of the quantization noise appears at the system output. Thus, $\Delta^2/12$ is a conservative estimate of the quantization noise power at the output of a PCM system.

Companding

Speech signals tend to have a large *crest factor* (the ratio of peak amplitude to rms amplitude) because they spend a relatively large amount of time at low amplitude and a small amount of time at their peak amplitude. This effect is compounded when more than one talker is producing the signals because some people speak loudly and others speak quietly. Because of this, the uniform quantizer, which has equally spaced zones as shown in Figure 6.20, does not perform as well as a quantizer with wider zones at high amplitudes and narrower zones at lower amplitudes.

The usual method for producing the desired nonuniform quantization characteristic is to first pass the message through a nonlinear device known as a *compressor*, which compresses the peak amplitudes. This is followed by a uniform quantizer. The input-output characteristic of a compressor is shown in Figure 6.25. Note from the figure that uniform zones at the output correspond to nonuniform zones at the input. At the lower amplitudes, where the message signal spends most of the time, the zones are narrow. At the high amplitudes, reached only occasionally by the signal, the zones are larger. Thus, the quantization error is less than for a uniform quantizer most of the time, and a net improvement results.

At the receiving end, the compressed message signal is recovered using a DAC with uniform zones. Then, the DAC output is passed through a nonlinear device known as an *expander*, which cancels the nonlinear effect of the compressor. The combined process is known as *companding*, a contraction of "compressing and expanding." A commonly used compression characteristic is the *μ-law characteristic* given by

$$m_c = (\text{sgn } m) \, \frac{\ln(1 + \mu|m|)}{\ln(1 + \mu)} \qquad \text{for } |m| \leq 1 \qquad (6.48)$$

where m is the input message value and m_c is the compressed output. This characteristic is used in the telephone system in the United States with $\mu = 255$.

Noise performance

A typical PCM communication system is shown in the block diagram of Figure 6.26. The message signal enters a *sample-and-hold circuit*, which samples the message signal and holds the sample value at its output while the analog-to-digital

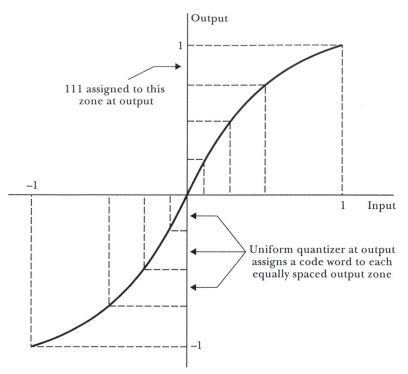

FIGURE 6.25
Nonlinear characteristic of compressor maps nonuniform input zones into uniform zones at the output. The compressor, followed by a uniform quantizer, is equivalent to a nonuniform quantizer with narrow zones for small input message values and wide zones for large input message values.

FIGURE 6.26
PCM communication system.

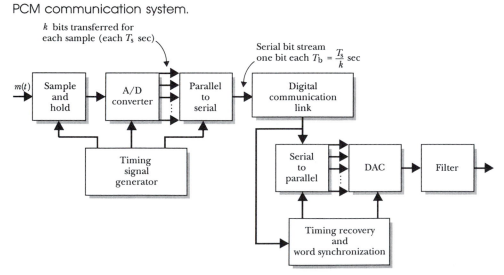

6 / PULSE MODULATION TECHNIQUES

(A/D) converter generates the corresponding code word. The k-bit code word is transferred to a *parallel-to-serial converter*, which places the bits into the digital communication link sequentially. The digital communication link could be any of a large variety of systems including baseband cable, radio using various modulation formats, optical fiber, or a dial-up telephone line, to mention a few possibilities. A large part of the remainder of this book is devoted to discussion of digital communication links of various kinds. At this point, we simply assume that the link conveys the bits at its input to the output with an occasional error. The received bits are collected into words by a *serial-to-parallel converter*. These words are then applied to a DAC, which generates an approximation to the message in the form of a flat-topped waveform that is applied to a filter to reconstruct the message. Due to the flat-topped nature of the waveform produced by the DAC, the recovery filter needs to have a reciprocal sinc transfer function to restore the high frequency message components to their original amplitude, as discussed in Section 6.1. The reconstructed message contains noise due to quantization at the sending end and to bit errors in the link.

Notice that it is necessary for the receiver to obtain the correct *word synchronization* so the received bits are divided into words at the proper boundaries. This is necessary so that the words applied to the DAC are the same words produced by the A/D converter. We consider this synchronization problem in more detail in Chapter 8.

For the code illustrated in Figure 6.20a, a bit error in the first bit of a code word causes the resulting sample value at the DAC output to be in error by ± 1 unit. An error in the second bit results in an error of $\pm \frac{1}{2}$ unit, and so on. (We assume that the output range of the DAC matches the ± 1 range of the input signal shown in Figure 6.20.) It can be shown, using probability concepts, that the resulting average noise power in the reconstructed signal due to these bit errors is

$$P_c = P_b \frac{4(1 + 2^{-2k})}{3} \tag{6.49}$$

where P_b is the probability of a bit error in the communication link. (It is assumed, as is often the case, that errors occur independently from bit to bit.)

The quantization noise and the bit error noise are caused by physically independent processes. Therefore, we expect that they are uncorrelated, and the total noise power at the system output is the sum of their separate contributions. This is given by

$$P_{no} = P_q + P_c \tag{6.50}$$

Substituting Equations 6.47 and 6.49 into Equation 6.50, we obtain

$$P_{no} = \frac{\Delta^2}{12} + P_b \frac{4(1 + 2^{-2k})}{3} \tag{6.51}$$

Substituting Equation 6.45 for Δ, this becomes

$$P_{\text{no}} = \frac{1}{3}[2^{-2k} + 4P_b(1 + 2^{-2k})] \tag{6.52}$$

The signal power at the system output is simply the power in the message signal. Since we have assumed that the peak value of the message is one, we denote the message power by P_{mn} as a reminder that the message was normalized. Thus, the output SNR is the ratio of the message power to the noise power given by Equation 6.52. The result is

$$\blacksquare \qquad \text{SNR}_o = \frac{3P_{\text{mn}}}{[2^{-2k} + 4P_b(1 + 2^{-2k})]} \tag{6.53}$$

Thus, the output SNR depends on the number of bits per code word k and the bit error probability P_b. Assuming that the error probability is low, the output SNR is set by the quantization noise. Then the SNR can be raised only by increasing the number of bits per code word. As we will see, increasing k increases the bandwidth required. Thus, PCM is a means for trading bandwidth to gain improved noise performance. PCM is similar to FM in this respect.

The probability of bit error P_b depends on many factors such as the type of modulation, the channel characteristics, and what type of error correction coding (if any) is used. These factors are discussed in detail in the following chapters. We show that if the best performing bit-by-bit modulation formats, known as *antipodal signal sets*, are used on an additive white gaussian noise channel without error-correction coding, the resulting error probability is

$$P_b = Q\left[\left(\frac{2P_r T_b}{N_0}\right)^{1/2}\right] \tag{6.54}$$

The function $Q(\cdot)$ is the integral of the tail of the gaussian probability density function (see Section 7.2), P_r is the power available at the receiver input terminals, T_b is the time slot for each bit, and N_0 is the effective noise PSD at the receiver input. Well-designed digital communication links often achieve an error probability on the order of the values predicted by Equation 6.54. However, as we will see, many factors affect the performance of actual systems.

Bandwidth requirements of PCM systems

The reciprocal of the time duration of each bit is the data rate on the link given by

$$R_b = \frac{1}{T_b} \qquad \text{bits/s}$$

The data rate for the communication link of Figure 6.20a is given by

$$R_b = kf_s \tag{6.55}$$

since k bits are generated for each message sample, and f_s is the sampling rate. If the minimum sampling rate to avoid aliasing is used, we have

$$R_b = 2kf_H \qquad (6.56)$$

The bandwidth required by a digital communication link is directly proportional to the data rate. For example, in Section 6.3, we found that pulses could be transmitted at a rate of $1/T$ pulses/s in a frequency band from dc to $1/(2T)$ using sinc pulses. These sinc pulses can be amplitude modulated with the binary data, so the two amplitudes are $\pm A$ (depending on whether the corresponding data bit is a 0 or a 1). Then, we can achieve transmission in a bandwidth equal to half of the data rate while still achieving the error probability given by Equation 6.54.

In the case of bandpass signals used in radio transmission, the bandwidth required depends on the modulation scheme used. We will see in Chapter 8 that by using a type of modulation known as *quadrature phase-shift keying*, we can achieve a bandwidth that is approximately equal to the data rate and achieve an error probability close to the value given by Equation 6.54. Thus, the bandwidth required by such a digital radio can be estimated as

$$B = 2kf_H \qquad (6.57)$$

We are now in a position to make a comparison of PCM using a bandpass communication link with the alternative analog schemes we have studied earlier.

EXAMPLE 6.2

Compare the output SNR of a $k = 8$ bit/sample PCM system with that of an FM system using the same bandwidth as a function of the ratio $P_r/N_0 f_H$. Assume a sinusoidal message. Also compare the output SNR and bandwidth requirements with those of DSB-SC and SSB systems.

SOLUTION
We assume that the sampling rate is chosen to be 50% higher than the minimum theoretical sampling rate. This higher sampling rate is often necessary in practical systems to be able to separate the alias terms from the desired message using practical filters. Therefore, the sampling rate is chosen as $3f_H$. The resulting data rate is

$$R_b = \frac{1}{T_b} = kf_s = 3kf_H$$

Substituting this into the expression for error probability given in Equation 6.54, we obtain

$$P_b = Q\left[\left(\frac{2P_r T_b}{N_0}\right)^{1/2}\right] = Q\left[\left(\frac{P_r}{N_0 f_H} \frac{2}{3k}\right)^{1/2}\right]$$

Once the error probability is known, the output SNR of the PCM system can be found from Equation 6.53. For a sinusoidal message, we have

$$P_{mn} = \frac{1}{2}$$

Substituting into Equation 6.53, we have

$$\text{SNR}_o = \frac{\frac{3}{2}}{[2^{-2k} + 4P_b(1 + 2^{-2k})]}$$

Using $k = 8$, the bit error probability can be computed from the previous equation for various values of $P_r/N_0 f_H$. Then the error probability can be used to compute the output SNR. The results are plotted in Figure 6.27. Notice that for high values of $P_r/N_0 f_H$, the output SNR of the PCM system is constant because of the quantization noise. At lower values of $P_r/N_0 f_H$, bit errors begin to occur, and the SNR falls off. Thus, the PCM system displays a threshold effect.

The bandwidth of the PCM system depends on the modulation format in use. (This is treated in detail in Chapter 8.) As we will see later, it is rea-

FIGURE 6.27
Comparison of output SNR for various systems.

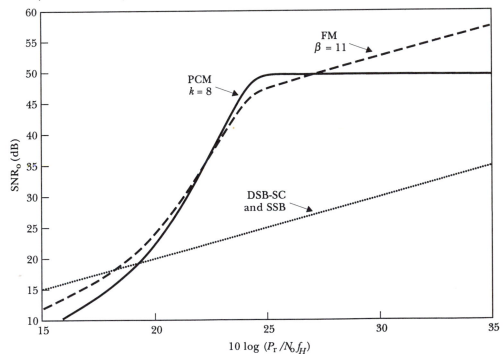

sonable to assume that the bandwidth is about equal to the data rate for this comparison. Therefore, the bandwidth is estimated as

$$B = R_b = 3kf_H = 24f_H$$

The bandwidth of the FM system can be estimated by the use of Carson's rule, given in Equation 4.27 as

$$B \simeq 2f_H[\beta + 1]$$

where we have used the modulation index β in place of the deviation ratio D because we are assuming a sinusoidal message. Equating the last two expressions for system bandwidth, we can solve for β. The result is $\beta = 11$.

The output SNR of an FM system is given in Equation 5.107. The results for the FM system are plotted for $\beta = 11$ in Figure 6.27.

The output SNR of the SSB and DSB-SC systems is given in Equation 5.71 as

$$SNR_o = \frac{P_r}{N_0 f_H}$$

This is also plotted in Figure 6.27.

Delta modulation

A digital communication system that is simpler than PCM is shown in Figure 6.28a. The message signal enters a comparator in which it is compared with the output of a circuit that is designed to predict the message waveform. The output of the comparator is high when the predicted value $m_p(t)$ is less than the actual message $m(t)$. On the other hand, the comparator output is low when the predicted value is greater than the actual message. The output of the comparator is connected to the input of a flip-flop acting as a sampler. At the end of each bit interval T_b, the clock signal causes the Q output of the flip-flop to change to the value present at the D input. Thus, the signal at the Q output of the flip-flop is a digital signal that is in the *high* state if the predicted value is lower than the message at the last clock pulse. The flip-flop output is low when the prediction is higher than the message. The output of the flip-flop is connected to the predictor block in the modulator. The flip-flop output is also the data sent over the link to the receiver. Based on the bit stream that appears at its input, the predictor adjusts its output to match the incoming signal. At the next sampling instant, a one or a zero appears at the input to the predictor, depending on whether its adjusted value was too low or too high, respectively.

Figure 6.28b shows example waveforms for a delta modulator. It is assumed that the logic levels of the flip-flop are $+V$ for a one and $-V$ for a zero. The predictor in this example consists of a simple integrator. The output

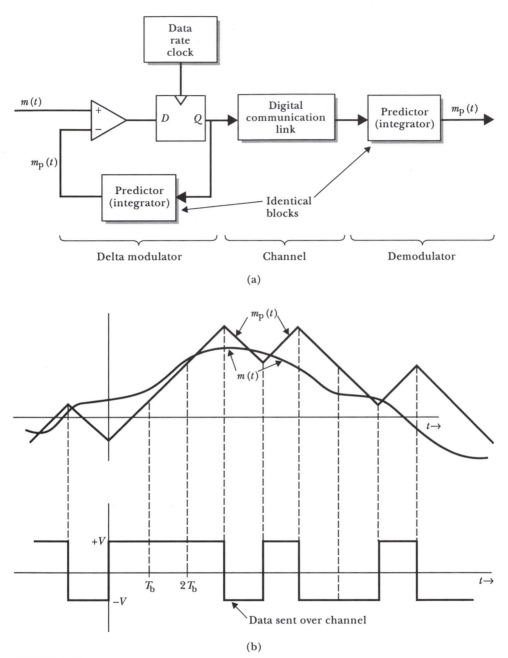

FIGURE 6.28
Delta modulation.

of the integrator ramps in the positive direction if its input is $+V$ and in the negative direction when the input is $-V$. Thus, the predicted message waveform ramps negative if it was too high at the last sampling instant and ramps positive if it was too low. Clearly, if the integrator responds quickly, and if the clock rate is high enough so that the direction of integration can change very often, the predicted waveform closely matches the actual message.

The receiver consists of a circuit that is identical to the predictor used in the modulator. Assuming that the prediction circuits are identical and that there are no bit errors in transmission, the waveform at the receiver output matches the predicted waveform at the input to the modulator. This scheme and variations of it are called *delta modulation* because only binary indications of the differences between the message and the predicted message are transmitted over the channel.

Many predictor circuits other than the simple integrator are possible. For example, we could use an integrator that increases its gain or rate of response if three or more bits in a row are the same, indicating that the prediction has fallen far behind the actual message and that the predictor must speed up. The gain would be reduced whenever the data bit changes its value.

The quality of the predicted message can be improved by increasing the sampling rate. However, this results in a higher bit rate on the channel and entails the use of more bandwidth. Therefore, it is desirable to find a predictor that can make good predictions even if the sampling rate is low. The ability to do this depends on the character of the message. Messages that are redundant and easy to predict require lower bit rates than highly variable messages.

The primary advantage of delta modulation is the extreme simplicity of the hardware required. In the example system of Figure 6.28, the integrator could be approximated by a simple lowpass RC circuit. Also, note that no word synchronization is required at the receiver as in PCM.

Now that we have illustrated some of the advantages of digital methods for communicating analog signals and made comparisons with analog modulation schemes, we will turn to the digital communication link, which conveys the bit stream to its destination. To make meaningful comparisons of the various modulation formats and error correction coding schemes, we need to use probability theory, which is the subject of the next chapter.

SUMMARY

1. A baseband signal that is limited to the frequency range from dc to f_H can be completely recovered from its samples if the sampling rate f_s is greater than $2f_H$. If the sampling rate is too low, alias terms, which are components of the signal that have been translated in frequency, appear in the recovered signal.

2. In principle, a signal can be recovered from its instantaneous samples by modulating an impulse train with the successive sample values and then passing the modulated impulse train through an ideal lowpass filter with a cutoff of f_H. A practical way of recovering the signal is to use the sample values to amplitude-modulate a flat-topped pulse train and then pass the result through a lowpass filter having a reciprocal sinc transfer function between dc and f_H.

3. Sampling can also be done by switching the waveform on and off, resulting in a sampled version that consists of short segments of the original signal. This is called *natural sampling*. The complete message can be recovered by lowpass filtering if the filter has constant gain in the passband. The required sampling rate is the same as for instantaneous samples.

4. Sampling rates must be higher than the minimum theoretical value to make implementation of the reconstruction filters feasible.

5. When there are components of a signal above the range of interest, it is necessary to eliminate them before sampling or take them into consideration in choosing the sampling frequency to avoid aliasing, which may create distortion in the desired frequency range.

6. Bandpass signals can sometimes be sampled at rates less than twice the highest frequency. When this is done, it is necessary to use a bandpass filter in recovering the signal.

7. Samples from a number of different messages can be interleaved in time and transmitted over a communication link. This is called time-division multiplexing (TDM).

8. The samples of a signal can modulate the amplitude of a pulse train (PAM), the pulse width (PWM), or the pulse position in time (PPM). Alternately, the samples can be quantized and each amplitude level can be represented by a digital code word, resulting in pulse code modulation (PCM).

9. Flat-topped and naturally sampled PAM signals have significant spectral content out to very high frequencies. It is desirable to limit the bandwidth of PAM signals by selection of the pulse shape and/or filtering. Unless filtering is carefully controlled, intersymbol interference (ISI) occurs.

10. The Nyquist criterion for zero ISI is a condition on the equivalent transfer function (which includes the transform of the transmitted pulse) of a PAM system so that the received pulses are zero at all of the adjacent sampling points.

11. The minimum bandwidth filter that meets the Nyquist criterion is an ideal lowpass filter with a cutoff frequency equal to one-half of the pulse rate. The received pulse is a sinc pulse. Practical systems must use filters with a more gradual roll-off and thus must use more bandwidth.

12. The overall equivalent transfer function in a PAM system must be divided among the transmitted pulse shape, the channel filtering, and the filter used at the receiver to limit noise.

13. When the added noise is white, the SNR of samples of the received PAM waveform is maximized by using a matched filter. The impulse response of a matched filter is the time-reversed version of the pulse shape received from the channel.

14. Theoretical considerations often ignore the realizability of filters, and nonzero impulse responses before $t = 0$ result. When the filters are to be implemented, a sufficient delay can be added to the impulse response to make them realizable. Circuit design approaches can then be applied to approximate the filters.

15. A minimum bandwidth PAM TDM system, using an ideal lowpass filter characteristic for the overall transfer function, uses a total bandwidth equal to the number of messages times the highest message frequency (assuming baseband messages, each with the same highest frequency limit). This is the same as the bandwidth requirement of an FDM system using SSB. The noise performance of the PAM TDM system in the presence of additive white noise is the same as SSB or DSB-SC.

16. Quantization is necessary when PCM is used because it is possible to represent only a finite number of amplitude levels with a fixed-length code word. Quantization error has the same effect as added noise. The amount of quantization noise can be reduced by using more quantization levels and longer code words, however, this leads to an increased use of bandwidth.

17. The overall SNR of signals that have a large crest factor, such as speech, can be improved by companding, which is equivalent to using nonuniform quantization zones.

18. PCM can be used to improve the output SNR of a communication system without the necessity of increasing the received (and therefore transmitted) power by expanding bandwidth. At high levels of received power, the output SNR is set by the quantization noise. As the received power decreases, a point is eventually reached at which a significant number of transmission errors occur, and the output SNR begins to decrease very rapidly. This is similar to the threshold effect experienced with FM systems.

19. Delta modulation is a technique for converting an analog message into a digital data stream, which is then used to predict the next message sample value. The difference between the prediction and the actual sample value determines the next character in the data stream. In its simplest forms, the hardware requirements of delta modulation systems are very modest.

REFERENCES

B. P. Lathi. *Modern Digital and Analog Communication Systems*. New York: Holt, Rinehart and Winston, 1983.

M. Schwartz. *Information Transmission, Modulation, and Noise*, 3d ed. New York: McGraw-Hill, 1980.

F. G. Stremler. *Introduction to Communication Systems*, 2d ed. Reading, Massachusetts: Addison-Wesley, 1982.

J. M. Wozencraft and I. M. Jacobs. *Principles of Communication Engineering*. New York: Wiley, 1965.

PROBLEMS

1. A message signal $m(t) = \cos(2000\pi t) + 3\cos(6000\pi t)$ is sampled by multiplication by an impulse train as shown in Figure 6.3. The weighted impulse train is then passed through an ideal lowpass filter in an attempt to recover the original message. The passband gain of the filter is unity, and the cutoff frequency is 3.5 kHz. Find an expression for the resulting output signal and identify any alias terms present if (a) the sampling rate is 10 kHz, (b) the sampling rate is 5 kHz.

2. A message signal with the amplitude spectrum shown in Figure P6.1 is to be sampled at 15 kHz. Sketch the amplitude spectrum of the sampled signal to scale if the sampled signal is (a) a weighted impulse train as shown in Figure 6.1c, (b) a flat-topped PAM waveform as shown in Figure 6.4c with $\tau = T_s/3$, (c) a naturally sampled waveform as shown in Figure 6.6c with $\tau = T_s/3$.

3. A message signal is sampled by multiplication with an impulse train as shown in Figure P6.2. The weighted impulse train is then passed through a filter with the impulse response $h_1(t)$ as shown. Sketch a typical input message waveform and the resulting output of the first filter. Find the frequency response of the second filter $H_2(f)$ so that the message appears at its output with no distortion. Assume that the sampling frequency is higher than twice the highest message frequency.

4. A bandpass signal is confined to the frequency range from 7 to 10 kHz. Find the allowed ranges of the sampling rate for this signal. Sketch the amplitude spectrum of a hypothetical message, the amplitude spectrum of the sampled signal, and the transfer function of a suitable recovery filter if the sampling rate is chosen in the center of the lowest range available.

5. In commercial FM stereo broadcasting, the message signals for the left and right channels are $m_l(t)$ and $m_r(t)$, respectively. These message signals are used to

FIGURE P6.1

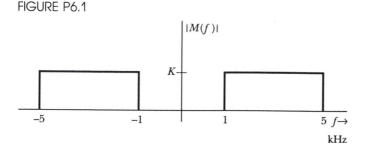

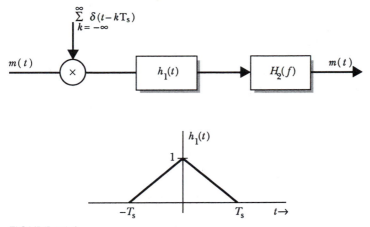

FIGURE P6.2

produce the signal

$$m(t) = \frac{1}{2}[m_l(t) + m_r(t)] + \frac{1}{2}[m_l(t) - m_r(t)]\cos(2\pi f_{sc}t) + A\cos(\pi f_{sc}t)$$

The first term on the right in brackets represents the monaural signal, the second term is the difference between the left and right channels DSB-SC modulated onto an $f_{sc} = 38$ kHz subcarrier, and the last term is a 19 kHz tone used by the receiver to regenerate the subcarrier at 38 kHz. The left and right signals are limited to a highest frequency of 15 kHz. Show that the system shown in Figure P6.3 can be used to generate this signal if the switch spends half of its time at each position and $t = 0$ falls in the center of an interval during which the switch is at the left signal position. Find the numerical value of the gain constant K of the filter. Hint: The system is linear (but time-varying), so superposition applies. Consider the case in which $m_l(t)$ is present and $m_r(t)$ is zero. Then sketch the

FIGURE P6.3

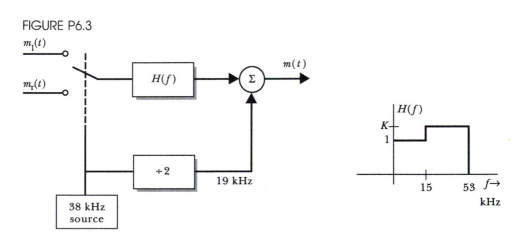

input to the filter and compare with Figure 6.6. Next, consider the case in which only the right signal is present.

6. Demonstrate that the individual left and right signals can be recovered from $m(t)$ in Problem 5 by instantaneous sampling at certain instants of time. Assume that the last term $[A \cos (\pi f_{sc} t)]$ has been removed before sampling. Suppose that the samples are obtained by sample-and-hold circuits that produce flat-topped PAM versions of the left and right signals, like the signal shown in Figure 6.5. Find and sketch to scale the recovery filter transfer functions needed to obtain the undistorted left and right signals from the sample-and-hold output signals. What is the consequence of simply amplifying and applying the sample-and-hold outputs to the speakers?

7. Consider the triangular PAM pulse train in Figure P6.4 derived from the samples of a message signal. This waveform can be produced by the system of Figure 6.4a with the proper choice of $h_g(t)$. Sketch $h_g(t)$ to scale. Find and sketch the amplitude response $H_r(f)$ that recovers the original message signal from the triangular PAM signal if the sampling frequency is four times the highest message frequency.

8. A sinusoid, $A \cos (2\pi f_H t)$, is sampled at $f_s = 4 f_H$ with one of the samples taken at $t = 0$. The samples are used to generate a flat-topped PAM signal as in Figure 6.5. Sketch the resulting PAM waveform to scale, find its Fourier series, plot the amplitude spectrum from the Fourier series coefficients, and verify that the result agrees with the spectrum of the flat-topped PAM signal given by Equation 6.11.

9. Consider a bandpass signal with a bandwidth of B and center frequency f_c (i.e., the frequency content extends from $f_c - B/2$ to $f_c + B/2$). If we can select the center frequency, what values could we choose so that the allowed sampling rate is as low as possible? What is the resulting sampling rate?

10. Three signals are to be transmitted using a TDM system similar to Figure 6.9. The highest frequency of $m_1(t)$ and $m_2(t)$ is 5 kHz and the highest frequency for $m_3(t)$ is 10 kHz. The sampling order is chosen to be $m_1, m_3, m_2, m_3, m_1, m_3, m_2$, and so on. Note that twice as many samples are taken from m_3 as from either m_1 or m_2. Sketch a diagram of the system. Determine the sampling rate for each signal so that in each case there is at least a 5 kHz guardband between the components to be passed by the recovery filters and the components to be rejected. Sketch typical message waveforms and the resulting transmitted waveform. Sketch typical amplitude spectra at the inputs to the recovery filters.

FIGURE P6.4

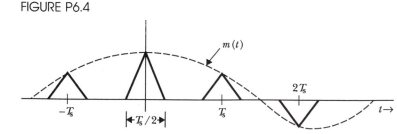

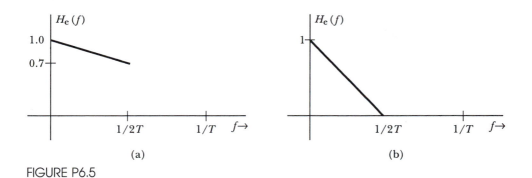

FIGURE P6.5

11. The Fourier transform of a pulse to be used in a TDM PAM system is shown in Figure P6.5a for the frequency range from 0 to $1/(2T)$. If the pulse is to have zero amplitude at integer multiples of T, sketch the complete transform for all frequencies. What is the amplitude of the pulse at $t = 0$? Repeat for the transform of Figure P6.5b.

12. Sketch the output signal versus time if the binary flat-topped pulse train shown in Figure P6.6 is applied to the input of the lowpass RC filter shown. Identify the ISI on the sketch.

13. Sketch the output waveform of a PAM system if the overall impulse response is the sinc pulse of Equation 6.20, and the message values are 0,0,0,1,0,0,1,1,0,0,0.

14. The pulse $p(t)$ shown in Figure P6.7 is added to white noise with a PSD, $S_n(f) = N_0/2$, and the sum is passed through a filter with the impulse response $h(t)$ shown. For $\tau = \frac{1}{2}T$, T, and $2T$, find the signal waveform at the output and

FIGURE P6.6

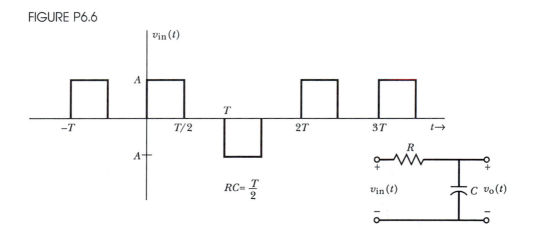

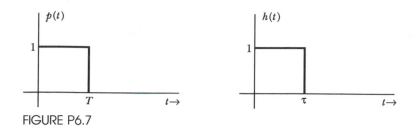

FIGURE P6.7

the peak signal power at the output. Find the noise bandwidth of the filter and the noise power at the output. Find the ratio of the peak signal power to the average noise power. Which value of τ gives the best SNR at the output? Sketch the impulse response of the matched filter for $p(t)$.

15. Consider the system shown in Figure P6.8. Sketch a typical waveform at the output of the pulse shaping filter and its amplitude spectrum. What is the spectral content of this PAM waveform in the vicinity of dc? Is it necessary for the channel to be dc-coupled for this PAM waveform? If the channel adds white noise and has constant gain at all frequencies, find the impulse response of the noise filter that maximizes the SNR at the output. Sketch the signal component of the waveform at the output of the noise-reduction filter resulting from the typical input waveform sketched earlier in this problem.

16. Twenty-five message signals, each with a highest frequency of 4 kHz, are to be sampled at the minimum rate to avoid aliasing and transmitted through a TDM PAM system. The overall transfer function is of the raised cosine type with a roll-

FIGURE P6.8

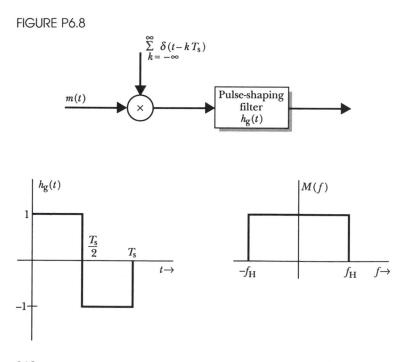

off factor of one-half. What is the highest frequency that the channel must pass (i.e., at what frequency does $|H_e(f)|$ become zero)? Repeat for roll-off factors of zero and one.

17. An audio signal with a crest factor (ratio of peak value to rms value) of 10 is to be quantized. If the SNR, after allowance for quantization noise, is to be 60 dB or greater, what is the minimum number of quantization levels that can be used assuming that uniform quantization is used? If the levels are to be represented by binary code words, what is the minimum number of bits per code word that can be used?

18. Twenty-five message signals, each with a highest frequency of 4 kHz, are sampled at the minimum sampling rate to avoid aliasing. The sample values are quantized to 256 levels represented by binary code words. The code words are interleaved in time and transmitted. What is the bit rate of the transmitted signal? What absolute bandwidth (i.e., the frequency for which $|H_e(f)|$ first becomes zero) is required if the overall channel response is of the raised-cosine type, with no ISI between bits and a roll-off factor of one-half? Repeat for roll-off factors of zero and one, and compare the results of this problem with Problem 16.

19. Threshold is often defined as the point at which the output SNR of a system falls 1 dB below the asymptote for high amounts of received power. Starting from Equation 6.53, determine the bit error probability at threshold for a PCM system. (Note that when the received power P_r is high enough, the error probability P_b is effectively zero. As P_r falls, P_b increases and the output SNR eventually falls.)

20. The bandwidth requirements of a PCM system can be reduced by grouping bits and transmitting each group with a single amplitude modulated pulse. For example, if two bits are taken at a time, pulses with four distinct amplitudes are transmitted. If raised-cosine shaping is used with a roll-off factor r of one-half, find the absolute bandwidth (see Problem 18) required to transmit binary data at a rate of 100 kbit/s if (a) 2 bits are transmitted at a time (4-level PAM), (b) 3 bits are transmitted at a time (8-level PAM).

21. A voice signal with a highest frequency of 3400 Hz is sampled at 8 kHz, and the samples are quantized to 256 levels. If a binary code such as the one in Figure 6.20a is used, what is the bit rate of this source? If these data are transmitted using binary PAM with raised-cosine pulses having a roll-off factor of one-half, what absolute bandwidth (see Problem 18) is required? If the data are to be transmitted with a highest frequency less than 20 kHz by the use of multilevel PAM (see Problem 20) using raised-cosine pulses with a roll-off factor of one-half, how many levels should the PAM signal have, assuming that it will be a power of two?

7

PROBABILITY, RANDOM VARIABLES, AND RANDOM PROCESSES

We will need the concepts of probability to discuss the performance of digital communication systems in the presence of noise. Probability theory is useful in making predictions about experiments in which the results are different every time the experiment is tried, even though the conditions are seemingly the same. For example, the result is different every time we measure the thermal noise voltage across a resistor in a specified bandwidth. Probability theory makes predictions about the fraction of the measurements that exceed a given value. This kind of consideration is useful in determining the bit-error rate of digital communication systems.

A *random variable* is a numerical value that is assigned to the outcome of an experiment, such as measuring the value of a noise voltage at a specific time. In a *random process*, a waveform is assigned to each outcome of an experiment. These concepts help us deal with random noise and message signals in communication systems. The value of a noise voltage determines whether a bit error will occur in the reception of a digital signal.

7.1
PROBABILITY

Random experiments

Probability is concerned with making predictions about random occurrences. Probability theory has its roots in the study of games of chance such as dice and card games, and many of the examples we will discuss are taken from that arena. However, our eventual aim is to obtain useful predictions about communication systems. The predictions made by probability theory are concerned with averages rather than the precise outcome of a particular trial of an experiment. For example, probability theory may predict that 0.1% of the bits transmitted through a particular system will be in error but does not try to predict which bits they will be.

A *random experiment* is one for which we cannot predict the *outcome*. Examples are the toss of a coin, for which the outcome is heads or tails, the toss of a die, for which the outcome is the number of spots on the face that comes up, or drawing a card from a deck of playing cards for which where there are 52 possible outcomes corresponding to the 52 cards. The set of all of the possible outcomes of a random experiment is called the *sample space* (S).

An *event* consists of a collection of outcomes of a random experiment. For example, if the experiment is the toss of a die, an event is that the number of spots that turns up is odd. In this case, the event contains three possible outcomes: one spot turns up, or three spots, or five spots. In the random experiment of drawing a card from a deck, an event is that the card is red, and another event is that the card is an ace. We will often use symbols to denote events. For example, we can let B denote the event that the card drawn from a deck is black and let A denote the event that the card is an ace.

Events that are combinations of other events are often of interest. The *union* of two events is the event consisting of all of the outcomes contained in either (or both) of the original events. We will denote the union of two events by A + B, read "A or B." If A denotes the event that a card drawn from a deck is an ace, and B denotes that the card is black, then C = A + B denotes the event that consists of all of the black cards plus the two red aces.

The *intersection* of two events consists only of the outcomes that are in *both* of the initial events and is denoted by AB, read "A and B." For example, in the card-drawing experiment, the event D = AB is the collection of all cards that are both aces and black. Thus D consists only of the two black aces.

The *complement* of an event A, denoted by A^c, is the collection of all outcomes not contained in A. Thus if A denotes the aces in a deck of 52 cards, then A^c denotes the other 48 cards.

It is possible, when combining events, to obtain a combination that contains no outcome. For example, if R denotes the event that a red card is drawn and B denotes the event that a black card is drawn, then the event RB contains

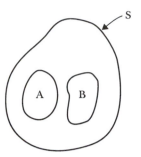

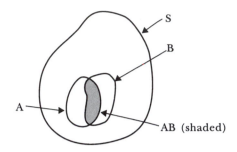

(a) A and B are disjoint, AB = ∅

(b) Intersection of events

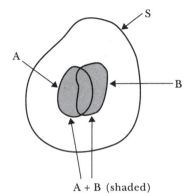

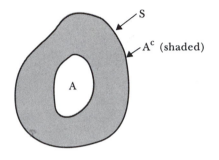

A + B (shaded)

(c) Union of events

(d) Complement of an event

FIGURE 7.1
Venn diagrams.

no outcome since no card is both red and black. We will call the event that contains no outcomes the *null event* and denote it by ∅. The event S, which contains all of the outcomes, is called the *certain event* because at least one outcome always occurs when the experiment is performed.

Graphic representations, known as *Venn diagrams*, are useful for illustrating the relationships among events. Venn diagrams illustrating the concepts of union, intersection, and complement are shown in Figure 7.1.

Probabilities and the axioms of probability

The *probability* of an event B, denoted by $P(B)$, is a number associated with the event. We will assign a probability to an event (often by an intuitive process) in such a way that it represents the fraction of the trials of an experiment for which we "expect" the event to occur. An intuitive (rather than a mathematical) definition of the probability of an event is the ratio of the number of times N_A that event A occurs, divided by the total number of times N that the experiment is performed, if the limit is taken for very large N.

$$P(A) = \lim_{N \to \infty} \frac{N_A}{N} \qquad (7.1)$$

We often use Equation 7.1 intuitively in assigning probability values to the events of a random experiment, but the equation does not constitute a mathematical definition.

Probability values are assumed to obey certain rules known as the *axioms of probability*. In equation form, these axioms for any two events A and B of a random experiment with sample space S are

- $$P(A) \geq 0 \qquad (7.2)$$

- $$P(S) = 1 \qquad (7.3)$$

- $$P(A + B) = P(A) + P(B) \qquad \text{if } AB = \varnothing \qquad (7.4)$$

Mathematicians have been led by the intuitive interpretation of Equation 7.1 to these axioms and other definitions in constructing probability theory, which is useful in modeling random experiments in the physical world. Certainly, if probability is to reflect the ratio of the number of times an event occurs to the number of trials, then Equation 7.2 must be true since an event cannot occur a negative number of times. The certain event S always occurs, so the probability assigned to it should be one. Finally, if A and B are mutually exclusive so that A never occurs when B does and vice versa, then the number of times that A + B occurs is the sum of the number of times A occurs plus the number of times B occurs; in equation form, we have

$$\frac{N_{A+B}}{N} = \frac{N_A}{N} + \frac{N_B}{N}$$

and Axiom 7.4 follows, using the intuitive definition of probability, if the limit is taken as N becomes very large.

Starting from the axioms, the following properties can be derived for any two events in a random experiment.

- $$P(A) \leq 1 \qquad (7.5)$$

- $$P(A) + P(A^c) = 1 \qquad (7.6)$$

- $$P(A + B) = P(A) + P(B) - P(AB) \qquad (7.7)$$

It can also be shown that $P(\varnothing) = 0$.

A mathematically valid *probability assignment* for experiments with a finite number of outcomes is to assign a positive probability to each outcome such that the sum of the probabilities assigned to all outcomes is unity. For example, if the experiment is the toss of a die, most of us would intuitively assign a probability of one-sixth to each of the six faces. There is no guarantee that this is the "correct" assignment, however. In making initial probability assignments, we must often be guided by intuition. Once a probability has been assigned to each outcome of an experiment that has a finite number of possible outcomes, the probability of an event is found by adding the probabilities of the outcomes included in it. For example, if the probability for each face of a die is one-sixth, then the probability of an even-numbered face coming up is one-half, since three faces are even.

Conditional probabilities

The probability of event A, *conditioned* on the fact that event B has happened, is defined to be

$$P(A|B) = \frac{P(AB)}{P(B)} \tag{7.8}$$

If $P(B) = 0$, the *conditional probability* $P(A|B)$ is undefined. The reason for this definition is that relative frequencies behave in this manner. After it is known that B has occurred in an experiment, the ratio of times that A occurs is

$$\frac{N_{AB}}{N_B} = \frac{N_{AB}/N}{N_B/N}$$

where N_{AB} is the number of times that both A and B occur in a trial of the experiment, N_B is the number of times event B occurs, and N is the number of trials of the experiment. If the limit of the last expression is taken as N becomes very large, and the intuitive definition of probability is used, the formal definition of conditional probability results.

Rearranging Equation 7.8, we obtain *Bayes' rule*:

$$P(AB) = P(A|B)P(B) = P(B|A)P(A) \tag{7.9}$$

Similarly for three events, we can write

$$P(ABC) = P(A|BC)P(BC) = P(A|BC)P(B|C)P(C) \tag{7.10}$$

Independent events

Two events are said to be *independent* if and only if

$$P(AB) = P(A)P(B) \tag{7.11}$$

Substituting this into Equation 7.8 results in

$$P(A|B) = \frac{P(AB)}{P(B)} = \frac{P(A)P(B)}{P(B)} = P(A) \tag{7.12}$$

for independent events. Similarly, we have

$$P(B|A) = P(B)$$

The intuitive notion of independence is that once we know that one of the events has occurred, we have exactly the same uncertainty about the occurrence of the other event as before. For example, if we toss two dice, a red one and a blue one, we would be inclined to assume that the event "one spot turns up on the red die" is independent of the event "five spots turn up on the blue die," since the dice presumably land in physically independent manners.

Total probability

Sometimes we encounter a situation in which an event B can occur in conjunction with a number of other events $A_1, A_2, \ldots, A_N$, such that the events $A_1 B, A_2 B, \ldots$, are all disjoint. If $B = A_1 B + A_2 B + \cdots + A_N B$, then it can be shown that

$$P(B) = \sum_{n=1}^{n=N} P(BA_n) = \sum_{n=1}^{n=N} P(A_n | B) P(B) \tag{7.13}$$

Equation 7.13 is called the theorem of *total probability*.

EXAMPLE 7.1

Two boxes, numbered 1 and 2, contain red and white balls as follows.

    Box 1: 5 red balls and 5 white balls
    Box 2: 1 red ball and 4 white balls

The experiment consists of selecting a box at random (take this to mean that they each have equal probability of being selected), selecting a ball at random from the box, and observing its color. Then the ball is replaced, the balls are mixed, another ball is selected at random from the same box, and its color is observed. Find the probability that the first ball is white, the probability that the second ball is white given that the first is white (but the box number is unknown), and the probability that box 2 was selected given that both balls selected are white.

SOLUTION

First, we must identify the outcomes of the experiment. An outcome consists of the number of the box selected, the color of the first ball, and the color of the second ball. There are a total of eight outcomes. We can denote a typical outcome by $B_1 R_1 R_2$, which indicates that box 1 was selected, the first ball was red, and the second ball was red. The event "box 1 was selected" denoted simply by B_1 consists of a collection of four outcomes, which are denoted by $B_1 R_1 R_2$, $B_1 R_1 W_2$, $B_1 W_1 R_2$, and $B_1 W_1 W_2$. Similarly, the event B_2 consists of four outcomes. We can assign a probability to each outcome as follows. First, using Equation 7.10, we have

$$P(B_1 R_1 R_2) = P(R_2 | R_1 B_1) P(R_1 B_1)$$
$$= P(R_2 | R_1 B_1) P(R_1 | B_1) P(B_1)$$

Given that we have selected box 1 and the fact that the first ball is replaced and the balls are mixed before the second draw, we expect the result of the second draw to be independent of the first draw. Thus, we assume that

$$P(R_2|R_1B_1) = P(R_2|B_1)$$

and so we can write

$$P(B_1R_1R_2) = P(R_2|B_1)P(R_1|B_1)P(B_1)$$

Now we find values for each of the terms on the right side of the last expression. There are two boxes, which are equally likely and mutually exclusive. (Only one box is selected in each trial.) Furthermore, the union of B_1 and B_2 is the certain event. Thus, we can write

$$P(B_1 + B_2) = P(S) = 1$$

$$P(B_1) + P(B_2) = 2P(B_1) = 1$$

This results in the, perhaps obvious, conclusion that

$$P(B_1) = P(B_2) = \frac{1}{2}$$

Since the first box contains 5 red and 5 white balls, we conclude that

$$P(R_1|B_1) = P(R_2|B_1) = \frac{5 \text{ red balls}}{10 \text{ total balls}} = \frac{1}{2}$$

Substituting values for $P(R_1|B_1)$, $P(R_2|B_1)$, and $P(B_1)$ into the equation for $P(B_1R_1R_2)$ results in

$$P(B_1R_1R_2) = 0.125$$

Continuing in similar fashion, we can find the probabilities of the remaining outcomes as

$$P(B_1R_1W_2) = 0.125$$

$$P(B_1W_1R_2) = 0.125$$

$$P(B_1W_1W_2) = 0.125$$

$$P(B_2R_1R_2) = 0.02$$

$$P(B_2R_1W_2) = 0.08$$

$$P(B_2W_1R_2) = 0.08$$

$$P(B_2W_1W_2) = 0.32$$

It can be verified that the probabilities assigned to the eight outcomes sum to one, as they should for a valid probability assignment. The probability of the event W_1 can be found by adding the probabilities of the outcomes that include a white first ball. The result is

$$P(W_1) = P(B_1W_1R_2 + B_1W_1W_2 + B_2W_1R_2 + B_2W_1W_2)$$
$$P(W_1) = P(B_1W_1R_2) + P(B_1W_1W_2) + P(B_2W_1R_2) + P(B_2W_1W_2)$$
$$= 0.65$$

(Note that this result is *not* the same as the ratio of the total number of white balls in both boxes to the total number of balls.) In exactly the same way, the probability of a white second ball can be found as

$$P(W_2) = 0.65$$

Considering the fact that the conditions are the same for the second draw as for the first, the fact that the probability of white on the second draw is the same as for the first draw is not surprising. The probability of white on the second draw, given white on the first, is

$$P(W_2|W_1) = \frac{P(W_1W_2)}{P(W_1)} = \frac{0.445}{0.65} = 0.685$$

Notice that $P(W_2|W_1)$ is not the same as $P(W_2)$, so W_1 and W_2 are not independent. Intuitively, this can be explained by observing that as soon as we learn that the first ball is white, the likelihood that box 2 was selected becomes greater, and therefore the probability of white on the second draw goes up. This would be much more obvious if the populations of the boxes was skewed more heavily, say 1000 to 1 in favor of red in one box and 1000 to 1 in favor of white in the other box. In such a case, a white first ball would be almost always followed by a white second ball. Note that, when the selection of the box is given, we have assumed that W_1 and W_2 are independent. Thus, we have a situation in which two events are dependent when not conditioned by another event but become independent when conditioned by another event. In equation form,

$$P(W_1W_2) \neq P(W_1)P(W_2)$$

but

$$P(W_1W_2|B_1) = P(W_1|B_1)P(W_2|B_1)$$

The probability of having selected box 2 given that both balls are white can be found as follows:

$$P(B_2|W_1W_2) = \frac{P(B_2W_1W_2)}{P(W_1W_2)} = \frac{0.32}{0.445} = 0.719$$

Thus we see that it is highly likely that box 2 was selected after it is known that two white balls have been drawn.

The binomial probability distribution

In binary digital communication systems, every bit transmitted has some probability p of being received in error, and errors on different bits are often independent events. The probability of k bit errors in an N-bit word is of interest. If a specific pattern of k errors in the N-bit word is specified, then, since we are assuming that errors on different bits are independent, the probability is

$$P(k \text{ errors in specified locations of an } N\text{-bit word}) = p^k(1 - p)^{N-k}$$

This is because the probability of k bit errors in the k specified locations has a probability of p^k, and the $N - k$ remaining bits have a probability of $(1 - p)^{N-k}$ of being received correctly.

If we are concerned with the probability of k errors without regard to the particular pattern of the errors, we must multiply the probability for a single specific pattern of errors by the number of patterns of k errors in an N bit word. This number can be shown to be

$$\binom{N}{k} = \frac{N!}{(N-k)!k!} \tag{7.14}$$

which is called the *binomial coefficient*. Therefore, the probability of k errors in any k locations of an N-bit word is given by

$$P(k \text{ errors in } N \text{ bits}) = \binom{N}{k} p^k(1 - p)^{N-k} \tag{7.15}$$

where p is the probability of a bit error. Equation 7.15, known as the *binomial probability distribution*, applies to any repeated experiment in which there are two complementary events and the result on each trial is independent of the other trials. For example, it gives the probability of k heads in N tosses of a coin if p is the probability of a head on each toss. Plots of the binomial probability distribution are shown in Figure 7.2 for several values of p and N. For large values of N, note that only values of k, such that k/N is close to p, have significant probability. This is the basis of the intuitive definition of probability as a relative frequency.

The Poisson probability distribution

Another interesting probability distribution is the *Poisson probability distribution*. This distribution often applies when the number of events, of some type, occurring in a given time interval is under consideration. The Poisson probability formula gives the probability of k events in an interval of time τ seconds long. It is given by

$$P(k \text{ events in } \tau \text{ seconds}) = \frac{(a\tau)^k \exp(-a\tau)}{k!} \tag{7.16}$$

The Poisson distribution can be derived by using two assumptions about the events. First, it is assumed that the numbers of events in nonoverlapping intervals are independent. Second, it is assumed that the probability of one event in a differential time interval $d\tau$ is given by $P(\text{one event in } d\tau) = a\,d\tau$, where a is a constant.

Many day-to-day examples approximate these assumptions, for example, the number of traffic accidents in a particular region for the interval between, say, 3:00 and 4:00 p.m. on weekday afternoons. The number of accidents between 3:00 and 3:30 p.m. and the number that occur between 3:30 p.m. and 4:00 p.m. seem, on an intuitive basis at least, to be independent. Intuition also indicates that the probability of an accident in a very short interval of time should be proportional to the length of the interval. Other examples of approximately Poisson processes are the number of fish caught in an hour of fishing and the number of bit errors in binary digital communication.

It can be shown that the parameter a in the Poisson distribution is the average number of events per unit of time. We will give a precise definition of the term *average*, as it is used in probability theory, in the next section, after we introduce some of the concepts associated with random variables.

7.2
RANDOM VARIABLES

Discrete versus continuous random variables

A *random variable* results when a finite real number is assigned to each outcome of an experiment. For example, if the experiment is the toss of a die, the possible outcomes are the six faces that can come up. A random variable associated with this experiment is the number of spots on the face that turns up. Several random variables can be associated with any experiment. In the toss of a die, we could define a second random variable by assigning the value 1 to an odd number of spots and the value 0 to an even number of spots; a third random variable could be the square of the number of spots.

Random variables can be either *discrete* or *continuous*. The examples associated with the toss of a die are all discrete random variables. An example of a continuous random variable is the measured value of the noise voltage across a resistor. (We assume in this discussion that voltage can be measured with infinite precision.) A continuous random variable can take on any value in some range, whereas a discrete random variable is restricted to, at most, a countably infinite set of values. An example of a discrete random variable associated with the measurement of a noise voltage would be one that is defined to be $+1$ if the measured value is positive and -1 if it is negative.

It is standard practice to use an upper-case letter to represent a random variable and the corresponding lower-case letter for the values that the random

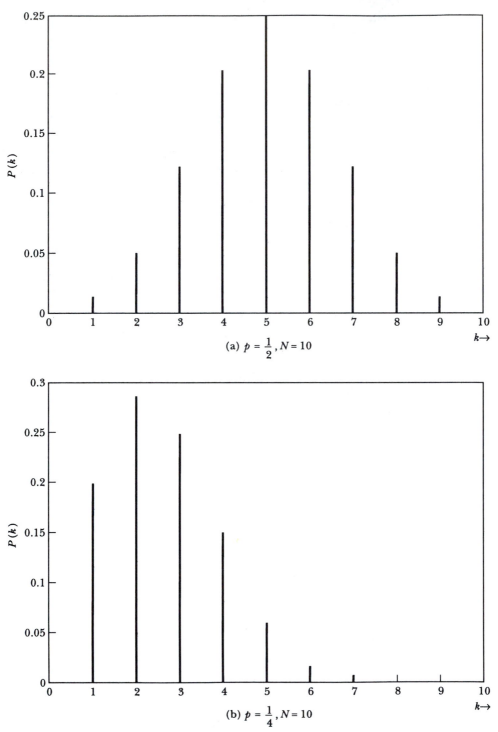

FIGURE 7.2
Binomial probability distribution.

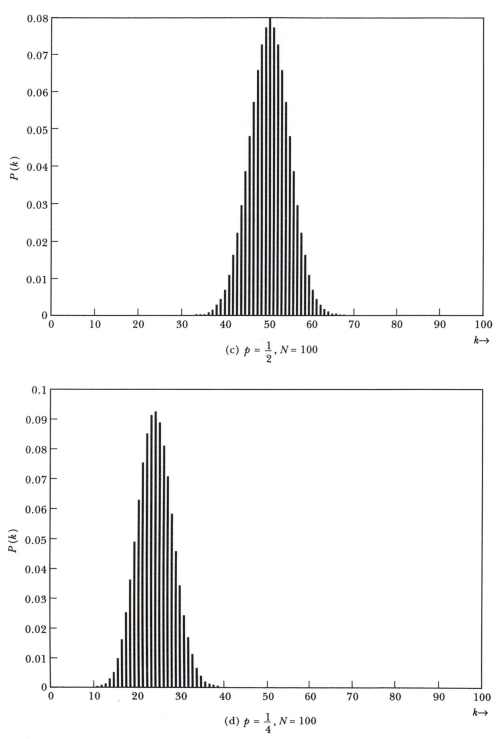

(c) $p = \frac{1}{2}$, $N = 100$

(d) $p = \frac{1}{4}$, $N = 100$

FIGURE 7.2 (*continued*)

variable takes on. Thus, if we use X to denote a random variable, then we use x to denote the values it takes on.

Cumulative distribution functions

The *cumulative distribution function* (CDF) for a random variable X is defined as

$$F_X(x) = P(X \le x) \qquad (7.17)$$

In words, the CDF of the random variable X is the probability of all of the outcomes of an experiment that result in a value for the random variable less than or equal to x. Suppose that the experiment under consideration is the toss of a fair die, and X is a random variable equal to the number of spots that turn up. The CDF for X is shown in Figure 7.3a. Note that the CDF for this

FIGURE 7.3
Cumulative distribution functions.

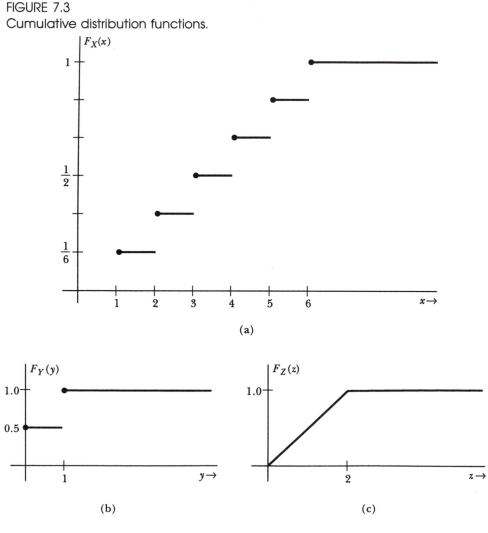

(a)

(b)

(c)

random variable is zero for x less than 1 because there are no possible outcomes of the experiment that result in values for X less than 1. When x is between 1 and 2 the value of the CDF is one-sixth because that is the probability of all of the outcomes that result in a value smaller than x. Suppose that Y is another random variable associated with the die toss that takes the value 1 if an odd number of spots turns up and the value 0 if an even number turns up. The CDF for Y is shown in Figure 7.3b.

These examples are discrete random variables. As an example of a continuous random variable, suppose that an experiment consists of selecting a point "at random" on the real line between the origin and two. The random variable Z is the distance to the point selected from the origin. A possible CDF for this continuous random variable is shown in Figure 7.3c.

CDFs have the following properties:

$$F_X(-\infty) = 0 \qquad (7.18)$$

$$F_X(\infty) = 1 \qquad (7.19)$$

$$F_X(a) \le F_X(b) \qquad \text{if } a < b \qquad (7.20)$$

$$\lim_{x \to a+} F_X(x) = F_X(a) \qquad (7.21)$$

The first property is due to the fact that some finite value is assigned to each outcome of the experiment, so there are no outcomes that result in arbitrarily large negative values. The second property follows from the fact that all outcomes result in finite values for the random variable. The third property shows that the CDF is nondecreasing because it is not possible to have an outcome with negative probability. The last property states that the CDF is continuous from the right. This means that the value of the CDF at a discontinuity is the same as the limit obtained when approaching the discontinuity from the right side. We have indicated this property by the heavy dots at each discontinuity in Figure 7.3a and b.

The probability that a random variable falls in an interval can be found from the CDF as follows:

$$P(a < X \le b) = P(X \le b) - P(X \le a)$$
$$P(a < X \le b) = F_X(b) - F_X(a) \qquad (7.22)$$

Note that a is assumed to be less than b.

Probability density functions

The *probability density function* (PDF) of a random variable is defined as the derivative of the CDF:

$$p_X(x) = \frac{dF_X(x)}{dx} \qquad (7.23)$$

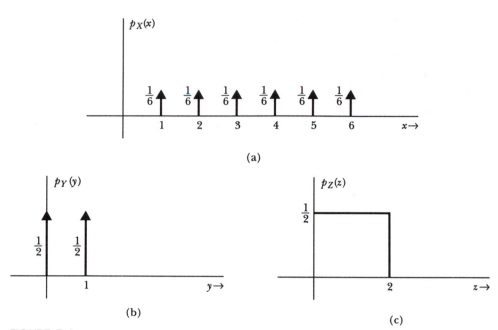

(a)

(b)   (c)

FIGURE 7.4

Probability density functions corresponding to the CDFs of Figure 7.3; X is the number of spots on a die, Y is zero for an even number of spots on a die and 1 for an odd number, and Z is a point selected "at random" between $z = 0$ and $z = 2$.

The PDFs corresponding to the CDFs of Figure 7.3 are shown in Figure 7.4. Note that the PDF contains an impulse at each point where the CDF has a discontinuity. The weight of the impulse is the same as the height of the discontinuity. An impulse in the PDF indicates that the value of x where the impulse occurs has a nonzero probability of occurrence. The height of the discontinuity in the CDF and the weight of the impulse are the same as the probability that X takes the value of x at the location of the impulse.

The PDF of a random variable is large for values of x that are likely to occur and small for unlikely values. Evidently, for random variable Z of Figure 7.4c, each point on the line between 0 and 2 is equally likely to be selected.

The CDF can be found from the PDF by integration:

$$F_X(x) = \int_{-\infty}^{x} p_X(\lambda)\, d\lambda \tag{7.24}$$

A dummy variable of integration λ has been used so that the limit and the variable of integration will not be confused.

The probability that the value of a random variable falls into a given interval can be found by integrating the PDF over the interval. For example,

$$P(a \le X \le b) = \int_{a}^{b} p_X(x)\, dx \tag{7.25}$$

Care should be taken if there are impulses in the PDF at the end points. If the end points of the interval are included, as in the last expression, then any area associated with impulses at a or b should be included in the integral. On the other hand, if we want to find $P(a < X < b)$, we perform the same integral but do not include areas associated with impulses at a or b.

Some properties of PDFs are given by

■
$$p_X(x) \geq 0 \qquad (7.26)$$

and

■
$$\int_{-\infty}^{\infty} p_X(x)\, dx = 1 \qquad (7.27)$$

The first property follows from the fact that the PDF is the derivative of the (nondecreasing) CDF. The second property is true because the integral of the PDF over all values is the probability that the random variable takes on a value between $-\infty$ and ∞. This probability is unity because the random variable assigns a finite value to every outcome of the experiment.

EXAMPLE 7.2

For the random variable X with the CDF shown in Figure 7.5a, find the PDF, $P(X < 1)$, $P(X \leq 1)$, and $P(X = 1.5)$.

SOLUTION
The PDF can be found from the given CDF by differentiation. The result is shown in Figure 7.5b. Note that, due to the discontinuity at $x = 1$ in the CDF, an impulse appears in the PDF at $x = 1$. The weight of the impulse is equal to the height of the discontinuity, which is $\frac{1}{4}$. The meaning of this is that $P(X = 1) = \frac{1}{4}$.

We can find $P(X < 1)$ by integrating the PDF from $-\infty$ to 1. The area of the impulse at $x = 1$ is not included since the point $x = 1$ is not included in the range for which we want to find the probability. The result is

$$P(X < 1) = \frac{1}{4}$$

To find $P(X \leq 1)$, we proceed in the same fashion, but in this case the area of the impulse is included in the integral since the point $x = 1$ is included in the range. The result is

$$P(X \leq 1) = \frac{1}{2}$$

Finally, we conclude that $P(X = 1.5)$ is zero because the area associated with an interval of zero length is zero unless an impulse occurs at the point

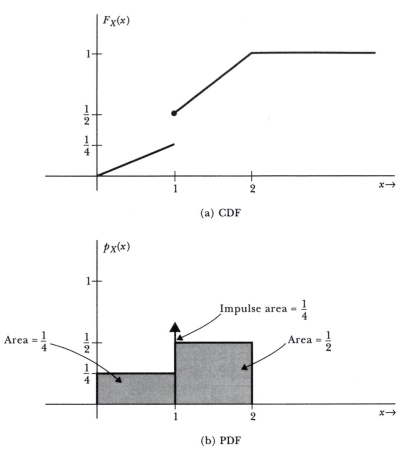

FIGURE 7.5
CDF and PDF of Example 7.2.

in question. Note that even though the probability that $X = 1.5$ is zero, it is not impossible for X to be equal to 1.5. Zero probability implies that the fraction of the trials in which such an event occurs is vanishingly small; however, the event could occur a few times in an infinite number of trials.

Joint random variables

When two random variables X and Y are associated with the same experiment, the *joint CDF* is defined to be

$$F_{XY}(xy) = P(X \le x \text{ and } Y \le y) \qquad (7.28)$$

The *joint PDF* is given by

$$p_{XY}(xy) = \frac{\partial^2 F_{XY}(xy)}{\partial x \, \partial y} \qquad (7.29)$$

Some important properties of the joint CDF and PDF are

$$F_{XY}(x,\infty) = F_X(x) \tag{7.30}$$

$$F_{XY}(\infty,y) = F_Y(y) \tag{7.31}$$

$$F_{XY}(\infty,\infty) = \int_{-\infty}^{\infty} \int_{-\infty}^{\infty} p_{XY}(x,y)\, dx\, dy = 1 \tag{7.32}$$

$$p_X(x) = \int_{-\infty}^{\infty} p_{XY}(x,y)\, dy \tag{7.33}$$

$$p_Y(y) = \int_{-\infty}^{\infty} p_{XY}(x,y)\, dx \tag{7.34}$$

The probability that the outcome of an experiment results in values of X and Y falling in some region of the x-y plane can be found by integrating the joint PDF over the region of interest.

If the joint PDF can be factored into the *marginal PDFs* of the separate random variables, given by

$$p_{XY}(x,y) = p_X(x)p_Y(y) \tag{7.35}$$

then the random variables are said to be *independent*.

The *conditional PDF* of a random variable X, given that the joint random variable Y takes the value y, is given by

$$p_{X|Y}(x,y) = \frac{p_{XY}(x,y)}{p_Y(y)} \tag{7.36}$$

provided that $p_Y(y)$ is not equal to zero. No definition is made for values of y where $p_Y(y)$ is zero. The conditional PDF gives the new PDF of X after it is learned that Y takes a particular value. $p_{Y|X}(x,y)$ is defined in a similar manner. Note that for independent random variables $p_{X|Y}(x,y) = p_X(x)$ and $p_{Y|X}(x,y) = p_Y(y)$. Thus, for independent random variables, the PDF of one random variable does not change when the value of the other random variable is learned.

EXAMPLE 7.3

The joint PDF of two random variables is given by

$$p_{XY}(x,y) = K \quad \text{if } 0 < x < 1 \text{ and } x < y < 1$$
$$= 0 \quad \text{otherwise}$$

as shown in Figure 7.6a. Find the value of K, the marginal PDFs of X and Y, and the conditional PDF $p_{X|Y}(x,y)$

SOLUTION
The value of the constant K can be found by integrating to find the volume under the joint PDF and setting it equal to unity, as required by Equation

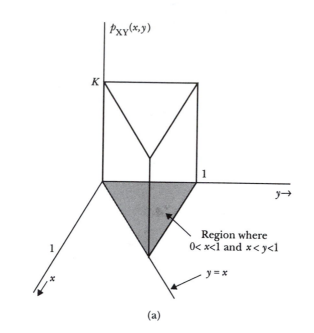

(a)

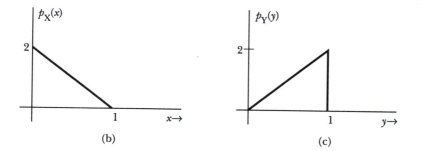

(b)

(c)

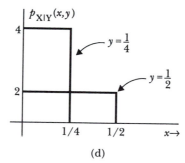

(d)

FIGURE 7.6
PDFs for Example 7.3.

7.32. This results in

$$\int_0^1 \int_0^y K \, dx \, dy = 1$$

The inside integral sign is for the integration with respect to x and the outside sign is for y. Integration and evaluation results in

$$K = 2$$

The marginal PDF for X can be found by integrating the joint PDF with respect to y as given by Equation 7.33. If x is less than zero or greater than unity, the integrand is zero for the entire range of y, and so we have

$$p_X(x) = 0 \qquad \text{for } x < 0 \text{ or for } x > 1$$

For values of x between 0 and 1, Equation 7.33 becomes

$$p_X(x) = \int_x^1 K \, dy$$
$$= K(1 - x) \qquad \text{for } 0 < x < 1$$

A plot of $p_X(x)$ is shown in Figure 7.6b. Note that the area under the plot of $p_X(x)$ is unity, as it should be for any PDF. In a similar fashion, it can be shown that the PDF of Y is given by

$$p_Y(y) = 0 \qquad \text{for } y < 0 \text{ or for } y > 1$$
$$= Ky = 2y \qquad \text{for } 0 < y < 1$$

A plot of $p_Y(y)$ is shown in Figure 7.6c.

The conditional PDF for X, given $Y = y$, is undefined for $y \leq 0$ or $y \geq 1$ since $p_Y(y)$ is zero in those ranges. From Equation 7.36 we can determine that

$$p_{X|Y}(x,y) = \frac{p_{XY}(x,y)}{p_Y(y)}$$
$$= \frac{K}{Ky} = \frac{1}{y} \qquad \text{for } 0 < x < 1 \text{ and } 0 < y < 1$$
$$= 0 \qquad \text{or undefined otherwise}$$

A plot of $p_{X|Y}(x,y)$ is shown versus x for several values of y in Figure 7.6d. Note from the joint PDF of Figure 7.6a that if $y = \frac{1}{4}$, then x ranges from 0 to $\frac{1}{4}$. This is borne out in the plot of the conditional PDF of Figure 7.6d.

Note that since $p_X(x)$ is not equal to $p_{X|Y}(x,y)$, the two random variables are dependent. This is intuitively obvious from the joint PDF shown in Figure 7.6a. For example, if it is known that Y is very small, then the combination of x and y must fall in the left-hand corner of the shaded area. As a result, X must also be very small.

Averages of a single random variable

The *expected value E* (also called the *mean value* or *average value*) of a random variable X is defined as

$$E(X) = m_x = \bar{x} = \int_{-\infty}^{\infty} x p_X(x)\, dx \qquad (7.37)$$

As we have indicated, several forms of notation for the expected value are in common use. If we perform an experiment a large number of times and average the values obtained for the random variable, the average is very close to the value computed from Equation 7.37 with high probability.

Sometimes we will be interested in the expected value of a function of a random variable. This is defined by

$$E[g(X)] = \int_{-\infty}^{\infty} g(x) p_X(x)\, dx \qquad (7.38)$$

The expected value of the nth power of the random variable $E[X^n]$ is called the n*th moment* of the random variable. The n*th central moment* is given by

$$E[(X - m_x)^n] \qquad (7.39)$$

The second central moment is called the *variance* of the random variable, defined by

$$\sigma_x^2 = \text{var}(X) = E[(X - m_x)^2] = E(X^2) - m_X^2 \qquad (7.40)$$

The square root of the variance, or σ_x, is called the *standard deviation* of the random variable. The variance (or equivalently the standard deviation) is a measure of how spread out the values of the random variable are likely to be for repeated trials of the experiment. A large variance indicates a wide spread of values. If the variance is zero, the random variable is almost always equal to its mean value.

The PDF of a discrete random variable consists of a sum of impulses. An impulse occurs at each value of the random variable that can occur, and the weight of the impulse is the probability. Examples were shown in Figure 7.4a and b. Therefore, the PDF of a discrete random variable can be written as

$$p_X(x) = \sum_{i=1}^{N} P(X = x_i)\delta(x - x_i) \qquad (7.41)$$

where x_i denotes the N discrete values that the random variable X takes on.

If the PDF of Equation 7.41 for a discrete random variable is substituted into Equation 7.38, the following result is obtained for the average of a function of a discrete random variable.

$$E[g(X)] = \sum_{i=1}^{N} g(x_i) P(X = x_i) \qquad (7.42)$$

EXAMPLE 7.4

Find the average number of spots that turns up on a fair die.

SOLUTION
The number of spots on a die is a random variable with the PDF shown in Figure 7.4a. The probability of each number is one-sixth. Substituting values into Equation 7.42, we have

$$E[X] = 1\left(\frac{1}{6}\right) + 2\left(\frac{1}{6}\right) + 3\left(\frac{1}{6}\right) + 4\left(\frac{1}{6}\right) + 5\left(\frac{1}{6}\right) + 6\left(\frac{1}{6}\right)$$

$$= 3\frac{1}{2}$$

Note that, as in this case, it is not necessary for the "expected value" to be one of the values the random variable takes on.

EXAMPLE 7.5

Find the expected value, the variance, and the standard deviation for the random variable Z whose PDF is shown in Figure 7.4c.

SOLUTION
The expected value can be found by substituting into Equation 7.37, resulting in

$$m_z = \int_0^2 \frac{1}{2} z\, dz = \frac{1}{4} z^2 \Big|_0^2 = 1$$

The second moment of Z is given by

$$E[Z^2] = \int_0^2 \frac{1}{2} z^2\, dz = \frac{1}{6} z^3 \Big|_0^2 = 1\frac{1}{3}$$

The variance is given by Equation 7.40 as

$$\sigma_z^2 = E[Z^2] - m_z^2 = \frac{1}{3}$$

Finally, the standard deviation is the square root of the variance.

$$\sigma_z = 3^{-1/2}$$

Averages of joint random variables

Averages of functions of joint random variables are computed from the definition:

■
$$E[g(X,Y)] = \int_{-\infty}^{\infty} \int_{-\infty}^{\infty} g(x,y) p_{XY}(x,y)\, dx\, dy \qquad (7.43)$$

The *covariance* of two random variables is defined as

■
$$\mu_{XY} = E[(X - m_x)(Y - m_y)] \qquad (7.44)$$

and the *correlation coefficient* is

■
$$\rho_{XY} = \frac{\mu_{XY}}{\sigma_X \sigma_Y} \qquad (7.45)$$

The correlation coefficient is a measure of potential linear dependence between the two random variables. It can be shown that the correlation coefficient is zero if the random variables are independent. The correlation coefficient is $+1$ if $Y = CX$ where C is a positive constant. If C is negative then the correlation coefficient is -1. Other values indicate a "loose" linear relationship between the random variables.

The expected value of a weighted sum of functions of random variables is equal to the weighted sum of the expected values. In equation form, this is

■
$$E\{ag_1(X) + bg_2(Y)\} = aE[g_1(X)] + bE[g_2(Y)] \qquad (7.46)$$

Thus the expected value operation is linear in the sense that superposition applies.

If two random variables are independent, then the expected value of a product of functions of the random variables is equal to the product of the expected values. This is given by

■
$$E\{g_1(X)g_2(Y)\} = \{E[g_1(X)]\}\{E[g_2(Y)]\} \qquad (7.47)$$

Note that this relationship does not hold in the general case. It should only be applied when X and Y are independent.

It can also be easily shown that the variance of the sum of independent random variables is the sum of the variances. In equation form, this is

■
$$\mathrm{var}(X + Y) = \mathrm{var}(X) + \mathrm{var}(Y) = \sigma_X^2 + \sigma_Y^2 \qquad (7.48)$$

Again note that this result requires the random variables to be independent.

The variance of a constant times a random variable is easily shown to be equal to the constant squared times the variance of the random variable. This is given by

$$\text{var}(aX) = a^2\sigma_X^2 \tag{7.49}$$

■

Transformations of random variables

A random variable that is defined as a function of other random variables is often of interest. This is often called a *transformation of random variables*. The CDF of the new random variable can be found by integrating the PDF of the old random variables. The PDF of the new random variable can then be found by differentiating its CDF.

EXAMPLE 7.6

A new random variable Y is defined as the square of an old random variable X. The PDF for X is shown in Figure 7.7a. Find the PDF of Y.

SOLUTION
First we find the CDF of Y, which is defined as

$$F_Y(y) = P(Y \leq y)$$

Since $Y = X^2$, it is not possible for Y to take on negative values. Thus, for negative values of y, the probability that Y is less than or equal to y is zero. Thus, we can write

$$F_Y(y) = 0 \qquad \text{for } y < 0$$

For positive values of y, the fact that $Y = X^2$ results in $|X| \leq y^{1/2}$ when $Y \leq y$. Thus, we can write

$$
\begin{aligned}
F_Y(y) &= P(Y \leq y) \\
&= P(|X| \leq y^{1/2}) \\
&= P(-y^{1/2} \leq X \leq y^{1/2})
\end{aligned}
$$

$$F_Y(y) = \int_{-y^{1/2}}^{y^{1/2}} p_X(x)\, dx$$

Using Figure 7.7a to find values for the integrand and appropriate limits results in

$$
\begin{aligned}
F_Y(y) &= \int_{-y^{1/2}}^{y^{1/2}} \frac{1}{4}\, dx = \frac{1}{2} y^{1/2} & \text{for } 0 \leq y \leq 1 \\
&= \int_{-1}^{y^{1/2}} \frac{1}{4}\, dx = \frac{1}{4} + \frac{1}{4} y^{1/2} & \text{for } 1 \leq y \leq 9 \\
&= \int_{-1}^{3} \frac{1}{4}\, dx = 1 & \text{for } 9 \leq y
\end{aligned}
$$

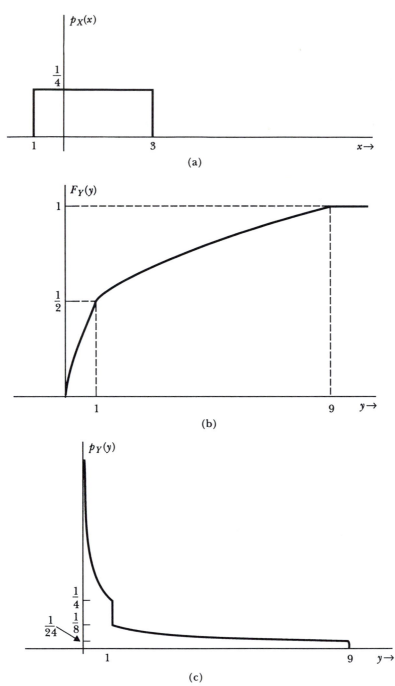

(a)

(b)

(c)

FIGURE 7.7
PDFs and CDF of Example 7.6.

The CDF of y is plotted in Figure 7.7b. The PDF for Y can now be found by differentiating the CDF with the results given by

$$p_Y(y) = 0 \qquad \text{for } y \leq 0$$

$$= \frac{1}{4} y^{-1/2} \qquad \text{for } 0 < y \leq 1$$

$$= \frac{1}{8} y^{-1/2} \qquad \text{for } 1 < y \leq 9$$

$$= 0 \qquad \text{for } 9 < y$$

The PDF for Y is plotted in Figure 7.7c.

Sum of random variables

Another transformation of random variables of interest is when a new random variable Z is formed by adding two old random variables X and Y. We now find an expression for the PDF of Z in terms of the joint PDF of X and Y. First, we note that the CDF of Z is given by

$$F_Z(z) = P(Z \leq z) \tag{7.50}$$

The condition $Z \leq z$ is equivalent to the condition that the values of x and y fall in the shaded region of Figure 7.8. The probability that x and y fall in the shaded region can be found by integrating the joint PDF of x and y over the shaded region. This results in

$$F_Z(z) = \int_{-\infty}^{\infty} \int_{-\infty}^{z-y} p_{XY}(x, y) \, dx \, dy \tag{7.51}$$

FIGURE 7.8
Range of Integration for Equation 7.51.

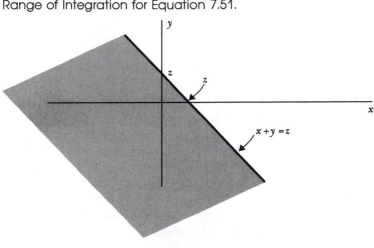

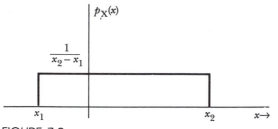

FIGURE 7.9
PDF of a uniform random variable.

where, as usual, the inside integral sign is for the first differential (dx in this case). Taking the derivative of Equation 7.51 with respect to z gives the PDF for Z as

$$p_Z(z) = \int_{-\infty}^{\infty} p_{XY}(z - y, y) \, dy \tag{7.52}$$

If the two random variables X and Y are independent, then their joint PDF can be factored, and we obtain

$$p_Z(z) = \int_{-\infty}^{\infty} p_X(z - y) p_Y(y) \, dy \tag{7.53}$$

Note that Equation 7.53 is a convolution integral. Thus, we see that when independent random variables are added, their PDFs are convolved to obtain the PDF of the sum. This result can be extended to more than two random variables by chaining the convolution operations. The result of convolving the first two PDFs is convolved with the third PDF and so forth.

The uniform PDF

The *uniform PDF* is shown in Figure 7.9. A uniform random variable has a range of values between x_1 and x_2 that are equally likely and has zero probability of being outside that range.

The gaussian PDF

A *gaussian* random variable has a PDF given by

$$p_X(x) = \frac{1}{(2\pi\sigma_x^2)^{1/2}} \exp\left(-\frac{(x - m_x)^2}{2\sigma_x^2}\right) \tag{7.54}$$

where, as before, m_x is the mean value of X and σ_x is the standard deviation. A plot of the gaussian PDF is shown in Figure 7.10. Note that the gaussian density function is the familiar bell curve centered at the mean value m_x and with a width determined by the standard deviation σ_x. The gaussian PDF is of importance because samples of many of the noise signals encountered in communication systems are gaussian random variables.

7 / PROBABILITY, RANDOM VARIABLES, AND RANDOM PROCESSES

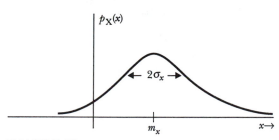

FIGURE 7.10
Gaussian PDF.

The Q function

The area under the "tail" of a gaussian PDF having a zero mean and unit variance as indicated in Figure 7.11 is defined as the Q *function*, given by

$$Q(z) = \int_z^\infty \frac{1}{(2\pi)^{1/2}} \exp\left(\frac{-x^2}{2}\right) dx \qquad (7.55)$$

A plot of the Q function is shown in Figure 7.12. The Q function is closely related to the *error function*, defined by

$$\mathrm{erf}\,(z) = \frac{2}{\pi^{1/2}} \int_0^z \exp\,(-x^2)\, dx \qquad (7.56)$$

The Q function in terms of the error function is given as

$$Q(z) = \frac{1}{2} - \frac{1}{2}\, \mathrm{erf}\left(\frac{z}{2^{1/2}}\right) \qquad (7.57)$$

An approximation for $Q(z)$ that is often sufficiently accurate for our purposes when $z > 3$ is given by

$$Q(z) \simeq \frac{1}{z(2\pi)^{1/2}} \exp\left(\frac{-z^2}{2}\right) \qquad (7.58)$$

FIGURE 7.11
Definition of the Q function.

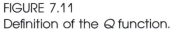

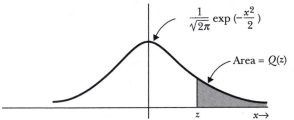

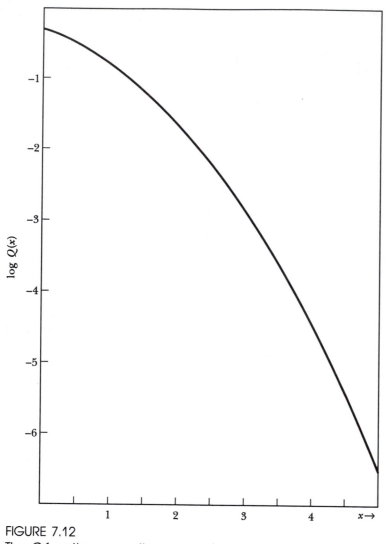

FIGURE 7.12
The Q function versus its argument.

The central limit theorem

The gaussian PDF often turns up when dealing with random physical quantities that result from the sum of many independent random quantities. For example, the scores earned by a class of students on an exam is often an approximately gaussian distribution.

The *central limit theorem* is an important mathematical result that, roughly speaking, states that a random variable resulting from the sum of many small independent contributions is approximately gaussian. Thus, it is not surprising that thermal noise, which results from the independent motion of many individual electrons in a resistor, turns out to be gaussian. As another example, note the similarity of the gaussian PDF to the plots of the binomial probability distribution in Figure 7.2 as N becomes large.

EXAMPLE 7.7

Three independent random variables, X_1, X_2, and X_3, that are uniformly distributed between -1 and $+1$ are added to form a new random variable Y. Find the PDF for Y and compare it with a gaussian PDF having the same mean and variance.

SOLUTION

The PDFs of each of the X random variables are the same. This PDF is shown in Figure 7.13a. The PDF of the sum of independent random variables is found by convolving the PDFs of the terms in the sum. First we can find the PDF of $Z = X_1 + X_2$ by convolving the uniform PDF in Figure 7.13a with itself. The result is shown in Figure 7.13b. Next the PDF of $Y = Z + X_3$ is found by convolving the PDF in Figure 7.13a with the PDF in Figure 7.13b. The result is

$$
\begin{aligned}
p_Y(y) &= 0 && \text{for } y \leq -3 \\
&= \frac{1}{16}(3+y)^2 && \text{for } -3 \leq y \leq -1 \\
&= \frac{1}{16}(6 - 2y^2) && \text{for } -1 \leq y \leq 1 \\
&= \frac{1}{16}(3-y)^2 && \text{for } 1 \leq y \leq 3 \\
&= 0 && \text{for } 3 \leq y
\end{aligned}
$$

This result is plotted as the solid line in Figure 7.13c.

The mean of Y can be found from the PDF for Y by the use of Equation 7.37 or from

$$E(Y) = E(X_1 + X_2 + X_3) = E(X_1) + E(X_2) + E(X_3) = 0$$

The variance of Y can be computed directly from the PDF for Y or by application of Equation 7.48, resulting in

$$\text{var}(Y) = \text{var}(X_1) + \text{var}(X_2) + \text{var}(X_3)$$

which is valid in this case because the X random variables are independent. Since the X random variables are identically distributed, they all have the same variance, given by

$$\text{var}(X_1) = E[(X_1)^2] - (m_{x_1})^2 = E[(X_1)^2] = \int_{-1}^{1} \frac{1}{2}(x_1)^2\, dx_1 = \frac{1}{3}$$

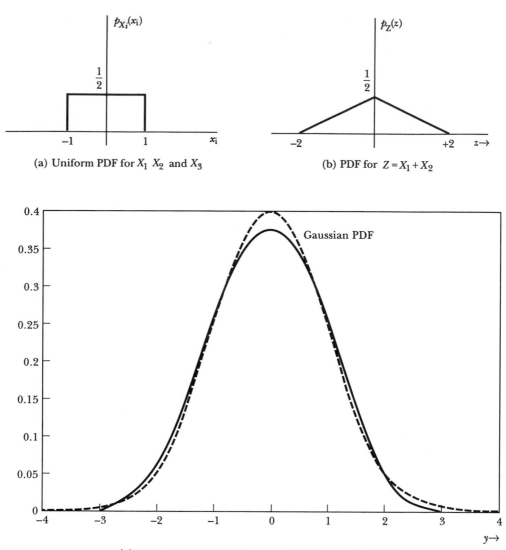

(a) Uniform PDF for X_1 X_2 and X_3

(b) PDF for $Z = X_1 + X_2$

(c) PDF of $Y = X_1 + X_2 + X_3$ compared with gaussian PDF

FIGURE 7.13
PDFs for Example 7.7.

Thus, the variance of Y is given by

$$\sigma_y^2 = 3 \text{ var }(X_1) = 1$$

A gaussian PDF with a mean of zero and unit variance is plotted in Figure 7.13c as the dotted line. Note that the effect of the central limit theorem is very noticeable in this case, and the PDF of the sum is nearly the same as the gaussian distribution.

Jointly gaussian random variables

Two random variables X and Y are *jointly gaussian* if they have a joint PDF given by

$$p_{XY}(x,y) = \frac{1}{2\pi\sigma_x\sigma_y(1-\rho_{xy}^2)^{1/2}} \exp\left[-\frac{(x-m_x)^2}{2\sigma_x^2(1-\rho_{xy}^2)} - \frac{(y-m_y)^2}{2\sigma_y^2(1-\rho_{xy}^2)} \right.$$
$$\left. + \frac{\rho_{xy}(x-m_x)(y-m_y)}{\sigma_x\sigma_y(1-\rho_{xy}^2)} \right]$$

where the symbols m_x, σ_x, and so on, have their usual meanings, and ρ_{xy} is the correlation coefficient as defined in Equation 7.45. When the correlation coefficient is zero, the last term in the exponent is zero, and the joint PDF can be factored into the product of gaussian PDFs for X and Y. Thus, when jointly gaussian random variables are uncorrelated (i.e., their correlation coefficient is zero), they are independent. This is not true in the general case for random variables that are not jointly gaussian.

It is possible for three or more random variables to be jointly gaussian, but the PDF is cumbersome to write unless matrix notation is used. We will not need this level of complexity in this book.

7.3
RANDOM PROCESSES

Specification of a random process

When a time waveform, called a *sample function*, is assigned to each outcome of a random experiment, we say that we have a *random process*, also called a *stochastic process*. A trivial example is to assign a deterministic waveform to each face of a die. The experiment is the toss of the die, the outcomes are the six faces of the die, and the six waveforms are the sample functions of the random process.

The collection of all of the possible sample functions is called the *ensemble*. We will encounter a random process when we consider a message source whose output we desire to transmit through a communication system. Every time we select one of the possible message sources, we are presented with a different sample function or message. Another example of a random process occurs when we use a communication system, and a different noise waveform is added to the transmitted signal every time the system is used. The noise waveform on each occasion will be selected from an infinite collection or ensemble of possible waveforms.

If we take a value from the sample functions at some time, we have a numerical value or random variable associated with each outcome of the experiment. In fact, if a number of values are measured at different times, a collection of random variables results. We denote a random process by upper

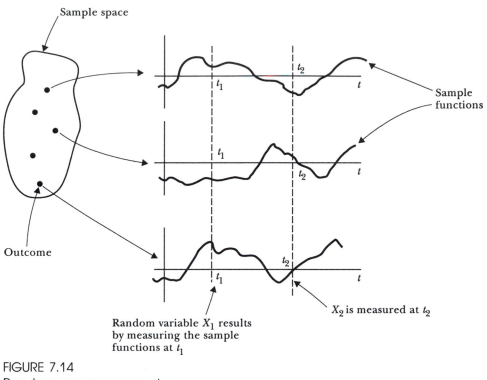

Sample space

Outcome

Sample functions

t_2

t_1

t

t_1

t_2

t

t_2

t_1

t

X_2 is measured at t_2

Random variable X_1 results by measuring the sample functions at t_1

FIGURE 7.14
Random process concepts.

case symbols such as $X(t)$. (In effect, a random process is a random variable that is a function of time, and the notation indicates the time dependence.) In addition, we will denote the value obtained from the random process by taking a measurement at t_1 by $X_1 = X(t_1)$. If a second measurement is taken at t_2, we have a second random variable $X_2 = X(t_2)$ and so forth. Figure 7.14 illustrates the concepts of the sample space of the random experiment, the outcomes of the experiment, the associated sample functions of the random process, and the random variables resulting from taking two measurements of the sample functions.

A *complete statistical description* of a random process is available if we can find the joint PDF of any number of measurements. This PDF is denoted by

$$p_{X_1 X_2 \cdots X_N}(x_1, x_2, \cdots, x_N;\, t_1, t_2, \cdots, t_N)$$

As usual, we have used upper-case symbols for random variables and the corresponding lower case symbols for the values that the random variables take on. It turns out, at least in principle, that it is possible to write the PDF of any order for a random process obtained by filtering thermal noise, but fortunately we will not need to do this for a large number of measurements. In general, the PDF is a function of the times $(t_1, t_2, \ldots)$ at which the measurements are taken.

Stationary random processes

An important concept is that of a *stationary random process*. For a stationary process, the joint PDF is not a function of the times at which the measurements are taken. However, even for a stationary process, the joint PDF may (and most probably, will) be a function of the intervals between the measurements. These intervals are $t_1 - t_2$, $t_1 - t_3$, $t_2 - t_3$, and so forth. We most often deal with the joint PDF for two measurements from a random process, and in this case we denote the interval between the two measurements as $\tau = t_2 - t_1$. Thus, for a stationary random process, the second-order PDF is denoted by

$$p_{X_1 X_2}(x_1, x_2, \tau)$$

As the notation indicates, this PDF does not depend on the time that the first measurement is taken if the interval between the two measurements is constant. In a similar fashion, the higher-order PDFs of a stationary process do not depend on the time that the first measurement is taken, but they do depend on the intervals between measurements. Figure 7.15a shows some of the sample functions of a nonstationary random process and Figure 7.15b shows some sample functions of a stationary process. For the most part, we will only deal with stationary processes.

Time averages versus ensemble averages

We are often interested in averages associated with a random process. The average value of many characteristics associated with a random process can be computed in two ways; either as a *time average* or as an *ensemble average*. For example, if we are interested in the average value of a random process $X(t)$, we can select one of the sample functions denoted by $x(t)$ and compute its time average, given by

$$\langle x(t) \rangle = \lim_{T \to \infty} \frac{1}{T} \int_{-T/2}^{T/2} x(t)\, dt \qquad (7.59)$$

where we have used the notation $\langle\ \rangle$ to denote the time-averaging process. As indicated in the equation, time averaging consists of integration followed by division by the interval of integration in the limit as the interval approaches infinity in both directions.

On the other hand, we could use the first-order PDF to compute an ensemble average of the random process, given by

$$E[X(t)] = m_x(t) = \int_{-\infty}^{\infty} x p_X(x, t)\, dx \qquad (7.60)$$

where we have denoted the time of the measurement as t rather than t_1 since only one measurement is under consideration. Notice that the ensemble average can be a function of time if the random process is nonstationary. For a stationary random process, the ensemble average of Equation 7.60 is not a function of time.

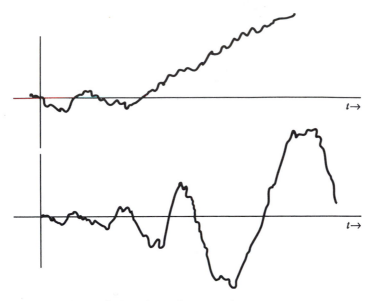

(a) Sample functions of a nonstationary process

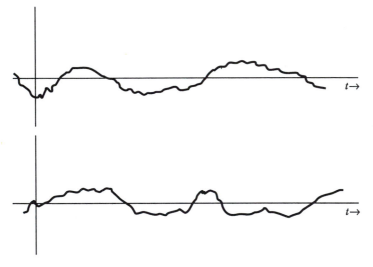

(b) Sample functions of a stationary process

FIGURE 7.15
Sample functions of random processes.

The time average of Equation 7.59 is not a function of time even if the process is nonstationary since the time variable drops out when the integration is performed. Notice, however, that the time average of Equation 7.59 may not be the same for different sample functions. (In this case, $\langle x(t) \rangle$ is a random variable.)

Ergodic random variables

An *ergodic random process* has the characteristic that time averages are the same for all sample functions and equal to the corresponding ensemble averages. For a process to be ergodic, it must be stationary. It can be a difficult mathematical exercise to prove that a random process is ergodic; and often in engineering we must assume ergodicity intuitively and proceed without the benefit of a formal proof.

EXAMPLE 7.8

A random process is generated by selecting an adjustable dc power supply and adjusting its output voltage "at random" to a value between zero and 10 V. The sample functions are constant values as indicated in Figure 7.16. Write expressions for the PDFs of first and second order for this process. Is this random process stationary? Is it ergodic? Compute the average voltage first as an ensemble average and second as a time average using sample functions 1 and 2.

SOLUTION

If we take a single measurement of the sample functions at t_1, the PDF of the resulting random variable does not depend on t_1 because the sample functions are constant with time, and the same value results regardless of the value of t_1. If we interpret the phrase "at random" to mean that all voltages

FIGURE 7.16
Some of the sample functions of the random process of Example 7.8.

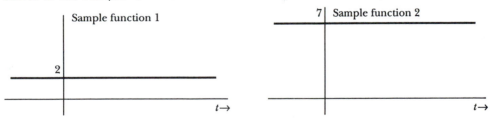

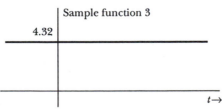

between zero and ten are equally likely, the first-order PDF is given by

$$p_{X_1}(x_1,t_1) = \frac{1}{10} \qquad \text{for } 0 < x_1 < 10$$

$$= 0 \qquad \text{otherwise}$$

A second measurement (from the same sample function) has exactly the same value as the first measurement since the sample functions are constant with time. Therefore, the conditional PDF of X_2, given X_1, is

$$p_{X_2|X_1}(x_1,x_2,t_1,t_2) = \delta(x_1 - x_2)$$

The meaning of this is simply that X_2 is equal to the given value for X_1 with a probability of one. Now we can write the second-order PDF as

$$p_{X_1 X_2}(x_1,x_2,t_1,t_2) = p_{X_2|X_1}(x_1,x_2,t_1,t_2)p_{X_1}(x_1,t_1)$$

$$= \frac{1}{10} \delta(x_1 - x_2) \qquad \text{for } 0 < x_1 < 10 \text{ and } 0 < x_2 < 10$$

$$= 0 \qquad \text{otherwise}$$

In a similar fashion, we could find higher-order PDFs, and like the second-order PDF, they would be independent of the measurement times t_1, t_2, and so on. Therefore, this random process is stationary. (Probably this fact should be intuitively obvious since the sample functions are constant with time.)

To compute the average value of the random process as an ensemble average, we use Equation 7.60 and the first-order PDF found above (with $t = t_1$). This results in

$$E[X(t)] = \int_0^{10} x\left(\frac{1}{10}\right) dx = 5$$

Since the voltages are equally likely to be any value between 0 and 10, the ensemble average of 5 is intuitively appealing. Notice that the ensemble average corresponds to the "real world" procedure of measuring the voltages of many sample functions at a given time and then averaging the measurements.

The time average can be calculated from Equation 7.61 for a given sample function. The result for sample function 1 is 2 V, and for sample function 2 it is 7 V. Notice that these values are different for each sample function and different from the ensemble average. Thus, this random process is not ergodic since the time averages and the ensemble average are not equal.

DC power, total power, and ac power of a random process

The ensemble average of a random process $m_x(t)$, given by Equation 7.60 is called the *mean function* of a random process. It is constant for an ergodic process and equal to the value computed as a time average from any of the sample functions. The mean value for an ergodic process can be interpreted as the *dc component* of the sample functions.

Other averages are of interest and can also be computed either as time or ensemble averages. For example, the average normalized power of a sample function can be computed as a time average:

$$\blacksquare \qquad P_{\text{avg}} = \langle [x(t)]^2 \rangle = \lim_{T \to \infty} \frac{1}{T} \int_{-T/2}^{T/2} [x(t)]^2 \, dt \qquad (7.61)$$

The corresponding ensemble average is given by

$$\blacksquare \qquad E[[X(t)]^2] = \int_{-\infty}^{\infty} x^2 p_X(x,t) \, dx \qquad (7.62)$$

For an ergodic process, the results of Equations 7.61 and 7.62 are the same. Note that Equation 7.62 is the second moment of a measurement of the random process. Thus, for an ergodic process we can interpret the second moment as the *total power* in the sample functions. Since m_x given by Equation 7.60 is the dc component of the sample functions, we can interpret m_x squared as the *dc power*. The variance of a measurement of the sample functions is given by Equation 7.40 as

$$\blacksquare \qquad \sigma_x^2 = E(X^2) - (m_x)^2 \qquad (7.63)$$

Evidently the variance is the total power in the sample functions minus the dc power. Thus, we can interpret the variance as the *ac power*. The standard deviation, which is the square root of the variance, is the rms value of the ac components in a sample function.

The autocorrelation function and power spectral density

Another average of random processes of importance is the autocorrelation function given as a time average in Equation 2.29, which is repeated here for convenience:

$$\blacksquare \qquad R_x(\tau) = \langle x(t)x(t+\tau) \rangle = \lim_{T \to \infty} \frac{1}{T} \int_{-T/2}^{T/2} x(t)x(t+\tau) \, dt \qquad (7.64)$$

Recall that the autocorrelation function is important because its Fourier transform is the power spectral density (PSD) of the time signal $x(t)$.

The autocorrelation function can also be computed as an ensemble average given by

$$R_x(t_1, t_1 + \tau) = E[X(t_1)X(t_1 + \tau)]$$

$$= \int_{-\infty}^{\infty} \int_{-\infty}^{\infty} x_1 x_2 p_{X_1 X_2}(x_1, x_2, t_1, t_2) \, dx_1 \, dx_2 \qquad (7.65)$$

where $t_2 = t_1 + \tau$. When the process is stationary, the second-order joint PDF inside the integral of Equation 7.65 depends only on $\tau = t_1 - t_2$. In this case, the autocorrelation function, computed as an ensemble average, is only a function of τ, and we can write

■
$$R_x(\tau) = E[X(t_1)X(t_1 + \tau)]$$
$$= \int_{-\infty}^{\infty} \int_{-\infty}^{\infty} x_1 x_2 p_{X_1 X_2}(x_1, x_2, \tau) \, dx_1 \, dx_2 \qquad (7.66)$$

If the process is ergodic as well as stationary, then the autocorrelation function computed as an ensemble average using Equation 7.66 is the same as the time average computed using Equation 7.64. Some authors use different symbols for the autocorrelation function computed as a time average and the autocorrelation function computed as an ensemble average. We use the same symbol for both functions because our main interest is in ergodic processes, in which the results are the same.

In our theoretical considerations, we often compute averages as statistical ensemble averages. When we make measurements of a dc component, average power, or PSD (using a spectrum analyzer), we must deal with a time average since it is not feasible to generate the ensemble of waveforms needed for an ensemble average. Thus, in theoretical work, it is often easiest to compute ensemble averages, whereas it is often the corresponding time averages that are of importance in the "real world". As long as we are dealing with ergodic processes, the results agree.

As we showed in Section 2.4, the Fourier transform of the autocorrelation function is the PSD of a signal. The PSD gives the distribution of the power in the signal as a function of frequency. The relationship between the autocorrelation function and the PSD is denoted by

■
$$R_x(\tau) \leftrightarrow S_x(f) \qquad (7.67)$$

If $X(t)$ is a random process at the input to a linear system with transfer function $H(f)$, and the random process resulting at the system output is $Y(t)$, then

■
$$S_y(f) = |H(f)|^2 S_x(f) \qquad (7.68)$$

This relationship was developed in Section 2.5 as Equation 2.38.

Gaussian random processes

A *gaussian random process* has PDFs of any order that are jointly gaussian. If the input to a linear system is a gaussian random process, then the output of the system is also a gaussian random process. It is possible to write the joint PDF of any order for a gaussian random process if the autocorrelation function is known. Thermal noise is a gaussian random process.

EXAMPLE 7.9

A 50 Ω resistor at a temperature of 290 K is in parallel with a 2 pF capacitor. Find the PSD of the noise signal across the terminals of the combination, the autocorrelation function of the noise signal, the first-order PDF of an instantaneous sample of the noise signal, and the probability that the noise sample value exceeds $V_t = 100 \ \mu V$.

SOLUTION

This circuit was analyzed in Example 5.5 in which the PSD of the output voltage was found to be

$$S_y(f) = \frac{2kTR}{[1 + (f/f_b)^2]}$$

where $k = 1.38 \times 10^{-23}$ J/K is Boltzmann's constant, T is the temperature of the resistor, and f_b is given by

$$f_b = \frac{1}{2\pi RC}$$

The autocorrelation function of the output signal is the inverse Fourier transform of the PSD given by

$$R_y(\tau) = \frac{kT}{C} \exp\left(-\frac{|\tau|}{RC}\right)$$

where we have made use of one of the entries in Table 2.2 to find the inverse transform.

We can see from the expression above for $S_y(f)$ that the PSD of the output voltage does not have an impulse at zero frequency. Therefore, $Y(t)$ does not contain a dc component. Thus, m_y is zero.

The second moment of a sample of the output voltage is the total power in $Y(t)$, given by

$$E(Y^2) = R_y(0) = \frac{kT}{C}$$

The variance of $Y(t)$ is given by

$$\sigma_y^2 = E(Y^2) - m_y^2 = \frac{kT}{C} = 1.98 \times 10^{-9} \text{ W}$$

This agrees with the total power in $Y(t)$ found in Example 5.5 by integrating the PSD over all frequencies. (Since there is no dc power, the total power is ac power.)

Since this filtered noise signal is gaussian, its PDF is given by Equation 7.54. With appropriate changes in notation this is

$$p_Y(y) = \frac{1}{(2\pi\sigma_y^2)^{1/2}} \exp\left(-\frac{y^2}{2\sigma_y^2}\right)$$

where we have made use of the fact that $m_y = 0$. The probability that the output voltage exceeds V_t can now be found by integrating the PDF. The integral is

$$P(Y > V_t) = \int_{V_t}^{\infty} p_Y(y)\, dy$$

$$P(Y > V_t) = \frac{1}{(2\pi\sigma_y^2)^{1/2}} \int_{V_t}^{\infty} \exp\left(-\frac{y^2}{2\sigma_y^2}\right) dy$$

If we make the substitution $x = y/\sigma_y$, the integral becomes

$$P(Y > V_t) = \int_{V_t/\sigma_y}^{\infty} (2\pi)^{-1/2} \exp\left(-\frac{x^2}{2}\right) dx$$

The last expression can be written in terms of the Q function defined in Equation 7.55 to obtain

$$P(Y > V_t) = Q\left(\frac{V_t}{\sigma_y}\right)$$

Substituting values and evaluating the Q function, we obtain

$$P(Y > 100\ \mu\text{V}) = Q(2.25) = 0.0123$$

Bandpass noise

In Section 5.1, a description of bandpass noise as a sinusoid with a slowly varying amplitude and phase was presented. Equation 5.9 for the sample function (we did not use the term *sample function* in Chapter 5) of a bandpass noise signal was given as

$$n(t) = a_n(t) \cos\left[2\pi f_c t + \theta_n(t)\right] \tag{7.69}$$

where $a_n(t)$ is the slowly varying amplitude and $\theta_n(t)$ is the slowly varying phase. We are now in a position to give an alternative explanation of why this description of narrowband noise is correct.

Consider the narrowband noise signal resulting from passing white noise with PSD $S_n(f) = N_0/2$ through an ideal bandpass filter. The PSD of the output noise signal is shown in Figure 7.17a. The autocorrelation function of the noise signal can be found by taking the inverse Fourier transform of the PSD, resulting in

$$R_n(\tau) = N_0 B \text{ sinc } (B\tau) \cos (2\pi f_c \tau) \tag{7.70}$$

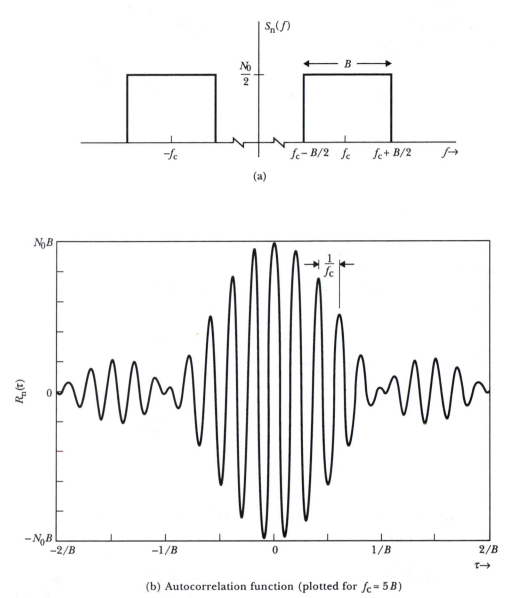

(a)

(b) Autocorrelation function (plotted for $f_c = 5B$)

FIGURE 7.17
PSD and autocorrelation function of bandpass noise.

A plot of $R_n(\tau)$ is shown in Figure 7.17b. Notice that the values of the auto-correlation function at $\tau = 1/f_c$ and $\tau = 2/f_c$ are nearly equal to $R_n(0)$. Thus, we can write

$$R_n(0) = E[n(t)n(t)] \simeq R_n\left(\frac{1}{f_c}\right) = E\left[n(t)n\left(t + \frac{1}{f_c}\right)\right]$$

This implies that $n(t) \simeq n(t + 1/f_c)$, which is exactly what we would expect for a sinusoidal signal with a frequency of f_c and a slowly varying amplitude and phase. This is because the noise signal is almost periodic for short intervals of time.

A sinusoid with a fixed amplitude and phase, given by

$$A \cos (2\pi f_c t + \theta)$$

has an autocorrelation function given by

$$\frac{A^2}{2} \cos (2\pi f_c \tau)$$

Notice that for small values of τ the sinc function in Equation 7.70 is approximately equal to unity; thus for small values of τ, the autocorrelation function of narrow-band bandpass noise has the same functional form as the autocorrelation function of a constant amplitude sinusoid. Thus, we again have the implication that bandpass noise is a sinusoid with a slowly varying phase and amplitude.

7.4
DIGITAL SIGNALS AND THEIR PSDs

In this section, we first illustrate the calculation of the autocorrelation function of one of the most common baseband binary digital signaling formats (the NRZ format) as an ensemble average. Next we take the Fourier transform of the autocorrelation function to find the PSD. Then we illustrate the calculation of the autocorrelation function as a time average. Finally, we present other signaling formats and their PSDs.

The autocorrelation function of the NRZ data signal as an ensemble average

The first signaling format to be considered is the *nonreturn-to-zero* (NRZ) format, illustrated in the waveforms of Figure 7.18. As indicated in the figure, the NRZ waveform has an amplitude of $+A$ for the entire bit interval T_b when the data bit is a 1. The NRZ waveform is $-A$ when the data bit is a 0.

Since we want to compute the autocorrelation function as an ensemble average, we must consider an ensemble of sample functions as indicated in Figure 7.18. We will consider the case in which the data sequence of 1s and 0s is random, with $P(1) = P(0) = \frac{1}{2}$, and the data are independent from bit interval to bit interval and from sample function to sample function. If the data sequence has a predominate number of 1s or 0s or if a recurring pattern of data symbols occurs frequently, the PSD is affected. The case of independent and equally likely data bits is often encountered in practice, so its consideration is of some importance. We will not consider other cases in detail.

7 / PROBABILITY, RANDOM VARIABLES, AND RANDOM PROCESSES

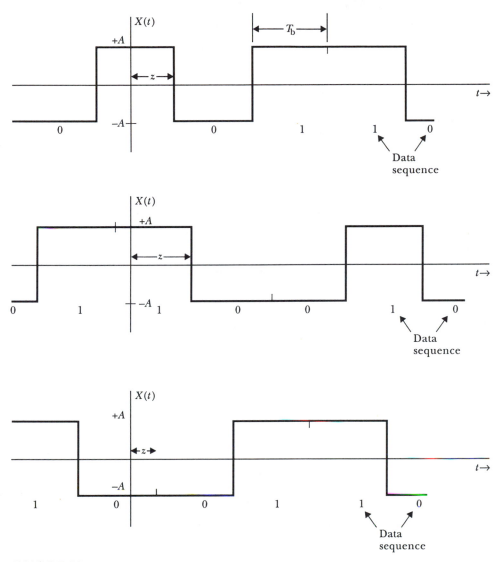

FIGURE 7.18
Sample functions from an ensemble of NRZ waveforms; Z is a random variable with a different value for each sample function.

If the bit transitions of all of the sample functions of the ensemble were time aligned, we would have a nonstationary random process. Then the auto-correlation function for this random process would be a function of both t_1 and t_2, whereas the autocorrelation function computed as a time average would only be a function of $\tau = t_2 - t_1$. Thus, the random process would not be ergodic, and we would not find the desired result from the ensemble average. Therefore, as indicated in Figure 7.18, the instant z of the first data-bit transition after $t = 0$ is different for each sample function. To obtain a

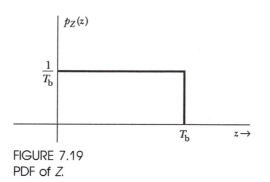

FIGURE 7.19
PDF of Z.

stationary (and hopefully ergodic) random process for consideration, we define Z to be a random variable that is uniformly distributed between 0 and T_b. Note that the sample functions change in amplitude at $t = z$ only if the data bit changes from 1 to 0 or from 0 to 1 at that point. The PDF for Z is shown in Figure 7.19.

The autocorrelation function as an ensemble average was given in Equation 7.66 as

$$R_x(\tau) = E[X(t_1)X(t_1 + \tau)] \qquad (7.71)$$

Since the process under consideration is stationary, the result is independent of the value of t_1, and we can assume for simplicity that $t_1 = 0$. Note that $X(0)$ can take on only two values: $+A$ and $-A$. Similarly, $X(\tau)$ takes on only two values.

Since we know that the autocorrelation function is an even function, we consider only positive values for τ in our analysis. Later, we will extend the results to negative values of τ by the relation

$$R_x(\tau) = R_x(-\tau) \qquad (7.72)$$

Now, since we consider only positive values of τ and have taken $t_1 = 0$, we can write

$$R_x(\tau) = E[X(0)X(\tau)] \qquad (7.73)$$

The outcomes resulting from selecting a sample function and measuring its value at $t = 0$ and $t = \tau$ can be divided into two mutually exclusive categories. First, there is the category in which there is no change to the next data bit between 0 and τ or, equivalently, in which z is greater than τ. Second, there is the category in which there is one or more bit transitions between 0 and τ or, equivalently, in which z is less than τ. If the average is found for each category, then the overall average for both cases is

$$R_x(\tau) = E[X(0)X(\tau)|Z > \tau]P[Z > \tau] + E[X(0)X(\tau)|Z < \tau]P[Z < \tau] \qquad (7.74)$$

where $E[X(0)X(\tau)|Z > \tau]$ denotes the average given that there is no bit transition between 0 and τ. (Notice that a bit transition does not result in a change in amplitude if the successive bits are both ones or both zeroes.) The meaning of the other notation is similar.

When there is no bit transition, we have either

$$X(0) = X(\tau) = +A \qquad \text{or} \qquad X(0) = X(\tau) = -A \qquad (7.75)$$

In this case, the product is always given by $X(0)X(\tau) = +A^2$. Thus

$$E[X(0)X(\tau)|\mathcal{Z} > \tau] = A^2 \qquad (7.76)$$

When there is a bit transition between 0 and τ, $X(0)$ and $X(\tau)$ are independent. In this case, the average of the product is the product of the averages. Therefore, we can write

$$E[X(0)X(\tau)|\mathcal{Z} < \tau] = E[X(0)|\mathcal{Z} < \tau]E[X(\tau)|\mathcal{Z} < \tau] \qquad (7.77)$$

Now, the value of $X(0)$ is independent of $\mathcal{Z}$ and τ, so we can write

$$E[X(0)|\mathcal{Z} < \tau] = E[X(0)] = 0 \qquad (7.78)$$

The average value of $X(0)$ is zero because $X(0)$ takes on the values $+A$ and $-A$ with equal probability. Substituting Equations 7.76, 7.77, and 7.78 into 7.74, we obtain

$$R_x(\tau) = A^2 P[\mathcal{Z} > \tau] \qquad (7.79)$$

Now, the probability that $\mathcal{Z}$ is greater than τ can be found by integrating the PDF of $\mathcal{Z}$ from τ to infinity. The result is given by

$$P(\mathcal{Z} > \tau) = \int_\tau^{T_b} \left(\frac{1}{T_b}\right) dz = 1 - \frac{\tau}{T_b} \qquad \text{for } 0 < \tau < T_b$$
$$= 0 \qquad \text{for } T_b < \tau \qquad (7.80)$$

Substituting this result into Equation 7.79 and extending the result to negative values of τ by replacing τ with $|\tau|$, we finally obtain the desired auto-correlation function, given by

$$R_x(\tau) = A^2\left(1 - \frac{|\tau|}{T_b}\right) \qquad \text{for } -T_b < \tau < T_b$$
$$= 0 \qquad \text{otherwise} \qquad (7.81)$$

The autocorrelation function of the NRZ data format is shown in Figure 7.20a.

The PSD of the NRZ signal

The PSD of the NRZ data format can be found by taking the Fourier transform of its autocorrelation function. The result is given by

$$S_x(f) = A^2 T_b \, \text{sinc}^2\,(f T_b) \qquad (7.82)$$

which is plotted in Figure 7.20b.

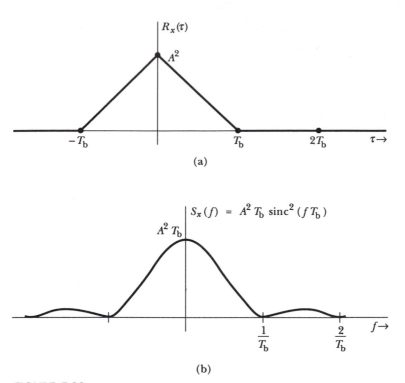

(a)

$$S_x(f) = A^2 T_b \text{ sinc}^2 (f T_b)$$

(b)

FIGURE 7.20
Autocorrelation function and PSD for the NRZ waveform with independent equally likely data.

To illustrate that the PSD is a function of the data sequence, we have plotted the PSD of an NRZ waveform with a data sequence consisting of alternating 1s and 0s in Figure 7.21 (this is a symmetrical square wave that is periodic and therefore has discrete frequency components). Note the difference between the spectrum for random data in Figure 7.20b and the alternating sequence in Figure 7.21.

FIGURE 7.21
PSD of NRZ waveform with alternating data sequence (101010 · · ·). Note that this is very different from the PSD for random data shown in Figure 7.20b.

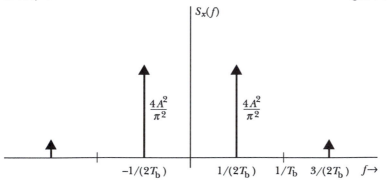

Autocorrelation function of the NRZ data format computed as a time average

Now we again obtain the autocorrelation function of the NRZ data signal by consideration of the time average, given by

$$R_x(\tau) = \lim_{T \to \infty} \frac{1}{T} \int_{-T/2}^{T/2} x(t)x(t + \tau)\, dt \qquad (7.83)$$

where $x(t)$ is a typical sample function of the random process. Notice that Equation 7.83 requires the sample function to be shifted by τ and multiplied by an unshifted version. Then the time average of the product is taken. This is similar to convolution except that no time reversal of the shifted function is required. By experience we find that when we correlate or convolve waveforms that are segments of constant value, the resulting correlation or convolution consists of segments of straight lines. The line segments in the result meet at values of the shift τ for which discontinuities in the shifted function pass by discontinuities in the unshifted function.

A typical sample function of the NRZ data signal is shown in Figure 7.22a, and the shifted version is shown in Figure 7.22b. Note that discontinuities of the shifted version pass by discontinuities in the unshifted version for $\tau = 0, \pm T_b, \pm 2T_b, \ldots$. Thus, the autocorrelation function consists of straight-line segments connecting the results for those values of τ. Therefore, we need to compute only the autocorrelation function for values of τ that are integer multiples of the bit interval T_b. First, we consider zero shift. In this case, we can write

$$x(t)x(t + 0) = x^2(t) = A^2 \qquad \text{for all } t$$

The time average of this is, of course, A^2. Thus $R_x(0) = A^2$, as indicated in Figure 7.20a.

Next, consider the case in which $\tau = T_b$. The shifted waveform $x(t + T_b)$ is shown in Figure 7.22c, and the product with the unshifted waveform is shown in Figure 7.22d. Note in Figure 7.22d that the product $x(t)x(t + T_b)$ is $+A^2$ in half of the intervals and $-A^2$ in the other half (if we consider a typical infinite sequence with independent data from bit to bit). Thus, the time average of the product waveform is zero, and we conclude that $R_x(T_b) = 0$. In a similar fashion, we conclude that the autocorrelation function is zero for all integer (except zero) multiples of T_b. These values are indicated by the heavy dots in Figure 7.20a. Finally, we find the intermediate values by connecting the values at the dots in Figure 7.20a by straight lines. As expected, the time average produces the same result for the autocorrelation function as the ensemble average.

We have presented this discussion of the autocorrelation function of the NRZ data waveform mainly to illustrate the principles of random process theory presented in the previous section. The autocorrelation functions and

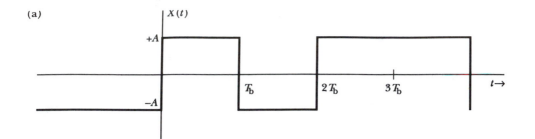

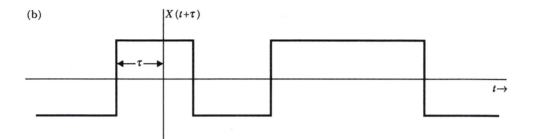

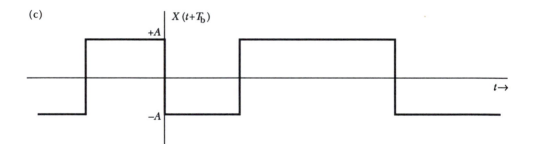

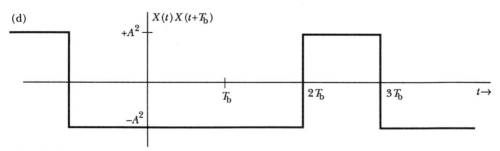

FIGURE 7.22
Sample function of an NRZ waveform, its shifted version, and the product of $x(t)$ and $x(t + T_b)$.

PSDs of many other data waveforms can be found by similar techniques. In some cases, special mathematical approaches must be invoked to achieve the result. A detailed discussion of these is not warranted in an introductory text, but some of the results are presented for digital data formats because the results are useful in their own right.

Signaling formats used for transmission of binary data

Figure 7.23 shows a number of signal formats for transmission of binary digital data. Sometimes these waveforms have been called *line codes*, but this is an unfortunate terminology because the term *code* is also used in other aspects of communication theory where it is more appropriate, such as in error correction coding and cryptology.

Figure 7.23a shows the assumed data sequence for the remaining parts of the figure. Part b shows the clock waveform, which is a square wave with a positive-going edge at the instants when the data changes. A clock waveform is often necessary to control the flow of data in digital systems. As we will see in our discussion of the reception of digital signals in the presence of noise, the clock waveform often must be recovered from the received noisy signal and used to synchronize the extraction of data from the noisy signal. Part c of the figure shows the NRZ waveform, which we have discussed earlier in this section.

Figure 7.23d shows a *biphase* data format, which is also known as a *Manchester code*. The biphase signal can be obtained by multiplying the NRZ waveform by the clock signal. This can be implemented by applying the clock signal and the NRZ signal to the inputs of an exclusive OR gate that is followed by an inverter, as indicated in Figure 7.24a. Note that when the logic levels are $\pm A$, the output of the exclusive OR and inverter combination is the scaled product of the inputs. This can be verified by examination of the outputs of the circuit given in the table in Figure 7.24b. Therefore, the biphase waveform consists of a cycle of the clock waveform when the data bit is a 1. The biphase signal is an inverted clock cycle when the data bit is a 0.

The *unipolar return-to-zero* (RZ) signal is shown in Figure 7.23e. The signal consists of a positive pulse that returns to 0 before the end of the data interval when the data bit is a 1. There is no pulse when the data bit is a 0. Because of the fact that the pulses are all positive, the unipolar RZ signal contains a discrete dc component. An RZ format is useful in minimizing intersymbol interference (ISI) because the response from one pulse has time to decay during the interval before the next pulse.

The *bipolar RZ* signal is shown in Figure 7.23f. It consists of a positive pulse when the data bit is a 1 and a negative pulse when the data bit is a 0. In either case, the pulse returns to 0 before the end of the data interval. When the 1s and 0s are equally likely, there is no discrete dc component in the bipolar RZ signal.

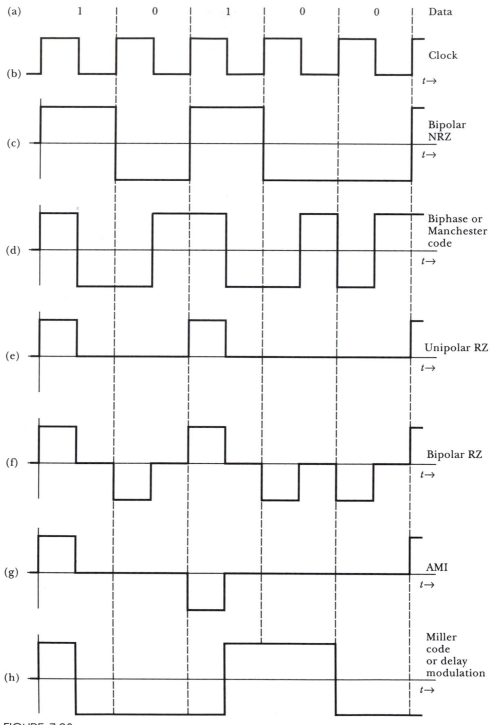

FIGURE 7.23
Binary signal formats.

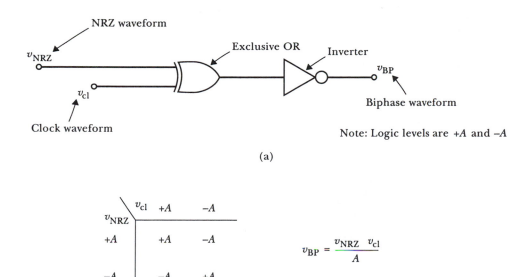

(a)

v_{NRZ} \\ v_{cl}	$+A$	$-A$
$+A$	$+A$	$-A$
$-A$	$-A$	$+A$

$$v_{BP} = \frac{v_{NRZ}\ v_{cl}}{A}$$

(b)

FIGURE 7.24
Logic circuit for producing the biphase data waveform.

The *alternate mark inversion* (AMI) data signal is shown in Figure 7.23g. In this signal, there are pulses only for 1s but they alternate in polarity. Because the signal contains an equal number of positive and negative pulses it does not contain a dc component.

The *Miller code* or *delay modulation* data format is shown in Figure 7.23h. The Miller code makes a transition in amplitude at the center of the data interval when the data bit is a 1. No transition occurs when the data bit is a 0 unless the next bit is also a 0. When the current bit is a 0 and the next bit is also a 0, a transition in amplitude is made at the end of the current bit.

Many other data formats and variations are discussed in the literature. Why are there so many? What are the characteristics of a "good" data format? There are many formats because the channel characteristics vary from application to application, and we are trying to satisfy many different objectives with our choices of data formats. For example, if the channel is ac coupled, we should not choose a format with a large dc component. Some of the points we should consider in selecting a data format are the following:

1.   The spectral characteristics. Does the frequency range for which the PSD of the format is largest match the passband of the channel? Is there a dc component?

2.   The immunity of the format to noise. As we will see in the next chapter, when the noise is white and gaussian, the best choice is to have the signal for a 0 be the negative of the signal for a 1. This

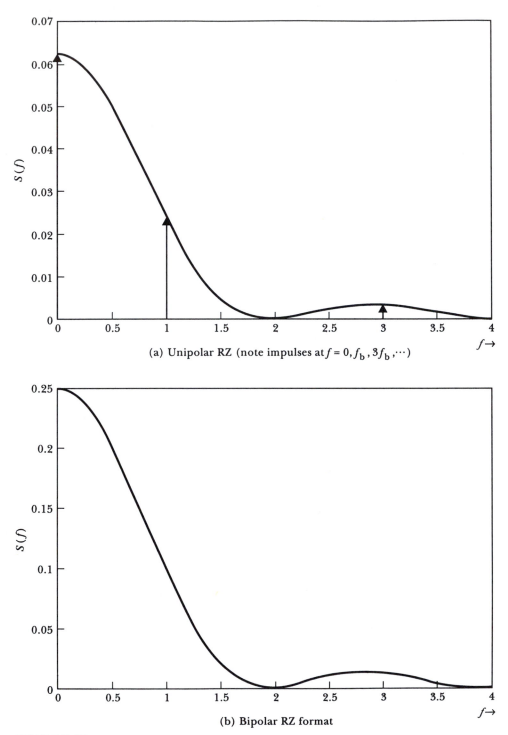

(a) Unipolar RZ (note impulses at $f = 0, f_b, 3f_b, \cdots$)

(b) Bipolar RZ format

FIGURE 7.25
PSDs for several signaling formats with random data. Plots are for normalized signals ($A = T_b = 1$).

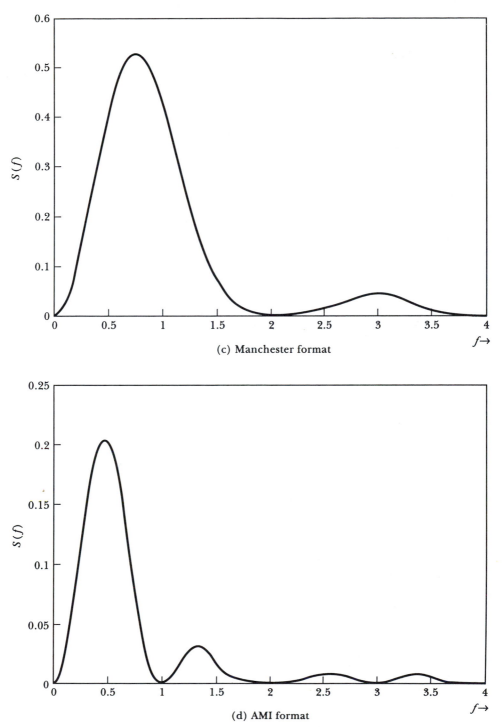

(c) Manchester format

(d) AMI format

FIGURE 7.25 (*continued*)

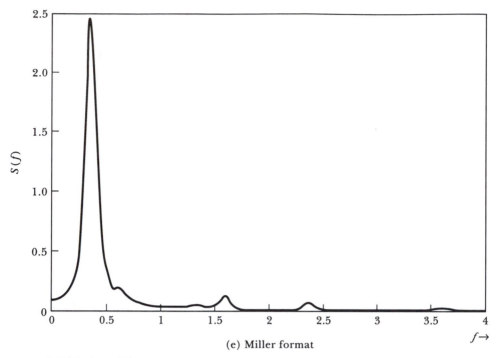

(e) Miller format

FIGURE 7.25 (*continued*)

$f \rightarrow$

condition is met by NRZ, biphase, and bipolar RZ but not by unipolar RZ, AMI, or the Miller code.

3. Recovery of the clock signal. Some formats, such as unipolar RZ, contain a discrete component at the clock frequency that can be recovered with a narrowband bandpass filter. The clock signal can be easily obtained from the bipolar RZ signal by the use of a full-wave rectifier.

4. Other factors that vary with the application.

The PSDs of the data formats we have discussed are shown in Figure 7.25. The PSD of the unipolar RZ format in Figure 7.25a is particularly interesting because it contains impulses at dc and at multiples of the clock frequency. These components are similar to the carrier component in amplitude modulation in that they do not convey information since they represent sinusoidal components with constant amplitude and phase.

SUMMARY

1. Probability theory is useful in making predictions about random experiments. The predictions are concerned with averages rather than the outcome of a particular trial of an experiment.

2. The collection of all outcomes of a random experiment is the sample space. Events are sets of outcomes. Combinations of events, including union, intersection, and complement can be illustrated by Venn diagrams. The null event contains no outcomes, and the certain event contains all of the outcomes of an experiment.

3. The probability of an event is a number assigned to each event that obeys the axioms of probability. The intuitive concept of probability is that it represents the relative frequency of the occurrence of the event when the experiment is repeated a large number of times.

4. Conditional probability gives the probability of an event after it is known that another event has occurred. Bayes' rule is a useful relation between conditional probabilities.

5. Events are independent if and only if the probability of their intersection is equal to the product of the probabilities of the separate events. If two events are independent, the probability of one event conditioned on the other is the same as its probability before conditioning.

6. The binomial probability distribution is concerned with the case in which the occurrence of two complementary events is of interest. An important example is a bit error versus no bit error. The binomial probability distribution gives the probability of k errors (in any k locations) of an $\mathcal{N}$-bit word when errors are independent from bit to bit.

7. If events occur at random times such that the numbers of events in non-overlapping time intervals are independent, and the probability of one event in a very short interval is proportional to the length of the interval, then the probability of k events in an interval τ is given by the Poisson probability distribution.

8. A random variable is a rule for assigning a finite real number to each outcome of a random experiment and the values that result. Random variables can be discrete, continuous, or a mixture of continuous and discrete.

9. The cumulative distribution function (CDF) of a random variable is defined as the probability that the random variable is less than or equal to the argument of the CDF. A number of important properties of the CDF have been given in the chapter.

10. The probability density function (PDF) of a random variable is defined as the derivative of the CDF with respect to its argument. The PDF of a discrete random variable contains impulses. The weight of each impulse is the probability that the random variable takes on the value where the impulse is located. The PDF of a continuous random variable is large for values that are most likely to occur and small for unlikely values. The area under a PDF is unity. Other properties of PDFs have been discussed.

11. Joint random variables result when two or more random variables are associated with an experiment. Joint random variables are characterized in terms of their joint CDF and joint PDF.

12. Random variables are independent if and only if their joint PDF can be factored into the marginal PDFs of the separate random variables.

13.   The conditional PDF of a random variable is the new PDF of the random variable after the value of a second random variable has been learned. When the random variables are independent, the conditional PDF is the same as the marginal PDF.

14.   The average or expected value of a random variable is the value that would result if the experiment could be performed an infinite number of times and the results were averaged. Expected values of a function of a random variable can be computed from the PDF. Some important averages are the mean and variance. Averages of functions of joint random variables can be computed in terms of the joint PDF. The expected value operation is linear. For independent random variables, the average of a product is the product of the averages. For independent random variables, the variance of the sum of several random variables is the sum of their variances.

15.   Transformations of random variables occur when new random variables are defined in terms of old random variables. When a new random variable is the sum of several independent random variables, the PDF of the new random variable is found by convolving the PDFs of the old random variables.

16.   The uniform and gaussian PDFs are of importance in the study of communication systems. The PDF of a gaussian random variable can be written if the mean and variance are known. The Q function is defined as the area under the tail of the PDF of a zero-mean, unit-variance gaussian random variable. Linear operations on jointly gaussian random variables result in gaussian random variables.

17.   The central limit theorem states that a random variable that results from the sum of many small independent contributions tends to become gaussian as the number of contributors becomes large.

18.   Jointly gaussian random variables that are uncorrelated are independent.

19.   In a random process, a function of time, called a sample function, is assigned to each outcome of the experiment. The collection of all of the sample functions is called the ensemble. A complete description of a random process is available if the joint PDF of any order can be written for measurements of the sample functions at specified time instants.

20.   A random process is stationary if its joint PDF of any order is dependent on the intervals between the measurements but not on the location of the first measurement.

21.   Averages such as the dc component, total power, autocorrelation function, or the PSD of a random process can be computed as either a time average or an ensemble average. Time averages are found from a single sample function and may, in general, depend on the particular sample function selected. Ensemble averages are computed from the joint PDF of the random process. When the random process is ergodic, the averages computed in either way are equal. A random process must be stationary to be ergodic.

22. For ergodic random processes, the following interpretations are valid. The mean value is the dc component of the sample functions. The second moment is the total power. The variance is the power in the ac part of the sample functions. The standard deviation is the rms value of the ac part of the sample functions. These quantities are often measured in the laboratory using voltmeters or power meters.

23. The autocorrelation function of an ergodic random process can be computed either as a time average or as an ensemble average. The Fourier transform of the autocorrelation function is the PSD of the sample functions of the random process. The PSD of the output of a linear system is the magnitude of the transfer function squared, times the PSD of the random process at the input.

24. A gaussian random process has a PDF of any order that is jointly gaussian. If the input to a linear system is a gaussian random process, the output is also a gaussian random process.

25. A large number of signaling formats are in use for baseband transmission of binary digital data. Some of the important parameters to consider in selecting a signaling format are its spectral characteristics in relation to those of the channel, the immunity of the format to noise, and recovery of the clock signal.

REFERENCES

W. B. Davenport, Jr. and W. L. Root. *An Introduction to the Theory of Random Signals and Noise.* New York: McGraw-Hill, 1958.

C. W. Helstrom. *Probability and Stochastic Processes for Engineers.* New York: Macmillan, 1984.

M. O'Flynn. *Probabilities, Random Variables, and Random Processes.* New York: Harper and Row, 1982.

A. Papoulis. *Probability, Random Variables, and Stochastic Processes*, 2d ed. New York: McGraw-Hill, 1984.

P. Z. Peebles, Jr. *Probability, Random Variables, and Random Signal Principles*, 2d ed. New York: McGraw-Hill, 1987.

J. M. Wozencraft and I. M. Jacobs. *Principles of Communication Engineering.* New York: Wiley, 1965.

PROBLEMS

1. Starting from the axioms of probability and using set theory concepts, prove the validity of Equations 7.5, 7.6, and 7.7. Also show that the probability of the null event is zero.

2. A random experiment consists of selecting an integer from 1 through 10. If E denotes the event that the result is even, O denotes the event that the number

is odd, and F denotes that the number is divisible by four, what outcomes are included in the events (a) F + O, (b)F^c, (c) EOF, and (d) EF?

3. A sample space consists of the cards in a deck of playing cards. Sketch a Venn diagram showing the relationships between the events consisting of the kings, the face cards, the red cards, and the black cards.

4. A random experiment consists of tossing two dice, a red one and a blue one. List all of the outcomes in the sample space of this experiment. Assuming that the number of spots that turns up on the red die is independent of the number that turns up on the blue die and that each face of a die is equally likely to turn up, find the probability of each outcome of the experiment. Find the probability that a total of seven spots turns up.

5. A random experiment consists of drawing a card from a deck of 52 playing cards. Assume that each card is equally likely to be drawn. Find the probability that the card drawn is black. Find the probability that the card drawn is an ace. Is the event "the card is an ace" independent of the event "the card is black?" Why or why not?

6. Two boxes contain colored balls as follows:

> Box 1: 8 red and 2 white
> Box 2: 2 red and 18 white

The experiment consists of selecting a box at random (equally likely) and then drawing a ball and observing its color. List all of the outcomes of the experiment and determine the probability of each outcome. What is the probability that a red ball is drawn? What is the probability that box 1 was selected, given that the ball is red?

7. Two boxes contain colored balls as in Problem 6. The experiment consists of selecting a box at random, drawing a first ball, observing its color, drawing a second ball from the same box (without replacing the first ball), and observing its color. List all of the outcomes of this experiment and determine the probability of each. Find the probability of the event "the first ball is white" and the event "the second ball is red." Are these two events independent? Why or why not?

8. Two dice are available. One is a "fair die" for which each face is equally likely to come up. The other is "loaded" so that the face with two spots has a probability of one-half and the other faces each have a probability of 0.1. One die is selected at random (equally likely) and tossed three times. If two spots come up on all three tosses, what is the probability that the loaded die was selected?

9. An electronic circuit contains 500 components. After 1000 h of operation, the probability that a given component has failed is 10^{-3}. Assume that the failures of various components are independent. If any of the components fail, the circuit will not function. Find the probability that the circuit is functioning after 1000 h. If two similar circuits are available, find the probability that at least one is functional after 1000 h.

10. The circuit of Problem 9 is redesigned so each component has a backup. The new circuit contains 1000 components consisting of 500 pairs, each original component and its backup. As long as one component in each pair has not failed, the circuit is operational. Find the probability that the circuit is operational after 1000 h.

11. Slick and Jack bet on a game in which they take turns tossing a fair coin until a head turns up. The first player to toss a head wins the bet. Slick al-

ways manages to get the first toss in each game. What is the probability that Slick wins a particular game?

12. N people are in a room. Find the probability that they all have birthdays on different days of the year. For simplicity assume that there are 365 days in a year (neglect the leap year possibility) and that a given person's birthday is equally likely to be any one of the 365 days. Also find the probability that at least two of the people in the room have the same birthday. Evaluate for several values of N.

13. Two events are mutually exclusive and independent. What can you conclude about the probability of at least one of the events?

14. A fair coin is tossed four times. How many sequences of heads and tails are possible? What is the probability of each sequence? How many sequences contain two heads? What is the probability of getting two heads in the four tosses?

15. A fair coin is tossed 50 times. Find (a) the probability that 25 heads turn up during the 50 tosses, (b) the probability that no heads turn up, and (c) the ratio of the probability in part a to that in part b and comment on the result.

16. A certain telephone exchange with a large number of customers experiences calls at the average rate of one per minute during the hours between 3:00 a.m. and 4:00 a.m. During this time period, it is often necessary to greatly reduce the call-handling capability (the maximum number of calls that the system can accommodate at a time) of the system to perform maintenance. Because of the emergency nature of many calls placed at this time of night, it is desired that the probability that the call-handling capability is exceeded at any particular instant in time is less than 10^{-6}. Assume that each call lasts for exactly two minutes. How many calls must the system be able to handle? It is reasonable to assume that the number of calls initiated in a time interval follows the Poisson distribution at this time of night. (This is not true earlier in the evening because many calls are synchronized with the commercials of popular television programs.)

17. A random experiment consists of drawing a card from a deck. An associated random variable takes the value 1 if the card is a heart, 2 if the card is a diamond, 3 if the card is a club, and 4 if the card is a spade. Sketch the CDF and PDF for this random variable.

18. A random variable has the PDF shown in Figure P7.1. Find the numerical value of the parameter K. Find and sketch to scale the CDF of this random variable. Find the probability that the random variable takes on a value less than 0.5. Find the probability that the random variable is positive.

FIGURE P7.1

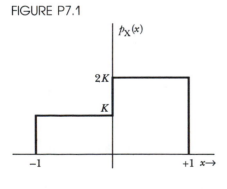

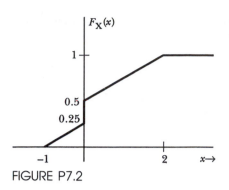

FIGURE P7.2

19. A random variable has the CDF shown in Figure P7.2. Find the PDF and sketch it to scale. Find $P(X < 0)$, $P(X = 0)$, and $P(X \le 0)$.

20. Joint random variables have $p_{XY}(x,y) = K$ for $0 < x < 1$ and $0 < y < 2$. $p_{XY}(x,y)$ is zero for all other values of x and y. Sketch $p_{XY}(x,y)$ versus x and y (similar to Figure 7.6a). Find the numerical value of K. Find the marginal PDFs, $p_X(x)$ and $p_Y(y)$, and sketch them to scale. Are X and Y independent? Explain. Find $p_{X|Y}(x,y)$, and sketch to scale versus x for several values of y. Find $P(X > Y)$.

21. Repeat Problem 20 if $p_{XY}(x, y) = Ky$ for $y > x > 0$ and $0 < y < 1$. $p_{XY}(x, y)$ is zero for other values of x and y.

22. A random variable is uniformly distributed between -2 and $+3$. Find its average value, variance, and standard deviation.

23. A discrete random variable takes on values of 0, 1, 2, and 3 with equal probability. Find its mean, variance, and standard deviation.

24. A random variable X is uniformly distributed between 0 and 2. A new random variable is defined as $Y = X^3$. Find and sketch to scale the PDF for Y.

25. Three independent random variables X_1, X_2, and X_3 are each uniformly distributed between -1 and $+1$. Two new random variables Y and Z are defined as $Y = aX_1 + (1 - a)X_2$ and $Z = (1 - a)X_2 + aX_3$. Where a is a constant, which can take on values from 0 to 1. Find the correlation coefficient of Y and Z, and comment on the result.

26. X is a random variable uniformly distributed from 2 to 3. Y is an independent random variable uniformly distributed from 0 to 2. Find and sketch to scale the PDF of $Z = X + Y$.

27. Θ is a random variable uniformly distributed from $-\pi$ to $+\pi$. If $X = \cos(\Theta)$ find the PDF of X and sketch to scale.

28. This is a continuation of the previous problem. If $Y = \sin(\Theta)$, find the PDF of Y and sketch to scale. Prepare a sketch showing the relationships between X, Y, and Θ. Write an expression for the conditional PDF, $p_{Y|X}(x, y)$. Are X and Y independent? Compute the correlation coefficient between X and Y. Comment.

29. A gaussian random variable has a mean value of 3 and a standard deviation of 2. Find the probability that the value of the random variable exceeds 9. Repeat for the probability that it is less than -5. The approximation for the Q function given in Equation 7.58 may be useful in evaluating the results.

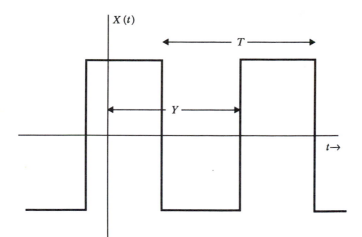

FIGURE P7.3

30. A random process consists of symmetrical square waves with a random "phase" as shown in Figure P7.3. The random variable Y is uniformly distributed between 0 and T.
    (a) Sketch several sample functions of this random process.
    (b) Write the first order PDF of the process.
    (c) Find the mean of the process as a time average and as an ensemble average. Are they the same?
    (d) Find the autocorrelation function of the process as a time average.

31. A random process has sample functions given by $A \cos (2\pi f_c t + \Theta)$ where A is a constant, and Θ is a random phase angle uniformly distributed between 0 and 2π.
    (a) Sketch several of the sample functions of this random process.
    (b) Find the first-order PDF of the process.
    (c) Find the autocorrelation function as a time average, then as an ensemble average. Are the results the same?
    (d) Find the PSD of the process.

32. A random process is given by $Z(t) = X \cos (2\pi f_c t) - Y \sin (2\pi f_c t)$, where X and Y are independent gaussian random variables with zero mean and a variance of σ^2; find (a) $E(Z)$, (b) $E(Z^2)$, (c) the first-order PDF for Z.

33. Show that the random process of the previous problem can be written as $Z(t) = R \cos (2\pi f_c t + \Theta)$. Find expressions for R and Θ in terms of X and Y. Determine the PDF for R. Determine the PDF for Θ.

34. A random process has the autocorrelation function shown in Figure P7.4. What is the total power in the random process? Find the PSD and sketch it to scale. What is the power in the dc component?

35. An ergodic gaussian random process has a dc component of 3 V and a total (normalized) power of 25 W. If the process is measured at some instant in time, write the PDF for the value obtained. Find the probability that the measured value exceeds 20 V.

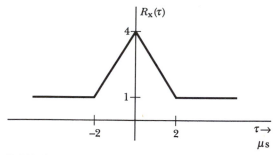

FIGURE P7.4

36. A typical electronic voltmeter, when used in the ac mode, processes the input signal $x(t)$ to obtain a reading as shown in Figure P7.5. As indicated in the figure, the input waveform is first passed through an ac-coupled buffer amplifier that removes any dc component from the input signal. Then the ac component of the input is passed to a full-wave rectifier circuit. The full-wave rectified output is passed to a circuit that finds its time average. The final reading is 1.111 times this average value. In summary, if the input signal $x(t)$, minus its average value, is denoted as $y(t)$, the reading is $[1.111 \langle |y(t)| \rangle]$ where $\langle \rangle$ denotes the time average.
   (a) If the input is $A + B \cos (2\pi ft)$, sketch the waveforms at various points in the diagram of Figure P7.5 and find the final reading.
   (b) If the input is a sample function from an ergodic gaussian random process with a mean of m and a variance of σ^2, find the reading. Note that for an ergodic process, the time average can be replaced by an ensemble average, so the reading is $[1.111 \, E(|Y(t)|)]$.
   The result of this problem shows that when a gaussian random process is measured with such a meter, the reading is not the true rms value of the ac part. However, the true rms value can be found by multiplying the reading by a "fudge factor."

37. A random process $Y(t)$ is obtained by passing white gaussian noise sample functions with a PSD of $S_x(f) = N_0/2$ through an ideal lowpass filter with the transfer function illustrated in Figure P7.6
   (a) Find the PSD of the filter output $Y(t)$. Sketch the PSD to scale.
   (b) Find the autocorrelation function of the filter output and sketch it to scale.
   (c) What is the minimum sampling rate for sampling the filter output so that the output waveform can be reconstructed from the samples without aliasing? Assume that the minimum sampling rate is used in the rest of this problem.

FIGURE P7.5
FIGURE P7.5
Functional block diagram of a typical electronic voltmeter in the ac mode.

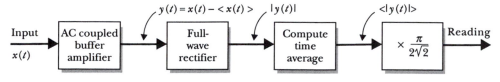

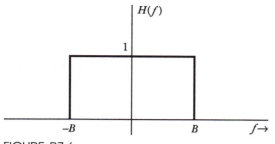

FIGURE P7.6

    (d)   What is the variance of the sample values? Find the covariance of any two sample values. Are different sample values independent? Why?

    (e)   Write the joint PDF for any two samples y_1 and y_2.

38. Find the autocorrelation function of the unipolar RZ format shown in Figure 7.23e for a random data sequence. Notice that except for a region around $\tau = 0$, the autocorrelation function is periodic. Decompose the autocorrelation function into a periodic part and an aperiodic part. Find the Fourier series of the periodic part and the Fourier transform of the aperiodic part. Compare the results with the PSD of Figure 7.25a.

8

DIGITAL MODULATION TECHNIQUES

In this chapter, we discuss bandpass digital communication techniques. The first section (8.1) presents signal-space concepts that enable a simple visualization of signals and noise as vectors, in a way similar to the use of phasors in circuit analysis. The second section presents optimum (in the sense of minimum error probability) receivers for several general classes of binary signals. The third section considers several commonly used coherent binary modulation formats. Coherent receivers recover the carrier (including its phase) embedded in the received signal and use the recovered carrier in processing the received signal. Section 8.4 presents noncoherent receivers, which are used when it is undesirable or impossible to recover the carrier phase. The fifth section presents M-ary modulation techniques, in which several bits are used at a time to modulate the carrier. Section 8.6 presents carrier recovery considerations for some of the modulation formats that require using the carrier in the demodulation process. This chapter concludes with a comparison of the performance of common digital modulation techniques with the limit imposed by the channel capacity equation and continues the design example of Section 5.6.

8.1
SIGNAL-SPACE CONCEPTS

In a binary digital communication system, the transmitter sends one of two possible waveforms $s_0(t)$ or $s_1(t)$, depending on whether the current data bit is a 0 or a 1. The channel adds noise to the transmitted waveform and presents the sum to the receiver. The receiver then faces the problem of deciding which waveform was transmitted, based on the noisy received waveform. As we will see, the optimum receiver (in the sense of minimum bit-error probability) processes the received waveform in a filter or a correlator to reduce the waveform to a single value that is compared to a threshold level to make the decision. In this section, we present the process for reducing the information in such received waveforms to a set of numbers that can be visualized as the components of a vector.

Generalized Fourier series

Consider the representation, known as a *generalized Fourier series*, for a signal $s_i(t)$, given by

$$s_i(t) = \sum_{j=1}^{N} s_{ij}\phi_j(t) \qquad 0 \leq t \leq T \tag{8.1}$$

where the s_{ij} are the *series coefficients*, and the $\phi_j(t)$ are *basis functions*. Notice that the first subscript of the coefficient refers to the signal being represented, and the second subscript is for the basis function. As indicated above, we assume that the representation is valid only for a time interval from 0 to T. An infinite number of terms may be required in the series, and N may be infinity.

We also require the basis functions to be *orthonormal*, which means that they satisfy the following condition:

$$\int_0^T \phi_i(t)\phi_j(t)\, dt = 1 \qquad \text{if } i = j$$
$$= 0 \qquad \text{if } i \text{ is not equal to } j \tag{8.2}$$

When the basis functions are orthonormal, it can be easily shown that the series coefficients are given by

$$s_{ik} = \int_0^T s_i(t)\phi_k(t)\, dt \tag{8.3}$$

This result can be derived starting from Equation 8.1 by multiplying both sides by $\phi_k(t)$, integrating from 0 to T, and using the orthonormality property. The operation of multiplying two time functions and integrating over an interval, as in Equation 8.3, is called *correlation*. Thus, the coefficients are obtained by correlating the signal $s_i(t)$ with each of the basis functions.

Signal-space vectors

We will find it very useful to consider the coefficients s_{ij} to be the components of an N-dimensional vector $\mathbf{S_i}$. The first component of the vector is s_{i1}, the second component is s_{i2}, and so forth. Thus, every signal that can be represented by Equation 8.1 has an associated vector. The space containing these vectors is called *signal space*.

Each of the basis functions can be represented as a (trivial) series expansion in which one of the coefficients is unity and the rest are zero. In equation form, we have

$$\phi_j(t) = \sum_{k=1}^{N} \phi_{jk}\phi_k(t)$$

where $\phi_{jj} = 1$ and all of the other coefficients are zero. Thus, the signal-space vector corresponding to a basis function is a unit vector pointing along one of the N coordinate axes in signal space.

EXAMPLE 8.1

Show that the set of basis functions of Figure 8.1a are orthonormal for the time interval from $t = 0$ to $t = T = 3$. Find the generalized Fourier series coefficients for the signal waveforms shown in Figure 8.1b and construct a signal-space diagram showing the basis function unit vectors and the vector representations for the signals.

SOLUTION

To demonstrate that the basis functions are orthonormal we must show that the integral

$$\int_0^T \phi_i(t)\phi_j(t)\, dt = \int_0^3 \phi_i(t)\phi_j(t)\, dt$$

is zero when i is different from j and is unity when $i = j$. Since the pulses in Figure 8.1a do not overlap in time, the product of two different basis functions is zero for all values of time. Therefore, the integral of the product of any two different basis functions is zero. For the case in which $i = j = 1$, the integral becomes.

$$\int_0^T \phi_i(t)\phi_j(t)\, dt = \int_0^T [\phi_1(t)]^2\, dt = \int_0^1 dt = 1$$

Similarly, the integrals of the squares of the other basis functions can be shown to be unity. Thus, we have shown that the basis functions are indeed orthonormal.

To represent $s_1(t)$ by the generalized Fourier series, we must find its coefficients. The first coefficient s_{11} is given by

$$s_{11} = \int_0^3 s_1(t)\phi_1(t)\, dt = \int_0^1 2\, dt = 2$$

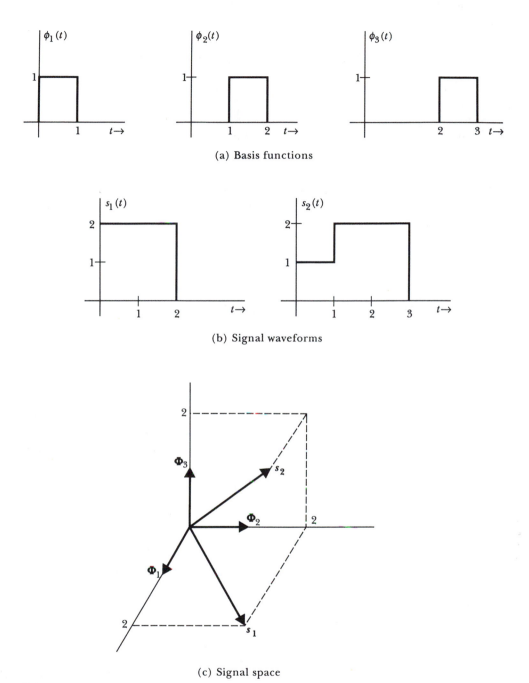

(a) Basis functions

(b) Signal waveforms

(c) Signal space

FIGURE 8.1
Signal-space example.

Similarly, the other coefficients of $s_1(t)$ can be found as

$$s_{12} = 2 \qquad \text{and} \qquad s_{13} = 0$$

The resulting signal-space vector associated with $s_1(t)$ is

$$\mathbf{S_1} = (s_{11}, s_{12}, s_{13}) = (2,2,0)$$

Similarly, the coefficients of $s_2(t)$ can be found as

$$s_{21} = 1, \qquad s_{22} = 2, \qquad \text{and} \qquad s_{23} = 2$$

and the vector corresponding to $s_2(t)$ is $\mathbf{S_2} = (1,2,2)$.

The vector representations of the basis functions themselves are

$$\mathbf{\Phi_1} = (1,0,0), \qquad \mathbf{\Phi_2} = (0,1,0) \qquad \text{and} \qquad \mathbf{\Phi_3} = (0,0,1)$$

The signal-space vector diagram is shown in Figure 8.1c.

Correlation as a dot product of signal-space vectors

In processing noisy signals in receivers for digital communications, we will often be concerned with the correlation of two signals, given by

$$\int_0^T s_i(t) s_j(t) \, dt \tag{8.4}$$

We now show that this correlation operation can also be accomplished by taking the dot product of the corresponding signal-space vectors. Substituting the generalized Fourier series expression for $s_i(t)$ and $s_j(t)$ in the correlation integral, we obtain

$$\int_0^T \left[\sum_{m=1}^N s_{im} \phi_m(t) \right] \left[\sum_{n=1}^N s_{jn} \phi_n(t) \right] dt$$

When the terms in the brackets are multiplied, a total of N^2 terms results. The terms that involve the product of different basis functions integrate to zero because of the orthonormality property of the basis functions. The terms that involve the product of a basis function with itself are then the only terms left. The result becomes

$$\int_0^T \sum_{m=1}^N s_{im} s_{jm} [\phi_m(t)]^2 \, dt$$

Interchanging the order of integration and summation and using the orthonormality property again results in

$$\sum_{m=1}^N s_{im} s_{jm}$$

which can be recognized as the *dot* or *inner product* of the signal-space vectors. (Recall that the dot product of two vectors can be computed by summing the products of corresponding components.) Thus, we have established the important result that the correlation operation gives the same result as the dot product of the signal-space vectors. In equation form, we have

$$\int_0^T s_i(t)s_j(t)\,dt = \sum_{m=1}^N s_{im}s_{jm} = \mathbf{S_i} \cdot \mathbf{S_j} \tag{8.5}$$

When a signal is correlated with itself, the correlation integral can be recognized as the normalized energy in the signal. This is also given by the dot product of the corresponding signal-space vector with itself. The dot product of a vector with itself is the length of the vector squared. Therefore, we can make the important conclusion that the length of a signal-space vector is equal to the square root of the energy in the signal. In equation form,

$$E_i = \int_0^T [s_i(t)]^2\,dt = \mathbf{S_i} \cdot \mathbf{S_i} = |\mathbf{S_i}|^2 = (\text{length of } \mathbf{S_i})^2 \tag{8.6}$$

EXAMPLE 8.2

Compute the energy of each of the signals in Figure 8.1b and the correlation of the signals by evaluating the time integrals of Equations 8.5 and 8.6. Repeat the computations by using the signal-space vector representations for these signals, which were found in Example 8.1.

SOLUTION
The energy in $s_1(t)$ is given by

$$E_1 = \int_0^3 [s_1(t)]^2\,dt = \int_0^2 4\,dt = 8$$

Alternately, the energy can be found from the dot product of $\mathbf{S_1}$ with itself, given by $\mathbf{S_1} \cdot \mathbf{S_1} = [2,2,0] \cdot [2,2,0] = 8$. In the same fashion, the energy in the second signal can be found by either procedure as $E_2 = 9$.

The correlation of the two signals can be computed using the integral of Equation 8.5 as

$$\int_0^3 s_1(t)s_2(t)\,dt = \int_0^1 2\,dt + \int_1^2 4\,dt = 6$$

Using the dot product of the corresponding signal vectors, we have

$$\mathbf{S_1} \cdot \mathbf{S_2} = [2,2,0] \cdot [1,2,2] = 6$$

Selection of the basis functions

In selecting a set of basis functions to use in constructing a signal space for some set of signals of interest, it is often convenient to make the selection so that the generalized Fourier series gives an exact representation with only a small number of terms. If we do not know the particular signals of interest in advance, it is necessary to use a large number of basis functions. An example of this was the exponential Fourier series, in which we had an infinite number of complex exponentials as the basis functions. A set of basis functions that can represent any function from some class over some interval of time is called a *complete set* for that class of functions on the given interval. For example, the complex exponentials of the ordinary Fourier series is complete for the class of periodic functions (which meet certain additional conditions known as the Dirichlet conditions) on the time interval from $-\infty$ to ∞. When we apply signal-space concepts to the signals used in digital communications, we are often able to identify, by inspection, a small set of suitable basis functions that are complete for the signal set of interest.

The Gram-Schmidt procedure

An alternative formal procedure for finding a small set of basis functions that can represent the members of a set of signals is the *Gram-Schmidt procedure*. In this procedure, we initially assume that the first signal can be represented by only the first basis function and that the second signal can be represented by a weighted sum of only the first two basis functions, and so forth. In equation form, these assumptions are

$$s_1(t) = s_{11}\phi_1(t) \tag{8.7}$$

$$s_2(t) = s_{21}\phi_1(t) + s_{22}\phi_2(t) \tag{8.8}$$

$$s_3(t) = s_{31}\phi_1(t) + s_{32}\phi_2(t) + s_{33}\phi_3(t) \tag{8.9}$$

and so forth.

The first equation in this set can be used to find the first basis function to within a multiplicative constant. In other words, the shape of the waveform for $\phi_1(t)$ is the same as the waveform shape of $s_1(t)$. Since the energy in a basis function is required to be unity by the orthonormality property of Equation 8.2, the amplitude of $\phi_1(t)$ can be determined. Once $\phi_1(t)$ is known, the first term on the right side of Equation 8.8 can be determined and subtracted from both sides. This reveals the waveform shape of the second basis function. The unit energy requirement establishes the amplitude of $\phi_2(t)$. Then the third equation in the set can be used to find $\phi_3(t)$ and so forth until all of the basis functions have been determined.

EXAMPLE 8.3

Find a set of basis functions that can be used to represent the signals of Figure 8.2a by use of the Gram-Schmidt procedure.

SOLUTION
Using Equation 8.7, we see that the shape of $\phi_1(t)$ is the same as the shape of $s_1(t)$. However, since the value of s_{11} is not known, the amplitude of $\phi_1(t)$

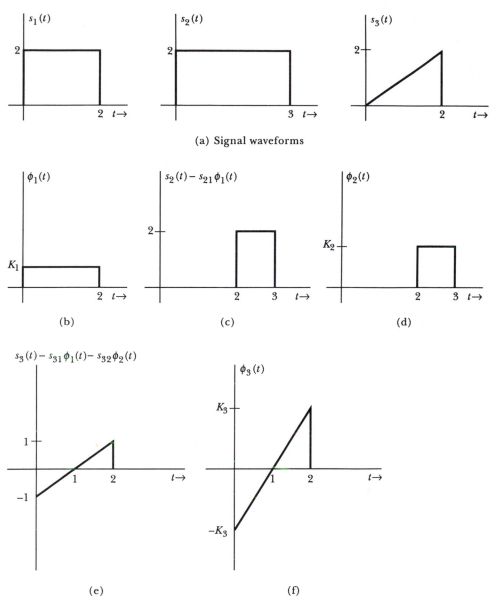

(a) Signal waveforms

(b)   (c)   (d)

(e)   (f)

FIGURE 8.2
Signals and basis functions of Example 8.3.

is not known. We denote the amplitude of $\phi_1(t)$ by K_1. Thus, at this point, we know that $\phi_1(t)$ takes the form shown in Figure 8.2b. Now we require the energy in $\phi_1(t)$ to be unity:

$$\int_0^T [\phi_1(t)]^2 \, dt = \int_0^2 (K_1)^2 \, dt = 2(K_1)^2 = 1$$

This yields $K_1 = 2^{-1/2}$, and the first basis function has been completely determined.

Now we begin working to find $\phi_2(t)$. To start, we can find the first coefficient for the second signal, given as

$$s_{21} = \int_0^T s_2(t)\phi_1(t)\,dt = \int_0^2 2^{1/2}\,dt = 2(2^{1/2})$$

Now we can rearrange Equation 8.8 to obtain

$$s_{22}\phi_2(t) = s_2(t) - s_{21}\phi_1(t)$$

where all of the quantities on the right side are given or have been determined. A plot of the right side of the last expression is shown in Figure 8.2c. Because of the last expression, we know that the waveshape of $\phi_2(t)$ is the same as the waveshape shown in Figure 8.2c except for an amplitude factor. At this point we know that $\phi_2(t)$ is as shown in Figure 8.2d. Now if we require the energy in $\phi_2(t)$ to be unity, we find that the amplitude of $\phi_2(t)$ is $K_2 = 1$. Thus, the first two basis functions have been completely determined.

Now we can use Equation 8.9 to determine $\phi_3(t)$. First, the coefficients s_{31} and s_{32} can be determined by use of Equation 8.3. The results are

$$s_{31} = 2^{1/2} \qquad \text{and} \qquad s_{32} = 0$$

Equation 8.8 can be rearranged to give

$$s_{33}\phi_3(t) = s_3(t) - s_{31}\phi_1(t) - s_{32}\phi_2(t)$$

where all of the terms on the right side have already been found. A plot of the right side of this last equation is shown in Figure 8.2e. Therefore, we know the shape but not the amplitude of $\phi_3(t)$, as indicated in Figure 8.2f. The unit energy requirement can be used to find that

$$K_3 = \left(\frac{3}{2}\right)^{1/2}$$

and $\phi_3(t)$ has been determined. Thus, we have found a set of basis functions that are complete for the three given signals.

When we use the Gram-Schmidt procedure, one of the basis functions may be zero. For instance, in the last example when we formed the expression

$$s_{33}\phi_3(t) = s_3(t) - s_{31}\phi_1(t) - s_{32}\phi_2(t) \tag{8.10}$$

the right side could have vanished. We find it convenient, when this happens, to drop that basis function from the equation set illustrated by Equations 8.7, 8.8, and 8.9 and to renumber the remaining basis functions consecutively.

Consideration of the Gram-Schmidt procedure shows that it is always possible to find a set of M (or fewer) basis functions that provide a complete basis set for any M given signals. Very often, we will be able to select a small set of functions by inspection of the signals to be represented without resorting to the formal procedure.

Representation of white gaussian noise in signal space

In addition to the transmitted signal, the receiver is presented with noise added in the channel. We will emphasize the case in which the noise is *white gaussian noise* (WGN). When WGN is represented by a generalized Fourier series, the series coefficients are random variables since they are different for each sample function of WGN. The generalized Fourier series for a sample function $n(t)$ of WGN is

$$n(t) = \sum_{k=1}^{\infty} n_k \phi_k(t) \qquad (8.11)$$

where we have indicated an infinite number of terms because an infinite number of basis functions are needed to make a complete set for WGN. The coefficients are given by

$$n_k = \int_0^T n(t) \phi_k(t) \, dt \qquad (8.12)$$

These coefficients are jointly gaussian random variables because the correlation integral used to find them is a linear operation. (Recall that linear operations on a gaussian random process produce gaussian results.)

We now consider the mean, variance, and covariance of the noise coefficients. The expected value of the coefficients is

$$E[n_k] = E\left[\int_0^T n(t) \phi_k(t) \, dt \right] \qquad (8.13)$$

Interchanging the operations of integration and expected value results in

$$E[n_k] = \int_0^T E[n(t)] \phi_k(t) \, dt \qquad (8.14)$$

Now $E[n(t)]$ is the dc component of the WGN sample function. Since we have assumed that the noise has a constant PSD (white noise), there is no impulse in the PSD at $f = 0$, and the dc component is zero. Therefore, the expected values of the coefficients of the noise sample function are zero.

Now we consider the covariance of the noise coefficients. The covariance was defined in Equation 7.44. Since the expected values of the coefficients are zero, their covariance is

$$\mu_{n_j n_k} = E(n_j n_k) \qquad (8.15)$$

Using Equation 8.12 to substitute for the coefficients, we obtain

$$E(n_j n_k) = E\left[\int_0^T n(x) \phi_j(x) \, dx \int_0^T n(y) \phi_k(y) \, dy \right] \qquad (8.16)$$

where x and y, rather than t, have been used as the dummy variables of integration because we will combine the integrals into a double integral. Combining the integrals and interchanging the expected value operation with the integration operations results in

$$E(n_j n_k) = \int_0^T \int_0^T E[n(x)n(y)]\phi_j(x)\phi_k(y)\, dx\, dy \qquad (8.17)$$

Now recall that the autocorrelation function of the noise is given by

$$R_n(\tau) = E[n(t)n(t + \tau)] \qquad (8.18)$$

Therefore, we can conclude that

$$E[n(x)n(y)] = R_n(x - y) \qquad (8.19)$$

Since the noise is white, the PSD of the noise is a constant:

$$S_n(f) = \frac{N_0}{2} \qquad (8.20)$$

The autocorrelation function of the noise is the inverse Fourier transform of the PSD. This results in

$$R_n(\tau) = \left(\frac{N_0}{2}\right)\delta(\tau) \qquad (8.21)$$

Substituting Equation 8.21 into 8.19 and the result into 8.17, we obtain

$$E(n_j n_k) = \int_0^T \int_0^T \left(\frac{N_0}{2}\right)\delta(x - y)\phi_j(x)\phi_k(y)\, dx\, dy \qquad (8.22)$$

Now using the integral property of the impulse function, we can carry out the integration with respect to x. The result is

$$E(n_j n_k) = \frac{N_0}{2}\int_0^T \phi_j(y)\phi_k(y)\, dy \qquad (8.23)$$

The orthonormality property of the basis functions given in Equation 8.2 can be applied to find

$$\begin{aligned} E(n_j n_k) &= \frac{N_0}{2} && \text{if } j = k \\ &= 0 && \text{if } j \text{ is not equal to } k \end{aligned} \qquad (8.24)$$

Equation 8.24 shows that the coefficients have a covariance of zero. Thus, the correlation coefficient of Equation 7.45 is zero, and the coefficients are uncorrelated. Since uncorrelated gaussian random variables are independent, the noise coefficients are independent. The variance of each coefficient is given by

Equation 8.24 when $j = k$. The result is

$$\text{var}\,(n_j) = \frac{\mathcal{N}_0}{2} \tag{8.25}$$

Thus, we have established the following result, which will be important in our consideration of digital communication: When WGN is represented by a generalized Fourier series, the coefficients are independent, zero-mean, gaussian random variables each having a variance of $\mathcal{N}_0/2$.

8.2
OPTIMUM RECEIVERS FOR BINARY SIGNALS

In this section, we derive receiver structures that are optimum, in the sense of minimum bit-error probability, for cases in which the received signal is one of two known waveforms plus WGN.

The system model

A conceptual block diagram of the transmitter, channel, and receiver is shown in Figure 8.3. The input data to the transmitter is a bit d, which can be either a 1 or a 0. We assume that the probabilities of the bit being a 1 or a 0 are known and denoted by $P(d = 1)$ and $P(d = 0)$. These probabilities are called the *a priori probabilities* since they are known to the receiver before a signal is received. We assume, except where noted otherwise, that the probability of a 1 and of a 0 are equal; so $P(d = 1) = P(d = 0) = \frac{1}{2}$.

When the data bit is presented to the transmitter, the transmitter selects one of two waveforms, either $s_1(t)$ or $s_0(t)$, depending on the data bit. The

FIGURE 8.3
Conceptual block diagram of binary communication system.

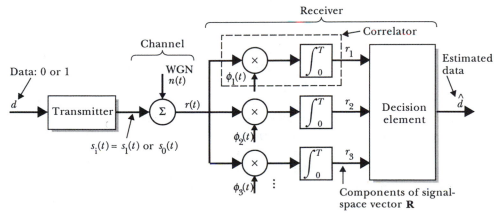

exact shape of these waveforms is assumed to be known to the receiver. The selected waveform is applied to the input of the channel, which adds a sample function from a white gaussian noise process and presents the sum to the receiver.

The function of the receiver is to use the received signal-plus-noise waveform to make an optimum decision about the transmitted data. The receiver decision is denoted as $\hat{d}$. As shown in Figure 8.3, the received signal-plus-noise $r(t)$ is first correlated with a set of orthonormal basis functions, $\phi_1(t)$, $\phi_2(t)$, and so forth. The correlator outputs are denoted as r_1, r_2, and so forth, as indicated in the figure. The decision about the transmitted data is based on these correlator outputs.

We assume that the first two basis functions form a complete set for the transmitted signals $s_0(t)$ and $s_1(t)$. As we have seen in the previous section, this can be achieved for any two signals by applying the Gram-Schmidt procedure. The remaining basis functions are assumed to form (with the first two basis functions) a complete set for the noise. As we will see shortly, only the outputs of the first two correlators are used in making the decision, so the exact specification of the basis functions numbered 3 and higher is not necessary.

The correlator outputs are the coefficients of a generalized Fourier series expansion for the received signal-plus-noise. Since we have assumed that a complete set of basis functions is used, the coefficients contain all of the information in the received signal. We can see that this is true because the original waveform can be reconstructed by summing the terms in the generalized Fourier series. Thus, the optimum decision can be based on the correlator outputs of Figure 8.3. These correlator outputs can be written as

$$r_1 = s_{i1} + n_1 \tag{8.26}$$

$$r_2 = s_{i2} + n_2 \tag{8.27}$$

$$r_3 = n_3 \tag{8.28}$$

$$r_4 = n_4 \tag{8.29}$$

$$\vdots$$

where s_{i1} is the first coefficient of the ith signal (i is 0 or 1), s_{i2} is the second coefficient of $s_i(t)$, and the n_j are the coefficients of the generalized Fourier series for the noise. Notice that only the first two correlator outputs contain terms due to the transmitted signal since we have selected the basis functions so that the first two alone are a complete set for the transmitted signals. As we showed in the previous section, the noise coefficients n_1, n_2, and so on, are independent, zero-mean, gaussian random variables.

The correlator outputs can be considered to be the components of a signal-space vector **R**, which contains all of the information in the received signal, and it is all that the decision device has available to make its decision

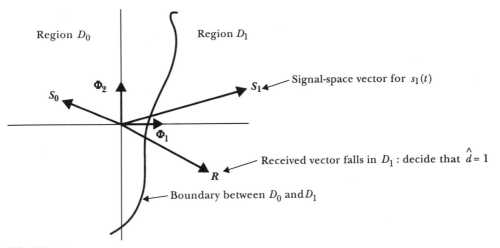

FIGURE 8.4.
Signal-space diagram showing transmitted signal vectors and decision regions (not optimum).

about the transmitted data. The decision is based on the location of this vector in signal space. As indicated in Figure 8.4, signal space can be divided into two regions: one where the decision is that a 0 was transmitted and the other where the decision is that a 1 was transmitted. These regions are labeled D_0 and D_1 in the figure. Therefore, the specification of the optimum (minimum error probability) decision device is simply a definition of the two *decision regions*. Next, we discuss how these optimum decision regions can be found.

The likelihood ratio test

If it were necessary to decide on the data before the signal was transmitted, the best decision would be always to select the data bit with the highest a priori probability. This would, of course, result in an unacceptably high error probability, but it would, nevertheless, represent the optimum decision strategy before signal transmission.

After the signal has been transmitted and the signal space vector $\mathbf{R}$ has been presented to the decision device, the probabilities change to the conditional probabilities denoted by $P(d = 1|\mathbf{R})$ and $P(d = 0|\mathbf{R})$. These are simply the new probabilities, given the value of the received signal-space vector. They are called the *a posteriori probabilities*. The optimum decision is to select the data bit that is most probable. Thus, the decision device must calculate the a posteriori probabilities and decide in favor of the data with the highest probability. Thus, the decision is $\hat{d} = 1$ if

$$P(d = 1|\mathbf{R}) > P(d = 0|\mathbf{R}) \tag{8.30}$$

and the decision is $\hat{d} = 0$ otherwise. (When the a posteriori probabilities are equal, either decision can be made and an error occurs in half of these cases.) This called the *maximum a posteriori* (MAP) decision criterion.

Equation 8.30 can be written as

$$\frac{p(\mathbf{R}|d=1)P(d=1)}{p(\mathbf{R})} > \frac{p(\mathbf{R}|d=0)P(d=0)}{p(\mathbf{R})} \tag{8.31}$$

where $p(\mathbf{R}|d=1) = p(r_1,r_2,r_3,\ldots|d=1)$ is the joint PDF of the components of the vector $\mathbf{R}$, given that the message bit is a 1. The meaning of the other notation is similar. Since $p(\mathbf{R})$ is positive, it can be canceled from both sides of Equation 8.31, and we can rearrange the expression to obtain

$$\blacksquare \qquad \lambda(\mathbf{R}) = \frac{p(\mathbf{R}|d=1)}{p(\mathbf{R}|d=0)} > \frac{P(d=0)}{P(d=1)} = \eta \tag{8.32}$$

The left side of the inequality, $\lambda(\mathbf{R})$, is known as the *likelihood ratio*, and the right side, η, is called the *threshold*. For equally likely data, the threshold value η is unity. Thus, the MAP decision process, which results in minimum error probability, consists of computing the value of the likelihood ratio for the received vector $\mathbf{R}$ and comparing the result with a threshold. If the likelihood ratio exceeds the threshold, the decision is in favor of a 1.

Decision regions in signal space

Now we will show that the MAP decision process can be simplified to an equivalent procedure that picks the signal vector closest to the received vector in signal space. The conditional joint PDF of the components of the received vector can be factored into the products of the PDFs of each of the components given by

$$\begin{aligned} p(\mathbf{R}|d=i) &= p(r_1,r_2,\ldots|d=i) \\ &= p(r_1|d=i)p(r_2|d=i)\cdots \end{aligned} \tag{8.33}$$

where $i = 0$ or 1. This is true because the generalized Fourier coefficients of the noise $n_1,n_2,\ldots$ are independent gaussian random variables. The components of the received vector are the noise coefficients plus the (nonrandom) signal coefficients as indicated by Equations 8.26 through 8.29. Therefore, the components of the received vector are also independent gaussian random variables. The variance of each of the noise coefficients was shown in the previous section to be $\mathcal{N}_0/2$. Thus, the PDF of the jth component of the received vector is

$$p(r_j|d=i) = (\pi\mathcal{N}_0)^{-1/2}\exp\left[-\frac{(r_j-s_{ij})^2}{\mathcal{N}_0}\right] \tag{8.34}$$

where $i = 0$ or 1 and, as before, s_{ij} is the jth coefficient of the ith signal. Notice that since the first two basis functions were chosen to be complete for the transmitted signals, s_{ij} is zero for j greater than two. Using Equations 8.33 and 8.34 to substitute for the PDFs in Equation 8.32 results in the following relation for the likelihood ratio:

$$\lambda(\mathbf{R}) = \frac{\exp\left[-(r_1 - s_{11})^2/N_0\right] \exp\left[-(r_2 - s_{12})^2/N_0\right]}{\exp\left[-(r_1 - s_{01})^2/N_0\right] \exp\left[-(r_2 - s_{02})^2/N_0\right]} > \eta = 1 \quad (8.35)$$

where the constants $(\pi N_0)^{-1/2}$ have canceled. Also notice that all of the factors for the PDFs of $r_3, r_4, \ldots$ have canceled. Thus, we have shown that the components of the received vector numbered three and higher are irrelevant to the decision. The threshold value η is unity since we have assumed that the data bits are equally likely to be 1 or 0.

Taking the natural logarithm of Equation 8.35 and rearranging, we obtain

$$(r_1 - s_{11})^2 + (r_2 - s_{12})^2 < (r_1 - s_{01})^2 + (r_2 - s_{02})^2 \quad (8.36)$$

where, as usual, the decision $\hat{d} = 1$ is made if the inequality is satisfied. Since the higher numbered components of the received vector $\mathbf{R}$ are irrelevant, we will ignore them and consider $\mathbf{R}$ to be a two-dimensional vector. The left side of Equation 8.36 can now be recognized as the square of the distance from the tip of the vector for $s_1(t)$ to the tip of $\mathbf{R}$. Similarly, the right side is the square of the distance from $\mathbf{R}$ to $\mathbf{S_0}$. Thus, the optimum decision rule is to decide in favor of the signal vector that is closest to the received vector. A signal-space diagram showing the transmitted signal vectors and the optimum decision regions is shown in Figure 8.5. Note that the boundary between the decision regions is the perpendicular bisector of the line from the tip of $\mathbf{S_0}$ to the tip of $\mathbf{S_1}$.

FIGURE 8.5
Signal-space diagram showing optimum decision regions for
$P(d = 1) = P(d = 0) = \frac{1}{2}$.

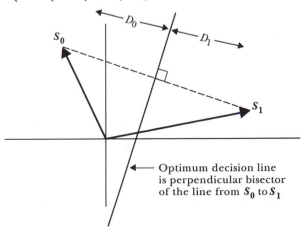

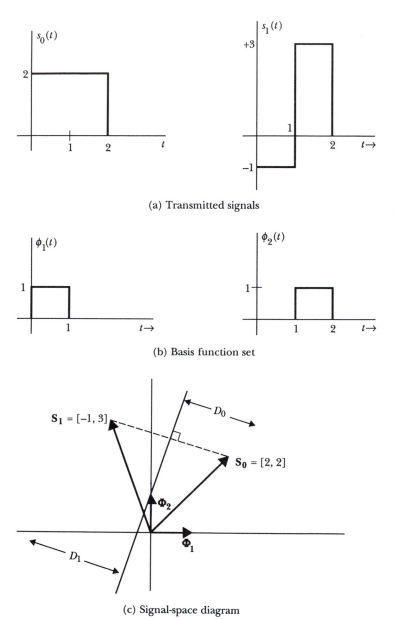

(a) Transmitted signals

(b) Basis function set

(c) Signal-space diagram

FIGURE 8.6
Signals and receivers for Example 8.4.

EXAMPLE 8.4

For the two signal waveforms shown in Figure 8.6a, find a suitable set of basis functions, draw the block diagram of an optimum receiver, and sketch a signal-space diagram showing the signal vectors and the optimum decision regions.

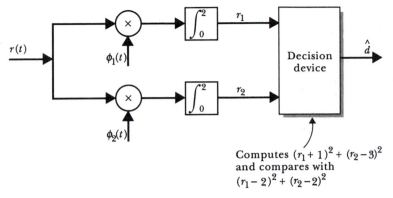

(d) Optimum receiver

Computes $(r_1 + 1)^2 + (r_2 - 3)^2$
and compares with
$(r_1 - 2)^2 + (r_2 - 2)^2$

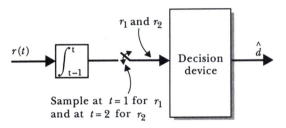

Sample at $t = 1$ for r_1
and at $t = 2$ for r_2

(e) Simplified version of the optimum receiver

FIGURE 8.6 (*continued*)

SOLUTION

We can find a set of two basis functions either by inspection of the signal waveforms or, if that fails, by use of the Gram-Schmidt procedure. For the waveforms shown, inspection yields the basis functions shown in Figure 8.6b. The coefficients of the generalized Fourier series can now be obtained. For example,

$$s_{01} = \int_0^T s_0(t)\phi_1(t)\,dt = \int_0^1 2\,dt = 2$$

Similarly, we can obtain $s_{02} = 2$, $s_{11} = -1$, and $s_{12} = 3$. Thus, the signal-space vectors are given by

$$\mathbf{S_0} = (2,2) \qquad \text{and} \qquad \mathbf{S_1} = (-1,3)$$

A signal-space diagram showing the signal vectors is shown in Figure 8.6c. The decision region where $\hat{d} = 0$, denoted as D_0, is shown in the figure. The decision region D_1 is also shown. The dividing line between the regions is the perpendicular bisector of the line between the tip of $\mathbf{S_0}$ and $\mathbf{S_1}$. Points on one side of the bisector are closer to $\mathbf{S_1}$, and points on the other side are closer to $\mathbf{S_0}$.

A block diagram of an optimum receiver similar to Figure 8.3 is shown in Figure 8.6d. Since the components produced by basis functions of index three and higher are irrelevant, they have not been included.

For this particular case, the receiver can be simplified by noting that the basis functions are unity (or zero), so the multipliers can be eliminated. Since r_1 is obtained by integrating $r(t)$ from $t = 0$ to $t = 1$ and r_2 is obtained by integrating over the nonoverlapping interval from $t = 1$ to $t = 2$, the receiver can be implemented with a single integrator, as shown in Figure 8.6e.

As indicated in the figure, the decision device computes the distance from $\mathbf{R}$ to $\mathbf{S_0}$ and the distance form $\mathbf{R}$ to $\mathbf{S_1}$. The decision is made in favor of the smallest distance.

The single correlator receiver

As we have indicated in Example 8.4, the optimum receiver for binary signals contains a maximum of two correlators, one for each relevant component of $\mathbf{R}$. We will now demonstrate that it is always possible to simplify the receiver for binary signaling so that only one correlator is required. Notice in Figure 8.7a that only the component of $\mathbf{R}$ that is perpendicular to the decision line, r_b, is of importance to the decision. The component of $\mathbf{R}$ that is parallel to the decision line, r_a, is irrelevant. The relevant component r_b can be obtained by taking the dot product of $\mathbf{R}$ with a unit vector $\boldsymbol{\Phi_b}$, directed perpendicular to the decision line. Since the dot product of two vectors corresponds to the correlation operation between the corresponding time signals, the relevant component r_b can be obtained from a correlator, as shown in Figure 8.7b. The figure shows that the optimum receiver using the relevant vector has a decision element consisting of a single comparator that compares r_b with a threshold. Figure 8.7c shows the conditional PDFs of the correlator output and the threshold level.

EXAMPLE 8.5

Continue Example 8.4 by finding the relevant unit vector, the corresponding time signal, and the value of the threshold used in the decision element.

SOLUTION
A vector pointing from $\mathbf{S_0}$ to $\mathbf{S_1}$ can be found by taking the difference between $\mathbf{S_1}$ and $\mathbf{S_0}$. From the vectors found in Example 8.4, we find that this is

$$\mathbf{S_1} - \mathbf{S_0} = (-1,3) - (2,2) = (-3,1)$$

The unit vector $\boldsymbol{\Phi_b}$ is simply a constant times the difference between the signal vectors. Thus, we have

$$\boldsymbol{\Phi_b} = K(\mathbf{S_1} - \mathbf{S_0}) = K(-3,1) = (-3K,K)$$

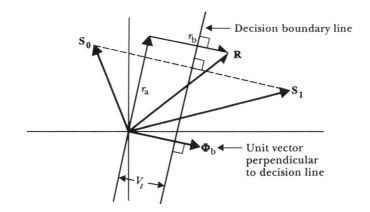

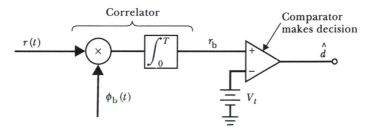

(a) Signal-space diagram showing received vector **R** resolved
into a component parallel to the decision line and
a component perpendicular to the decision line

(b) Optimum receiver using a single correlator

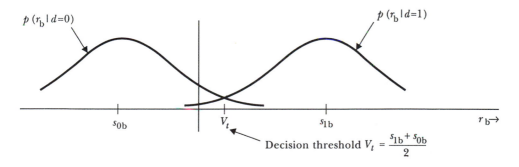

(c) PDFs of the correlator output

FIGURE 8.7
Single correlator receiver.

Since $\boldsymbol{\Phi_b}$ is a unit vector, we require that the sum of the squares of its components equal unity. This results in

$$(-3K)^2 + (K)^2 = 1 \quad \text{or} \quad K = 10^{-1/2} = 0.316$$

Thus, the relevant unit vector is

$$\boldsymbol{\Phi_b} = [-0.949, 0.316]$$

Since the components of the signal-space vector are the coefficients of the generalized Fourier series for the time waveform, we can find $\phi_b(t)$ as

$$\phi_b(t) = -0.949\phi_1(t) + 0.316\phi_2(t)$$

The basis functions are shown in Figure 8.6b. The resulting waveform of $\phi_b(t)$ is shown in Figure 8.8a.

FIGURE 8.8
Single correlator receiver of Example 8.5.

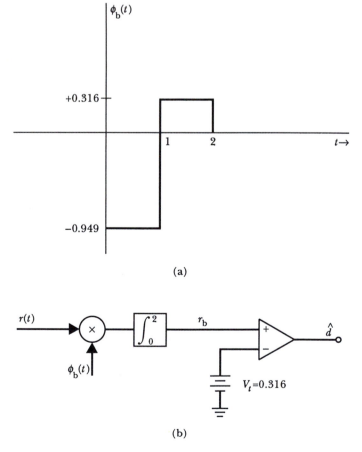

(a)

(b)

The optimum receiver using the relevant basis function is shown in Figure 8.8b. When $s_0(t)$ is transmitted, the output of the correlator in the absence of noise is

$$s_{0b} = \int_0^2 s_0(t)\phi_b(t)\,dt = \mathbf{S_0} \cdot \mathbf{\Phi_b} = -1.265$$

Similarly, the correlator output when $s_1(t)$ is transmitted is $s_{1b} = 1.897$. The optimum threshold is midway between these values, or $V_t = 0.316$.

Error performance

Now we will derive an expression for the error probability for the optimum receiver. Figure 8.9a shows a signal-space diagram including the signal vectors $\mathbf{S_0}$ and $\mathbf{S_1}$. The length of the line from the tip of $\mathbf{S_0}$ to the tip of $\mathbf{S_1}$ is denoted as ℓ. The unit vector $\mathbf{\Phi_b}$, pointing in the direction from the tip of $\mathbf{S_0}$ to the tip of $\mathbf{S_1}$, is also shown in Figure 8.9. Notice that the length of the line from $\mathbf{S_0}$ to $\mathbf{S_1}$ is given by

$$\ell = |s_{1b} - s_{0b}| \tag{8.37}$$

where s_{0b} is the component of $\mathbf{S_0}$ in the direction of the unit vector $\mathbf{\Phi_b}$ given by

$$s_{0b} = \mathbf{S_0} \cdot \mathbf{\Phi_b} \tag{8.38}$$

Similarly, s_{1b} is the component of $\mathbf{S_1}$ in the direction of $\mathbf{\Phi_b}$. (Note that for the case illustrated in Figure 8.9, s_{0b} is a negative value and s_{1b} is positive.)

The conditional probabilities of the relevant component of the received vector r_b, are shown in Figure 8.9b. Note that when a 0 is transmitted, the mean value of r_b is s_{0b} and when a 1 is transmitted the mean value of r_b is s_{1b}. In either case, the variance of r_b is $\mathcal{N}_0/2$. This is because, as shown in Section 8.1, every component of a signal-space vector derived from WGN is gaussian with a variance of $\mathcal{N}_0/2$, and r_b is the component of the noise in the direction of $\mathbf{\Phi_b}$ plus the nonrandom signal component s_{ib}.

When a 1 is transmitted, the probability of the decision being in error is the area under $p(r_b|d=1)$ to the left of the decision line. Similarly, when a 0 is transmitted, the probability of error is the shaded area to the right of the decision line. When the probabilities of a 1 or a 0 are the same, so that the decision line is midway between $\mathbf{S_0}$ and $\mathbf{S_1}$, these error probabilities are the same. (We only consider the case of equally probable 1s and 0s in detail.) Therefore, the unconditional error probability is the same as the probability of error when a zero is transmitted, which is given by

$$P(E) = P(E|d=0) = \int_{V_t}^{\infty} p(r_b|d=0)\,dr_b \tag{8.39}$$

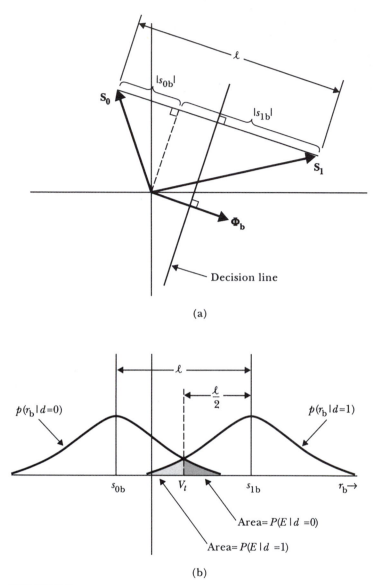

(a)

(b)

FIGURE 8.9
Signal-space diagram illustrating the distance ℓ between signals and the conditional PDFs of the relevant received vector component r_b.

Now, as we have seen, when a 0 is transmitted, r_b is a gaussian random variable with a mean of s_{0b} and a variance of $N_0/2$. Therefore, we can write Equation 8.39 as

$$P(E) = \int_{(s_{0b}+s_{1b})/2}^{\infty} (\pi N_0)^{-1/2} \exp\left[-\frac{(r_b - s_{0b})^2}{N_0}\right] dr_b \qquad (8.40)$$

Now, if the change of variables given by

$$\frac{x^2}{2} = \frac{(r_b - s_{0b})^2}{\mathcal{N}_0} \qquad (8.41)$$

is used in Equation 8.40, the result can be shown to be

■ $$P(E) = \int_{\ell/(2\mathcal{N}_0)^{1/2}}^{\infty} (2\pi)^{-1/2} \exp\left(\frac{-x^2}{2}\right) dx = Q\left[\frac{\ell}{(2\mathcal{N}_0)^{1/2}}\right] \qquad (8.42)$$

where the Q function was defined in Equation 7.55.

To summarize our results to this point, we have found several alternatives for the optimum receiver for binary digital signals in the presence of WGN shown in Figures 8.3 and 8.7b. The error probability of these receivers is given by Equation 8.42.

EXAMPLE 8.6

If the signals of Example 8.4 are used to communicate equally likely 1s and 0s, and any of the optimum receivers found in Examples 8.4 and 8.5 are used for reception, find the probability of error when the channel adds WGN with $\mathcal{N}_0/2 = 1$ W/Hz.

SOLUTION
The probability of error is given by Equation 8.42 for any of the optimum receivers that we have found for the specified signal set. First, we must find the signal-space distance ℓ between $\mathbf{S_0}$ and $\mathbf{S_1}$. This can be found by several methods. We could obtain ℓ from the signal-space diagram of Figure 8.6c by geometrical considerations. Alternately, it can be found as

$$\ell = |\mathbf{S_1} - \mathbf{S_0}| = |(-1,3) - (2,2)| = |(-3,1)| = [(-3)^2 + (1)^2]^{1/2}$$
$$= 10^{1/2} = 3.16$$

Since the square of a length in signal space is equal to the energy in the corresponding time signal, we can also find ℓ from the relation

$$\ell^2 = \int_0^2 [s_1(t) - s_0(t)]^2 \, dt = 10$$

Substituting the value found for ℓ into Equation 8.42, we find the probability of error as

$$P(E) = Q\left[\frac{\ell}{(2\mathcal{N}_0)^{1/2}}\right] = Q(1.58) = 5.69 \times 10^{-2}$$

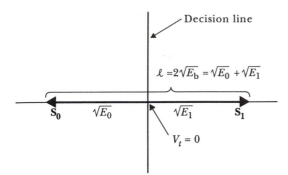

FIGURE 8.10
Signal-space diagram for antipodal signals.

Antipodal signal sets

So far we have considered the general case for the transmitted waveforms and the corresponding configuration of the signal vectors. Now we will consider several specific configurations of signal-space vectors that are often of interest. The first of these occurs when the signal $s_0(t)$ is the negative of the signal $s_1(t)$. In this case, the signal-space vectors for the two signals are of equal length and point in opposite directions, as indicated in Figure 8.10. This case is called an *antipodal signal set*. Notice that the line from the tip of $\mathbf{S_0}$ to the tip of $\mathbf{S_1}$ points in the same direction as $\mathbf{S_1}$. Therefore, the unit vector, $\mathbf{\Phi_b}$, perpendicular to the decision line is in the same direction as $\mathbf{S_1}$, and the waveform for $\phi_b(t)$ is the same as the waveform of $s_1(t)$.

We denote the energy in $s_1(t)$ as E_1 and the energy in $s_0(t)$ as E_0. The average energy per bit transmitted is given by

$$\blacksquare \qquad E_b = E_0 P(d = 0) + E_1 P(d = 1) \qquad (8.43)$$

For the antipodal signal set under consideration, we have

$$E_1 = E_0 = E_b \qquad (8.44)$$

Recall from our discussion of signal-space vectors in Section 8.1 that the length of a signal-space vector is the square root of the energy in the signal. This is indicated in Figure 8.10. For antipodal signals, we can conclude that the length ℓ from $\mathbf{S_0}$ to $\mathbf{S_1}$ is given by

$$\ell = 2(E_b)^{1/2} \qquad (8.45)$$

Substituting Equation 8.45 into 8.42, we find the probability of error for optimum reception of antipodal signals as

$$\blacksquare \qquad P(E) = Q\left[\left(\frac{2E_b}{N_0}\right)^{1/2}\right] \qquad \text{(antipodal signals)} \qquad (8.46)$$

This is plotted in Figure 8.11.

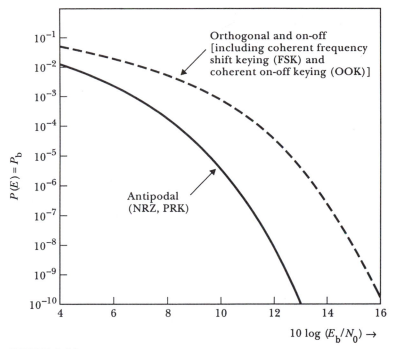

FIGURE 8.11
Bit-error rate (BER) curves for antipodal and orthogonal signal sets.

Usually, it is desirable to design a communication system for the minimum possible average transmitter power. This in turn makes it desirable to minimize the length of the signal-space vectors $\mathbf{S_0}$ and $\mathbf{S_1}$ since their length is proportional to the signal energies. At the same time, we want to minimize the bit-error probability $P(E)$. The error probability decreases rapidly as the distance ℓ between $\mathbf{S_0}$ and $\mathbf{S_1}$ increases. Therefore, it is desirable to maximize the distance between the signal vectors while minimizing the lengths of the vectors. Clearly, this optimum situation is achieved by the antipodal signal set. Certainly, for signal vectors of any specified length, the maximum distance is obtained between their tips when they point in opposite directions. It can be shown that, for a given total length of the two vectors, the average signal energy is minimum when the vectors are of equal length (provided that the signals are equally likely).

An optimum receiver for antipodal signals is shown in Figure 8.12. Notice from the signal-space diagram of Figure 8.10 that the threshold is zero, provided again that $P(d = 1) = P(d = 0) = \frac{1}{2}$. This is an advantage in a practical receiver because the attenuation in the channel is often not known; so the actual received signal amplitudes are unknown. Since the threshold is zero regardless of the signal amplitude, it is not necessary to know the received

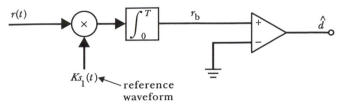

FIGURE 8.12
Optimum correlation receiver for antipodal signals.

amplitude to set the threshold. For the same reason, it is also not necessary to control the amplitude of the reference signal used in the correlation receiver shown in Figure 8.12. For some other signal sets, the optimum threshold depends on the received signal amplitude.

FIGURE 8.13
Signal-space diagram for orthogonal signals.

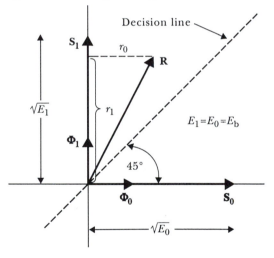

(a)

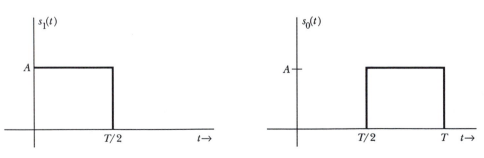

(b)

Orthogonal signal sets

Figure 8.13a is a signal-space diagram for the *orthogonal signal set*, showing the signal-space vectors at right angles. (We have assumed that the signals have equal energy.) For signal vectors at right angles, we have

$$\mathbf{S_0} \cdot \mathbf{S_1} = 0 \qquad (8.47)$$

Since the dot product corresponds to a correlation operation of the time signals, this is equivalent to the condition

$$\int_0^T s_0(t)s_1(t)\, dt = 0 \qquad (8.48)$$

Examples of orthogonal signal sets are shown in Figure 8.13b, c, and d. Figure 8.13a also shows the decision line. Note that the decision can be based on a comparison of the components of the received vector. If r_0 is greater than

FIGURE 8.13 (*continued*)

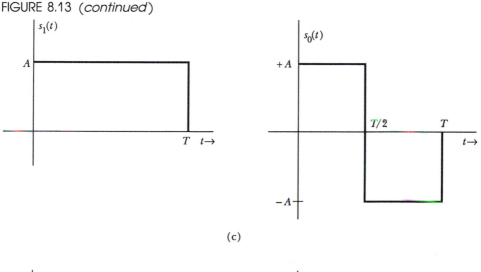

(c)

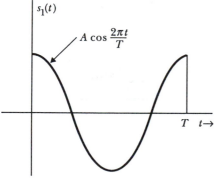
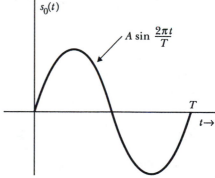

(d)

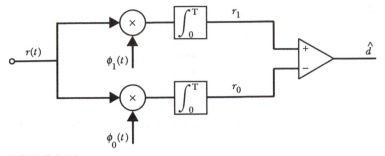

FIGURE 8.14
Optimum receiver for orthogonal signals.

r_1, then the decision is made in favor of a 0. (Note that we found it convenient to number the basis vector starting from 0 in this case, with $\mathbf{\Phi_0}$ in the same direction as $\mathbf{S_0}$.) One version of the optimum receiver using two correlators is shown in Figure 8.14. Of course, it is also possible to find a receiver with a single correlator for this signal set.

Since we have assumed that the energy in the two orthogonal signals is the same, the average energy, given by Equation 8.43, is equal to the energy in either signal:

$$E_b = E_1 = E_0 \qquad (8.49)$$

From the signal-space diagram of Figure 8.13a, we can easily determine that the distance between the signals is given by

$$\ell = (2E_b)^{1/2} \qquad (8.50)$$

Substituting Equation 8.50 into 8.42, we find the error probability for the orthogonal signal set as

$$\blacksquare \qquad P(E) = Q\left[\left(\frac{E_b}{N_0}\right)^{1/2}\right] \qquad \text{(orthogonal signals)} \qquad (8.51)$$

This is plotted in Figure 8.11. Note by comparing Equations 8.51 and 8.46 that the orthogonal signal set requires an E_b twice as high as the antipodal signal set to achieve the same error probability. Thus, the antipodal signal set is 3 dB more efficient in the use of signal power than the orthogonal set.

On-off signal sets

Another signal configuration of importance occurs when one of the signals is zero. We refer to this case as the *on-off signal set* because a signal is turned *on* when the message bit is, say, a 1 and *off* if the message bit is a 0. The signal-

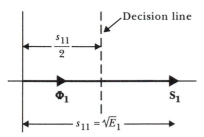

FIGURE 8.15
Signal-space diagram for on-off signals.

space diagram for this case is shown in Figure 8.15. The figure also shows the decision line and a unit vector $\mathbf{\Phi_1}$ in the direction of $\mathbf{S_1}$. Notice that, since the unit vector is perpendicular to the decision line, the relevant component of the received vector is the component in the direction of $\mathbf{\Phi_1}$.

As usual, the length of the signal vector $\mathbf{S_1}$ is the square root of its energy. Since the energy in the other signal is zero, the distance between the tips of the signal vectors is equal to the length of $\mathbf{S_1}$:

$$\ell = (E_1)^{1/2} \tag{8.52}$$

The average energy in a signal set is given by Equation 8.43. In the present case, we have $E_0 = 0$, and Equation 8.43 yields

$$E_b = \frac{1}{2} E_1 \tag{8.53}$$

where we have assumed equally likely data. Equations 8.52 and 8.53 can be solved for the signal-space distance ℓ in terms of the average energy per bit E_b. If the result is substituted into Equation 8.42, we find the probability of error for the on-off signal set as

$$P(E) = Q\left[\left(\frac{E_b}{\mathcal{N}_0}\right)^{1/2}\right] \tag{8.54}$$

Note that this is the same as the bit-error probability for the orthogonal signal set given by Equation 8.51. Thus, both the on-off and orthogonal signal sets perform 3 dB poorer than the antipodal signal set, which is optimum for transmission of binary data one bit at a time.

A correlation receiver for the on-off signal set is shown in Figure 8.16a. Notice that the threshold level depends on the amplitude of the received signal; so some mechanism must be provided in a practical implementation to adjust the threshold as the received-signal level varies.

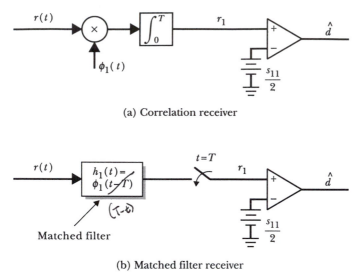

(a) Correlation receiver

(b) Matched filter receiver

FIGURE 8.16
Optimum receivers for on-off signals.

Matched-filter receivers

An important variation of any of the correlation receivers shown in this section is possible. The correlation operation of the received signal $r(t)$ with the reference waveform $\phi_1(t)$ is given by

$$r_1 = \int_0^T r(t)\phi_1(t)\, dt \qquad (8.55)$$

It can be easily shown, using the definition of the convolution operation, that the same result can be obtained from

$$r_1 = [r(t) \otimes \phi_1(T-t)]\Big|_{t=T} \qquad (8.56)$$

where $\otimes$ denotes the convolution operation. Since the output of a linear system is given by the convolution of the input with the impulse response of the system, the relevant component r_1 can be obtained by passing the received signal through a filter and sampling the output at $t = T$. The impulse response of the filter is given by

$$h_1(t) = \phi_1(T-t) \qquad (8.57)$$

The resulting filter is called a *matched filter*, said to be matched to the waveshape of $\phi_1(t)$. Since the waveshape of $\phi_1(t)$ is the same as the waveshape of $s_1(t)$, the filter is also matched for $s_1(t)$. A matched-filter receiver for the on-off signal set is shown in Figure 8.16b. The matched filter and sampler replaces the correlator.

356

Now we will give an alternative derivation of the error performance of the matched-filter receiver for the on-off signal set using the noise bandwidth concept discussed in Section 5.1. The noise equivalent bandwidth of the matched filter can be found by substituting the impulse response of Equation 8.57 into Equation 5.8. The result is

$$B_n = \frac{1}{2|H_0|^2} \int_0^\infty [\phi_1(T-t)]^2 \, dt \tag{8.58}$$

where H_0 is the nominal gain of the filter. Since the energy in the basis function $\phi_1(t)$ is unity, the value of the integral in Equation 8.58 can be easily shown to be unity. (Recall that we have assumed that $s_1(t)$, and therefore $\phi_1(t)$, is zero outside of the time range from 0 to T.) Thus, the noise bandwidth of the filter is

$$B_n = \frac{1}{2|H_0|^2} \tag{8.59}$$

The noise waveform at the input to the matched filter of Figure 8.16b is a sample function from a WGN process. The noise power at the output of the filter can be computed by substituting the noise bandwidth just found into Equation 5.5. The result is

$$P_n = |H_0|^2 B_n N_0 = \frac{N_0}{2} \tag{8.60}$$

We could have anticipated this result because the power in a noise waveform is the variance of a sample of the noise. In Section 8.1 we showed that the variance of components of WGN, such as n_1, is $N_0/2$. Since n_1 is a sample of the noise at the filter output, we could have concluded that the noise power at the filter output is also $N_0/2$.

The conditional PDF of the matched-filter output (at $t = T$) is shown in Figure 8.17. The probability of error when the transmitted bit is a 1 is the shaded area indicated below the threshold. By integrating the PDF, the error probability can be found as

$$P(E) = Q\left[\frac{s_{11}}{2}\left(\frac{2}{N_0}\right)^{1/2}\right] = Q\left[\left(\frac{E_b}{N_0}\right)^{1/2}\right] \tag{8.61}$$

where s_{11} is the filter output due to the signal at $t = T$. [s_{11} is also the component of $\mathbf{S_1}$ in the direction of $\mathbf{\Phi_1}$ and the square root of the energy in $s_1(t)$.]

The matched-filter receiver is, like the correlation receiver, optimum in minimizing the error probability. The probability of error is minimum when the argument of the Q function in Equation 8.61 is maximum. Thus, the matched filter maximizes the ratio

$$\frac{2(s_{11})^2}{N_0}$$

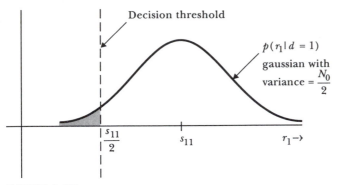

FIGURE 8.17
Conditional PDF of the matched-filter output at $t = T$.

where s_{11} is the signal amplitude at the filter output at $t = T$. In fact, s_{11} is the peak value of the signal component at the matched-filter output for all time instants. Therefore, the square of s_{11} is the peak signal power at the filter output. $N_0/2$ is the average noise power at the output of the matched filter. Thus, the matched filter maximizes the ratio of the peak signal power to average noise power at the filter output. Of all possible filters, the matched filter is optimum when we want to maximize the peak SNR at the filter output for a known pulse shape in the presence of additive WGN. Because of this fact, matched filters are useful when we want to detect the presence of pulses of a known shape that are immersed in noise. For example, radar systems often use matched-filter techniques in detecting return echos.

Now that we have presented the principles of optimum receivers for binary digital signals and some of the common signal configurations, we will turn to a detailed discussion of specific digital modulation formats.

8.3
EXAMPLES OF BINARY COMMUNICATION WITH KNOWN SIGNALS

We have discussed many of the concepts in binary communication systems when the transmitted waveform is completely known in advance to the receiver designer. We will now show how these principles apply to several commonly used digital signaling formats.

Baseband nonreturn-to-zero system

First, we consider a baseband system using the NRZ waveform described in Section 7.4. A diagram of such a system is shown in Figure 8.18a. A typical transmitted NRZ waveform is shown in Figure 8.18b. Notice that, in each bit interval, the transmitted waveform for a 0 is the negative of the waveform for

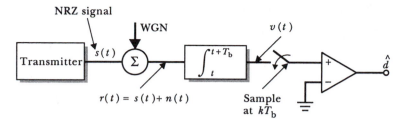

NRZ signal

WGN

$v(t)$

Transmitter — $s(t)$ — Σ — $\int_{t}^{t+T_b}$ —

$r(t) = s(t) + n(t)$

Sample at kT_b

$\hat{d}$

(a) Conceptual baseband system

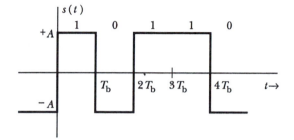

(b) Typical NRZ waveform

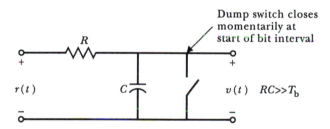

Dump switch closes momentarily at start of bit interval

R

$r(t)$

C

$v(t)$ $RC >> T_b$

(c) An implementation of the integrate-and-dump circuit

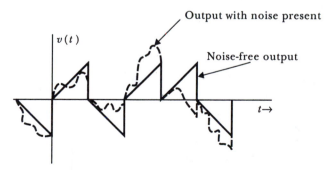

Output with noise present

$v(t)$

Noise-free output

(d) Output of integrate-and-dump circuit

FIGURE 8.18
NRZ baseband communication system.

a 1. Therefore, the NRZ waveform is an example of an antipodal signal set.

The correlation receiver for antipodal signals, shown in Figure 8.12, first multiplies the received waveform by the basis function $\phi_1(t)$ and then integrates the product over the bit interval. Since the transmitted waveform is constant during the bit interval, the basis function is also constant. Therefore, the multiplier in the correlation operation can be eliminated. Thus, the optimum receiver simply integrates the received waveform over each bit interval and decides in favor of a 1 if the result is positive or a 0 if the integral is negative. A conceptual diagram of the receiver is also shown in Figure 8.18a.

One method for implementing the integrator is shown in Figure 8.18c. The received signal-plus-noise is applied to the input of the circuit. The switch closes for an instant at the beginning of each bit interval to discharge the capacitor to zero. If the time constant of the RC circuit is much longer than the bit interval, it can be shown that the output voltage at the end of a bit interval is approximately the integral of the input over the past bit interval. The voltage at the end of each bit interval is sampled and compared with zero to make the data decision. This circuit is known as an *integrate-and-dump circuit* because it integrates for a bit interval and then the capacitor is "dumped" by the discharge switch. The waveform across the capacitor, corresponding to the received NRZ waveform of part b of the figure, looks like the solid line in Figure 8.18d when no noise is added in the channel; a typical waveform at the output of the integrate-and-dump circuit in the presence of channel noise resembles the dotted line.

An alternative to the integrate-and-dump receiver is the matched-filter receiver shown in Figure 8.19a. The impulse response of the matched filter appropriate for the NRZ waveform is shown in Figure 8.19b. The amplitude K of the impulse response is arbitrary and does not affect the final decision because the threshold is zero.

The transfer function of the filter can be found by taking the Fourier transform of the impulse response. The result is

$$H_m(f) = KT_b \operatorname{sinc}(fT_b) \exp(-j\pi T_b f) \qquad (8.62)$$

FIGURE 8.19
Matched-filter receiver for NRZ baseband communication.

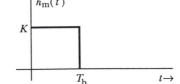

(a) Matched-filter receiver

(b) Impulse response of matched filter

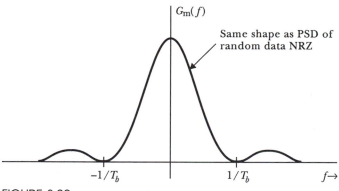

FIGURE 8.20
Power gain of matched filter versus frequency.

The power gain of the matched filter is simply the square of the magnitude of the transfer function, given by

$$G_m(f) = |H_m(f)|^2 = (KT_b)^2 \operatorname{sinc}^2 (fT_b) \tag{8.63}$$

This is plotted in Figure 8.20.

Comparison of the power gain of the matched filter with the PSD of the NRZ waveform shown in Figure 7.20b reveals that they have the same form. This is another reason for the term *matched filter*. Also notice that the matched filter for the NRZ signal has a lowpass characteristic. The matched filter is often approximated in practice by various lowpass filter circuits.

The output waveform of the matched filter can be found by convolving its impulse response with the input waveform. A typical noise-free input is shown in Figure 8.21a, and the corresponding matched-filter output is shown in part b of the figure. Part c shows the output of a simple *RC* lowpass filter approximation to the matched filter.

The eye pattern

The *eye pattern* of a filtered digital waveform is obtained by displaying the waveform on an oscilloscope that is triggered by the data clock signal. The pattern superimposes all of the transient responses of the filtered data signal. The eye pattern of the matched-filter output for the NRZ waveform is shown in Figure 8.21d; that for the *RC* filter output is shown in part e of the figure. Generally, the optimum sampling point is at the widest opening of the eye.

When the filter output at the sampling instant depends only on the value of the last data bit, the eye pattern indicates this fact because all of the transients pass through just two points at the sampling instant. This is the case for the matched-filter output of Figure 8.21b and d. At the end of a bit interval the simple *RC* lowpass filter output depends on several of the past data bits. This results in *intersymbol interference (ISI)*, which we discussed in Chapter 6.

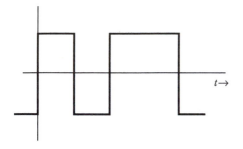

(a) NRZ waveform

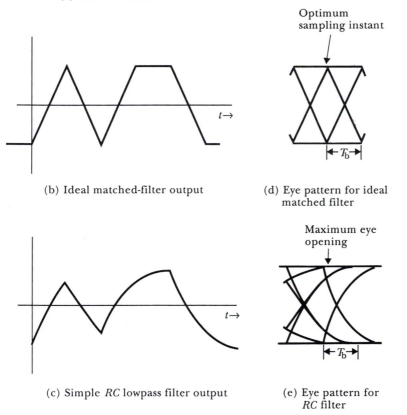

(b) Ideal matched-filter output

(c) Simple *RC* lowpass filter output

Optimum sampling instant

(d) Eye pattern for ideal matched filter

Maximum eye opening

(e) Eye pattern for *RC* filter

FIGURE 8.21
Filtered NRZ waveforms and corresponding eye patterns.

The fact that ISI exist in the *RC* filter output is apparent in the eye pattern of Figure 8.21e. The eye pattern takes more than two values at the sampling instant.

The eye pattern provides a convenient way to judge the performance of a digital receiver. The width of the opening reflects on the error performance since a greater distance between the levels yields a smaller error probability. The optimum sampling instant is at or very near the widest opening. The tolerence of the system to timing errors in the sampling can be judged by observing the length of time that the pattern stays at its widest opening. As we have illustrated, the presence of ISI is also apparent in the pattern. ISI can result from the use of poor approximations to the matched filter or from channel effects such as multipath. ISI usually leads to degradation in system performance.

A practical implementation of the NRZ receiver

A more practical diagram of a receiver for NRZ signals in the presence of WGN is shown in Figure 8.22. The attenuated NRZ signal plus channel noise is first amplified by a low noise amplifier to provide a high-level input to the matched filter and clock recovery circuit. The effective value of N_0 used in error probability calculations should include both the channel noise and the noise added in the receiver. The noise performance of the receiver is primarily determined by the low noise amplifier if its gain is high. As we have shown in Chapter 5, N_0 can be determined from knowledge of the noise added by the channel and the noise figure of the receiver.

The matched filter shown in Figure 8.22 is an approximation to the ideal matched filter. The transfer function of the ideal matched filter given in Equation 8.62 is a sinc function with its first null at the data rate f_b. The values of the elements in the approximate matched filter should be chosen to approximate the frequency response of the ideal matched filter. For example, the parallel combination of L and C_1 should be selected to be resonant at f_b to provide the null in the transfer characteristic.

The order of the decision element (comparator) and the sampling function (flip-flop) have been interchanged in Figure 8.22 compared with most of

FIGURE 8.22
Practical implementation of matched-filter receiver for NRZ baseband signaling.

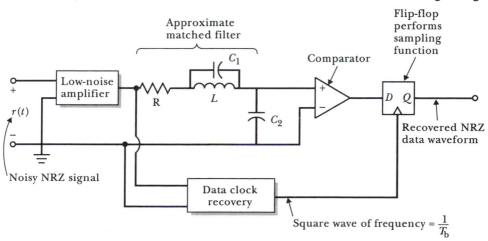

the previous diagrams. The output of the approximate matched filter is compared with zero by the comparator. The output of the comparator is a logic level indicating whether the filter output is positive or negative. The data clock recovery circuit provides a square wave at the data rate. If the flip-flop is the positive-going edge-triggered type, the positive-going edge of the square wave should be time-aligned with the sampling instant. In practice, this can be achieved by displaying the data clock and the output of the matched filter on a dual trace oscilloscope that is triggered by the data clock. The data recovery circuitry can then be adjusted to align the positive going edge of the data clock signal with the widest opening of the eye pattern. The details of the data clock recovery function will be discussed in Section 8.6.

Since the NRZ waveform is an antipodal signal set in each bit interval, and the sampled output of the ideal matched filter depends only on one bit, the error performance of an ideal receiver for NRZ signal in the presence of WGN was given in Equation 8.46. The ideal error performance was plotted in Figure 8.11, but the error performance of an actual system is usually worse, due to a number of possible causes. These include the fact that the matched filter is often only an approximation, that bandlimiting of the transmitted signal or multipath effects in the channel may cause ISI, and that the recovered data clock signal often has significant timing jitter due to noise. The performance of an actual system is displaced to the right of the ideal *bit-error rate* (BER) curve of Figure 8.11. (The term *bit-error rate* is often used with the same meaning as bit-error probability.) The amount that it is displaced to the right typically ranges from a few tenths to several decibels for systems that nominally meet the WGN assumption and are well designed.

Phase reversal keyed system

Now we consider a bandpass communication system in which the data sequence is used to modulate a sinusoidal carrier. The particular modulation scheme is known either as *phase reversal keying* (PRK) or as *biphase shift keying* (BPSK). In this type of digital modulation, the transmitted signal is a sinusoid of constant amplitude and 0° phase during the bit intervals for which the message bit is a 1. When the message bit is a 0, the phase of the transmitted sinusoid is reversed to 180°. This type of modulation can be generated by multiplying a sinusoidal carrier by an NRZ data waveform. Thus the PRK signal is given by

$$\blacksquare \qquad g_{\text{PRK}}(t) = A_c d(t) \cos{(2\pi f_c t)} \qquad (8.64)$$

where $d(t)$ is the NRZ data waveform that we assume takes on values of ± 1. Since the signal for a 0 is the negative of the signal for a 1, PRK produces an antipodal signal set in each bit interval.

Figure 8.23a shows a conceptual diagram of a PRK system. In the modulator, the NRZ data waveform is multiplied by the carrier to produce the PRK signal. A typical NRZ data waveform is shown in Figure 8.23b. Part c

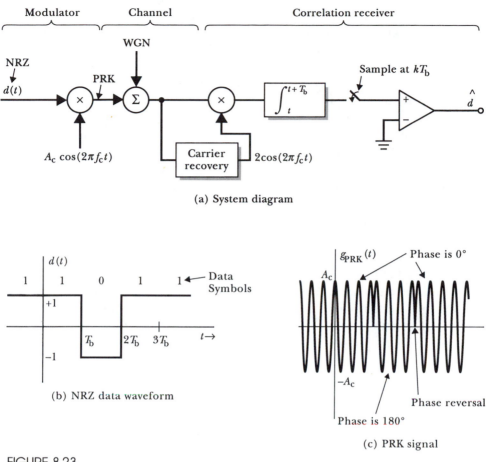

(a) System diagram

(b) NRZ data waveform

(c) PRK signal

FIGURE 8.23
PRK communication system.

of the figure shows the corresponding PRK signal. The channel adds a sample function from a WGN process to the PRK signal and presents the sum to the receiver. Since the signal set is antipodal, a version of the optimum receiver is shown in Figure 8.12. The reference waveform is the transmitted signal when a 1 is transmitted. In other words, the reference waveform is the carrier at 0°. This carrier must be available at the receiver with exactly the same frequency and phase as at the transmitter (allowing for doppler and phase shift in the channel). Usually, the carrier reference is derived from the received signal. We will discuss the details of carrier recovery briefly in Section 8.6.

In the absence of noise, the output of the multiplier in the receiver of Figure 8.23a is given by

$$[A_c d(t) \cos 2\pi f_c t] \times (2 \cos 2\pi f_c t) = A_c d(t) + A_c d(t) \cos (4\pi f_c t) \quad (8.65)$$

in which we have used the identity for cosine squared to obtain the right side. This is the input to the integrator. If, as is often the case, the carrier frequency is high compared with the data rate, the double carrier frequency term on the right side of Equation 8.65 does not cause a significant contribution to the output of the integrator. Thus, the effective part of output of the multiplier is an NRZ data waveform plus noise. (The PRK system can be viewed as a DSB-SC system in which the data waveform is modulated onto the carrier by the multiplier at the sending end and demodulated by the multiplier in the receiver.)

The integrator in Figure 8.23a can be implemented as an integrate-and-dump circuit, shown previously in Figure 8.18c. When the carrier frequency is high, the output waveform from the integrate-and-dump circuit for the PRK system is nearly the same as the waveform shown in Figure 8.18d for the baseband system. Alternately, the integrator in the PRK system can be implemented as a matched filter for the NRZ waveform as shown in Figure 8.19.

The PSD of the PRK signal is the same as the PSD of the NRZ data waveform except that it is translated to be centered at the carrier frequency. A plot of the PSD of the PRK signal is shown in Figure 8.24. Notice that the null-to-null bandwidth of the PRK signal is twice the data rate.

As we pointed out in Section 8.2, the correlators in optimum receivers can be replaced by matched filters. The impulse response of the matched filter is a time-reversed and delayed version of the signal for which it is matched. When the message bit is a 1 in the interval from $t = 0$ to $t = T_b$, the transmitted signal for this single data bit is

$$
\begin{aligned}
s_1(t) &= A_c \cos\ (2\pi f_c t) & 0 < t < T_b \\
&= 0 & \text{otherwise}
\end{aligned}
\tag{8.66}
$$

The impulse response of the matched filter for this signal is

$$
\begin{aligned}
h_m(t) &= s_1(T_b - t) \\
&= A_c \cos\ [2\pi f_c(T_b - t)] & 0 < t < T_b \\
&= 0 & \text{otherwise}
\end{aligned}
\tag{8.67}
$$

FIGURE 8.24
PSD of PRK signal with random data. (The PSD is identical for negative frequencies.)

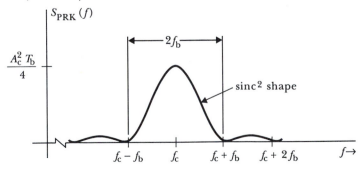

The signal $s_1(t)$, matched-filter impulse response, matched-filter transfer function, and the filter output due to the single data pulse are shown in Figure 8.25. Notice that the frequency response of the matched filter in this case is a bandpass characteristic. The shape of the transfer magnitude is a sinc function centered at the carrier with a null-to-null bandwidth equal to twice the data rate. We may have anticipated this form of the transfer function because it results in a power gain characteristic that matches the PSD of the PRK signal. Notice that the output of the matched filter corresponding to a single bit reaches its maximum amplitude at the end of the bit interval. Since the output of the filter is a modulated sinusoid at the carrier frequency, the peak amplitude is of very short duration, and the sampling instant must occur within a small fraction of a carrier cycle from the correct point. The matched-filter receiver for the PRK signal is shown in Figure 8.26a, but this receiver is seldom used due to the necessity for tight control of the sampling instant. Instead, a receiver that converts the output of the filter to baseband in a coherent demodulator is used in practice. The diagram of this receiver is shown in Figure 8.26b. The received signal-plus-noise is first passed through the matched filter. The output of the filter is multiplied by the recovered carrier. The product is passed through a lowpass filter to eliminate the components in the vicinity of twice the carrier frequency. When the carrier frequency is high compared with the data rate, the resulting (noise-free) waveform presented to the sampler is identical to the waveform of Figure 8.21b. The sampling point for this baseband waveform does not require the extreme accuracy that the matched-filter output does. The sampling point must only be controlled to within a small fraction of the bit interval rather than a small fraction of the carrier cycle.

Coherent on-off keying

On-off keying (OOK) is a form of modulation in which the transmitted signal is a constant amplitude sinusoid during intervals when the message bit is a 1 and the modulated signal is zero when the message bit is a 0. Two variations of OOK modulation are possible. In *coherent OOK* the phase of the sinusoid is the same each time a sinusoidal pulse occurs. This can be achieved by having a continuously running carrier oscillator that is switched to the output by the modulator when the message bit is a 1. In the other version of OOK, known as *noncoherent OOK*, the phase of the transmitted sinusoidal pulses are random variables that are independent from pulse to pulse. Noncoherent OOK can result if the carrier oscillator is turned *on* and *off* by the data. Each time the oscillator is turned on, it can start with a different phase angle.

When the transmitted signal is coherent so that the phase is the same each time the signal is turned on, the receiver can recover the original carrier for use in a correlation receiver. Such a receiver, which uses the recovered carrier, is called a *coherent receiver*. Receivers that do not recover the carrier, known as *noncoherent receivers*, are possible for OOK signals. We will discuss

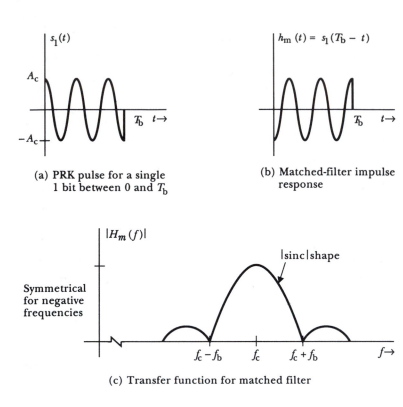

(a) PRK pulse for a single
1 bit between 0 and T_b

(b) Matched-filter impulse
response

$|H_m(f)|$

$|sinc|$ shape

Symmetrical
for negative
frequencies

$f_c - f_b$ f_c $f_c + f_b$ $f \rightarrow$

(c) Transfer function for matched filter

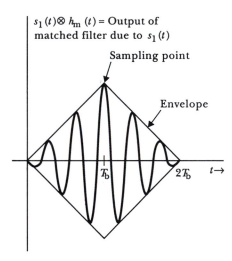

$s_1(t) \otimes h_m(t) =$ Output of
matched filter due to $s_1(t)$

Sampling point

Envelope

T_b $2T_b$ $t \rightarrow$

(d) Matched-filter output

FIGURE 8.25
Bandpass matched filter for PRK.

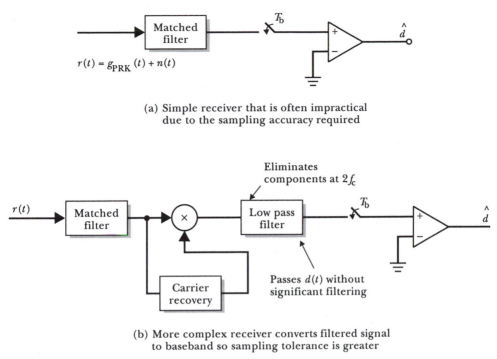

(a) Simple receiver that is often impractical
due to the sampling accuracy required

(b) More complex receiver converts filtered signal
to baseband so sampling tolerance is greater

FIGURE 8.26
PRK receivers using a bandpass matched filter.

noncoherent receivers in the next section. In this section, we consider only coherent reception of coherent OOK signals.

A diagram of a communication system using coherent OOK modulation and a coherent receiver is shown in Figure 8.27a. The input signal to the modulator is assumed to be $\frac{1}{2}d(t) + \frac{1}{2}$ where $d(t)$ is an NRZ waveform taking values of ± 1. A plot of $\frac{1}{2}d(t) + \frac{1}{2}$ is shown in Figure 8.27b for a typical data sequence. The corresponding coherent OOK signal is shown in Figure 8.27c. Note that every time the signal is turned on by the presence of a 1 in the data stream, the phase is the same as the continuation of the previous pulse indicated by the dotted line. This phase continuity is the characteristic that causes this OOK signal to be coherent. An expression for the coherent OOK signal is

$$ g_{\text{OOK}}(t) = A_c \left[\frac{1}{2} + \frac{1}{2}d(t) \right] \cos{(2\pi f_c t)} \tag{8.68} $$

Notice that the expression contains an unmodulated carrier component. In fact, it can be easily shown that exactly half of the power is in the carrier component. The other term can be recognized as a PRK signal. Thus, the coherent OOK signal consists of PRK plus a carrier component. Coherent OOK can be viewed as an AM signal just as PRK can be viewed as a DSB-SC signal. Recall that adding a carrier component to DSB-SC produces AM. The

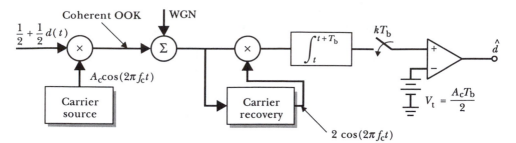

(a) System diagram

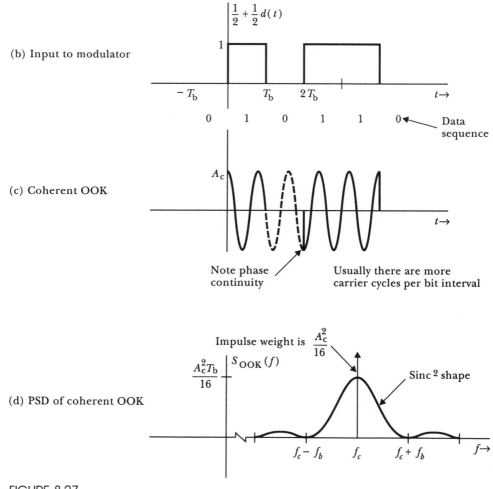

(b) Input to modulator

(c) Coherent OOK

Note phase continuity

Usually there are more carrier cycles per bit interval

(d) PSD of coherent OOK

FIGURE 8.27
Coherent OOK communication system.

PSD of the coherent OOK signal is the same as the PSD of PRK with an impulse added at the carrier frequency. This is shown in Figure 8.27d.

The coherent OOK signal is an example of an on-off signal set, as discussed in Section 8.2, with

$$s_1(t) = A_c \cos (2\pi f_c t)$$

An optimum correlation receiver for on-off signals was shown in Figure 8.16. The receiver shown in Figure 8.27a is a version of this, except that the amplitude of the reference (recovered carrier) has not been adjusted to make it a unit energy basis function. The amplitude of the reference is unimportant as long as the threshold is set appropriately. The proper setting for the threshold is exactly halfway between the output of the integrator when a 1 is transmitted and the output when a 0 is transmitted, under noise-free conditions in both cases.

Since OOK modulation is an example of an on-off signal set, the BER performance of the optimum receiver is shown in Figure 8.11. Notice that coherent OOK is exactly 3 dB poorer in performance (more transmitter power is required) than PRK. This can be attributed to the fact that the OOK signal has exactly half of its power in an unmodulated carrier component, whereas PRK has all if its power in the side frequencies. Because the coherent receiver for OOK signals is as complex as the receiver for PRK, there is no hardware advantage for OOK. Thus, coherent OOK systems are seldom used. We have presented the coherent OOK system mainly to contrast it with the noncoherent approach discussed in the next section.

If we should want to use a coherent receiver for OOK, it is necessary to insure that the transmitted signal is coherent since the carrier recovery system obtains the carrier phase by averaging the phase of a large number of received pulses. As we will see, a noncoherent receiver is not sensitive to the phase of the transmitted signal. Therefore, when a noncoherent receiver is in use, the transmitted signal can be either coherent or noncoherent.

8.4
NONCOHERENT RECEIVERS

In this section, we present optimum receivers for several noncoherent binary signaling schemes. As we pointed out in the previous section, a signal is said to be noncoherent if the carrier phase is a random variable that is independent from bit interval to bit interval and is uniformly distributed from 0 to 2π.

OOK

We consider the case of noncoherent OOK first. The noncoherent OOK signal in the bit interval from 0 to T_b is given by either

$$s_0(t) = 0 \tag{8.69}$$

or

$$s_1(t) = A_c \cos (2\pi f_c t + \theta) \qquad 0 < t < T_b$$
$$= 0 \qquad \qquad \qquad \text{for other values of } t \qquad (8.70)$$

depending on whether the message bit is a 1 or a 0. The phase angle θ is a random variable uniformly distributed from 0 to 2π. (We assume throughout our discussion that $2f_c T_b$ is an integer or is very large compared with unity.) The fact that the received signal contains an unknown random variable other than the data value is a new situation compared with that discussed in Section 8.2. In the previous work, we assumed that the received signal was completely known except for the data.

The energies in the two signals $s_0(t)$ and $s_1(t)$ are

$$E_0 = 0 \qquad \text{and} \qquad E_1 = \frac{A_c^2 T_b}{2}$$

The average energy per bit can be computed from Equation 8.43 as

$$E_b = E_0 P(d = 0) + E_1 P(d = 1)$$

Assuming that the message bit is equally likely to be a 1 or a 0, this reduces to

$$E_b = \frac{A_c^2 T_b}{4} \qquad (8.71)$$

To develop the optimum receiver structure, we expand the received signal-plus-noise in a generalized Fourier series as we did for the known signal in Section 8.2. First, notice that by application of the identity for the cosine of the sum of two angles, we can write

$$s_1(t) = A_c \cos (2\pi f_c t + \theta)$$
$$s_1(t) = A_c \cos \theta \cos (2\pi f_c t) - A_c \sin \theta \sin (2\pi f_c t) \qquad (8.72)$$

We choose the basis functions so that the first two are given by

$$\phi_1(t) = \left(\frac{2}{T_b}\right)^{1/2} \cos (2\pi f_c t) \qquad 0 < t < T_b$$
$$= 0 \qquad \qquad \qquad \text{otherwise} \qquad (8.73)$$

and

$$\phi_2(t) = -\left(\frac{2}{T_b}\right)^{1/2} \sin (2\pi f_c t) \qquad 0 < t < T_b$$
$$= 0 \qquad \qquad \qquad \text{otherwise} \qquad (8.74)$$

It can be shown that the basis functions given in Equations 8.73 and 8.74 are indeed orthonormal if $2f_c T_b$ is either an integer or is very large compared with unity. This implies that the number of half carrier cycles per bit interval is either an integer or very large. Often the carrier frequency is much higher than the data rate, so there is a large number of cycles per bit.

Notice that when the message bit is a 1, the expansion of the signal given in Equation 8.72 is the orthonormal series expansion for the basis functions we have selected. By comparing Equation 8.72 with 8.73 and 8.74, we can see that the orthonormal series coefficients are given by

$$s_{11} = \left(\frac{T_b}{2}\right)^{1/2} A_c \cos \theta = (2E_b)^{1/2} \cos \theta \qquad (8.75)$$

and

$$s_{12} = \left(\frac{T_b}{2}\right)^{1/2} A_c \sin \theta = (2E_b)^{1/2} \sin \theta \qquad (8.76)$$

We have used Equation 8.71 for the average energy per bit to obtain the expressions on the far right sides. Since the signal for a 0 data bit is zero, the coefficients for $s_0(t)$ are

$$s_{01} = s_{02} = 0 \qquad (8.77)$$

Since both signals are completely represented by the first two basis functions, the coefficients of any higher-order basis functions are irrelevant, just as in the known signal case. This is because the higher-order coefficients of the received signal depend only on the noise and are statistically independent of the first two coefficients.

At this point, we could formally develop the optimum receiver structure by use of the likelihood ratio test, as we did in Section 8.2 for the known signal case. The reader who is interested in a more formal development is referred to Whalen. We will use an intuitive approach instead.

For a particular known signal phase, denoted by $\theta = \theta_1$, the signal-space diagram of the two signals is shown in Figure 8.28a. Notice that the phase angle of the signal $s_1(t)$ is also the angle of the vector $\mathbf{S_1}$ in signal space. For the choice of basis functions we have made, the signal-space diagram effectively becomes a phasor diagram (except for scaling of the lengths of the phasors).

Since the phase is random and uniformly distributed from 0 to 2π, the signal-space vector $\mathbf{S_1}$ can end at any point on the circle shown in Figure 8.28b. Because of the random parameter θ, included in the noncoherent signal, the signal does not map into a point in signal space as was the case for completely known signals. Clearly, since the signal-space diagram is circularly symmetric, the optimum decision line is also a circle as indicated in Figure 8.28b. Thus, the optimum receiver considers only the length ℓ of the received vector in making the decision about the transmitted data.

The block diagram of an optimum receiver for noncoherent OOK is shown in Figure 8.29. The received signal is correlated with the two basis functions to obtain r_1 and r_2. The basis functions are two phases of the carrier

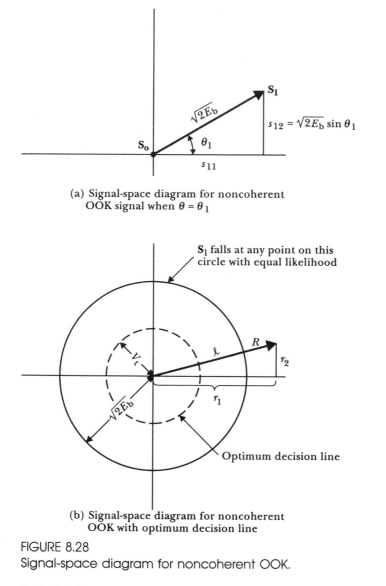

(a) Signal-space diagram for noncoherent
OOK signal when $\theta = \theta_1$

(b) Signal-space diagram for noncoherent
OOK with optimum decision line

FIGURE 8.28
Signal-space diagram for noncoherent OOK.

FIGURE 8.29
An optimum receiver for noncohorent OOK signals.

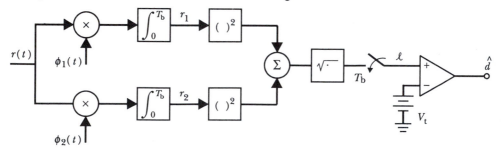

8 / DIGITAL MODULATION TECHNIQUES

that are 90° apart. The length of the received signal is then computed as

$$\ell = [(r_1)^2 + (r_2)^2]^{1/2} \qquad (8.78)$$

As shown in Figure 8.29, the value of ℓ is then compared with the threshold to make the decision.

For a given signal phase, the components of the received vector r_1 and r_2 are gaussian random variables because we have assumed that the channel noise is gaussian and the components are obtained from the channel signal by a linear operation. The length of the received vector ℓ is not gaussian because it is obtained from r_1 and r_2 through the use of nonlinear operations as indicated by Equation 8.78. It can be shown (see Whalen, for example) that the conditional PDFs of ℓ are given by

$$p(\ell|d=0) = \frac{2\ell}{N_0} \exp\left(-\frac{\ell^2}{N_0}\right) \qquad\qquad 0 < \ell < \infty$$

$$= 0 \qquad\qquad\qquad\qquad\qquad \ell < 0 \qquad (8.79)$$

and

$$p(\ell|d=1) = \frac{2\ell}{N_0} \exp\left(-\frac{\ell^2 + 2E_b}{N_0}\right) I_0\left(\frac{2\ell(2E_b)^{1/2}}{N_0}\right) \qquad 0 < \ell$$

$$= 0 \qquad\qquad\qquad\qquad\qquad\qquad\qquad\qquad \ell < 0 \qquad (8.80)$$

where $I_0(\cdot)$ is the *modified Bessel function* of the first kind and order zero, defined by

$$I_0(x) = \frac{1}{2\pi} \int_{-\pi}^{\pi} \exp\left(x \cos \phi\right) d\phi \qquad (8.81)$$

Equation 8.79 is known as a *Rayleigh PDF*, and Equation 8.80 is called the *Rician PDF*. Typical plots of these PDFs are shown in Figure 8.30. When the message bits are equally likely to be a 1 or a 0, the optimum threshold value occurs at the point where these PDFs intersect, as indicated in Figure 8.30.

FIGURE 8.30
Conditional PDFs of ℓ.

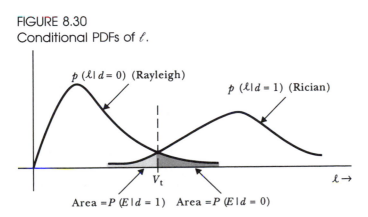

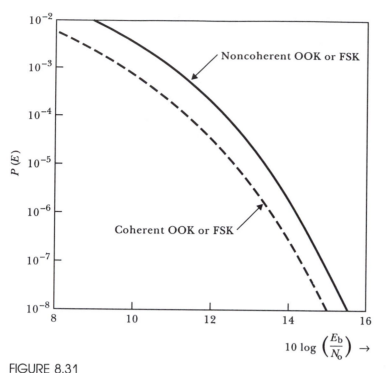

FIGURE 8.31
Bit-error probability curves for noncoherent OOK and FSK compared with coherent OOK and FSK.

The conditional probabilities of error are the shaded areas indicated in the figure.

Under high SNR conditions, the optimum threshold is approximately half of the signal output when the message bit is a 1. For the receiver of Figure 8.29 this threshold is

$$V_t = \left(\frac{E_b}{2}\right)^{1/2} \tag{8.82}$$

The resulting probability of error is given by the approximate expression

$$\blacksquare \qquad P(E) \simeq \frac{1}{2} \exp\left(-\frac{E_b}{2N_0}\right) \qquad \text{(noncoherent OOK)} \tag{8.83}$$

This is plotted in Figure 8.31, which also shows the error probability for coherent OOK. Notice that the degradation in BER performance of a non-coherent system compared with a coherent system is very small at high SNR.

Now we demonstrate that the optimum receiver of Figure 8.29 has the much simpler equivalent shown in Figure 8.32. This receiver contains a band-pass matched filter followed by an envelope detector. The output of the enve-lope detector is sampled at the end of the bit interval, and the value obtained

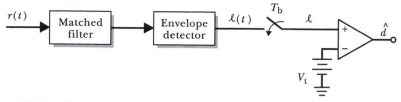

FIGURE 8.32
Simpler optimum receiver for noncoherent OOK.

is compared with the threshold. Now we show that the output of the envelope detector, sampled at $t = T_b$, is the same as the value ℓ in the more complex receiver of Figure 8.29.

The impulse response of the matched filter in Figure 8.32 is given by

$$h_m(t) = \left(\frac{2}{T_b}\right)^{1/2} \cos\left(2\pi f_c t + \phi\right) \qquad 0 < t < T_b$$
$$= 0 \qquad\qquad\qquad\qquad\qquad \text{otherwise} \qquad (8.84)$$

where ϕ is an arbitrary phase angle. The output of this filter due to its input signal $r(t)$ is given by the convolution integral:

$$v_o(t) = \int_{t-T_b}^{t} r(x) \left(\frac{2}{T_b}\right)^{1/2} \cos\left[2\pi f_c(t - x) + \phi\right] dx \qquad (8.85)$$

Using the identity for the cosine of the difference of two angles, we can write the last equation as

$$v_o(t) = \left[\left(\frac{2}{T_b}\right)^{1/2} \int_{t-T_b}^{t} r(x) \cos\left(2\pi f_c x\right) dx\right] \cos\left(2\pi f_c t + \phi\right)$$
$$+ \left[\left(\frac{2}{T_b}\right)^{1/2} \int_{t-T_b}^{t} r(x) \sin\left(2\pi f_c x\right) dx\right] \sin\left(2\pi f_c t + \phi\right) \qquad (8.86)$$

This can be written as

$$v_o(t) = r_1(t) \cos\left(2\pi f_c t + \phi\right) + r_2(t) \sin\left(2\pi f_c t + \phi\right) \qquad (8.87)$$

where $r_1(t)$ and $r_2(t)$ are defined as

$$r_1(t) = \left[\left(\frac{2}{T_b}\right)^{1/2} \int_{t-T_b}^{t} r(x) \cos\left(2\pi f_c x\right) dx\right] \qquad (8.88)$$

and

$$r_2(t) = \left[\left(\frac{2}{T_b}\right)^{1/2} \int_{t-T_b}^{t} r(x) \sin\left(2\pi f_c x\right) dx\right] \qquad (8.89)$$

Notice that the value of $r_1(T_b)$ is the same as the output of the top correlator in the receiver of Figure 8.29, which is denoted simply by r_1. Similarly, the value $r_2(T_b)$ is the same as the output of the bottom correlator, r_2.

The expression for the matched-filter output given in Equation 8.87 can be written as

$$v_o(t) = \{[r_1(t)]^2 + [r_2(t)]^2\}^{1/2} \cos\left\{2\pi f_c t + \phi - \tan^{-1}\left[\frac{r_2(t)}{r_1(t)}\right]\right\} \quad (8.90)$$

The output of the envelope detector is the amplitude of the last expression, which is given by

$$\ell(t) = \{[r_1(t)]^2 + [r_2(t)]^2\}^{1/2} \quad (8.91)$$

At $t = T_b$ the last expression becomes

$$\ell(T_b) = [(r_1)^2 + (r_2)^2]^{1/2} = \ell \quad (8.92)$$

Therefore, the output of the envelope detector at $t = T_b$ is the same value ℓ produced in the receiver of Figure 8.29, and thus, the simple receiver of Figure 8.32 is equivalent.

Comparison of the noncoherent receiver of Figure 8.32 with the coherent OOK receiver of Figure 8.27a shows that the noncoherent receiver is much simpler. This is mainly because the coherent receiver must contain a carrier recovery subsystem whereas the noncoherent receiver does not. As we have shown in the comparison of BER performance in Figure 8.31, the noncoherent system performs nearly as well as the coherent OOK system. Therefore, when OOK is used, the receivers used are almost always of the noncoherent type. Since the noncoherent receiver is not responsive to the received signal phase, it can be used with either coherent or noncoherent signals. The BER performance is exactly the same in either case. For the coherent receiver to be useful, the signal must be coherent because the carrier recovery subsystem averages the phase of past data pulses to obtain the phase of the recovered carrier.

The PSD of the noncoherent OOK signal is different from the spectrum of the coherent OOK signal. The PSD of the coherent OOK signal was shown in Figure 8.27d. Coherent OOK has half of its power in a pure carrier term, resulting in an impulse at the carrier frequency. The PSD of the noncoherent OOK signal, in which the phase in each data interval is assumed to be a different independent random variable uniformly distributed from 0 to 2π, is shown in Figure 8.33. Notice that the noncoherent OOK signal does not contain an impulse at the carrier frequency. All of the power of the noncoherent OOK signal is contained in the side frequencies. In fact, the PSD of noncoherent OOK has the same form as the PSD of the PRK signal shown in Figure 8.24.

Even though the power of the noncoherent OOK signal is all in the sidebands, the noncoherent receiver is not able to use the power as effectively as the coherent receiver in making the data decisions. This is due to the fact that the coherent receiver can predict the phase of the signal; so it only has to be responsive to signal and noise components in a single direction in signal space.

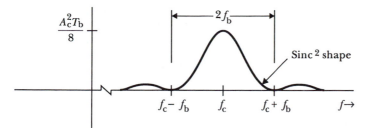

FIGURE 8.33
PSD of noncoherent OOK does not contain a discrete carrier component.

On the other hand, the noncoherent receiver does not predict the carrier phase, and therefore it must be responsive to signal and noise components in an entire plane in signal space. Thus, a second noise component potentially can influence the decision, and as a result, more errors are made by the noncoherent receiver.

Frequency shift keying

A *frequency shift keyed* (FSK) signal consists of a sinusoid with a frequency that can take on two values, f_0 and f_1, depending on the data. As in the case of OOK, FSK signals can be either coherent or noncoherent. In coherent FSK, the phases of the sinusoids at each of the two frequencies are continuous from bit interval to bit interval. In noncoherent FSK, the phase changes randomly between bit intervals and is uniformly distributed from 0 to 2π. The coherent FSK signal for the bit interval between 0 and T_b is given by

$$s_0(t) = A_c \cos (2\pi f_0 t) \qquad (8.93)$$

or by

$$s_1(t) = A_c \cos (2\pi f_1 t) \qquad (8.94)$$

depending on the data bit. Notice that we have assumed that the phases of the signals are known and equal to zero for both frequencies. If the frequency difference between the two signals, $f_0 - f_1$, is such that $(f_0 - f_1)T_b$ is either an integer or very large, it can be easily shown that the FSK signal set forms an orthogonal signal set.

An optimum receiver for coherent FSK signals is shown in Figure 8.34a. The received signal-plus-noise is first split and passed through filters that are matched for the signals of Equations 8.93 and 8.94. The outputs of the matched filters are split and applied to the input of carrier recovery subsystems and mixers. The carrier recovery subsystems retrieve the two carrier frequencies with the same phase as at the transmitter (allowing for any phase shift in the channel and filters). The output of each mixer is passed through a lowpass

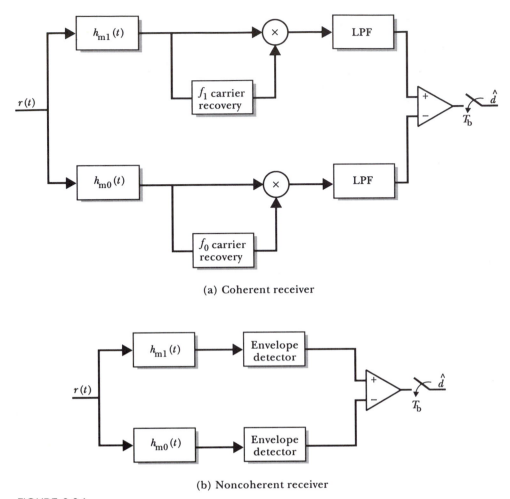

(a) Coherent receiver

(b) Noncoherent receiver

FIGURE 8.34
FSK receivers.

filter that passes the difference frequency components and rejects the sum frequencies in the vicinity of twice the carrier frequencies. The outputs of the lowpass filters are applied to a comparator that forms a decision in favor of the largest signal. The output of the comparator is sampled at the end of each bit interval to recover the data. The BER performance of coherent FSK is the same as coherent OOK as indicated in Figure 8.31.

The noncoherent FSK signal set is the same as the coherent signal set given in Equations 8.93 and 8.94, except that an independent random phase is included in each expression. An optimum receiver for noncoherent FSK is shown in Figure 8.34b. Notice that the noncoherent FSK receiver is much simpler than the coherent receiver since the carrier recovery functions are not required. The envelope detectors are very simple compared with the carrier recovery subsystems.

The probability of error of the optimum noncoherent FSK receiver is given by

$$\blacksquare \qquad P(E) = \frac{1}{2} \exp\left(-\frac{E_b}{2N_0}\right) \qquad \text{(noncoherent FSK)} \qquad (8.95)$$

This is the same as the approximate error probability of noncoherent OOK as shown in the plot of Figure 8.31. Because of the small difference in performance between the coherent receiver for FSK and the noncoherent receiver, the more complex coherent receiver is seldom used. When the noncoherent receiver is in use, whether the transmitted signal maintains phase continuity between bit intervals or not makes no difference to the BER performance.

Differential phase shift keying

Noncoherent approaches do not apply to PRK because the information is contained solely in the transmitted phase. If a random phase were included in the signal, as in noncoherent OOK or FSK, the information would be totally obscured. However, the *differential phase shift keying* (DPSK) approach provides much of the hardware simplicity of noncoherent approaches while achieving a BER performance close to that of PRK. In DPSK, the phase of the filtered received signal at the end of each bit interval is compared with the phase at the end of the previous interval to make the data decision. Thus, the phase for each bit provides the phase reference for the next bit. This approach is termed differentially coherent because it is the difference in phase between adjacent data intervals, which is used by the receiver to recover the data.

Usually, if a differentially coherent receiver is to be used, the transmitted signal is *differentially encoded*. The diagram of a modulator using differential encoding is shown in Figure 8.35. It is assumed that the message $d(t)$ is an NRZ

FIGURE 8.35
DEPSK modulator.

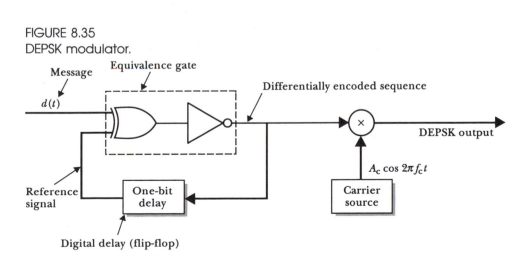

TABLE 8.1
Truth table for the equivalence gate

Message	Reference	Encoded output
0	0	1
0	1	0
1	0	0
1	1	1

waveform with logic levels of ± 1. Differential encoding is performed by the equivalence gate, which consists of an exclusive OR gate followed by an inverter. The inputs to the exclusive OR gate are the data and the differentially encoded output delayed by one bit. The truth table for the equivalence gate is given in Table 8.1.

During intervals in which the message bit is a 1, the resulting transmitted signal maintains the same phase as the preceding interval. In intervals in which the current message bit is a 0, the transmitted phase is inverted compared with the phase of the preceding interval. A typical message sequence, the differentially encoded sequence, and the phase of the transmitted signal are illustrated in Table 8.2. Since the output of the differential encoder at any instant depends on the output during the preceding bit interval, an initial state of the reference bit must be given to determine the encoded sequence. In Table 8.2 we have assumed that the initial reference bit is a 1. If a 0 is assumed for the initial reference bit, the entire encoded sequence is inverted. However, the eventual recovered data at the DPSK receiver output will be the same. This is due to the fact that the receiver recovers the data by comparing the phase for the current bit with the phase for the preceding bit.

A DPSK receiver is shown in Figure 8.36. Assuming that the channel adds WGN and that the data decision must be based on the segment of the received signal for the current data and one preceding bit, this receiver is optimum. In the receiver of Figure 8.36, the received signal-plus-noise is first

TABLE 8.2
Example of differential encoding and the
transmitted phase

Data sequence	1 1 0 1 0 0 1 1
Encoded sequence	⬈1 1 1 0 0 1 0 0 0
Transmitted phase	0 0 0 π π 0 π π π
	Initial arbitrary reference bit

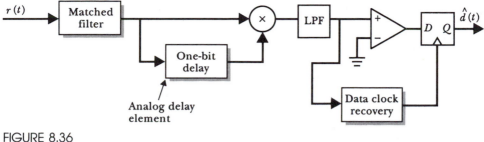

FIGURE 8.36
DPSK receiver.

passed through a matched filter with an impulse response given by

$$h_m(t) = A \cos (2\pi f_c t + \theta) \qquad 0 < t < T_b$$
$$= 0 \qquad\qquad\qquad \text{otherwise} \qquad (8.96)$$

where both A and θ can take any convenient values with no effect on the error performance (other than $A = 0$, of course). For each bit, this filter produces the maximum possible SNR at the end of the bit interval.

The output of the matched filter is split and applied to the inputs of both a delay element and a multiplier. We assume that the delay is one bit interval and that this is also equal to an integer number of carrier cycles. This implies that the carrier frequency is a multiple of the data rate.

In the absence of noise, the output signal from the matched filter reaches steady-state conditions at the end of a bit interval because the duration of the impulse response is T_b. Therefore, the duration of a transient due to a phase reversal at the beginning of the interval is also T_b. Thus, the phase of the matched-filter output depends only on the received phase during the preceding data interval. Phase reversals of the input signal due to the data result in phase reversals of the output signal sampled at the ends of bit intervals.

The filter output and its delayed version are applied to the multiplier. Since both of these signals are sinusoids at the carrier frequency, the product contains low-frequency terms and terms in the vicinity of twice the carrier frequency. The twice-carrier frequency terms are rejected by the lowpass filter following the multiplier. At the end of a bit interval, the input signals to the multiplier are either in phase or out of phase, depending on whether the message bit for the current interval is a 1 or a 0, respectively. Multiplication of sinusoids of the same phase results in a positive valued low-frequency component at the input to the comparator. When the inputs to the multiplier are out of phase, a negative value results at the input to the comparator. Thus, in the absence of noise, the comparator output is the same as the current message bit.

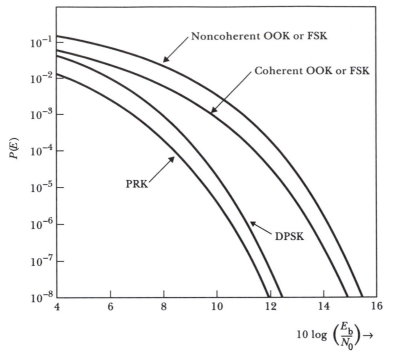

FIGURE 8.37
Bit-error probability curves for various modulation formats.

The data clock signal is recovered from the output of the lowpass filter and used to trigger the flip-flop, which implements the sampling function at the end of each bit interval.

The bit-error probability of the DPSK receiver can be shown to be given by

$$P(E) = \frac{1}{2} \exp\left(-\frac{E_b}{N_0}\right) \quad \text{(DPSK)} \qquad (8.97)$$

A plot of this as well as the BER curves for several of the other systems we have considered is shown in Figure 8.37. Notice that the performance of DPSK is only about 1 dB worse than that of PRK at an error probability of 10^{-4}. DPSK performance is exactly 3 dB better than noncoherent FSK. This can be verified by comparing Equation 8.97 with 8.95.

In our discussion so far, we have assumed that the delay in the DPSK receiver is an integer number of carrier cycles. However, the receiver performs just as well if the delay is an odd multiple of half carrier cycles, provided that an inverter is added after the flip-flop. When the delay is an odd multiple of half carrier cycles, signals with the same phase at the output of the matched filter result in out-of-phase inputs to the multiplier due to the phase inversion in the delay. Thus, the data decisions are reversed.

The DPSK receiver of Figure 8.36 seems to be quite simple compared with the coherent PRK receiver of Figure 8.23a because the delay element replaces the relatively more complex carrier recovery subsystem. However, as we have noted, the delay element is required to provide a phase shift that is a multiple of exactly 180° at the carrier frequency. This can be a difficult requirement when the carrier is high compared with the data rate. As a result, DPSK may not be as attractive to the designer as the carrier recovery subsystem.

8.5
M-ARY MODULATION TECHNIQUES

In this section, we present some of the commonly used digital modulation techniques in which more than two message symbols are used. Very often in practice, the number of symbols M is a power of two, and we can use each of the M data symbols to represent a group of binary symbols. For example, if M is eight, each data symbol can represent three bits of a binary data stream.

Quadrature phase-shift keying

In *quadrature phase-shift keying* (QPSK), there are four data symbols, and the modulated signal is a sinusoid with four possible phases which are 90° apart. Most often, each of the four data symbols represents two bits of a binary data stream. A QPSK modulator for this case is shown in Figure 8.38a. The input is assumed to be a binary stream with a new bit every T_b seconds. The input data are split into two parallel binary data signals by the *demultiplexer* (DMUX). The demultiplexer may, for example, separate the input data bits in the even-numbered bit intervals into the output signal $d_i(t)$ and the odd bits into $d_q(t)$. The bit intervals for each of the demultiplexed data streams are $2T_b$. A typical data signal and the corresponding demultiplexed signals are shown in Figure 8.38b. As indicated in the figure, we assume that the data signals are of the NRZ type, taking values of ± 1.

The demultiplexed data streams are then PRK modulated onto quadrature versions of the carrier. The two PRK signals are summed to produce the QPSK signal, given by

$$\blacksquare \qquad g_{qpsk}(t) = d_i(t)A_c \cos(2\pi f_c t) - d_q(t)A_c \sin(2\pi f_c t) \qquad (8.98)$$

Provided that $4f_c T_b$ is either very large or an integer, the basis functions for a generalized Fourier series representation of this QPSK signal can be chosen to be time segments of the quadrature carriers with a duration of $2T_b$. (The provision concerning $4f_c T_b$ is necessary to insure that the basis functions

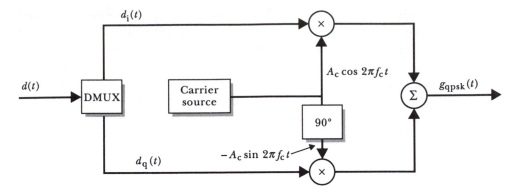

(a) Modulator

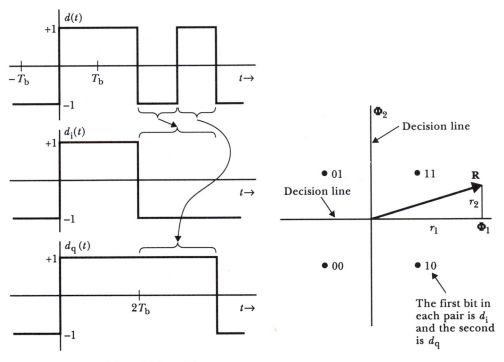

(b) Input data and demultiplexed data streams

(c) Signal-space diagram

FIGURE 8.38
QPSK modulator [if $d_q(t)$ is staggered by T_b relative to $d_i(t)$ this becomes OQPSK].

are indeed orthogonal.) For the interval from 0 to $2T_b$, these basis functions are given by

$$\phi_1(t) = (T_b)^{-1/2} \cos(2\pi f_c t) \qquad 0 < t < 2T_b$$
$$= 0 \qquad\qquad\qquad\qquad \text{otherwise}$$

386

and

$$\phi_2(t) = -(T_b)^{-1/2} \sin (2\pi f_c t) \qquad 0 < t < 2T_b$$
$$= 0 \qquad\qquad\qquad\qquad \text{otherwise}$$

As we have seen, for this choice of basis functions, the signal-space diagram becomes a scaled phasor diagram. A signal-space diagram for the QPSK signal is shown in Figure 8.38c. Notice that the QPSK signal takes on four phases 90° apart, one for each of the four combinations of data bits.

The optimum receiver for QPSK resolves the received noisy signal into a vector in the two-dimensional signal space and then decides in favor of the closest signal. The resulting decision lines fall on the coordinates axes, as indicated in the signal-space diagram of Figure 8.38c. For example, when the received vector falls into the upper right-hand quadrant, the decision is in favor of the 11 combination of data bits. Notice that the decision concerning the first bit of each pair can be made simply by comparing the first component r_1 of the received vector with zero. Similarly, the decision about the second bit can be made by comparing r_2 with zero.

An optimum receiver for QPSK, in the presence of additive WGN, is shown in Figure 8.39. The received signal-plus-noise is correlated with the basis functions (quadrature versions of the carrier) to obtain the two components r_1 and r_2 of the relevant received vector. Actually, since the threshold is zero, it is not necessary to use a specific amplitude for the basis functions (i.e., the basis functions do not have to be normalized). Each component is then compared with zero to make the decision concerning the corresponding data bit. Finally, the recovered data is *multiplexed* (**MUX**) to obtain the composite output data stream.

FIGURE 8.39
QPSK receiver (by changing the sampling instants for $\hat{d}_q$ to $-T_b, T_b, 3T_b, \ldots$, this becomes an OQPSK receiver).

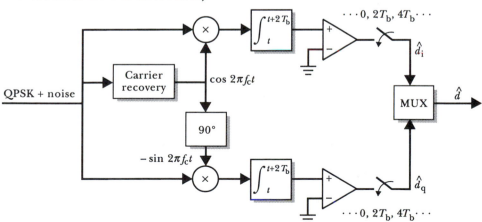

Notice that the upper branch of the QPSK receiver of Figure 8.39 is identical to a PRK receiver such as the one shown in Figure 8.23a. The upper branch responds only to the cosine (or in-phase) term of the QPSK signal of Equation 8.98. Similarly, the lower branch of the QPSK receiver responds only to the sine component of the QPSK signal. Therefore, the QPSK system can be considered to be two parallel PRK systems that operate without mutual interference because of the quadrature relationship between the carriers. Thus, the bit-error probability of the QPSK system is identical to that of PRK, given by

$$\blacksquare \qquad P(E) = P_b = Q\left[\left(\frac{2E_b}{\mathcal{N}_0}\right)^{1/2}\right] \qquad \text{(QPSK or PRK)} \qquad (8.99)$$

where, as before, E_b is the received signal energy per data bit and $\mathcal{N}_0/2$ is the PSD of the noise including both the received noise and any contribution from the receiver circuits. Also, we have introduced the notation P_b to stand for the bit-error probability.

Sometimes in QPSK systems the probability of error for the 4-ary symbols is of interest. For the correct decision to be made concerning each symbol, the decisions for both components of the received signal must be correct. In other words, for 4-ary symbols representing two bits each to be correct, both of the bits must be received correctly. Thus, the probability of correctly receiving a symbol is given by

$$P\text{ (symbol correct)} = (1 - P_b)^2 \qquad (8.100)$$

The probability of symbol error is thus given by

$$P_s = 1 - (1 - P_b)^2$$
$$\blacksquare \qquad P_s = 2P_b - P_b^2 \qquad (8.101)$$

Note that when the probability of a bit error is small, the probability of symbol error is very nearly twice the probability of bit error.

The PSD of the QPSK signal is the sum of the PSDs of the separate in-phase and quadrature terms of Equation 8.98. Each of these components is a PRK signal carrying half of the data bits. Therefore, the PSD of QPSK is the sum of the PSDs of two PRK signals, each with a bit interval of $2T_b$. The PSD of the QPSK signal is shown in Figure 8.40. Notice that the null-to-null bandwidth of the QPSK signal is f_b. The PSD of a PRK signal, shown in Figure 8.24 has a null-to-null bandwidth of $2f_b$. Therefore, QPSK achieves a two-fold savings of bandwidth compared with PRK while providing equal BER performance. As we will see in a later section of this chapter, QPSK has some disadvantages with respect to complexity of the carrier recovery function and sensitivity to carrier phase errors.

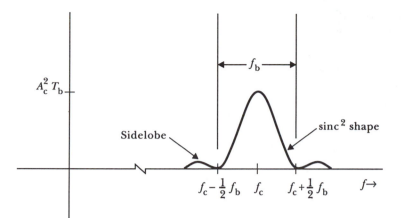

FIGURE 8.40
Power spectral density of QPSK or OQPSK.

Offset quadrature phase-shift keying

The *offset quadrature phase-shift keying* (OQPSK) signaling scheme is identical to QPSK except that the two demultiplexed data signals $d_i(t)$ and $d_q(t)$ are offset in time by T_b so that the transitions of one signal occur in the middle of the data intervals for the other signal. This method is also called *staggered quadrature phase-shift keying* (SQPSK). Except for modifications needed to offset the data clocks in the transmitter and receiver, the systems are the same. The spectral characteristics and bit-error probability of OQPSK are the same as those of QPSK.

The PSD of both QPSK and OQPSK contain power in the sidelobes that can cause significant interference for systems operating in adjacent frequency bands. This is because the sinc-squared function contains significant sidelobes when the transmitter power is high. The power in these sidelobes is associated with the sudden changes in the phase of the modulated signal caused by the data transitions. This is similar to the higher harmonics of a square wave, which are due to the discontinuities in the square wave. To minimize interference with adjacent systems, the spectra of many digital transmitters are limited by bandpass filtering to eliminate much of the power in these sidelobes. Typical waveforms of bandlimited QPSK and OQPSK signals are shown in Figures 8.41a and b. In the bandlimited QPSK signal, the amplitude decays to zero and then builds up with the opposite phase when both data streams change at the same time. The OQPSK signal can change by, at most, 90° at each transition because only one data signal changes at a given instant due to the time offset of the data signals. Therefore, the amplitude variations of bandlimited OQPSK are not as severe as those of the bandlimited QPSK signal. This is indicated in the waveform sketches of the figures.

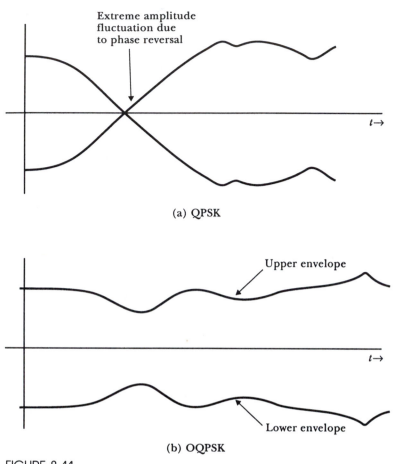

Extreme amplitude
fluctuation due
to phase reversal

$t\rightarrow$

(a) QPSK

Upper envelope

$t\rightarrow$

Lower envelope

(b) OQPSK

FIGURE 8.41
Envelopes of bandlimited signals.

Often, it is desirable to be able to amplify signals to the power levels required for radio transmission with efficient class C amplifiers. To obtain high efficiency, class C amplifiers are designed so that the active elements behave as switches rather than linear amplification elements. As a result, such amplifiers tend to produce a constant output level independent of the amplitude of the input signal. Such nonlinear amplifiers can produce frequency components in the output signal that are not present in the input. In the case of the bandlimited OQPSK signal, since the amplitude variations are relatively small, class C amplification can be used without severe regeneration of spectral sidelobes. On the other hand, the bandlimited QPSK signal contains relatively large amplitude fluctuations, and when these are converted into a constant amplitude signal by a class C amplifier, the spectral sidelobes reappear. Thus, OQPSK has an advantage in that it is not as necessary to filter the signal after amplification as it is for QPSK. This can be significant, for example, in the case of satellite transmitters, in which efficient amplifiers and simplicity are of primary importance. For such applications a *constant envelope property* is important for modulated signals. Bandlimited OQPSK has a more nearly constant envelope than bandlimited QPSK and is, therefore, more desirable in these cases.

Minimum shift keying

The *minimum shift keying* (MSK) modulation technique can be viewed as a special case of OQPSK with shaped data pulses. An equation for the MSK signal based on this point of view is

$$
\blacksquare \qquad g_{\text{msk}}(t) = A_c d_i(t) \cos\left(\frac{\pi f_b t}{2}\right) \cos(2\pi f_c t)
$$

$$
+ A_c d_q(t) \sin\left(\frac{\pi f_b t}{2}\right) \sin(2\pi f_c t) \qquad (8.102)
$$

where $d_i(t)$ is an NRZ waveform taking values of ± 1 with potential data transitions at $-T_b, T_b, 3T_b, \dots$. The quadrature data sequence $d_q(t)$ is a similar NRZ sequence with potential data transitions at $-2T_b, 0, 2T_b, \dots$. Notice that the MSK signal is the same as OQPSK except for the inclusion of the *data pulse shaping* terms, $\cos(\pi f_b t/2)$ and $\sin(\pi f_b t/2)$. Figure 8.42 shows typical in-phase and quadrature data sequences as well as the amplitudes of the in-phase and quadrature terms of the MSK signal. Notice that when a data signal makes a transition, the amplitude of the corresponding component of the MSK signal is zero, due to the pulse shaping terms. Therefore, no phase or amplitude discontinuity occurs in the MSK signal due to data transitions. This is in contrast to conventional OQPSK, in which phase discontinuities of $\pm 90°$ occur at data transitions. As we will see, this phase continuity leads to a PSD with reduced sidelobes.

Notice that since $d_i(t)$ and $d_q(t)$ take on values of ± 1, using the identity for the cosine of the sum or difference of angles, we can also write Equation 8.102 as follows:

$$
\blacksquare \qquad g_{\text{msk}}(t) = A_c d_i(t) \cos\left[2\pi\left(f_c \pm \frac{1}{4} f_b\right)t\right] \qquad (8.103)
$$

where the plus sign applies for intervals in which $d_i = -d_q$, and the minus sign applies if $d_i = d_q$. In this form, the MSK signal can be seen to be a form of *continuous phase frequency shift keying* (CPFSK). The continuous phase characteristic is not apparent from Equation 8.103, because the fact that $d_i(t)$ is present as a multiplier gives the impression that phase reversals will occur on data transitions; however, when $d_i(t)$ changes, the sign inside the argument of the sinusoid also changes, and the result is no net phase change. Also notice from Equation 8.103 that MSK has the desirable constant envelope feature.

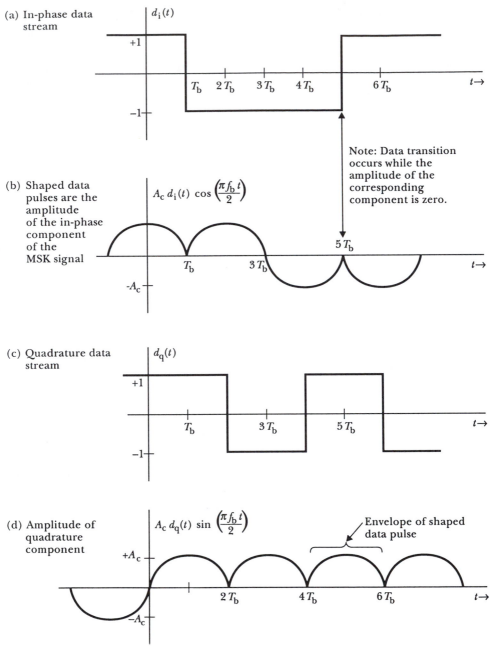

(a) In-phase data stream

$d_i(t)$

+1

T_b $2T_b$ $3T_b$ $4T_b$ $6T_b$ $t \rightarrow$

−1

Note: Data transition occurs while the amplitude of the corresponding component is zero.

(b) Shaped data pulses are the amplitude of the in-phase component of the MSK signal

$A_c\, d_i(t)\, \cos\left(\dfrac{\pi f_b\, t}{2}\right)$

T_b $3T_b$ $5T_b$ $t \rightarrow$

$-A_c$

(c) Quadrature data stream

$d_q(t)$

+1

T_b $3T_b$ $5T_b$ $t \rightarrow$

−1

(d) Amplitude of quadrature component

$A_c\, d_q(t)\, \sin\left(\dfrac{\pi f_b\, t}{2}\right)$

Envelope of shaped data pulse

$+A_c$

$2T_b$ $4T_b$ $6T_b$ $t \rightarrow$

$-A_c$

FIGURE 8.42
Components of the MSK signal.

Equation 8.103 can also be written as

$$g_{msk}(t) = A_c \cos \left[2\pi f_c t + \phi(t) \right] \qquad (8.104)$$

where $\phi(t)$ contains all of the phase variation due to the data. Notice that in intervals of time when the plus sign applies in Equation 8.103, the value of $\phi(t)$ increases linearly with time with a slope of 90° per bit interval. When the minus sign applies, the phase decreases at the rate of 90° per bit interval. Figure 8.43a shows the possible variations of $\phi(t)$ versus time. If we assume that the phase is zero at $t = 0$, the phase can either increase to $+90°$ (if $d_i = -d_q$) or decrease to $-90°$ (if $d_i = d_q$) in the first bit interval. In the next bit interval, a new data combination occurs, and the phase can again take either an upward path or downward path. This diagram is called a *phase trellis*. (This figure is constructed assuming that the bit immediately prior to $t = 0$ is a 0.)

Figure 8.43b shows the phase versus time for a specific data sequence in which the phase at $t = 0$ is taken as zero. Notice, as we have already observed, that the phase decreases in intervals in which $d_i = d_q$, and the phase increases when $d_i = -d_q$.

MSK modulators similar to the QPSK (or OQPSK) modulator of Figure 8.38a can be constructed by including additional multipliers in each arm to multiply by the pulse shaping terms $\cos(\pi f_b t/2)$ and $\sin(\pi f_b t/2)$. However, a much simpler method of producing MSK is to filter a PRK signal. This approach is known as *serial MSK* because the modulator does not demultiplex the data into parallel streams. The block diagram of a serial MSK modulator based on this approach is shown in Figure 8.44a. The composite data stream is first PRK modulated onto a carrier with a frequency of $f_c - \frac{1}{4}f_b$. The carrier at $f_c - \frac{1}{4}f_b$ is known as the serial carrier, whereas f_c is the carrier frequency used in the parallel approach. The resulting PRK signal is passed through a conversion filter with an impulse response given by

$$h_c(t) = \frac{\pi}{T_b} \cos \left[2\pi \left(f_c + \frac{1}{4} f_b \right) t \right] \qquad 0 < t < T_b$$
$$= 0 \qquad\qquad\qquad\qquad\qquad \text{otherwise} \qquad (8.105)$$

Notice that when two or more bits of the composite data sequence are the same, the filter output reaches steady state at the end of the first bit because the duration of the transient response of the conversion filter is T_b. In this steady-state condition, the input to the filter is a sinusoid at a frequency of $f_c - \frac{1}{4}f_b$, and the output is a sinusoid at the same frequency. This corresponds to a downward arm in the trellis diagram of Figure 8.43. When the input data bit changes, a transient occurs in the output of the filter. If f_c is much larger than f_b, it can be shown that the resulting filter output is a constant amplitude sinusoid at a frequency of $f_c + \frac{1}{4}f_b$. This corresponds to an upward arm in the trellis diagram (Amoroso and Kivett).

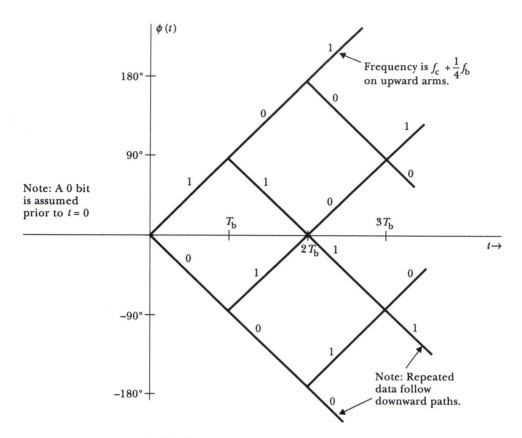

(a) Trellis diagram of possible phase paths versus time

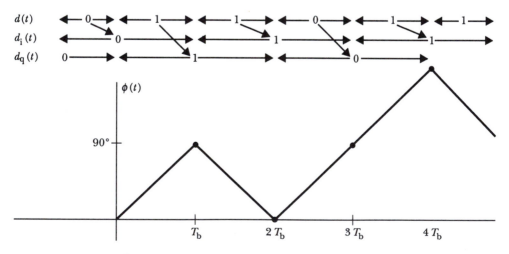

(b) Phase path for a specific data sequence

FIGURE 8.43
Phase versus time for MSK signals.

(a) Modulator block diagram

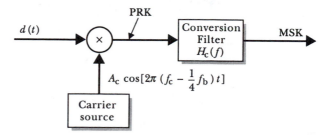

$$A_c \cos[2\pi\,(f_c - \tfrac{1}{4}f_b\,)\,t]$$

(b) PSD of PRK signal

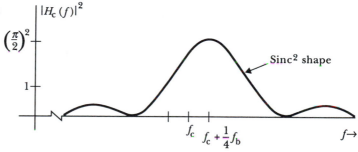

(c) Power gain of conversion filter

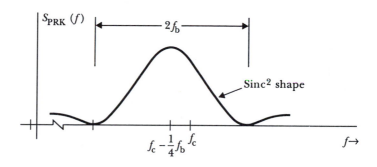

(d) PSD of MSK signal (product of b and c)

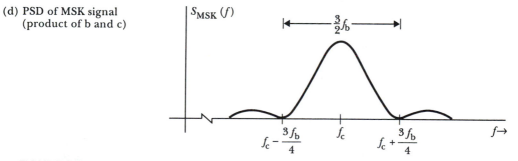

FIGURE 8.44
Serial MSK modulator.

The PSD of the PRK signal at the input to the conversion filter is shown in Fig. 8.44b. The transfer function of the conversion filter is the Fourier transform of the impulse response. The power gain of the conversion filter is the magnitude squared of the transfer function given by

$$|H_c(f)|^2 = \left(\frac{\pi}{2}\right)^2 \text{sinc}^2\left[\left(f - f_c - \frac{1}{4}f_b\right)T_b\right]$$
$$+ \left(\frac{\pi}{2}\right)^2 \text{sinc}^2\left[\left(f + f_c + \frac{1}{4}f_b\right)T_b\right] \qquad (8.106)$$

(As usual, we have assumed that f_c is much greater than f_b.) The power gain of the conversion filter is plotted versus frequency in Figure 8.44c. The PSD of the MSK signal is the product of the PSD of the PRK signal at the input to the conversion filter and the power gain of the conversion filter. For positive frequencies this is given by

$$S_{\text{msk}}(f) = \left(\frac{\pi A_c}{4}\right)^2 T_b \text{sinc}^2\left[\left(f - f_c - \frac{1}{4}f_b\right)T_b\right]\text{sinc}^2\left[\left(f - f_c + \frac{1}{4}f_b\right)T_b\right]$$
$$(8.107)$$

This is plotted in Figure 8.44d. Notice that the null-to-null bandwidth of the MSK signal is $3f_b/2$. This is 50% more than the null-to-null bandwidth of the QPSK or OQPSK signal. However, the sidelobes of the MSK signal decay much faster than the sidelobes of the QPSK signal; so the bandwidth advantage is not clearly in favor of either modulation format.

An optimum receiver for MSK in the presence of additive WGN is shown in Figure 8.45. The noisy signal is first passed through a matched filter to maximize the SNR at the end of each bit interval. The impulse response of the matched filter is a time reversed and delayed version of one of the data pulses

FIGURE 8.45
Serial MSK demodulator.

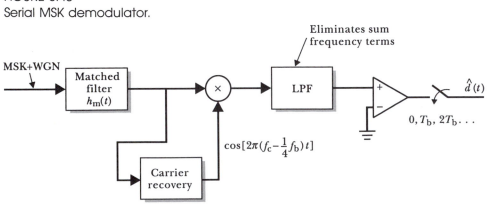

of the MSK signal given in Equation 8.102. This impulse response is given by

$$h_{\mathrm{m}}(t) = \sin\left(\frac{\pi f_{\mathrm{b}} t}{2}\right) \sin\left(2\pi f_{\mathrm{c}} t + \theta\right) \qquad 0 < t < 2T_{\mathrm{b}}$$

$$= 0 \qquad\qquad\qquad\qquad \text{otherwise} \qquad (8.108)$$

where θ is an arbitrary phase angle whose effect is cancelled by the tracking action of the carrier recovery subsystem. The carrier at a frequency of $f_{\mathrm{c}} - \frac{1}{4} f_{\mathrm{b}}$ is recovered and multiplied by the filtered MSK signal. It can be shown that the filtered MSK signal is either in phase with the recovered carrier or exactly out of phase at the sampling instants depending on the current data bit (Amoroso and Kivett). Therefore, the difference frequency components are either positive or negative at the sampling instants. The lowpass part of the product is passed to a comparator by which the decisions are made. The output of the comparator is sampled at the end of each bit interval to recover the final output data.

M-ary phase-shift keying

In *M-ary phase-shift keying* the transmitted signal is a sinusoidal carrier with *M* equally spaced phases depending on the data symbols. Most often, the number of phases is a power of two, and a bit pattern is associated with each phase. The signal-space diagram for 8-ary PSK is shown in Figure 8.46.

Note that the decision region for each phase is a sector. When errors are made, the error usually involves the selection of one of the two sectors immediately adjacent to the correct sector. Because of this, the best assignment of bit patterns to the phases is one in which there is only a single bit that is different in adjacent sectors. A sequence of bit patterns in which adjacent patterns differ in only a single bit is known as a *Gray code*. A Gray code assignment is shown for the 8-ary signal-space diagram in Figure 8.46. The advantage of using the Gray code with *M*-ary PSK is that most errors fall in the adjacent sector, and only a single bit is then in error. If, for example, we were to assign 111 to a sector adjacent to the sector for 000, then an error between these sectors would lead to three bit errors instead of one.

The Gray code for 4-bit words is shown in Table 8.3. Notice the Gray code for $k + 1$ bits can be obtained by repeating the code for k bits in reverse order and then adding a 1 to the first half of the sequence and a 0 to the second half of the sequence.

When a Gray code is used and the received SNR is high, the probability of bit error is related to the probability of symbol error by the approximate expression

$$P_{\mathrm{b}} \simeq \frac{1}{k} P_{\mathrm{s}} \qquad (8.109)$$

where k is the number of bits per symbol.

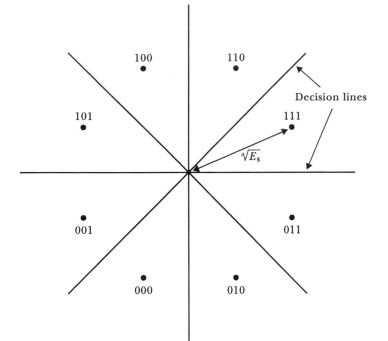

FIGURE 8.46
Signal-space configuration for 8-ary PSK with Gray code data assignment.

TABLE 8.3
Gray code for 4-bit words

1 1 1 1
1 1 1 0
1 1 0 0
1 1 0 1
1 1 0 0 1
1 0 0 0 ──3-bit Gray code
1 0 1 0
1 0 1 1

0 0 1 1
0 0 1 0
0 0 0 0
0 0 0 1
0 1 0 1 ──3-bit code in reverse order
0 1 0 0
0 1 1 0
0 1 1 1

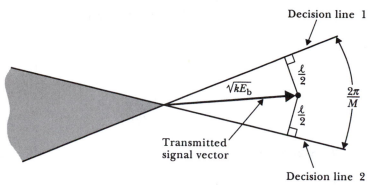

FIGURE 8.47
Decision regions for *M*-ary PSK.

Because of the symmetry of the signal-space diagram for *M*-ary PSK, the probability of a symbol error is the same regardless of the symbol (phase) transmitted. A signal-space diagram showing a transmitted signal and the associated decision region is shown in Figure 8.47. Since we have assumed that the signal represents k data bits, and the received signal energy per bit is E_b, the energy for each symbol is given by

$$E_s = kE_b \qquad (8.110)$$

The length of the signal-space vector is the square root of this energy as indicated in Figure 8.47.

When the particular signal phase indicated in this figure is transmitted, the probability of a symbol error is the probability of the received vector falling outside the sector containing the transmitted phase. This is approximately equal to the sum of the probability of falling above decision line 1 and the probability of falling below decision line 2. Actually, the symbol error is slightly less than this sum because the sum includes the probability of falling in the shaded area twice. However, when the SNR is high, the probability of falling in the shaded area is very small compared with the symbol error probability. Thus, we can write the approximate equation

$$P_s \simeq P(\text{falling above line 1}) + P(\text{falling below line 2}) \qquad (8.111)$$

Because of symmetry, we can see that the probability of crossing line 1 is the same as the probability of crossing line 2. Therefore, we have

$$P_s \simeq 2P(\text{falling above line 1}) \qquad (8.112)$$

In the discussion leading up to Equation 8.42, we found that when a decision line is located a distance $\ell/2$ from a transmitted signal, the probability of noise placing the received vector on the other side of the line is given by

$$P(\text{falling above line 1}) = Q\left(\frac{\ell}{\sqrt{2N_0}}\right) \qquad (8.113)$$

From Figure 8.47, we can see that the distance from the transmitted signal to the decision line is given by

$$\frac{\ell}{2} = (kE_b)^{1/2} \sin\left(\frac{\pi}{M}\right) \tag{8.114}$$

Solving the last equation for ℓ and substituting into Equation 8.113 and then into 8.112 results in the following expression for the symbol error probability:

$$P_s \simeq 2Q\left[\sin\left(\frac{\pi}{M}\right)\left(\frac{2kE_b}{N_0}\right)^{1/2}\right] \tag{8.115}$$

Substituting into Equation 8.109 results in the bit-error probability when the Gray code is used to assign binary words to the phases:

$$P_b \simeq \left(\frac{2}{k}\right)Q\left[\sin\left(\frac{\pi}{M}\right)\left(\frac{2kE_b}{N_0}\right)^{1/2}\right] \tag{8.116}$$

This is plotted in Figure 8.48 for $M = 8$, 16, and 32. The case for $k = 1$ and $M = 2$ corresponds to PRK, and the exact result in Equation 8.46 was used to produce the plot instead of the approximate Equation 8.116. The case for $k = 2$ and $M = 4$ corresponds to QPSK, which has the same bit-error probability as PRK.

When the phases of M-ary PSK are equally likely, it can be demonstrated that the PSD of the signal is shown in Figure 8.49. Notice that as the number of bits per symbol, k, becomes larger, the null-to-null bandwidth becomes smaller.

FIGURE 8.48
Bit-error performance of M-ary PSK.

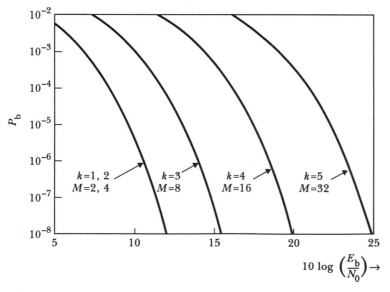

8 / DIGITAL MODULATION TECHNIQUES

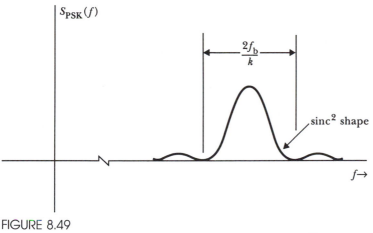

FIGURE 8.49
PSD for *M*-ary PSK with equally likely symbols ($M = 2^k$).

Notice from Figure 8.48 that, as the number of phases used by *M*-ary PSK increases beyond four, the received signal energy per bit must increase to achieve a given BER. The advantage of *M*-ary PSK for large *M* is that the bandwidth requirements are reduced as *M* is increased. As often occurs in communication systems, we have an opportunity to trade increased power for smaller bandwidth.

Quadrature amplitude modulation

In *quadrature amplitude modulation* (QAM), message symbols are used to determine the amplitudes of quadrature carriers. The two amplitude modulated signals are added together to produce the QAM signal. A signal-space diagram for a 4 × 4 QAM signal is shown in Figure 8.50. This diagram is just one example of many possible QAM signal configurations that can be obtained by using various combinations of amplitudes for the in-phase and quadrature carriers. Notice that the data symbols affect both the amplitude and phase of the transmitted signal. It can be demonstrated that high-order QAM can be used to trade increased power for a reduction in bandwidth just as *M*-ary PSK can.

8.6
TIMING CONSIDERATIONS

Several levels of time synchronization are necessary in digital communication systems. In coherent bandpass communication systems, the carrier recovery subsystem in the receiver must synchronize the locally generated carrier phase with the carrier phase imbedded in the received signal. This may be different from the carrier phase at the transmitter due to transmission delay and Doppler shift caused by motion of the transmitter or receiver. Many receivers also

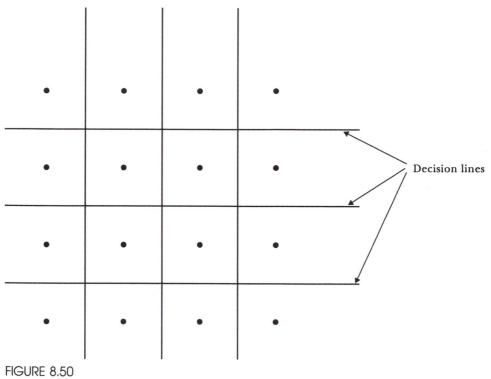

Decision lines

FIGURE 8.50
Signal-space configuration for 4 × 4 QAM.

require bit or symbol synchronization to properly establish the correct sampling instant for data decisions. Often, in practical communication systems, the recovered bit stream is used at the destination in the form of groups of bits called *words*. Then it is necessary to provide word synchronization so that the word boundaries are correctly determined. Often the data words from several sources are TDM multiplexed creating *frames*, and frame synchronization is then necessary so that the words can be routed to the appropriate destination at the receiving end. In this section, we discuss some of the possibilities for providing this synchronization.

Carrier synchronization

The simplest form of carrier recovery subsystem is possible when the modulation format in use results in a discrete carrier component in the transmitted signal. For example, coherent OOK with random data has half of its power contained in a carrier frequency component. This is illustrated in the PSD shown in Figure 8.27. For such a modulation format, the carrier can be recovered by using a bandpass filter centered at the carrier frequency with a bandwidth narrow enough so noise and the data modulation cause an acceptably small variation in phase and amplitude at the output of the filter. Alternately, a PLL with a suitably narrow bandwidth can be used to lock onto the carrier component.

It may be necessary to provide a mechanism to insure that the data stream does not contain data sequences that adversely affect the carrier recovery subsystem. For example, if a long sequence of zeroes occurs often, an OOK signal will be in the *off* state for long periods and the input to the bandpass filter or PLL disappears. This results in poor performance of the carrier recovery subsystem, such as frequent disappearance of the carrier at the filter output or loss of lock in the PLL. Since long sequences of zeroes and other nonrandom sequences often occur in typical data sequences, it is often necessary to *scramble* the data sequence before transmission and *descramble* it at the receiver. Scrambling is a method for insuring that the data sent on the link has nearly random properties. It can be achieved by modulo-two adding a binary scrambling sequence to the data. Descrambling is achieved by modulo-two adding the scrambling sequence a second time. In Section 11.1, we discuss methods for generating *pseudorandom binary sequences* suitable for use in scrambling.

Carrier recovery for modulation formats that do not contain a discrete carrier component can often be based on nonlinear operations that generate the carrier or one of its harmonics. For example, if the PRK signal given in Equation 8.64 as

$$g_{\text{PRK}}(t) = A_c d(t) \cos (2\pi f_c t)$$

is passed through a square-law device, we obtain

$$[g_{\text{PRK}}(t)]^2 = A_c^2 d^2(t) \cos^2 (2\pi f_c t)$$

Since the data signal $d(t)$ is assumed to be an NRZ waveform taking values of ± 1, this equation can be written as

$$[g_{\text{PRK}}(t)]^2 = \frac{1}{2} A_c^2 + \frac{1}{2} A_c^2 \cos (4\pi f_c t)$$

where the identity for cosine squared has been used. The second harmonic of the carrier can be separated from any noise terms that may be present by the use of a bandpass filter or PLL. Passing the second harmonic of the carrier through a divide-by-two frequency divider results in the desired carrier term. Notice that this method of carrier recovery is nearly identical to carrier recovery for DSB-SC discussed in Section 3.1.

One of the problems, known as *phase ambiguity*, associated with carrier recovery for PRK is that the divide-by-two operation can produce the carrier with either of two phases that are 180° apart. This is due to the random starting value of the internal state of the divide-by-two circuit. Thus, the demodulator can produce either the correct data stream or an inverted version of it. This same problem occurred with DSB-SC as discussed in Section 3.1.

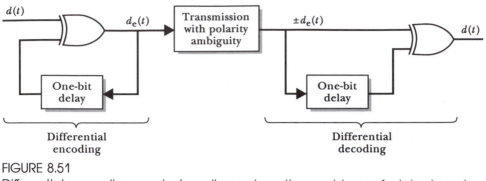

FIGURE 8.51
Differential encoding and decoding solves the problem of data inversion ambiguity.

The problem of inversion of the data in transmission can be easily solved by differentially encoding the data stream before the modulator and differential decoding at the receiver output. The diagram of the circuits for differential encoding and decoding are shown in Figure 8.51. It can be verified that the correct data appear at the output of the decoder regardless of whether the link inverts or not. A slight performance penalty is incurred by differential encoding and decoding since each isolated bit error in the link is converted into two successive bit errors in the decoder. Thus, the bit-error rate is very nearly doubled.

A second problem associated with carrier recovery for PRK occurs when the PRK signal has been passed through a bandpass filter before it reaches the square-law device. If, as is often the case, the filter passes only the main lobe of the spectrum and eliminates most of the side lobes, the PRK signal reaching the square-law device has amplitude fluctuations. A similar condition occurs for filtered QPSK and OQPSK signals as shown earlier in Figure 8.41. The amplitude fluctuations of the PRK signal occur at each data transition where the filtered PRK signal diminishes to zero amplitude before building up with the opposite phase. These amplitude fluctuations result in similar amplitude fluctuations of the second harmonic component at the output of the square-law device. As a result, it turns out that the output of the square-law device contains spectral lines at $2f_c \pm f_b$ as well as the desired line at $2f_c$. When the frequency uncertainty is comparable to or larger than the bit rate f_b, there is a strong possibility that the PLL will lock to one of the lines at $2f_c \pm f_b$ rather than the desired line at $2f_c$. In that case, the output of the frequency divider is at a frequency of $f_c \pm \frac{1}{2}f_b$ and the demodulator does not operate properly. This problem is called *false lock*.

Alternative carrier recovery schemes similar to the Costas PLL discussed in Section 3.1 are also possible for PRK. These approaches also suffer from the phase ambiguity and false lock phenomena discussed above. Phase ambiguity is inherent in a PRK signal since it is impossible to determine by examining a PRK signal which phase represents 0s in the data and which represents 1s if the data bits are random. The false lock problem is associated with filtering the signal ahead of the nonlinearity and can also occur when using the Costas approach.

Carrier recovery for QPSK or OQPSK can be achieved by the use of a fourth-order nonlinearity. QPSK and OQPSK signals take on four different phases spaced 90° apart. If one of these phases is selected as the reference phase of 0°, the other phases are $\pm 90°$ and 180°. When the fourth harmonic is created in the nonlinearity, these phases are multiplied by four, resulting in 0°, $\pm 360°$, and 720°. All of these are equivalent to 0°, so the fourth harmonic is a pure unmodulated tone that can be separated from the noise by a band-pass filter or PLL. Passing the fourth harmonic through a divide-by-four frequency divider produces the recovered carrier.

The recovered carrier for QPSK or OQPSK with random data can be any of four different phases, depending on the initial internal state of the divide-by-four circuits. Thus, a four-valued phase ambiguity exists for these modulation formats. The ambiguity is inherent in the signal, and any method for retrieving the carrier from the received signal incurs the problem for random data. Only one of the possible phases produces the in-phase data and quadrature data with the correct polarity at the corresponding outputs in the demodulators of Figure 8.39.

As we have noted, bandlimiting of OQPSK or QPSK results in amplitude modulation of the signal. This leads to amplitude modulation of the fourth harmonic and the production of spectral lines at frequencies other than the desired line at four times the carrier. For QPSK, this can lead to false lock at $f_c \pm f_b/8$, where f_b is the total data rate in the in-phase data stream plus the quadrature stream. Since the false lock frequency is closer to the desired carrier for QPSK than for PRK, false lock is potentially a more serious problem for QPSK.

Because of frequency uncertainty and receiver misalignment, the recovered carrier may have a static phase error. Due to noise and data modulation, the recovered carrier may also have significant phase jitter. Both static and dynamic phase errors lead to degraded performance in the receiver. The various modulation schemes differ in their tolerance to phase errors. For example, it can be shown that PRK and MSK are much less sensitive to phase errors than QPSK or OQPSK.

Symbol synchronization

All of the receivers presented in this chapter have included a mechanism for sampling the output of a matched filter or correlator at the end of each symbol interval so the decisions are made at the instant with the best SNR. Synchronization of the sampling instant is called symbol or bit synchronization. Many of the waveforms available for deriving this synchronization take the

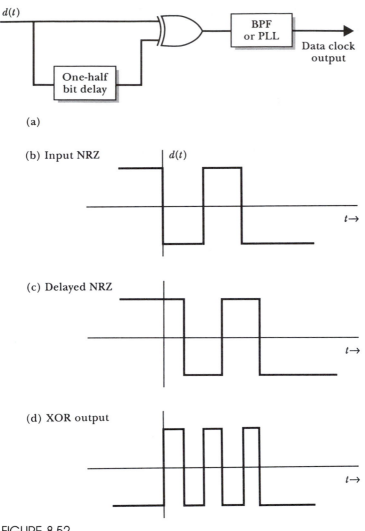

FIGURE 8.52
Bit synchronizer for NRZ signals.

form of a filtered NRZ waveform with a rate equal to the symbol rate. We will now demonstrate that a component at the data clock frequency can be generated by passing a filtered NRZ waveform through a nonlinearity.

Consider the system shown in Figure 8.52a with an NRZ waveform applied to the input. The input is split into two paths with one path applied to a delay element with a delay of one-half bit. The output of the delay element and the original input NRZ signal are applied to an exclusive OR gate. A typical NRZ signal at the system input is shown in part b of the figure, the delayed signal is shown in part c, and the output of the exclusive OR is shown in part d. Notice that each time the input signal changes value, the output of

FIGURE 8.53
Bit synchronizer for NRZ signals.

exclusive OR contains a high pulse that is one-half bit wide. For random data on the input waveform, the probabilities of a transition or of no transition are each one-half. The occurrences of transitions are independent from interval to interval. Thus, the probability of a positive pulse in the output of the exclusive OR gate is one-half for each data interval and is independent from interval to interval. Comparison of the output waveform of the system of Figure 8.52a with the unipolar RZ data format of Figure 7.23e shows that the XOR output is a unipolar RZ data waveform minus a dc component. The power spectrum of the unipolar RZ waveform with random data is shown in Figure 7.25a. Notice that this PSD contains a line at the data clock frequency f_b. Therefore, as indicated in Figure 8.52a, the data clock waveform can be recovered from the output of the XOR by the use of a bandpass filter or PLL.

A line at the clock frequency can be created from an NRZ waveform by passing the waveform through a linear filter followed by a nonlinearity. This fact can be explained as follows. As a result of the convolution of the input with the impulse response of the filter, the output of a linear filter contains a sum of delayed versions of the input signal. When this filtered signal is passed through a nonlinearity with a second-order term, products of the delayed components of the input NRZ waveform are formed. This product operation is the function performed by the XOR gate in the system of Figure 8.52a. Therefore, a linear filter followed by a nonlinearity performs the same functions as this system, and the clock frequency term can be separated by the use of a bandpass filter or PLL. A symbol clock recovery subsystem based on this approach is shown in Figure 8.53. Since filtered NRZ waveforms frequently occur in digital receivers, symbol synchronization can often be based on this approach.

Word synchronization

The symbols in a digital data stream are often used in groups at the receiving end. For example, if digitized voice signals are created by an 8-bit analog-to-digital converter, the bits are used in 8-bit words to reconstruct the signal at the receiving end with a digital-to-analog converter. When the message is English text, the characters are often represented by groups of bits. As we will see later, one approach for error correction coding results in code words. Therefore, it is often necessary to separate the received bits into the appropriate words before they are useful.

When TDM is in use, it may also be necessary to identify the words that should be routed to a particular destination. When this is the case, the words are often grouped in frames. For example, if N digitized voice signals are to be multiplexed, a frame may consist of a string of interleaved words from the N signals. The first word in the frame is from the first voice signal, the second word from the second voice signal, and so forth.

Word and frame synchronization can be obtained by placing a known synchronization word into the data stream periodically. For example, a particular pattern could be inserted at the beginning of each frame. A system for recognizing the synchronization word and producing both frame and word synchronization is shown in Figure 8.54. It is assumed that each frame consists of N words and the first word in each frame is a known pattern given by

$$s_1 s_2 s_3 \cdots s_w$$

where each s_i is a logic value, 0 or 1, and w is the number of bits per word. Inverted s_i values are also input to the XOR gates as indicated in the figure. The incoming data stream is passed through the shift register. When the syn-

FIGURE 8.54
Word and frame synchronizer.

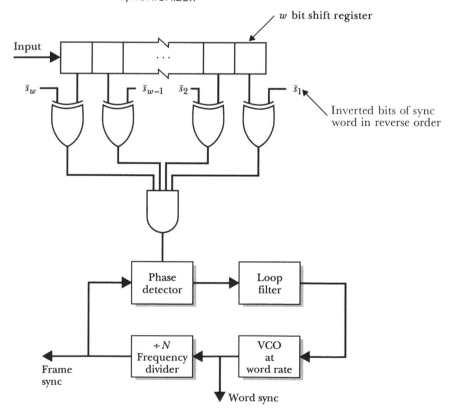

chronization word in the data stream fills the shift register, the outputs of each of the XOR gates are high, and the output of the AND gate is high. When any pattern except the synchronization word fills the shift register, the output of the AND gate is low. Thus, the output of the AND gate is a pulse train with a pulse each time the synchronization word occurs. The synchronization word occurs periodically at the beginning of each frame, so the output of the AND gate contains a component that is periodic at the framing frequency. In addition, the synchronization word may occur at random in the data words, but as long as the data bits are random, the periodic component is synchronized to the beginning of a frame. The PLL can lock onto this component and provide both the frame synchronization and word synchronization since the word rate is a multiple of the frame rate.

Notice the similarity of the shift register and logic gate combination of Figure 8.54 to a matched filter. As the received data stream moves through the shift register, it is combined with a replica of the synchronization word by the XOR gates. The outputs of the XOR gates are then passed through a logic circuit that produces a high output when each bit of the stored replica of the sync word matches the content of the shift register. In an analog matched filter, which is often used for detecting pulses of known shape in white noise, the impulse response is a time-reversed replica of the pulse shape of interest. The output of the analog matched filter is the convolution of the input signal with this time-reversed replica. As the input signal is moved past the replica, the product of corresponding points is formed and the integral of the product is taken at each point. In the absence of noise, the output of the matched filter reaches its peak value when the time-reversed impulse response and the input signal are aligned.

8.7
COMPARISON OF DIGITAL MODULATION TECHNIQUES

In this section, we will compare selected digital modulation techniques with each other and to the theoretical limit set by the Shannon-Hartley theorem with respect to bandwidth and power requirements. Then we will compare the hardware complexity of the modulation schemes. Finally, we will consider the use of digital PCM techniques in the communication system design example of Section 5.6.

Performance compared to the Shannon-Hartley limit

The *Shannon-Hartley theorem*, introduced in Chapter 1, gives the capacity of a communication channel as

$$C = B \log_2 \left(1 + \frac{S}{N} \right) \tag{8.117}$$

where C is the maximum capacity of the channel in bits of information per second, B is the bandwidth of the channel in hertz, S is the signal power at the output of the channel, and N is the noise power at the output of the channel. The theorem applies when the noise is gaussian. In theory it is possible to find signaling schemes for which the probability of bit error is as small as desired provided that the data rate is less than the channel capacity.

For a digital communication system operating at capacity, the bit rate f_b is equal to the capacity C. Also, the received signal power is equal to the product of the energy per bit E_b and the bit rate. In equation form, this is

$$S = E_b f_b$$

The total noise power in the channel bandwidth due to WGN is given by

$$N = N_0 B$$

where, as usual, $N_0/2$ is the PSD of the noise. Substituting these results into Equation 8.117 results in

$$C = f_b = B \log_2 \left(1 + \frac{E_b f_b}{N_0 B} \right)$$

This equation can be rearranged to obtain

$$\frac{E_b}{N_0} = [(2)^{f_b/B} - 1] \frac{B}{f_b} \qquad (8.118)$$

This relation gives the Shannon-Hartley limit for digital communication systems in terms of the ratio of bit rate to channel bandwidth and the ratio of bit energy to noise PSD. A plot of Equation 8.118 is shown in Figure 8.55. We will refer to this figure as the *bandwidth-efficiency plane*.

Notice that when the available bandwidth is much greater than the bit rate of the system (f_b/B is much smaller than 1), the Shannon-Hartley limit indicates that arbitrarily low error rates are possible if the ratio E_b/N_0 is greater than -1.6 dB. Systems designed to approach this limit are said to be *power limited* since they expend large amounts of bandwidth in an attempt to reduce the received energy per bit E_b. Space probe communication links to earth are often designed as power-limited systems. The long distances result in large attenuation, and it is difficult to provide high values of E_b/N_0 by increasing the transmitter power. On the other hand, the desired data rates are often small, and the available bandwidth is large; so the ratio f_b/B can easily be made small.

When bandwidth is scarce and transmitter power can be easily increased to provide a high value of E_b/N_0, systems are designed to operate in the *bandwidth-limited* region of the bandwidth-efficiency plane. Point-to-point terrestrial microwave communication systems are often designed to operate in the bandwidth-limited region because high data rates are desired, bandwidth is

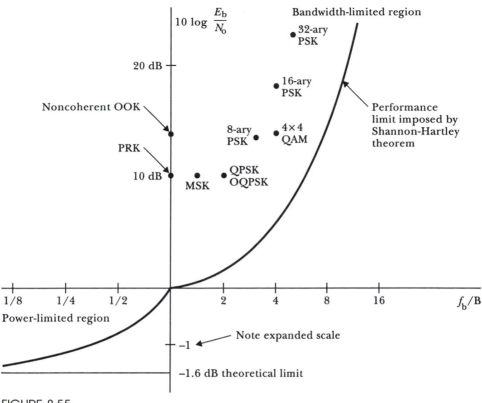

FIGURE 8.55
The bandwidth-efficiency plane.

limited, and high transmitter power is easily achieved. High-order modulation schemes such as M-ary PSK and M-ary QAM can be used when operation in the bandwidth-limited region is desired.

Estimates of the parameters of several of the digital modulation techniques we have studied are shown in Figure 8.55. These values were obtained for a bit-error probability of $P_b = 10^{-6}$. The values of E_b/N_0 required for this error rate were calculated using the equations for error probability developed in this chapter, except for the 4×4 QAM case, which is treated in the problems at the end of the chapter. If a lower error probability is required, the points plotted will move upward. The amount of this movement is roughly 1 dB for an order of magnitude reduction in the error probability.

The bandwidth required by each of the systems plotted in Figure 8.55 was estimated to be one-half of the null-to-null bandwidth of the PSD of the modulated signal with random data. This represents a fairly high degree of bandlimiting and would typically result in several decibels of performance degradation, which has not been accounted for in the figure. In other words, the actual performance of the digital modulation formats is typically either several decibels above or significantly to the left of the points plotted. Nevertheless, the points plotted give an indication of the comparison between the modulation formats and with the Shannon-Hartley limit.

All of the modulation schemes we have presented so far give performance that is significantly poorer than the Shannon-Hartley limit. For example, a well-designed PRK system with an error probability of 10^{-6} is about 12–13 dB above the theoretical limit. (This figure takes into account several decibels of degradation due to bandlimiting.) Recall, however, that an antipodal signal set is optimum for transmission of one bit at a time, and PRK is antipodal. The use of QPSK, MSK, or OQPSK reduces the bandwidth and provides operation closer to the theoretical limit; however, a significant improvement is still promised by the Shannon-Hartley limit. We will see that coding provides a method for approaching this theoretical limit more closely and requires that the data bits be transmitted in combination rather than one bit at a time.

Comparisons of hardware complexity

Digital communication techniques can also be compared on the basis of hardware complexity. In general, such a comparison must be somewhat subjective since the detailed requirements of a particular system shade the degree of complexity of each of the approaches in a different way. Also, the state of the art of electronic hardware is constantly changing. This affects the relative cost of the various elements in a system. Nevertheless, some general conclusions can be drawn. Certainly a noncoherent system without the need for carrier recovery is less complex than a coherent system. Figure 8.56 reflects the author's judgement of the relative complexity of various systems.

Design example continued

At this point, we will use the results of this chapter and Chapter 6 to continue the design example of Section 5.6 by considering the use of digital PCM. Much of the work done in Section 5.6 for analog systems also applies to a digital system. This includes the choice of the operating frequency, the estimated system noise temperature, and selection of the antennas.

FIGURE 8.56
Relative complexity of various communication systems.

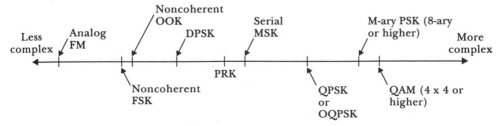

The first step is to establish the sampling rate for the message signal. The highest message frequency was given in the design example as 5 kHz. We saw in Chapter 6 that the minimum theoretical sampling frequency is twice the highest message frequency for baseband message signals. In practice, the sampling rate must be picked somewhat higher than the theoretical limit so that the antialias and reconstruction filters can be implemented without undue difficulty. On the other hand, we do not want to pick the sampling rate too high because this increases the data rate of the system. Ultimately, the required transmitter power is proportional to the data rate. This is because operation below a given bit-error probability sets a lower limit on the ratio E_b/N_0, and, assuming that the noise level is fixed, this requirement sets a lower limit on the received energy per bit E_b. This in turn requires the transmitter power to increase in proportion to the data rate since the received power is the product $E_b f_b$. Thus, we must pick a sampling rate higher than twice the highest message frequency but not too much higher since a primary consideration in the design example is to minimize transmitter power. Perhaps a good choice would be $f_s = 12$ kHz.

The output SNR of a digital PCM system was given in Equation 6.53 as

$$\text{SNR}_o = \frac{3P_{m_n}}{[2^{-2k} + 4P_b(1 + 2^{-2k})]} \qquad (8.119)$$

where P_{m_n} is the power in a normalized version of the analog message, k is the number of bits per code word used to represent the samples, and P_b is the bit-error probability of the communication link. In Section 5.6, we found that P_{m_n} was 0.01. Since the probability of bit error is a strong function of the received power level, it is important to design the system with some margin. Thus we should design to make the second term in the denominator of Equation 8.119 negligible. The output SNR for the design is required to be 40 dB. Neglecting the second term in the denominator and using the values for P_{m_n} and SNR_o we can solve for the number of bits used to code each sample as $k = 9.17$. Since k must be an integer, we select the next highest integer value or $k = 10$. When, as we have assumed in this calculation, the bit-error probability is very small, the noise level is set by quantization error.

Now if we choose the bit-error probability to be 10^{-8}, the second term in the denominator of Equation 8.119 is negligible, and the output SNR is determined primarily by the quantization error. If the error probability becomes less than this value, the output SNR quickly falls below the desired design value. This happens very quickly as the received signal power falls because the error rate is a very strong function of E_b/N_0. As the received power is increased beyond the value required for an error probability of 10^{-8}, an additional safety margin is gained, but the output SNR does not improve significantly because of quantization noise. Since we are interested in a nominal design and would presumably include a safety margin later in refining the design, we use a bit-error probability of 10^{-8}.

For this system, an important consideration is minimal complexity of the transmitter circuits and minimum transmitter power, whereas the use of bandwidth is unimportant. Among the digital modulation techniques we have studied, PRK is probably the best. QPSK or OQPSK would require more complexity in the modulator and provide no better error performance. The factor-of-two bandwidth savings possible with these modulation formats is not attractive for the application at hand. For signaling one bit at a time, PRK provides optimum performance. As we will see later, coding can provide an improvement in error performance, but we will not consider its use at this time. Thus, we proceed assuming that uncoded PRK will be used.

To achieve an error probability of 10^{-8} with PRK, we can find from Equation 8.46 that E_b/N_0 must be 12 dB. This applies for an ideal system with no bandlimiting, no phase errors in the carrier recovery subsystem, and so forth. A more realistic figure would be 14 dB for a well-designed system with the bandwidth limited to the main spectral lobe.

In Section 5.6, we obtained an estimate of the received noise PSD as

$$N_0 = 1.38 \times 10^{-21} \text{ W/Hz}$$

The energy per bit E_b must be 14 dB (approximately a ratio of 25) greater than this value or $E_b = 34.6 \times 10^{-21}$ J. The data rate of the system is the number of bits per sample, $k = 10$, times the number of samples per second, $f_s = 12$ kHz or $f_b = 120$ kHz. The required values of the received power can now be found as $P_r = E_b f_b = 4.15 \times 10^{-15}$ or -143.8 dBW. This compares with $P_r = -144.4$ dBW for the analog FM system. Thus, in this case, the performance of analog FM and PCM are comparable. The use of coding would give the digital PCM approach a clear advantage but at the expense of added complexity. A detailed design tradeoff study would be necessary to determine which system best meets the overall mission objectives.

SUMMARY

1. *M* known signals can always be represented by a generalized Fourier series with *M* or fewer basis functions. A suitable set of basis functions can be found either by inspection or by use of the Gram-Schmidt procedure.

2. A signal can be represented by a vector in signal space where the components of the vector are the coefficients of the generalized series representation. Correlation of signals consists of multiplying the signals and integrating the product with respect to time. The correlation of two signals can also be computed by taking the dot product of their signal-space vectors. The length of a signal-space vector is the square root of the energy in the signal.

3. When white gaussian noise (WGN) with a PSD of $\mathcal{N}_0/2$ is partially represented in a general Fourier series, the coefficients of the series are independent gaussian random variables, each with zero mean and a variance of $\mathcal{N}_0/2$. Thus, WGN is represented by a vector in signal space whose components are independent gaussian random variables.

4. Optimum receivers for digital signals in the presence of WGN can base their decisions on the signal-space components of the received signal. Components of the received noise that arise from basis functions that are not necessary in representation of the transmitted signals are irrelevant to the decision and therefore do not need to be retrieved by the receiver.

5. The maximum a posteriori (MAP) decision criterion is to select the transmitted signal that is most probable, given the received signal and the a priori probabilities of the messages. The MAP criterion leads to minimum-error probability. Application of the MAP criterion results in a receiver that computes the likelihood ratio for the received signal vector and compares the result with a threshold. When the messages are equally likely, the optimum receiver chooses the message signal closest to the received signal vector in signal space. The optimum receiver for binary signals can be implemented with a single correlator or matched filter.

6. The minimum probability of error between two equally likely signals in WGN depends on the distance ℓ between the signals in signal space and the noise level $\mathcal{N}_0$. The error-probability is given by Equation 8.42.

7. An antipodal signal set occurs when one signal is the negative of the other. The antipodal signal set is optimum in the sense of minimizing error probability for a given signal energy when the signals are equally likely. Thus, antipodal signals are optimum for bit-by-bit transmission (i.e., only one bit is associated with each signal). In the next chapter, we will see that better bit-error performance can be obtained by transmitting bits in combination using coding.

8. Orthogonal signal sets have the property that correlation of one signal with another in the set results in a value of zero. In the binary case, orthogonal signals are 3 dB inferior to antipodal signals. In the binary on-off signal set, one of the signals is zero. The binary on-off signal set attains the same error performance as the binary orthogonal set.

9. Optimum receivers often contain correlators, which can be replaced by matched filters. The impulse response of the matched filter is a time-reversed and delayed version of the correlator reference signal. The output of the matched filter is sampled at the end of the signaling interval, and the resulting value is the same as the correlator output at the end of the interval. The output of the correlator and the equivalent matched filter are not necessarily the same except at the end of the interval. The filter is said to be matched to the signal for which the impulse response is a time-reversed and delayed version. The matched filter produces the maximum ratio of peak signal power to average noise power that can be attained by any linear filter.

10. Baseband communication with an NRZ waveform is an example of anti-podal signals consisting of rectangular pulses. The correlation receiver can be implemented with an integrate-and-dump circuit.

11. The eye pattern of a filtered digital waveform can be obtained by displaying the waveform on an oscilloscope triggered by the data clock. The optimum sampling instant occurs at the maximum opening of the eye pattern. The presence of intersymbol interference can be seen clearly in the eye pattern of a binary system because the eye pattern takes on more than two values at the sampling instant.

12. Phase reversal keying (PRK) results when the phase of a sinusoidal carrier takes one of two values $180°$ apart depending on the data symbol. PRK is an antipodal signal set and can be viewed as DSB-SC modulation of the carrier by an NRZ waveform. The PSD of PRK takes the form of a sinc-squared function centered at the carrier frequency with a null-to-null bandwidth equal to twice the bit rate of the system (assuming equally-likely independent data values).

13. A coherent receiver recovers the carrier imbedded in the received signal and uses it in processing the received signal to produce data decisions. Thus, the coherent receiver treats the transmitted signal as a completely known waveform including the carrier phase. A noncoherent receiver treats the carrier phase as a random variable and does not attempt to estimate phase by retrieving the imbedded carrier. For a coherent receiver to be useful, the transmitted carrier phase should be the same (except for data modulation) in each signaling interval. In this case, the signal is said to be coherent. If a noncoherent receiver is in use, it is not necessary to control the carrier phase from interval to interval. When the phase changes randomly over the full range from 0 to 2π, the signal is said to be noncoherent.

14. An on-off keyed (OOK) signal is one in which the data turn the carrier *on* or *off*. The signal can be either coherent or noncoherent, depending on the behavior of the phase each time the signal is turned on. Coherent receivers perform only slightly better than noncoherent receivers, but the carrier recovery function adds significantly to the complexity of the receiver. Thus, coherent receivers are seldom used for OOK. The PSD of the coherent OOK signal contains an impulse with half of the total power at the carrier frequency. The remaining power is distributed according to a sinc-squared function centered at the carrier frequency with a null-to-null bandwidth of twice the data rate. Noncoherent OOK does not have a discrete carrier component, but instead has all of its power distributed in the form of a sinc-squared function centered at the carrier frequency and with a null-to-null bandwidth of twice the data rate. Coherent OOK performs 3 dB poorer than PRK. Noncoherent OOK performs only slightly poorer than coherent OOK. The chief advantage of OOK is the simplicity of the noncoherent receiver.

15. Frequency shift keying (FSK) occurs when the data change the frequency of the transmitted signal. FSK signals can be either coherent or non-

coherent. The error performance of binary FSK is nearly identical to that of OOK.

16. Differential phase shift keying (DPSK) signals have a phase reversal at the start of the interval when the message bit is a 0 and have a constant phase between intervals when the message bit is a 1. A method for demodulating DPSK is to compare the phase of the signal between bit intervals. This receiver has an error performance nearly as good as that of the coherent PRK receiver. The advantage of DPSK is that the receiver is simpler and sometimes easier to implement than a coherent PRK receiver since the carrier recovery function is not needed.

17. *M*-ary modulation techniques use more than two transmitted signals. Often, bits from a binary bit stream are grouped and different bit patterns are assigned to each of the *M* signals.

18. In quadrature phase shift keying (QPSK), the transmitted signal is a sinusoid that can take on any of four phases, 90° apart. It is often produced by separating the input bit stream into two parallel paths that are PRK modulated onto quadrature carriers. QPSK has the same bit-error performance as PRK but uses only half of the bandwidth.

19. Offset QPSK (OQPSK) is the same as QPSK except that one of the parallel data streams is offset by one bit interval (half a symbol) from the other. The phase of the OQPSK signal can change instantaneously by at most 90° whereas the QPSK signal can change by 180°. The advantage of OQPSK as compared with QPSK is that it does not have as large amplitude fluctuations when bandlimited. As a result, bandlimited OQPSK does not experience as much sidelobe regrowth when amplified in hard limiting amplifiers.

20. Minimum shift keying (MSK) has a phase that can either increase or decrease linearly in time by 90° during each bit interval. Because MSK does not have phase discontinuities like PRK, QPSK, or OQPSK, its spectral sidelobes decrease more rapidly with frequency. The main lobe has a null-to-null bandwidth of one and one-half times the data rate. Serial MSK modulators and demodulators give the bandwidth and constant envelope advantages of OQPSK with only slightly more complexity than a PRK system.

21. *M*-ary PSK is a sinusoidal signal that can take on *M* phases 360°/*M* apart. PRK is a special case with $M = 2$, and QPSK is a special case with $M = 4$. When $M = 2^k$, the null-to-null bandwidth of the *M*-ary PSK signal is two times the bit rate f_b, divided by k. As *M* increases beyond four, the error performance becomes poorer. Thus, high-order *M*-ary PSK is a technique for trading increased transmitter power for less bandwidth. Use of the Gray code in assigning bit patterns to the phases reduces the bit-error rate.

22. Quadrature amplitude modulation (QAM) is produced by adding quadrature carriers that have been amplitude modulated. QPSK is a special case in which there are two amplitudes (one is the negative of the other)

for each carrier. Higher-order QAM provides a method for trading increased transmitter power for reduced bandwidth.

23. Carrier recovery in coherent receivers can be accomplished by use of a narrowband filter or a PLL if the modulation format produces a discrete component at the carrier frequency. Suppressed carrier modulation formats often produce a discrete component at a harmonic of the carrier frequency when passed through a nonlinearity. The carrier can then be obtained by use of a bandpass filter or PLL followed by a frequency divider. Modified Costas loops are also useful for carrier recovery and demodulation of suppressed carrier modulation formats.

24. Data clock recovery for NRZ waveforms can be accomplished by filtering the NRZ waveform followed by a nonlinearity, which produces a discrete component at the bit rate. A bandpass filter or PLL then can recover the data clock signal.

25. Word and frame synchronization is often based on the periodic insertion of a known word into the data sequence. At the receiver the periodic occurrence of this word can be used to generate a spectral line at the frame rate.

26. Digital modulation techniques can be compared with respect to error performance for a given received SNR, the bandwidth required, and the complexity of the modulator and demodulator. No single modulation format is best in all situations.

27. Antipodal signals such as PRK provide optimum error performance when signaling is done on a bit-by-bit basis. However, the Shannon-Hartley theorem indicates that significant improvement is possible. We will see that realization of this improvement can be provided by coding for which groups of data bits are used to specify the transmitted signal at any instant.

REFERENCES

F. Amoroso and J. A. Kivett. "Simplified MSK Signaling Technique." *IEEE Trans. Commun.* COM-25, (April 1977): 433–41.

F. Amoroso. "The Bandwidth of Digital Data Signals." *IEEE Communications Magazine* (November 1980): 13–24.

C. R. Ryan, A. R. Hambley, and D. E. Vogt. "760 Mb/s Serial MSK Microwave Modem." *IEEE Trans. Commun.* COM-28, (May 1980): 771–7.

S. Pasupathy. "Minimum Shift Keying: A Spectrally Efficient Modulation." *IEEE Communications Magazine* (July 1979): 14–22.

B. Sklar. "A Structured Overview of Digital Communications-A Tutorial Review-Part I." *IEEE Communications Magazine* (August 1983): 4–17.

B. Sklar. "A Structured Overview of Digital Communications-A Tutorial Review-Part II." *IEEE Communications Magazine* (October 1983): 6–21.

A. D. Whalen. *Detection of Signals in Noise.* New York: Academic Press, 1971.

J. M. Wozencraft and I. M. Jacobs. *Principles of Communication Engineering.* New York: Wiley, 1965.

R. E. Ziemer and R. L. Peterson. *Digital Communications and Spread Spectrum Systems.* New York: Macmillan, 1985.

PROBLEMS

1. Find the values of K and Δf such that the two signals $\phi_1(t) = K \cos(2\pi f_c t)$ and $\phi_2(t) = K \cos[2\pi(f_c + \Delta f)t]$ are orthonormal on the interval from 0 to T. Assume that $f_c T$ is very large and that Δf is much less than f_c.

2. Find the coefficients of the generalized Fourier series for the signals shown in Figure P8.1a using the basis functions shown in Figure P8.1b. Construct the signal-space diagram showing the vector representations for $s_1(t)$, $s_2(t)$, and $s_3(t)$. Compute the correlation of each pair of signals as a time integral and as a dot product of the signal space vectors. Find the energy in each signal by use of the time integral and by finding the squared length of the signal-space vector.

3. Use the Gram-Schmidt procedure to find an alternate set of basis functions for the signals of Figure P8.1a.

4. Two signals $s_1(t)$ and $s_2(t)$ have energies $E_1 = 25$ and $E_2 = 100$, respectively. The energy in the sum of the signals is $E_s = 75$. Sketch a signal-space diagram of the signals showing their lengths and the angle between the vectors.

5. Two signals $s_1(t)$ and $s_2(t)$ have energies $E_1 = 50$ and $E_2 = 100$, respectively. What is the maximum energy that could be contained in the sum of these two signals? What is the relationship of the directions of the signal-space vectors for maximum energy? Repeat for minimum energy in the sum.

6. WGN with a PSD given by $S_n(f) = N_0/2 = 1.5$ is added to the signal $s_1(t)$ of Figure P8.1a, resulting in a signal $r(t) = s_1(t) + n(t)$, where $n(t)$ is the sample function of WGN. Are the basis functions shown in Figure P8.1b a complete set for all of the possible waveforms that $r(t)$ can have? Explain. The coefficients of a generalized Fourier series for $r(t)$ using the basis functions of Figure P8.1b are random variables. Find the PDFs for the coefficients r_1, r_2, and r_3 and sketch to scale.

7. Three dice are in a box; two are fair dice, and the other is loaded so that the face with six spots comes up half of the time whereas the other five faces come up one-tenth of the time each. A die is selected at random from the box and thrown.
    (a) What is the a priori probability of having selected the loaded die before the outcome of the throw is known?
    (b) What is the a posteriori probability of having selected the loaded die, given that the throw comes up six spots? Find the a posteriori probability, given that the throw comes up other than six spots. If the MAP decision criterion is used to decide whether the die selected is loaded, what decision is made in each case?
    (c) Suppose that the die is thrown a second time. Find the a posteriori probability that the loaded die was selected if the throw comes up with six spots both times. Repeat if six spots comes up once. Repeat if six spots comes

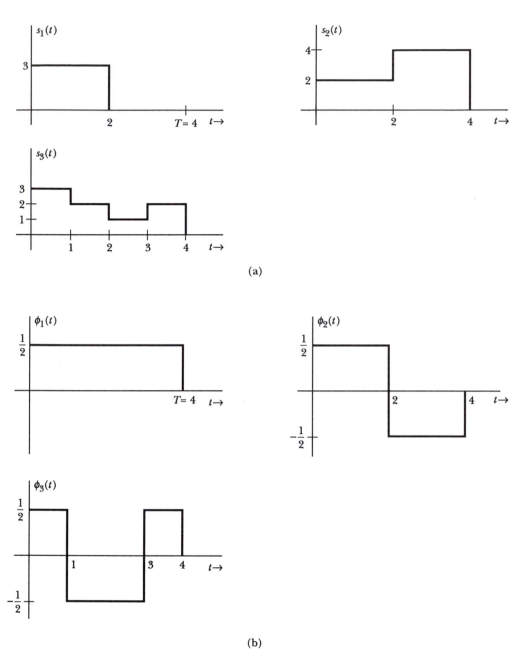

FIGURE P8.1
(a) Signal waveforms and (b) basis functions.

up on neither throw. What is the MAP decision rule to decide if the die is loaded for each experimental outcome?

8. In the simple communication system illustrated in Figure P8.2a, a single bit d is communicated by sending a constant value of -3 if $d = 0$ and a value of $+5$ if $d = 1$. The a priori probabilities are $P[d = 0] = \frac{1}{3}$ and $P[d = 1] = \frac{2}{3}$. A random

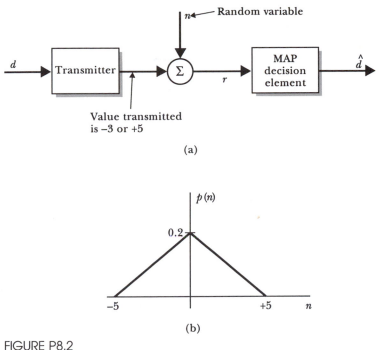

(a)

(b)

FIGURE P8.2
(a) System diagram and (b) PDF of random variable n.

variable with the PDF shown in Figure P8.2b is added to the transmitted value, and the resulting value r is presented to a MAP decision device. Find the conditional PDFs $p(r|d = 1)$ and $p(r|d = 0)$ and sketch them to scale. What is the threshold for the likelihood ratio test of Equation 8.32? Find the MAP decision rule for deciding on d based on the received value. What is the resulting probability of error?

9. Repeat Problem 8 if the a priori probabilities are equal.

10. A binary communication system transmits one of the two equally likely signals shown in Figure P8.3a. Using the basis functions of Figure P8.3b, find the signal-space vectors for the transmitted signals. Draw the signal-space diagram showing the optimum decision regions. Find the probability of error if the noise added in the channel is WGN with a PSD given by $S_n(f) = 0.25$.

11. For the signals of Problem 10, sketch the block diagram of an optimum receiver using only a single correlator. Show the threshold level and the reference waveform for the correlator. Sketch the noise-free output waveform from the integrator when $s_1(t)$ is transmitted and when $s_0(t)$ is transmitted. What are the output values at the end of the signaling interval? Where is the threshold in relation to these values?

12. Sketch the block diagram of a receiver for the signals of Problem 10 using a single matched filter. Sketch the impulse response of the matched filter. Sketch the noise-free output waveform of the matched filter when $s_1(t)$ is transmitted and when $s_0(t)$ is transmitted. What are the values of the filter output at the sampling instant in each case? Where is the threshold in relation to these values?

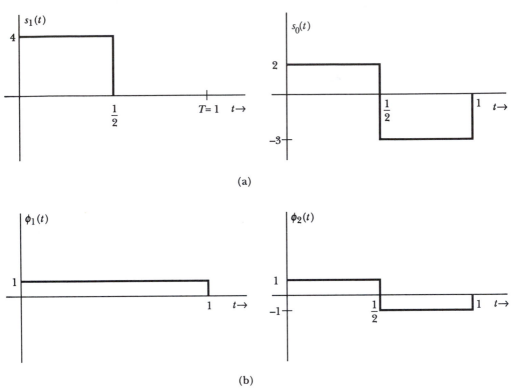

FIGURE P8.3
(a) Signal waveforms and (b) basis functions.

13. Select a set of basis functions to represent the four signals of Figure P8.4 either by inspection or by use of the Gram-Schmidt procedure. Sketch the signal-space diagram and show the optimum decision regions if the signals are equally likely and are used for communication in the presence of WGN.

14. For each of the binary signal formats shown in Figure 7.23, (a) sketch the waveform(s) that can occur in the bit interval between 0 and T_b when the message bit is a zero; repeat when the message bit is a one; (b) for the signals of part (a) select a set of basis functions by inspection and construct a signal-space diagram of the signal sets; and (c) classify the signal sets of part (b) as antipodal, orthogonal, on-off, or none of these.

15. An NRZ waveform is passed through a filter with the impulse response shown in Figure P8.5a. Sketch a typical NRZ waveform and the corresponding output waveform of the filter. Sketch the eye pattern for this filtered signal. Does this pattern display intersymbol interference? Repeat for the impulse response of Figure P8.5b.

16. A communication system using PRK has an effective system noise temperature T_{syst} of 150 K and a data rate of 100 kbit/s. Find the received signal power required so that the bit error rate is 10^{-6} if no allowance is made for degradations due to bandlimiting or other causes. Hint: Recall from Chapter 5 that $\mathcal{N}_0 = kT_{syst}$, where k is Boltzmann's constant.

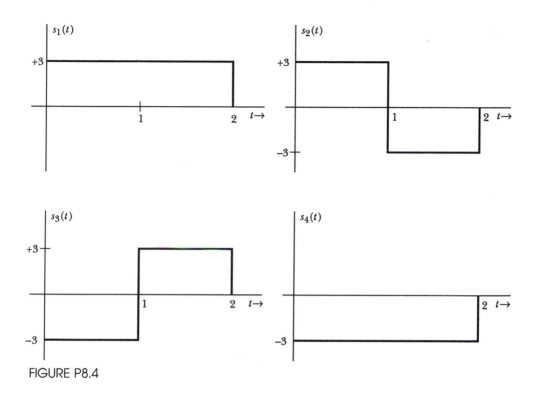

FIGURE P8.4

17. Consider the PRK communication system of Figure 8.23 with a static phase error in the recovered carrier. In other words, the carrier used in the correlator is $2 \cos{(2\pi f_c t + \theta)}$, where θ is the static phase error. Find the output of the correlator as a function of θ at the end of a bit interval if the message bit is a one. Find the bit-error probability of the receiver as a function of θ. Plot the result versus θ for $E_b/N_0 = 10$ and θ ranging from $0°$ to $30°$.

18. A PRK communication system is designed to operate with a bit-error probability of 10^{-5}. By what factor must the transmitter power be increased if the system is changed to DPSK? Repeat for a change from PRK to coherent OOK, noncoherent OOK, or QPSK.

FIGURE P8.5

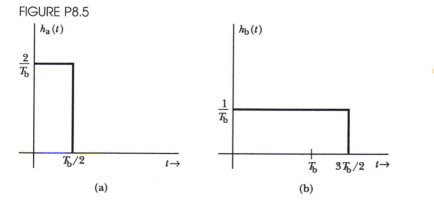

(a)   (b)

19. Find the percentage of the total power of a PRK signal that is contained in the main spectral lobe. Assume a random data sequence. Hint: Use a computer or calculator program to integrate the area of the main lobe of the PSD shown in Figure 8.24.

20. Consider the DPSK receiver of Figure 8.36 and recall that for proper operation the delay must be very close to one bit-interval and also be a multiple of the carrier period. Suppose that a DPSK signal with a data rate of 10 kbit/s and a carrier frequency of 100 MHz is to be demodulated. What should the delay of the delay element be? How many carrier cycles are contained in this delay? What is the phase shift of the delay element at the carrier frequency? Show that if the delay is (slightly) too long or short, so that the net phase shift is 90°, the noise-free signal at the output of the lowpass filter becomes zero. The consequence of this is that the demodulator would base its decisions entirely on noise. What net phase error in the delay would result in a 1 dB loss in the noise-free signal at the filter output? What percentage error in the delay does this represent? Comment on the practicality of the receiver for this combination of data rate and carrier frequency.

21. Consider the signal-space diagram of the QPSK signals and the decision lines shown in Figure 8.38. The effect of a static phase error, denoted by θ, in the recovered carrier is to rotate the decision lines by θ as indicated in Figure P8.6. The effect of this rotation for the in-phase data is to increase the distance to the pertinent decision line for two signals and decrease the distance for the other two signals. Thus, the bit-error probability for the in-phase data will be increased for two signal points and decreased for the other two. Find an expression for the probability of bit error in the in-phase data as a function of θ for each of the four cases. (Hint: The error probability is related to the distance to the decision line by Equation 8.42, where $\ell/2$ is the distance from the signal to the decision line.) Now, using the fact that each of the four signals is equally likely, find an expression for the average probability of bit error for the in-phase data. Because

FIGURE P8.6

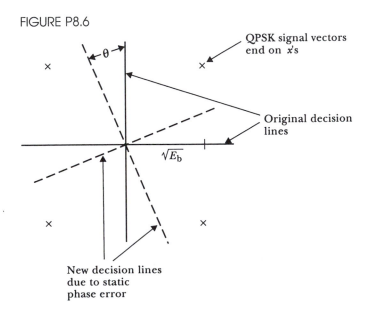

QPSK signal vectors end on x's

Original decision lines

$\sqrt{E_b}$

New decision lines due to static phase error

of symmetry, the result also holds for the quadrature data. Plot the result for $E_b/N_0 = 10$ and θ ranging from $0°$ to $30°$. It is interesting to compare this result with the result of Problem 17. The conclusion is that QPSK is much more sensitive to static phase error than PRK.

22. Find the percentage of the total power contained in the main spectral lobe of the MSK signal by use of numerical integration techniques (computer or calculator) using the PSD shown in Figure 8.44. It is interesting to compare this with the result for PRK (which is the same as for QPSK) found in Problem 19.

23. Consider the signal-space diagram for 4×4 QAM shown in Figure P8.7. Four bits can be assigned to each signal point. Use the two-bit Gray code to assign the first two bits to the in-phase component. Assign the second two bits to the quadrature component also using a Gray code. Verify that the signals in adjacent decision zones differ by only one bit for this assignment. What is the approximate relationship between the bit-error probability and the symbol-error probability for this assignment? Give another assignment of bit patterns to signal points that is bad for minimizing bit-error probability and discuss why the assignment is worse than the Gray code assignment just found.

24. Find the energy of each of the signals of the 4×4 QAM signal set shown in Figure P8.7 in terms of the signal-space distance ℓ. If each of the signals is equally likely, find the average symbol energy. Then find the average energy per bit in terms of ℓ.

FIGURE P8.7
Signal set for 4×4 QAM.

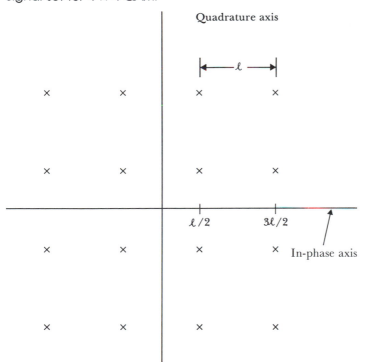

25. If the signals in the 4×4 QAM signal set shown in Figure P8.7 are received in the presence of additive WGN and the signals are equally likely, find an estimate of the symbol-error probability for each of the signals in terms of ℓ and $\mathcal{N}_0$ by adding the probabilities that the received signal-plus-noise crosses each of the decision lines bounding the corresponding decision region. Notice that this estimate is an overbound because the probability of falling in certain regions is added in twice. (However, under conditions of high SNR, it is a close bound.) Then compute the average symbol-error probability. Then use the result of Problem 24 to express the average symbol-error probability in terms of E_b and $\mathcal{N}_0$. Finally, use the results of Problem 23 to find the bit-error probability for the optimum Gray code assignment. Evaluate to find the value of $E_b/\mathcal{N}_0$ required for a bit-error probability of 10^{-6}.

26. Consider the differential encoding and decoding scheme for circumventing the problem of data inversion in the transmission system shown in Figure 8.51. For the data sequence 11001010011, find the output of the differential encoding subsystem, assuming that the output of the encoder was a 1 prior to the start of the given sequence. Show that the original sequence is recovered by the differential decoder regardless of whether there is an inversion in the channel or not.

27. Draw the block diagram of a carrier recovery scheme for 8-ary PSK using a frequency multiplier. Discuss the function of each block in the system. This system has a potential false lock problem because the phase modulation is partly converted to amplitude modulation when bandlimited. This amplitude modulation results in lines at the multiple of the carrier frequency plus and minus the symbol rate. The system can lock to one of these rather than the desired line at a multiple of the carrier frequency. What frequency does the carrier recovery system produce at its output in the false locked condition? This error in the recovered carrier frequency causes the decision lines to rotate with time. By how much do they rotate during a symbol period? Notice that the signal points can fall in the center of a decision sector at the end of each symbol in the false locked condition, just as they do for the correct lock. Thus, any system that is designed to correct the recovered carrier phase by looking for accumulated phase drift during each symbol period potentially suffers from this same false lock condition.

9

ERROR
CONTROL
CODING

As we have seen, the effect of channel noise in digital communication systems is to create bit errors in the output of the demodulator. One method for reducing the error rate is to increase the received energy per bit E_b, either by increasing the transmitter power or by reducing the data rate. If neither of these methods is practical, coding can be used to reduce the error rate. *Coding* is the addition of redundant symbols to the data stream. These redundant symbols can be used at the receiving end to detect errors in the data or actually to correct them.

If the data stream consists of equally likely independent bits, then every possible pattern of ones and zeroes can occur. An error simply changes one data pattern into another, so its presence cannot be detected at the receiver. Adding redundant symbols that are not independent of the data creates recognizable patterns of ones and zeroes, called *code words*. Then the presence of errors is apparent at the receiver whenever the received pattern does not match one of the possible patterns of data plus the redundant bits. Thus, one of the key ingredients of error-control coding is the use of *redundancy*.

Another key ingredient of any good coding scheme is *noise averaging*, which is based on the fact that the fraction of errors in a code word becomes more predictable as the code word becomes longer. This is the familiar idea that the natural frequency of an event approaches the probability of the

427

event when the number of trials becomes very large. If we toss a coin three times, we are not surprised if it comes up heads every time. However, if we toss the coin a thousand times, we are very surprised if it comes up heads every time. Similarly, we may find a large percentage of errors in a short code word, but we expect the ratio of errors in a long code word to be closer to the error probability of the channel. Thus, if a channel has an error probability of 1%, we use a sufficiently long code word so that we expect to find no more than 2% of the received symbols in error, and we would design enough redundancy into the code sequences so that the correct sequence can always be recognized even with 2% of the symbols in error. In this way, we can construct codes to decrease errors in various communication channels.

Two approaches to using redundancy and noise averaging to produce good codes for error detection and correction are *block coding*, which will be discussed in the first section of the chapter, and convolutional coding, discussed in the second section. In block coding, blocks of data bits are transformed into blocks of code symbols or code words. In *convolutional coding*, data bits are shifted into a shift register; and after each shift, several code symbols are produced by modulo-two summing some of the data bits in the register. Thus, block coding segments the data into blocks, and convolutional coding is a continuous operation on the data stream.

Various types of communication channels can benefit from the use of coding. Until now, we have concentrated our attention on the additive white gaussian noise (AWGN) channel. In an AWGN channel, errors occur randomly. Whether a bit is in error or not is independent of an error in the preceding bit or the following bit. In other channels, this may not be the case. For example, a military communication system may encounter a burst noise jammer that is designed to disrupt communications by transmitting randomly timed bursts of high-power noise. In this case, errors tend to occur in bursts when the jammer is *on* rather than being randomly distributed. Another example is the compact disk, which conveys the data bits of digitized music from the recording studio to the pickup head of our player. Scratches or dirt on the surface of the disk create bursts of errors similar to those of a pulsed noise jammer. We will discuss the application of coding to various types of channels with a primary emphasis on the AWGN channel in Section 3 of this chapter.

Coding schemes for correcting errors at the receiving end are known as *forward error correction* (FEC). An alternative technique is simply to use the code to detect errors and request retransmission when errors are detected. This approach is known as *automatic repeat query* (ARQ) and will be discussed in the last section of the chapter. The main advantage of ARQ techniques lies in the fact that error detection is much simpler than error correction. The disadvantages of ARQ include the need for two-way communication, the uneven flow of data when errors occur, and the additional complexity of the transmitter, which must be designed to repeat transmissions on request.

9.1
BLOCK CODES

In a block code, a sequence of k data bits is combined with redundant bits to make a block of n code symbols. The resulting code is called an (n,k) block code.

Single parity check codes

One of the simplest block codes, commonly used for detecting errors in computer data, is the *single parity check code*. In this code, a single redundant bit is made by adding the data bits of a data word modulo-two fashion, and appending the resulting check bit to the data word. Modulo-two addition is defined by

$$0 \oplus 0 = 1 \oplus 1 = 0$$

and

$$1 \oplus 0 = 0 \oplus 1 = 1$$

If the data blocks are eight bits long, we obtain a (9,8) block code. Some of the code words for this code are the following:

```
0000  0000  0
0000  0001  1
1111  1111  0
1010  0000  0
1110  0000  1
```

data bits   redundant parity bit

We have put the spaces in the code words to improve legibility. Sometimes, this code is said to have *even parity* because each of the code words has an even number of 1s. A single error in transmission of the code word becomes obvious because the number of 1s in a word with a single error is odd. However, it is not possible to determine which bit is in error with a single parity bit. For example, if we receive the erroneous code word

$$0011 \quad 0011 \quad 1$$

it is obvious that at least one error has occurred because the word contains an odd number of ones, but it is not possible to determine where the errors have occurred. A valid code word can be obtained by changing any one of the nine symbols in the word.

Sometimes, code words with *odd parity* in which the redundant bit is selected to result in an odd number of 1s in the code words is used, and the receiver then detects an error when even parity occurs.

Code rate

The *rate* R_c of a block code is the ratio of the number of data bits in a code word k to the total number of symbols per code word n, given by

$$R_c = \frac{k}{n} \tag{9.1}$$

The code rate is a measure of the efficiency of a code. The $n - k$ redundant symbols must be transmitted through the system along with the k data bits, increasing the number of symbols that must be transferred from the transmitter to the receiver. For codes with otherwise equal performance, the one with the highest rate is generally the most desirable. The rate of the single parity check code discussed above is $\frac{8}{9}$.

Repetition codes

Another simple code is the *repetition code* in which a code word contains a single data bit that is repeated n times. For $n = 5$, the two code words are

$$00000 \quad \text{and} \quad 11111$$

If this code is used for error correction, up to two errors can be corrected. For example, if 00111 is received, the most likely transmitted sequence is 11111, and the most likely data bit is a 1. If the repetition code is used for error detection, then any number of errors up to $n - 1$ can be detected. If 11111 is transmitted and four errors occur, so that 01000 is received, the four errors can be detected; but if error correction is attempted, the word is incorrectly determined to be 00000. In general, a code can be used to detect more errors that it can correct, and error detection is simpler to perform than error correction.

As we have discussed, the performance of codes becomes better as the code words become longer, due to noise averaging. As the code words become longer, the fraction of code symbols in error becomes more and more likely to be close to the error probability of the channel. Thus, incorrect decoding due to a large fraction of errors in a code word becomes unlikely. For the repetition code, as the code words become longer, the code rate R_c becomes smaller, indicating that the number of code symbols that must be transmitted per data bit is increasing. As we will see, the result is that the repetition code does not provide any performance improvement for the AWGN channel. Thus, it is necessary to use more complex codes for which more than one data bit is combined in each code word. However, the repetition code provides a significant performance improvement for the case of interference by a pulsed noise jammer.

Simple multiple parity check codes

A simple code that is capable of correction of a single error can be constructed by arranging the data bits $d_1, d_2, \ldots, d_k$ in a rectangular array. A redundant parity check bit is computed for each row and column in the array by modulo-

TABLE 9.1
A single error correcting code
based on multiple parity checks

d_1	d_2	d_3	r_1
d_4	d_5	d_6	r_2
d_7	d_8	d_9	r_3
c_1	c_2	c_3	

two addition of the data bits in the row or column. An example is shown in Table 9.1 for nine data bits arranged into a square array. Six parity check bits are computed, one for each column or row of the array, resulting in a 15-bit code word. This is an example of a (15,9) block code.

For the code of Table 9.1 we have

$$r_1 = d_1 \oplus d_2 \oplus d_3$$
$$r_2 = d_4 \oplus d_5 \oplus d_6$$
$$\vdots$$
$$c_3 = d_3 \oplus d_6 \oplus d_9$$

Therefore, the top row of the array, including the parity symbol r_1 must contain an even number of 1s for a correctly received code word. The received words can be decoded by checking the parity of the three rows and three columns. If the parity is even in all six cases, it is assumed that the word has been received correctly. If a data bit is received in error, parity is odd for the corresponding row and column, and the error can be corrected. If a single parity check symbol experiences an error in transmission, then only that parity check fails, and the data bits should be assumed to be correct. Thus, a single error can always be corrected. Some thought will show that double errors and some triple errors can be detected but not corrected.

Hamming distance

We will find it convenient to denote code words in matrix form as a row vector whose elements are the code symbols. Thus, a code word $\mathbf{C_i}$ can be written in terms of its successive symbols $c_{i1}, c_{i2}, \ldots$ as

$$\mathbf{C_i} = (c_{i1}, c_{i2}, \ldots, c_{in}) \qquad (9.2)$$

Note that c_{ij} is the jth symbol of the ith code word. Usually, we assume that the data bits are the first k symbols in the code word. When the data bits appear directly in the code words, the code is said to be *systematic*.

The *weight* of a code word, denoted by $w(\mathbf{C_i})$, is the number of 1s in the word. For example, we have

$$w(0,0,1,1,0,1,1) = 4$$

and

$$w(1,1,1,1,0,1) = 5$$

The *Hamming distance* d_{ij} between two code words $\mathbf{C_i}$ and $\mathbf{C_j}$ is the number of symbols in which the two words are different. This distance is the weight of the bit-by-bit modulo-two sum of the words. Thus, the Hamming distance is given by

■ $$d_{ij} = w(\mathbf{C_i} \oplus \mathbf{C_j}) \qquad (9.3)$$

For example, the Hamming distance between the two code words

$$\mathbf{C_1} = (1,1,0,0) \qquad \text{and} \qquad \mathbf{C_2} = (1,0,1,0)$$

is

$$d_{12} = w(\mathbf{C_1} \oplus \mathbf{C_2}) = w(0,1,1,0) = 2$$

We could have found this result by noting that the two code words are different in the middle two symbols.

Error detection and correction

One approach to decoding the received blocks is to determine which of the possible code words is closest to the received sequence in Hamming distance. This is called the *maximum likelihood decoding algorithm* because it results in the code word with the highest a posteriori probability of having been transmitted, given the received sequence (assuming, as we always will, that the probability of error for a given bit is less than one-half).

The error correction and detection capability of a code is a function of the minimum Hamming distance between any pair of code words. For example, suppose that the minimum Hamming distance for a code is five. This situation is illustrated in Figure 9.1, in which the heavy dots represent two

FIGURE 9.1
Diagram showing two code words $\mathbf{C_i}$ and $\mathbf{C_j}$ that are 5 units of Hamming distance apart.

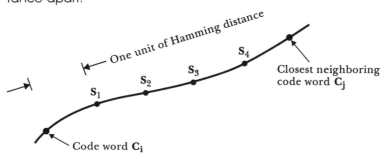

of the closest code words and the light dots represent sequences that are not valid code words. If code word $\mathbf{C_i}$ is transmitted and a single error occurs, the received sequence $\mathbf{S_1}$ is one unit of Hamming distance away, as indicated in the figure. Similarly, two errors could result in the received sequence represented by $\mathbf{S_2}$, and so on. For a minimum distance of five units, the decoder can correct one or two errors if error correction is desired, because two errors result in a sequence closer to the correct code word than to any other code word. However, up to four errors can be detected by a distance five code. Note that five errors can transform a code word into its nearest neighbor and thereby remain undetected. Alternatively, for a minimum distance of five, a single error can be corrected and up to three errors can be detected. Note that in this case, four errors can result in improper "correction" to a different code word, resulting in errors in the decoder output. Thus, for a code with a minimum distance between closest code words of five Hamming units, we have the following options at the receiver:

1. Correct all single and double errors. Three or more errors then result in improper "correction" and possible errors in the decoded data.
2. Correct single errors and detect two or three errors. In this case, four or more errors may result in decoding errors.
3. Detect one through four errors. In this case, five or more errors may not be detected.

Similar options exist for codes with other values of minimum distance.

In general, if a code is required to correct t or fewer errors, the minimum distance between code words must be

$$d_{\text{min}} = 2t + 1 \qquad (9.4)$$

If a code is required to detect the presence of l or fewer errors, the minimum distance must be at least

$$d_{\text{min}} = l + 1 \qquad (9.5)$$

A conceptually simple decoder for a given block code is to use a read only memory (ROM), in which each possible n-symbol sequence that is received serves as the address, and the word stored in each location of the memory is the data sequence for the code word closest to the address sequence. In this case, the received sequence is used simply to retrieve the decoded data from the memory. This is satisfactory for short code words but quickly becomes impractical for the longer code words needed to provide improved performance by noise averaging. The number of memory words required is 2^n, where, as usual, n is the number of symbols in a code word. For example, a 20-bit code length requires approximately one million words of memory whereas a 30-bit code length requires approximately one billion words of memory.

Clearly, a simpler decoder is required for practical realization with longer code words. We will see that this can be achieved by structuring the code.

Linear codes

A great reduction in decoder complexity can be achieved by requiring the code to be linear. A *linear code* is one in which the bit-by-bit modulo-two sum of any two code words is also a code word. Linear codes are also known as *group codes* because the code words form a mathematical entity known as a group. A consequence of linearity is that the code always contains the all-zero code word because the bit-by-bit modulo-two sum of a code word with itself is all zeroes and is required to be a code word by linearity. Another important consequence is that the minimum distance between any two distinct code words is the weight of the code word with the fewest number of 1s. This is because we determine distance by summing (bit-by-bit modulo-two, of course) the two code words that form a third code word whose weight is the distance between the first two code words. Furthermore, the distance from any code word to its nearest neighbor is the same as the distance from any other code word to its nearest neighbor. Thus, the error performance is the same for all of the code words, and we can consider transmission of any convenient code word in evaluating performance. Therefore, in theoretical considerations we often assume that the all-zero code word is transmitted. This is a significant simplification in the analysis of error performance of codes. Note that this simplification is not usually possible for nonlinear codes.

A simple example of a linear code consists of the following code words:

$$00000$$
$$01101$$
$$10110$$
$$11011$$

Notice that this code contains the all-zero code word and that the sum of any two code words is also a code word. The minimum weight of a nonzero code word is three, which is the same as the minimum distance from any code word to its nearest neighbor.

Generator and parity check matrices

The *generator matrix* for a linear code is a k row by n column matrix consisting of the code words corresponding to the data sequences containing a single 1. Usually, we arrange the rows of the generator matrix so that the left portion forms an identity matrix. For example, the generator matrix for a certain (6,3) systematic linear code is

$$\mathbf{G} = \begin{bmatrix} 100011 \\ 010110 \\ 001101 \end{bmatrix} \qquad (9.6)$$

The code word **C** for any data sequence $\mathbf{D} = [d_1, d_2, \ldots, d_k]$ can be found by forming the matrix product

$$\mathbf{C} = \mathbf{DG} \tag{9.7}$$

where modulo-two addition is used. As an example, we can find the eight code words for the generator matrix of Equation 9.6 as

$$
\begin{array}{l}
000000 \\
001101 \\
010110 \\
011011 \\
100011 \\
101110 \\
110101 \\
111000
\end{array} \tag{9.8}
$$

Notice that, as required for a linear code, the sum of any two code words is also a code word. Also, the minimum weight and therefore the minimum distance of this code is three. Thus, the code can be used to correct one error. If only error detection is wanted, this code can be used to detect one or two errors. Each code word consists of the three data bits plus three parity check bits. The general code word is given by

$$
\mathbf{C} = \mathbf{DG} = [d_1, d_2, d_3] \begin{bmatrix} 100011 \\ 010110 \\ 001101 \end{bmatrix}
$$

$$
\mathbf{C} = [\underbrace{d_1, d_2, d_3}_{\text{data}}, \underbrace{d_2 \oplus d_3,\ d_1 \oplus d_2,\ d_1 \oplus d_3}_{\text{redundant parity check bits}}] \tag{9.9}
$$

Thus, the code word consists of the data bits plus redundant bits. Each redundant bit forms a parity check with some combination of the data bits.

The *parity check matrix* for a code is a matrix that can be used to check each of the parity relations of a code word. The parity check matrix **H** has n columns and $n - k$ rows, one row for each redundant parity check bit in the code word. The right-hand $n - k$ columns of the parity check matrix consist of an $n - k$ by $n - k$ identity matrix. Each of the first k bits of each row is set to one only if the data bit corresponding to the column is included in the parity check corresponding to the row. For example, the parity check matrix for the code words of Equation 9.9 is given by

$$
\mathbf{H} = \begin{bmatrix} 011100 \\ 110010 \\ 101001 \end{bmatrix} \tag{9.10}
$$

Notice that this parity check matrix has one row for each of the three redundant check bits of the code word given in Equation 9.9. The right-hand portion of the **H** matrix forms an identity matrix. The first element of row 1 is 0 because the first data bit is not included in the first parity check bit (which is the fourth bit of the code word). The second bit of row 1 is a 1 because the second data bit is included in the first parity bit. We proceed in this fashion, finally reaching the third bit of the last row, which is a 1 because the third data bit is included in the last parity check bit.

Due to the manner in which the parity check matrix is formed, the product of the parity check matrix with the transpose of any valid code word results in an $n - k$ element column matrix with all zero elements. To illustrate this for the parity check matrix of Equation 9.10 and the seventh code word in the list given in 9.8, we have

$$\mathbf{H}\mathbf{C}^{\mathrm{T}} = \begin{bmatrix} 011100 \\ 110010 \\ 101001 \end{bmatrix} \begin{bmatrix} 1 \\ 1 \\ 0 \\ 1 \\ 0 \\ 1 \end{bmatrix} = \begin{bmatrix} 0 \\ 0 \\ 0 \end{bmatrix}$$

As usual, modulo-two arithmetic is used in the calculation.

In general, the generator matrix for a systematic linear code can be written in the form

$$\mathbf{G} = [\mathbf{I_k}|\mathbf{P}] \tag{9.11}$$

where $\mathbf{I_k}$ is a k by k identity matrix and $\mathbf{P}$ is a k row by $n - k$ column matrix. The rows of the generator matrix consist of the code words for data sequences containing a single 1. The corresponding parity check matrix can be shown to be given by

$$\mathbf{H} = [\mathbf{P^T}|\mathbf{I_{n-k}}] \tag{9.12}$$

where $\mathbf{P^T}$ is the transpose of $\mathbf{P}$ and $\mathbf{I_{n-k}}$ is an $n - k$ by $n - k$ identity matrix.

The syndrome

The parity check matrix is very useful in providing a simplified method for error detection and error correction with a linear code. When a code word is transmitted through the channel, the effect of the errors is to add a 1 to some of the code symbols, which are then received in error. Thus, the received pattern is given by

$$\mathbf{R} = \mathbf{C} \oplus \mathbf{E} \tag{9.13}$$

where **C** is the transmitted code word and **E** is a row vector that contains a 1 in each position where a symbol error occurs. If the parity check matrix is used to check the received pattern, the resulting matrix is called the *syndrome*, given by

■ $\quad$ $\mathbf{S} = \mathbf{HR}^{\mathrm{T}} = \mathbf{H}[\mathbf{C}^{\mathrm{T}} \oplus \mathbf{E}^{\mathrm{T}}] = \mathbf{HC}^{\mathrm{T}} \oplus \mathbf{HE}^{\mathrm{T}} = [000 \cdots 0]^{\mathrm{T}} \oplus \mathbf{HE}^{\mathrm{T}}$

$$(9.14)$$

Finally, we obtain

■ $\quad\quad\quad\quad\quad\quad\quad\quad \mathbf{S} = \mathbf{HE}^{\mathrm{T}} \quad\quad\quad\quad\quad\quad (9.15)$

From this result, we see that the syndrome depends only on the error pattern. If no errors occur, the syndrome is zero in every element. Thus, if only error detection is required, we need only to compute the syndrome and check to see if it is zero. If the syndrome is zero, presumably no errors have occurred. (At least, no error patterns that the code is capable of detecting have occurred.) When error correction is required, the syndrome indicates the pattern of the errors. For example, if only a single error has occurred, the syndrome is the same as the column of the parity check matrix corresponding to the error location. To illustrate this, suppose that, for the code of Equations 9.8 and 9.10, the code word

$$[110101]$$

is transmitted but received with an error in the third bit as

$$[111101]$$

The syndrome can then be computed as

$$\mathbf{S} = \mathbf{HR}^{\mathrm{T}} = \begin{bmatrix} 011100 \\ 110010 \\ 101001 \end{bmatrix} \begin{bmatrix} 1 \\ 1 \\ 1 \\ 1 \\ 0 \\ 1 \end{bmatrix} = \begin{bmatrix} 1 \\ 0 \\ 1 \end{bmatrix}$$

Note that the syndrome pattern matches the third column of the parity check matrix, as it should when an error occurs in the third bit.

Clearly, from this result we can see that to construct a single error correcting code, the necessary condition is for each column of the parity check matrix to be unique and nonzero. Certainly, if a column of the parity check matrix is zero, an error in the corresponding location of the code word results in an all-zero syndrome and therefore goes undetected. Also, if two columns of the parity check matrix are the same, then an error in either location is indistinguishable from an error in the other location.

Hamming codes

Hamming codes are single error correcting codes with an integer number, p of parity checks, $n = 2^p - 1$ and $k = n-p$. The columns of the parity check matrix of a Hamming code are all of the nonzero bit patterns of p bits each.

EXAMPLE 9.1

Write the parity check matrix for a Hamming code with $p = 3$ parity checks, find the corresponding generator matrix, list all of the code words, and verify that the minimum distance is consistent with the ability to correct single errors. What happens if errors occur in locations 1 and 3 and the decoder "corrects" the errors?

SOLUTION

The parity check matrix contains three rows, one for each of the parity check bits in the code. The right-hand three columns form an identity matrix. With three bits, seven different nonzero bit patterns can be produced. Three of these appear in the right-hand columns of the parity check matrix, and the remaining four can be placed in any order in the first four columns of the matrix. Therefore, one possibility for the parity check matrix is

$$\mathbf{H} = \begin{bmatrix} 1101100 \\ 1011010 \\ 0111001 \end{bmatrix}$$

This is the parity check matrix for a $(7,4)$ Hamming code. The generator matrix can be found by the use of Equations 9.11 and 9.12. The generator matrix is a $k = 4$ by $n = 7$ matrix. The first four columns of the generator matrix are a four-by-four identity matrix, and the last three columns are the transpose of the first four columns of the parity check matrix. The resulting generator matrix is

$$\mathbf{G} = \begin{bmatrix} 1000110 \\ 0100101 \\ 0010011 \\ 0001111 \end{bmatrix}$$

The list of code words can be found by multiplying the generator matrix by each of the 16 data patterns of four bits each. In general, this is given by

$$\begin{aligned} \mathbf{C} &= [d_1, d_2, d_3, d_4]\mathbf{G} \\ &= [d_1, d_2, d_3, d_4, d_1 \oplus d_2 \oplus d_4, d_1 \oplus d_3 \oplus d_4, d_2 \oplus d_3 \oplus d_4] \end{aligned}$$

The list of code words is

0000000	1000110
0001111	1001001
0010011	1010101
0011100	1011010
0100101	1100011
0101010	1101100
0110110	1110000
0111001	1111111

The first four bits of each code word are the data bits, and the remaining three bits are the parity check bits. Notice that the minimum weight of any of the code words is three. Therefore, the minimum distance of this linear code is three, which is the required minimum distance for single error correction. Notice that the all-zero code word is included and the sum of any two code words is also a code word.

If errors occur in locations one and three, the result of the syndrome calculation is

$$\mathbf{S} = \mathbf{H}\mathbf{E}^{\mathrm{T}} = \mathbf{H}[1010000]^{\mathrm{T}} = \begin{bmatrix} 1 \\ 0 \\ 1 \end{bmatrix}$$

This syndrome pattern matches column 2 of the parity check matrix, so an error correction decoder would improperly "correct" bit 2 of the received sequence, resulting in errors in locations 1, 2, and 3 at the decoder output. Thus, the coding and decoding process can result in more errors than when no coding is used. However, the decoder would correct all of the single errors; and since these are more likely than double errors when the probability of error in transmission is small, the coding and decoding can reduce the overall error probability.

Cyclic codes

The structure in linear codes allows the decoding process to be simplified by the use of the syndrome to identify the error pattern. Additional code structure in the form of a requirement for the code words to be cyclic can further reduce the complexity of the coding and decoding process while still allowing the construction of codes with good distance properties. A *cyclic code* is one for which a cyclic shift of a symbol from the beginning of a code word to the end produces another code word in the code. For example, the following set of code words form a cyclic code:

$$\begin{aligned} & 0000 \\ & 0101 \\ & 1010 \\ & 1111 \end{aligned}$$

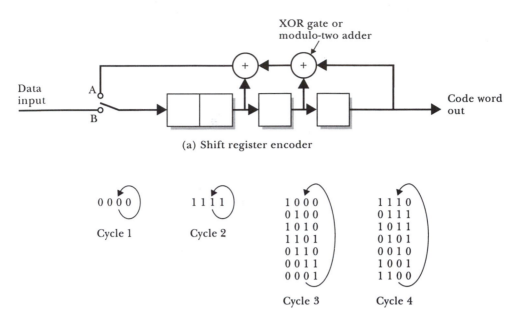

(a) Shift register encoder

0 0 0 0
Cycle 1

1 1 1 1
Cycle 2

Cycle 3
1 0 0 0
0 1 0 0
1 0 1 0
1 1 0 1
0 1 1 0
0 0 1 1
0 0 0 1

Cycle 4
1 1 1 0
0 1 1 1
1 0 1 1
0 1 0 1
0 0 1 0
1 0 0 1
1 1 0 0

(b) Cycles of encoder states with switch in position A

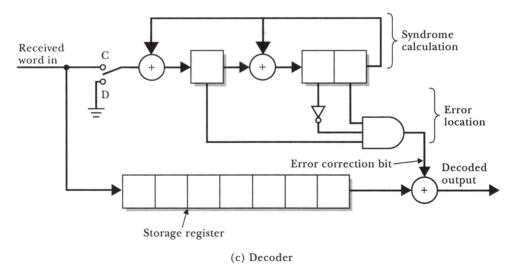

(c) Decoder

FIGURE 9.2
Encoder and decoder for Example 9.2.

This code is cyclic because when a symbol is moved from the beginning to the end of a code word another code word (or the same word) in the set is obtained. In fact, if the first two bits are the data, this is a linear cyclic systematic code with a minimum distance of two.

The study of cyclic codes involves the theory of polynomials with binary coefficients and a branch of mathematics known as Galois field theory. The use of this theory provides many interesting and useful results for block coding. The reader who is interested will find additional information in the references. The following example illustrates some of the simplifications that are possible with cyclic codes.

EXAMPLE 9.2

Show that the shift-register encoder of Figure 9.2a produces a cyclic code by listing the code words. Find the generator matrix and the parity check matrix. Illustrate by example that the decoder computes the syndrome of the received sequence and corrects single bit errors.

SOLUTION

With the switch in position A, the state of the encoder shift register cycles through one of the four cycles indicated in Figure 9.2b. For example, if the shift register starts in the state 1000 (where the right-hand bit corresponds to the right-hand cell of the shift register), the bit fed back to point A by the modulo-two adders (these are the same as exclusive OR gates) is a 0; so the next state is 0100. On the next shift, the bit fed back by the adder is a 1, and so the next state is 1010. After seven shifts the state returns to the starting value. As indicated in Figure 9.2b, there are two cycles of length seven and two cycles of unit length for this shift register encoder.

The encoder operates by feeding the data into the register with the switch in position B. Therefore, if the data sequence is 0111 (with, as usual, the first bit in time given first), the state of the shift register is 1110 after shifting the data into the encoder register. Note that the order of the bit pattern in the register is the reverse of the data pattern because the first bit shifted into the circuit ends up in the right-hand stage. The switch is then placed in position A, and the register is shifted seven times to produce the output code word. For a data sequence of 0111, the starting state is the top state of cycle 4 shown in Figure 9.2b. The output code word is the right-hand bit of the shift register state for seven shifts or 0111010. Note that the data sequence is the first four bits of the code word, and the last three bits are sums of some of the data bits.

In general, if the shift register is loaded with the data, the shift register state starts as

$$d_4, d_3, d_2, d_1$$

After one shift, the state becomes

$$d_1 \oplus d_2 \oplus d_3, d_4, d_3, d_2$$

On the next shift, the state becomes

$$d_2 \oplus d_3 \oplus d_4, d_1 \oplus d_2 \oplus d_3, d_4, d_3$$

Shifting one more time produces

$$d_1 \oplus d_2 \oplus d_4, d_2 \oplus d_3 \oplus d_4, d_1 \oplus d_2 \oplus d_3, d_4$$

At this point, the register contains the last four bits of the output code word. The output code word is given by the successive content of the right-hand stage of the register. This is given by

$$d_1, d_2, d_3, d_4, d_1 \oplus d_2 \oplus d_3, d_2 \oplus d_3 \oplus d_4, d_1 \oplus d_2 \oplus d_4$$

The code words can now be listed:

0000000	1000101
0001011	1001110
0010110	1010011
0011101	1011000
0100111	1100010
0101100	1101001
0110001	1110100
0111010	1111111

Inspection of this list shows that a cyclic shift of any code word results in another code word.

The generator matrix can be constructed by listing the code words with a single 1 in the data sequence in proper order so that an identity matrix results in the left four columns. The result is

$$\mathbf{G} = \begin{bmatrix} 1000101 \\ 0100111 \\ 0010110 \\ 0001011 \end{bmatrix}$$

Using Equations 9.11 and 9.12, we can find the parity check matrix as

$$\mathbf{H} = \begin{bmatrix} 1110100 \\ 0111010 \\ 1101001 \end{bmatrix}$$

Notice that the columns of this parity check matrix include all of the possible nonzero three-bit combinations. Therefore, the code is a Hamming code, but it has the additional property of being cyclic, whereas the Hamming code of Example 9.1 was not cyclic. Cyclic codes have the advantage that they can be generated and the syndrome can be calculated using simple feedback shift register circuits.

The decoder of Figure 9.2c operates in the following manner. Starting with the registers in the all-zero state, the received word (with possible channel errors) is shifted into the circuit. After seven shifts, the received code word has been shifted into the circuit, the syndrome register contains the syndrome for the received sequence, and the storage register contains the received word. Then the switch is moved to position D so that zeroes are shifted into the circuit on each shift, and the circuit is shifted seven more times. If there was no error in the received sequence, the content of the syndrome circuit is zero, the error correction bit remains as a 0, and the word in the storage register is shifted out without change. However, if a single error has occurred, the content of the syndrome register is not zero, and the error location circuit produces a 1 when the erroneous bit is in the last stage of the storage register. Thus, the output of the decoder is corrected.

We will now give an example of the operation of the circuit. Suppose that the transmitted code word is 0001011 but is received with an error in the fourth location as 0000011. The syndrome register is in the all-zero state before the received word is shifted into the circuit. Therefore, the syndrome register stays in the all-zero state until the sixth received bit is shifted into the circuit. After the sixth bit, the state of the syndrome register is 100. The seventh bit of the received sequence is also a 1, so the state of the syndrome register is 110 after the complete word is received. Now, as the decoded word is shifted out of the register with the switch in position D, we have the following results:

Syndrome register	Correction bit	Storage register	Decoded output bit	
110	0	1100000	0 ←	register states after
011	0	0110000	0	receiving the
111	0	0011000	0	word 0000011
101	1	0001100	1 ←	
100	0	0000110	0	Erroneous bit
010	0	0000011	1	corrected
001	0	0000001	1	

Thus, the output of the circuit is the corrected code word. It can be shown that the circuit always corrects single errors in a received code word.

In general, decoders for cyclic codes consist of a syndrome calculation circuit consisting of a feedback shift register, a circuit for error location, and a register to store the received word until the errors can be corrected. The block diagram of this decoder architecture is shown in Figure 9.3. For a simple code like the Hamming code, the error location circuits are fairly simple. However, for long multiple error correcting codes, the error location process can be rather complex. The interested reader should see the references for more information on this subject.

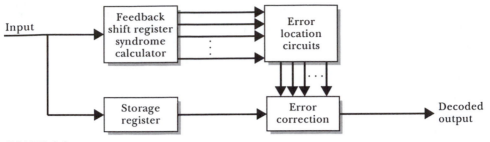

FIGURE 9.3
Typical decoder architecture for cyclic block codes.

9.2
CONVOLUTIONAL CODES

In block coding, the information bits are divided into blocks of k bits that are then encoded into longer blocks of n code symbols. Convolutional coding, on the other hand, does not divide the information bits into blocks but uses the continuing stream of data bits to produce a stream of code symbols. As in the case of block codes, there are more bits in the coded result than in the input data stream since redundancy is a necessary ingredient of any successful error control coding scheme.

A simple convolutional encoder

An example of a simple convolutional encoder is shown in Figure 9.4. In this encoder, information bits are shifted into the shift register one at a time. After each data bit is shifted into the register, three code symbols are produced by

FIGURE 9.4
Rate $\frac{1}{3}$ convolutional encoder.

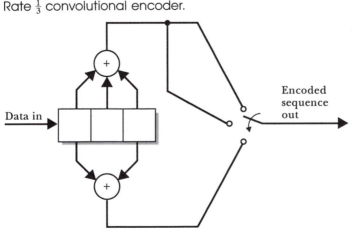

moving the switch through its contacts from top to bottom to route the code symbols produced by the modulo-two adders to the output. For example, assuming that the content of the register is 000 before any data arrive, when a data bit that happens to be a 1 is shifted into the register, the content of the register becomes 100. Then the outputs of both adders are 1, and the three code symbols at the output are 111. If another 1 enters the register, the register content becomes 110, and the top adder output is a 0, whereas the bottom adder output is a 1. The output switch samples the adder outputs from the top to the bottom contact resulting in a second output code group of 001.

The tree diagram

A simple way of portraying the outputs of a convolutional encoder is the *tree diagram*. The tree diagram for the encoder of Figure 9.4 is shown in Figure 9.5. When the data input to the encoder is a 0, the operation follows an upward branch of the tree whereas an input of 1 causes operation along a downward branch. For example, the input data sequence 0110 would cause operation from the start at point A along an upward branch to point B since the first data bit is a 0. Then, since the next bit is a 1, operation would follow a downward path to point C and so forth to points D and E. The resulting output code bits are written along the branches of the tree diagram. Thus, for the data sequence, 0110, (and assuming that the encoder register starts in the all-zero state), the output code bits are 000 111 001 001. We have grouped the output bits in groups of three simply to emphasize that three output symbols are produced for each input bit for this particular convolutional encoder.

The encoder of Figure 9.4 is called a rate $\frac{1}{3}$ coder because it produces three code symbols for each data bit. We could change the encoder into a rate $\frac{1}{2}$ convolutional encoder simply by eliminating one of the switch contacts. A rate $\frac{1}{4}$ encoder could be constructed by adding another modulo-two adder and switch contact so that four code symbols are produced on each shift. Another possibility would be to use the encoder as shown in Figure 9.4 but to shift in two data bits before sampling the adder outputs to produce the three code symbols. Then we would have a rate $\frac{2}{3}$ coder.

The trellis diagram

Tree diagrams such as the example in Figure 9.5, are unnecessarily complicated when long data sequences are transmitted, because the number of branches doubles at every step but the encoder register has a small number of states, which recur as we progress through the tree. An alternative portrayal of the convolutional coding process is the *trellis diagram*, shown in Figure 9.6 for the encoder of Figure 9.4. In this trellis diagram there are four possible states at each step. Each state corresponds to the contents of the two left-hand stages of

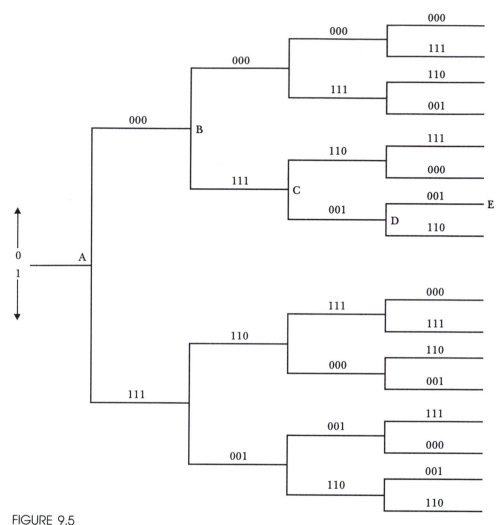

FIGURE 9.5
Tree diagram for the encoder of Figure 9.4.

the shift register. If the left stages contain the 00 pattern, then the encoder is in the top state of the trellis diagram, and so forth for the other three possible contents. When a new data bit enters the register, the state changes and three code symbols are emitted. For example, if the coder is in state 00 (as at point A in Figure 9.6) and a 1 enters, the new state is 10 (the left bit of the state designation corresponds to the left-most stage of the register), and the three code symbols emitted are 111. Thus, the coder state has moved from point A to point B on the trellis diagram. Now, if a second 1 enters the encoder, the state changes to 11 and the code symbols are 001. Notice that after step 2 in the trellis diagram, the pattern of state transitions and code bits repeats without any further increase in complexity.

State

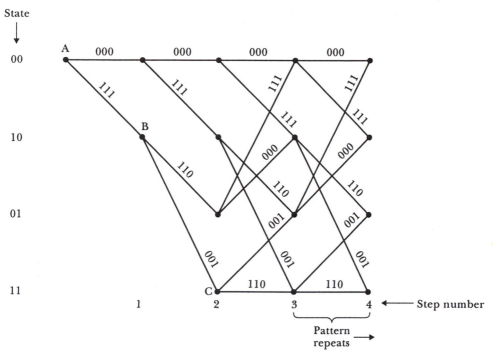

FIGURE 9.6
Trellis diagram for the encoder of Figure 9.4.

Constraint length

If the number of bits in the encoder shift register were increased from three to four, then the number of states in the trellis diagram would double from four to eight because the state of the encoder would depend on the content of the left three stages. In this case, the complexity of the trellis would increase. Thus, the complexity of a convolutional code depends on the length of the shift register (assuming that all of the stages are connected to at least one of the adders). The effective number of shift register stages is sometimes called the *constraint length* of the encoder, though this terminology is not standard in the literature.

As in the case of block codes, an important measure of the ability of a convolutional code to correct errors is the Hamming distance between various code sequences produced by the possible input data sequences. Codes that have greater distances between the closest output sequences usually have better performance. For a given code rate and shift register length, a number of possible connections to the modulo-two adders are possible. The connections that produce convolutional codes with the best distance properties have been found by exhaustive computerized searches and are listed in the literature. This procedure for finding good convolutional codes is in contrast to block codes in which good codes are usually found by theoretical approaches.

The name *convolutional code* derives from the similarity of the shift and add procedure for computing the code symbols to the shift and integrate process in convolution of analog signals. As in the case of analog convolution, convolutional encoding is linear, but in the sense of modulo-two addition. In other words, if data sequence D_1 produces an output code sequence C_1, and similarly, if data sequence D_2 produces the output C_2, then the data sequence $D_1 \oplus D_2$, which is the bit-by-bit modulo-two sum of the two input data sequences, will produce a coder output given by $C_1 \oplus C_2$. The Hamming distance between any two code sequences is the weight of their sum, which in turn is the weight of another code sequence because of the linearity property. The implication of this fact is that the distance between any two code sequences is the same as the distance between the all-zero sequence and some other sequence. Thus, the set of distances from the all-zero sequence to its neighbors is the same as the set of distances from any other sequence to its neighbors. Therefore, we can evaluate a convolutional code by considering only the transmission of the all-zero sequence. This fact makes the evaluation of convolutional coders much simpler, just as the linearity property simplifies the analysis of block codes.

The Viterbi algorithm

The optimum approach (in the sense of smallest error probability for random data) to decoding a convolutional code is to find the data sequence that produces a code sequence that is closest in Hamming distance to the received sequence. The *Viterbi algorithm* is a practical procedure for finding this closest sequence. The Viterbi algorithm can best be explained by use of the trellis diagram such as the one in Figure 9.6. A record is kept of the path closest to the received sequence for each of the nodes at each step in the trellis. Each time the algorithm is extended by one step in the trellis, two paths potentially end on each of the nodes in the new step. The best of this pair of paths is retained for each node and the other is discarded. Then the paths are extended to the next step, and the best path at each node is retained. This process continues until only one path remains in the early steps of the trellis and then the best data sequence is known. Thus, the algorithm keeps track only of the best path ending on each node in the current step and the accumulated distance for each of the paths.

As an example of Viterbi decoding, consider the case in which the data sequence is all zeroes and the encoder is as shown in Figure 9.4. Then the transmitted sequence is also all zeroes. For our example, we assume that the first group of three code bits contains two errors and that the rest of the code symbols are received correctly; so the received sequence is 101 000 000 000 000 and so on. Notice that if a decision on the first data bit were to be made after receiving only the first three code symbols, an erroneous decision in favor of a 1 would be made, since a 1 results in 111 as the first three code symbols whereas a 0 results in 000 as the code group. As we will see in this example, the Viterbi algorithm delays the decision to the tenth step in the decoding process. As a result, it makes the correct decision.

Figure 9.7a illustrates the progression of the Viterbi algorithm for the first three steps into the trellis diagram. The trellis diagram and the accumulated distance between the received sequence and the code sequence along each path up to the second step is shown. For example, the code sequence leading to the bottom node of step 2 is 111 001, which is two units from the received sequence 101 000. The distances at the other nodes in step 2 are found by similar comparisons with the received sequence.

Once the distances to each of the nodes in step 2 have been found, the algorithm extends the calculation to step 3. The possible extensions of the trellis to step 3 are shown as dotted lines in Figure 9.7a. Notice that two potential paths end on each of the nodes of step 3. At this point, the accumulated distance is found for each pair and only the shortest distance path is retained. For example, for the two paths ending at the top node of step 3, we find that the distance for the upper path is 2, and the distance for the lower path is 6. Therefore, the top path ending on the top node (state 00) of step 3 is retained. This is indicated by the solid line ending on the top node of step 3 in Figure 9.7b, along with the accumulated distance of 2 units. We can safely discard the lower path ending on the top node of step 3 because the upper path is better at this step, and we know that distance cannot decrease as we go forward in the trellis. Thus, any path extending forward from this node, which includes the top path up to this node, is better than the option including the lower path to this node. The best path leading to a node in the current step is called a *survivor path*. Thus, the solid upper path leading to the top node of step 3 shown in Figure 9.7b is the survivor path for the top node of step 3.

Next, consider the potential survivor paths ending on the state 10 node (second node from the top) of step 3, shown as dotted lines in Figure 9.7a. The upper path ending on this node has an accumulated distance of 5 and the lower path has a distance of 3. Therefore, the lower path is the survivor. This is indicated by the solid line ending on state 10 of step 3 in Figure 9.7b. The process of finding the survivor path for each node proceeds through all of the nodes of step 3. The results are indicated by the solid lines. Notice that the survivor path ending on node 10 of step 2 does not extend to step three. Thus a survivor at one step may be eliminated at a later step. The Viterbi algorithm defers data decisions until only one survivor remains for each data bit.

The algorithm now extends the paths to step 4. At the 01 node in step 4 a tie between paths occurs: Both paths ending on the 01 node of step 4 have an accumulated distance of 5 units. Either path may be selected without loss of optimality of the procedure. When they occur, we will (arbitrarily) resolve ties in favor of the lower branch.

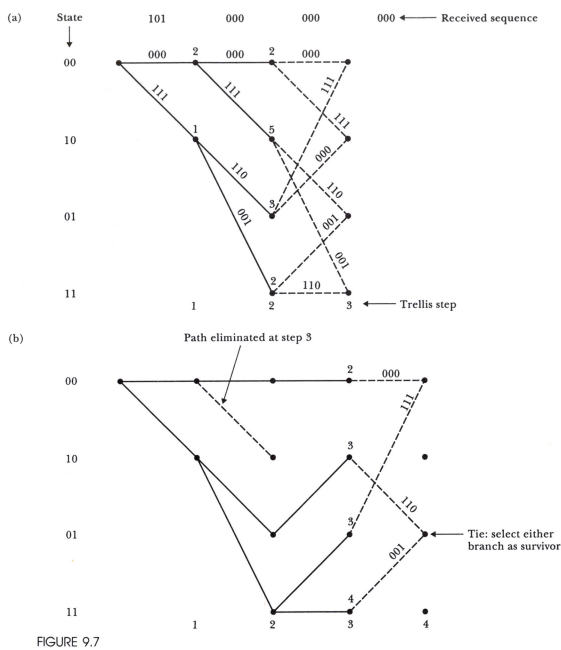

(a)

(b)

FIGURE 9.7
First four steps of the Viterbi algorithm.

If the Viterbi algorithm is extended one step at a time, the survivor paths and accumulated distances are obtained as shown in Figure 9.8.

Finally, when step 10 is reached, the lower paths in the first seven steps are eliminated. At this point, the algorithm can make the decision that the first seven data bits are zeroes corresponding to the data sequence for the

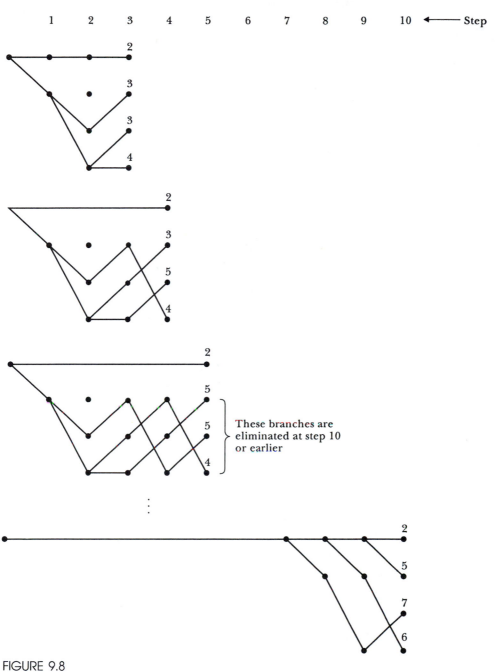

FIGURE 9.8
Illustration of the Viterbi algorithm.

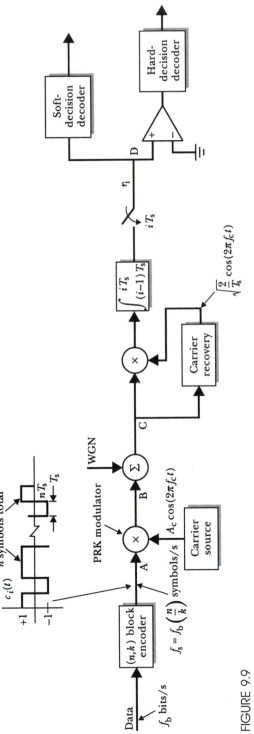

FIGURE 9.9

Communication system using an (n,k) block code on an AWGN channel with PRK modulation.

single survivor path in that span of the trellis. As the algorithm extends further, additional data bits are eventually decoded.

The Viterbi algorithm is widely used because it is practical at high data rates with constraint lengths less than about ten. As we will see, Viterbi decoding has a significant advantage compared to block codes in that it can be easily modified to make full use of the correlator output of a digital receiver by using soft decisions. This point will be explained in the next section. The primary disadvantage of the Viterbi algorithm is that the complexity grows exponentially with the constraint length. Other suboptimum but less complex decoding procedures can be used for longer constraint lengths. The interested reader will find a discussion of these in the literature.

9.3
EXAMPLES OF CODING APPLICATIONS

Communication systems use many modulation schemes, and noises of various characters cause the performance of any given code to vary. For example, we might use an additive white gaussian noise (AWGN) channel with various modulation types. On the other hand, we might need a code to cope with bursts of errors in digitized music caused by scratches on the surface of a compact disk. In the AWGN case, symbol errors are dispersed randomly through the received sequence whereas in the compact disk the errors tend to occur in groups or bursts. In the AWGN case, we can increase the transmitter power to achieve a lower error rate but not in the case of errors caused by scratches on a compact disk. The performance improvement provided by a coding scheme depends greatly on the system to which it is applied. In this section, we consider a few examples to demonstrate the performance improvements that can be achieved with coding. Since we have emphasized digital communication through the AWGN channel up to this point, we start with a consideration of coding applied to this channel with the use of PRK modulation.

A typical AWGN system

Figure 9.9 shows the block diagram of the system to be analyzed. The system uses an (n,k) block code and PRK modulation. The coded and modulated data signal is passed through the AWGN channel, and the resulting noisy received signal is presented to the receiver. As usual in many of the digital receivers we have considered, the first step in the receiver is to correlate the received signal with a set of orthonormal basis functions that are complete for the transmitted signal set. (We give a precise definition of the basis set later.) This process results in the components of the signal-space vector for the received signal. The correlator outputs resulting at point D in the figure are then used to determine the most likely data block that entered the coder at the transmitter.

Hard decisions versus soft decisions

We will consider two decoder approaches. Thus, the correlator output is split at point D into two paths feeding the two different decoders. In one case, the outputs of the correlator (these are analog numbers that represent the components of the received vector) are fed directly to the decoder, which calculates the signal-space distance to each of the transmitted signals for each possible code word. This decoder is known as a *soft-decision* decoder. In the other case, the outputs of the correlator are passed through a decision device that makes binary decisions on the individual code symbols. These decisions are then passed to the decoder. In this case, the input to the decoder is a binary sequence consisting of individual decisions about each symbol in the transmitted code word, some of which may be in error. The decoder operates by finding the code word closest to the received sequence in Hamming distance as discussed in Section 9.1. In this case, the decoder operates on "hard" decisions about each code symbol and is known as a *hard-decision decoder*. We will see that the soft-decision decoder performs better and can typically produce a given error performance with about 2 dB less received E_b/N_0 than is required for the hard-decision decoder. On the other hand, the hard-decision decoder is simpler to implement.

The input to the encoder in Figure 9.9 is a data bit stream that, as usual, we assume is a sequence equally likely to be 0 or 1 in each bit interval and independent from bit to bit. The data rate of the input is f_b bits per second. Each block of k input bits results in a block of n binary symbols at the encoder output. A typical signal for one code word is shown in the figure, where it is assumed that the logic levels of the NRZ waveform are $+1$ and -1 for convenience. The symbol rate f_s of the encoder output is higher than the data rate, due to the redundant symbols added by the encoder. The data bit rate f_b is related to the code symbol rate f_s by the expression

$$f_b = \frac{k}{n} f_s = R_c f_s \tag{9.16}$$

where, as before, $R_c = k/n$ is the rate of the block code.

As we implied, the operation of the communication link is repetitive in time. A block of k data bits is collected in an input buffer in the encoder while the n code symbols produced by the last data block are output to the modulator. Similarly, the decoder decodes the last received code word signal while the next is being received and stored. In our analysis, we focus on the operations for the block of data that results in the transmitted signal between $t = 0$ and $t = nT_s$, where $T_s = 1/f_s$ is the time for each code symbol.

The encoder output multiplies the sinusoidal carrier in the modulator resulting in a PRK signal, which is given by

$$s_i(t) = A_c c_i(t) \cos\left(2\pi f_c t\right) \tag{9.17}$$

where A_c is the peak carrier amplitude, $c_i(t)$ is the NRZ coded waveform for the ith data pattern, as shown in Figure 9.9, and f_c is the carrier frequency. Since we have assumed that the encoder output signal $c_i(t)$ takes values of ± 1, the modulated signal $s_i(t)$ has a phase of $0°$ when the current code symbol is a 1 and has a phase of $180°$ when the symbol is a 0. Thus, the code symbols are PRK modulated onto the carrier. There are $M = 2^k$ possible code words, and therefore the transmitted signals can be visualized as a set of M vectors in signal space. As we have seen for the AWGN channel, the error performance is dependent on the distances between these signal-space vectors. A high-performance code results in large distances between the signal vectors.

Since the transmitted signals consist of segments of the carrier with a phase of either $0°$ or $180°$, a complete set of orthonormal basis functions for this signal set can be selected to be segments of the unmodulated carrier with unit energy. These are given by

$$
\begin{aligned}
\phi_1(t) &= \left(\frac{2}{T_s}\right)^{1/2} \cos\left(2\pi f_c t\right) && 0 < t < T_s \\
&= 0 && \text{otherwise} \\
\\
\phi_2(t) &= \left(\frac{2}{T_s}\right)^{1/2} \cos\left(2\pi f_c t\right) && T_s < t < 2T_s \\
&= 0 && \text{otherwise} \\
\\
\phi_n(t) &= \left(\frac{2}{T_s}\right)^{1/2} \cos\left(2\pi f_c t\right) && (n-1)T_s < t < nT_s \\
&= 0 && \text{otherwise}
\end{aligned}
\tag{9.18}
$$

These basis functions are shown in Figure 9.10. The amplitude $(2/T_s)^{1/2}$ was selected so that the basis functions have unit energy as required for an orthonormal set. (We will assume throughout our discussion that $2f_c T_s$ is either very large or an integer; this assumption is necessary for the basis functions given to have unit energy.) The basis functions are orthogonal since the time intervals in which they are nonzero do not overlap. Each of the transmitter output signals can be written in terms of these basis functions as

$$
s_i(t) = c_{i1}(E_s)^{1/2}\phi_1(t) + c_{i2}(E_s)^{1/2}\phi_2(t) + \cdots + c_{in}(E_s)^{1/2}\phi_n(t) \tag{9.19}
$$

where $c_{i1} = \pm 1$ is the first symbol of the ith code word, c_{i2} is the second symbol, and so on. (Note that we find it convenient at times to take the binary symbols of code words to be ± 1 rather than 0 and 1.) E_s is the energy associated with the transmitted signal for each code symbol, given by

$$
E_s = \frac{1}{2}(A_c)^2 T_s \tag{9.20}
$$

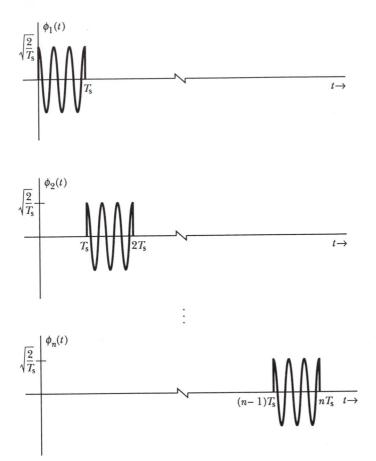

FIGURE 9.10
Basis function set.

The total energy associated with each code word is given by

$$E = nE_s = kE_b \qquad (9.21)$$

where E_b is the energy available to the receiver for each data bit. Thus, we have the following relationship between the energy per data bit and the energy per code symbol:

$$\blacksquare \qquad E_s = \frac{k}{n} E_b = R_c E_b \qquad (9.22)$$

where R_c is the code rate. Notice that the energy available for each code symbol is less than the energy available per data bit. As a result, when coding is added to a system without any change in E_b, the error rate for the code symbols is (much) higher than for the data bits before coding is added. The error correction properties of the code must be powerful enough to more than overcome this increased error rate if the code is to provide an advantage.

A performance bound for the hard-decision decoder

First, we consider the performance of the hard-decision decoder. The error probability of the code symbols at the output of the decision device is given by Equation 8.46 since the PRK modulation produces an antipodal signal set for each code symbol. This error probability is given by

$$P_s = Q\left[\left(\frac{2E_s}{N_0}\right)^{1/2}\right] \tag{9.23}$$

where E_b in Equation 8.46 has been replaced by E_s because it is the code symbol energy that pertains for the error probability of each code symbol.

We assume that the hard-decision decoder can correct t symbol errors per code word and makes a decoding error if more than t symbol errors occur per code word. Therefore, the probability of a word error in decoding is the probability that $t + 1$ or more symbol errors occur in a code word. The errors occur randomly from symbol to symbol, and therefore the probability of errors in a code word is a binomial probability distribution as given in Equation 7.15. The resulting word-error probability is given by

$$P_w = \binom{n}{t+1} P_s^{t+1}(1 - P_s)^{n-t-1} + \binom{n}{t+2} P_s^{t+2}(1 - P_s)^{n-t-2} + \cdots \tag{9.24}$$

When the SNR is high so that the probability of a symbol error is small, this expression for the word-error probability can be approximated by the first term. Also, we have $(1 - P_s)^{n-t-1} \simeq 1$. Thus, we have the approximate expression

$$P_w \simeq \binom{n}{t+1} P_s^{t+1} \tag{9.25}$$

When word errors occur, the occurrence of data-bit errors is related to the structure of the code and depends on which particular word errors are most likely. For example, suppose that the code word for the 0000 data sequence is decoded in error so that the code word for the data sequence 0100 is selected by the decoder. In this case, only one of the four data bits is in error. On the other hand if the decoder selects the data sequence 1111, then all of the data bits are in error. Certainly, if a word error occurs, at least one of the data bits is in error. Thus, the probability of error for a given bit P_b is bounded by

$$\frac{P_w}{k} \leq P_b \leq P_w \tag{9.26}$$

Most of the time when a word error occurs, the erroneous word selected by the decoder is one of the words closest to the transmitted word. Therefore, the erroneous word usually contains $d_{\min}$ symbol errors. Thus, we can estimate the probability that a given data bit is in error if a word error has occurred as d_{min}/n. Thus, we have

$$P_b \simeq \frac{d_{min}}{n} P_w \qquad (9.27)$$

Substituting Equation 9.23 into 9.25 and the result into 9.27, we find an approximate expression for the bit-error probability given by

$$P_b \simeq \frac{d_{min}}{n} \binom{n}{t+1} \left\{ Q\left[\left(\frac{2R_c E_b}{N_0}\right)^{1/2}\right] \right\}^{t+1} \qquad \text{(hard-decision decoder)} \qquad (9.28)$$

We will consider some examples and the implications of this result, but first we will derive the corresponding result for the soft-decision decoder.

A performance bound for the soft-decision decoder

The soft-decision decoder is presented with the components of the received vector (these are the correlator outputs) and finds the code word that results in the closest transmitted signal in signal space. The received signal vector is denoted by

$$\mathbf{R} = (r_1, r_2, \ldots, r_n) \qquad (9.29)$$

where r_1 is the correlator output for the first code symbol interval, r_2 is the output for the second symbol, and so forth. The soft-decision decoder computes the distance to each of the possible transmitted code word signals and chooses the code word with the smallest distance. Thus, the decoder computes the minimum of

$$(\ell_i)^2 = (\mathbf{R} - \mathbf{S_i}) \cdot (\mathbf{R} - \mathbf{S_i}) \qquad (9.30)$$

where ℓ_i is the distance from the received signal to the ith code-word signal. This can be expanded as

$$(\ell_i)^2 = \mathbf{R} \cdot \mathbf{R} - 2\mathbf{R} \cdot \mathbf{S_i} + \mathbf{S_i} \cdot \mathbf{S_i} \qquad (9.31)$$

The first term on the right in the last expression is the energy of the received signal, and the last term is the energy of the transmitted code-word signal. Both of these terms are the same for every distance calculation, because the energy in all code word signals is the same due to the fact that PRK modulation does not change the energy in a signal. Thus, it is necessary for the decoder to compute only the middle term of Equation 9.31 and find the minimum value. Dropping the factor of two and the minus sign, the decoder computes the maximum of

$$\mathbf{R} \cdot \mathbf{S_i} \qquad (9.32)$$

As shown by Equation 9.19, all of the components of the signal-space vector for the code-word signals have the same magnitude, and the components corresponding to the code symbols that are 1 are positive, whereas the components corresponding to 0 are negative. Thus, an equivalent procedure for the soft-decision decoder is to find the maximum of

$$\mathbf{R} \cdot \mathbf{C_i} \tag{9.33}$$

where $\mathbf{C_i}$ is a vector with components that are $+1$ or -1 depending on whether the corresponding code bit is a 1 or 0, respectively. Thus, the calculations performed by the soft-decision decoder amount to computing a sum (with some negative terms) of the correlator outputs for each possible code word. These values are then compared, and the largest sum determines the best choice.

This decoding algorithm seems fairly simple. However, when k is large, so that there are many code words, the amount of computation can be prohibitive, and simpler but poorer performing decoding algorithms must be used. For block codes, the known decoding approaches often are suited only for hard-decision decoding. As we will see, hard-decision decoders have a poorer performance than soft-decision decoders for a given code. Viterbi decoding of convolutional codes can be easily modified to make use of soft decisions and, therefore, often has an advantage over block codes in the AWGN channel.

Now that we have described the computations performed by the optimum soft-decision decoder, we will estimate the error performance of the decoder. Assuming that $\mathbf{S_i}$ represents the transmitted signal, the probability that the decoder erroneously selects the code-word signal $\mathbf{S_j}$ depends on the signal-space distance ℓ_{ij}, between the two signals. This distance can be computed from the transmitted time domain signals as

$$[\ell_{ij}]^2 = \int_0^{nT_s} [s_i(t) - s_j(t)]^2 \, dt \tag{9.34}$$

The integrand of this expression is zero for symbol intervals in which the two code symbols are the same and is a sinusoid with an amplitude of $2A_c$ in intervals in which the code symbols are different. If we denote the Hamming distance between the code words as d_{ij}, Equation 9.34 can be written as

$$[\ell_{ij}]^2 = d_{ij} \int_0^{T_s} (2A_c \cos 2\pi f_c t)^2 \, dt = d_{ij} 2A_c^2 T_s \tag{9.35}$$

As usual, we assume that the double frequency term in the integral can be ignored due to the assumption that $2f_c T_s$ is large or an integer. Using Equation 9.20, we can write this as

$$(\ell_{ij})^2 = 4d_{ij}E_S \tag{9.36}$$

An important implication of this result is that code words selected to have a large Hamming distance result in a large distance between signal-space vectors when PRK modulation is used. This is not always the case when other modulation schemes are used, particularly higher-order modulation schemes in which more than one code symbol is associated with each modulator output signal.

The error probability of selecting $\mathbf{S_j}$ when $\mathbf{S_i}$ is transmitted is given by substituting the distance between the signals ℓ_{ij} into Equation 8.42, resulting in

$$P_2 = Q\left[\frac{\ell_{ij}}{2}\left(\frac{2}{N_0}\right)^{1/2}\right] \tag{9.37}$$

where we have used the notation P_2 because it is the probability of the decoder confusing a pair of code-word signals. Substituting Equation 9.22 into 9.36 and the result into 9.37 gives

$$P_2 = Q\left[\left(\frac{2E_b d_{ij} R_c}{N_0}\right)^{1/2}\right] \tag{9.38}$$

We can place an upper bound on this error probability by replacing the actual distance d_{ij} between the two code words by the minimum Hamming distance between code words $d_{\min}$, resulting in

$$P_2 \leq Q\left[\left(\frac{2E_b d_{\min} R_c}{N_0}\right)^{1/2}\right] \tag{9.39}$$

Now the soft-decision decoder must compare the received signal with all 2^k code-word signals, and there is a possibility of making an error each time one of the signals other than the one transmitted is considered. Thus, the decoder has $2^k - 1$ opportunities to make a word error in decoding each received signal. To obtain an upper bound on the error probability for decoding the code word, we use the *union bound*. The union bound for the probability of the union of two events A and B is based on the following relationship:

$$P(A + B) = P(A) + P(B) - P(AB)$$

Since $P(AB)$ is always equal to or greater than zero, we can write

$$P(A + B) \leq P(A) + P(B) \tag{9.40}$$

In words, the union bound states that the probability of the union of two or more events is less than or equal to the sum of the probabilities of the separate events. The word-error event consists of the union of the error events consisting of picking one of the $2^k - 1$ erroneous code words in favor of the correct code word. Thus, the total word-error probability is less than or equal to the sum of the probabilities of picking each of the $2^k - 1$ erroneous code words. Each of these $2^k - 1$ probabilities is over bounded by the expression in Equation 9.39. Thus, we have the following bound on the probability of the decoder selecting an erroneous code word:

$$P_w \leq [2^k - 1]Q\left[\left(\frac{2E_b d_{min} R_c}{N_0}\right)^{1/2}\right]$$ (9.41)

Just as in the case of the hard-decision decoder, word errors do not always result in all of the data bits associated with that code word being in error. Therefore, we know that the bit-error probability is less than the word-error probability. Substituting Equation 9.41 into the approximate relationship of Equation 9.27, we have an approximate bound on the bit-error probability of the soft-decision decoder given by

■ $$P_b \lesssim \frac{d_{min}}{n}(2^k - 1)Q\left[\left(\frac{2E_b d_{min} R_c}{N_0}\right)^{1/2}\right]$$ (optimum soft-decision decoder) (9.42)

At this point it may be instructive to compare the performance of the hard- and soft-decision decoders using a simple code from Section 9.1.

EXAMPLE 9.3

Compute the bit-error probability for a system having $E_b/N_0 = 10$ if (a) PRK with no coding is used, (b) a (15,11) single error correction Hamming code is used with a hard-decision decoder, (c) the same code as in part b is used with a soft-decision decoder.

SOLUTION
(a) The bit-error probability of the uncoded system can be computed from Equation 8.46. The result is

$$P_b = Q\left[\left(\frac{2E_b}{N_0}\right)^{1/2}\right] = Q[(20)^{1/2}] = 3.88 \times 10^{-6}$$

(b) For an approximate upper bound on the bit-error probability of the system using the $k = 11$, $n = 15$, $d_{min} = 3$, $t = 1$ Hamming code, we use Equation 9.28 to find

$$P_b \simeq \frac{d_{min}}{n}\binom{n}{t+1}\left\{Q\left[\left(\frac{2R_c E_b}{N_0}\right)^{1/2}\right]\right\}^{t+1}$$

$$\simeq \frac{3}{15}\binom{15}{2}\left\{Q\left[\left(2\left(\frac{11}{15}\right)\frac{E_b}{N_0}\right)^{1/2}\right]\right\}^2 \simeq 2.88 \times 10^{-8}$$

Thus, the simple Hamming code with hard-decision decoding reduces the bit-error probability by more than two orders of magnitude compared with the uncoded system. (The actual performance would be better than this because the computation is an upper bound on the error probability.)

(c) Carrying out the corresponding computation for the soft-decision decoder using Equation 9.42, we have

$$P_b \lesssim \frac{d_{\min}}{n} (2^k - 1) \, Q\left[\left(\frac{2d_{\min} R_c E_b}{\mathcal{N}_0} \right)^{1/2} \right]$$

$$P_b \lesssim \frac{3}{15} (2047) \, Q\left\{ \left[2(3) \left(\frac{11}{15} \right) \frac{E_b}{\mathcal{N}_0} \right]^{1/2} \right\}$$

$$P_b \lesssim 6.76 \times 10^{-9}$$

Thus, the Hamming code with soft-decision decoding reduces the bit error by almost three orders of magnitude compared with the uncoded system. Plots of the bit-error probability for all three systems are shown in Figure 9.11.

Coding gain

The distance between the bit-error rate curves for an uncoded system and a coded system at a given error probability is called the *coding gain* of the code in use. Coding gain is the amount that using the code allows the transmitter

FIGURE 9.11
Performance bounds for (15,11) Hamming code.

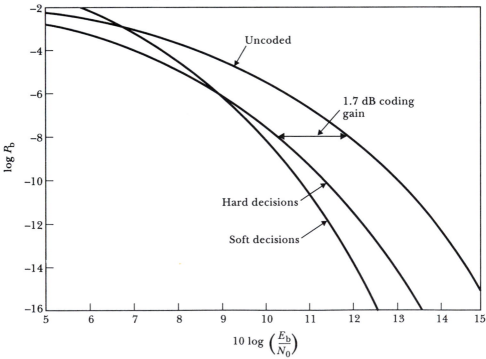

power to be reduced while maintaining a given bit-error probability. As indicated in Figure 9.11, the estimated coding gain of the (15,11) Hamming code for a PRK system on an AWGN channel is 1.7 dB for the hard-decision decoder at an error probability of 10^{-8}. This coding gain is slight because of the simplicity of the Hamming code. More complex multiple-error correction codes achieve higher coding gains.

Coding gain depends on the code, the modulation in use, the character of the noise, and the error probability at which it is measured. In general, coding gain becomes larger as the SNR of the system is increased (or equivalently, as the error probability is reduced). For block codes on the AWGN channel with PRK modulation and a hard-decision decoder under high SNR conditions, the coding gain can be shown to be given by

■ $$G = 10 \log \left[R_c (t + 1) \right] \text{ dB} \qquad (9.43)$$

Under the same conditions, the gain with a soft-decision decoder can be shown to be

■ $$G = 10 \log \left[R_c d_{\min} \right] \text{ dB} \qquad (9.44)$$

Since the error correction capability t of a block code is approximately half of the minimum distance $d_{\min}$, this indicates that under high SNR conditions the soft-decision decoder provides about 3 dB more coding gain than the hard-decision decoder. However, in most practical systems, operated under lower SNR, the difference in coding gain is less. The typical difference in coding gain between the soft- and hard-decision decoders for practical systems is approximately 2 dB.

Typical coding gains for the AWGN channel with PRK modulation range from a few to perhaps 8 dB for even fairly powerful codes. Thus, the achievable coding gain for the AWGN channel is not spectacular. Next, we will consider the application of coding to a communication system in which the interference is due to a pulsed noise jammer, and we will find that even a simple repetition code can provide impressive performance improvements. Before we consider the pulsed jammer, notice that, in the AWGN channel, the repetition code does not change the performance when a soft-decision decoder is used. This is because, for PRK modulation, the transmitted signals are exactly the same when the code is in use as when the data are transmitted without coding. The two code words for a (3,1) repetition code are 111 and 000; so the two transmitted signals are a sinusoid with a phase of either $0°$ or $180°$ for the entire code word, and these are the same signals as when the data is PRK modulated onto the carrier without coding. If a repetition code is used with the AWGN channel and a hard-decision decoder, a performance degradation occurs because the decision device and hard-decision decoder represent a suboptimum processor for the received signal. (See problem 14.)

Coding performance against a pulsed noise jammer

A *pulsed noise jammer*, encountered by military communication systems, disrupts the communications of its adversary by transmitting a high-power pulsed noise signal in the frequency band used by the communications system. We assume that the jamming signal has an average power spectral density of $N_j/2$ on a two-sided frequency basis. If the bandwidth of the communication system is B, then the total average power due to the jammer is

$$P_j = N_j B \qquad (9.45)$$

Note that the required jammer power is proportional to the bandwidth used by the communicator. Therefore, the jammer power required to disrupt communications can often be increased by choosing communication signals that occupy a large bandwidth. This *antijamming* (AJ) technique is the basis of *spread spectrum communication* techniques, which will be discussed in a later chapter.

The pulsed noise jammer tries to maximize the error rate of the communication system by transmitting a high-power level part of the time and no noise the rest of the time. We denote the fraction of time that the jammer is on by ρ. If the jammer power in the *on* state is P_{jon}, then the average power is given by

$$P_j = P_{jon}\rho \qquad (9.46)$$

Substituting for P_j using Equation 9.45 and rearranging, we obtain

$$\frac{P_{jon}}{B} = \frac{N_j}{\rho} \qquad (9.47)$$

Thus, the PSD of the jammer noise during the time that the jammer is *on* is N_j/ρ.

When it is on, we assume that the jammer signal behaves like WGN as far as its influence on the receiver is concerned. Then if the transmitter uses PRK modulation without coding, the bit-error probability is given by Equation 8.46 with N_0 replaced by the noise power spectral density of the jammer in the *on* state N_j/ρ. This results in

$$P_{bon} = Q\left[\left(\frac{2E_b\rho}{N_j} \right)^{1/2} \right] \qquad (9.48)$$

We assume that the residual noise level is negligible and the error rate is zero when the jammer is off. This is reasonable when the communication system has been designed to overcome the effects of a high-power jamming signal. Therefore, the average bit-error probability is given by

$$P_b = \rho P_{bon} = \rho Q\left[\left(\frac{2E_b\rho}{N_j} \right)^{1/2} \right] \qquad (9.49)$$

This is plotted for various values of ρ in Figure 9.12. Notice that for each value of E_b/N_j, the error probability is maximized for a different value of ρ. Thus, the jammer would try to adjust ρ to produce this maximum error rate. Under the condition that the argument of the Q function is larger than about two, the Q function can be approximated by Equation 7.58. Using this approximation in Equation 9.49, we find

$$P_b \simeq \left(\frac{\rho N_j}{4\pi E_b}\right)^{1/2} \exp\left(\frac{-\rho E_b}{N_j}\right) \tag{9.50}$$

The optimum value of ρ (from the jammer's point of view) can be found by differentiating this expression with respect to ρ, setting the result to zero, and solving. The result is

$$\rho_{\text{opt}} \simeq \frac{N_j}{2E_b} \tag{9.51}$$

Substituting this result into Equation 9.50 results in the maximum error probability:

$$P_{b\,\text{max}} \simeq \frac{N_j}{2E_b(2\pi e)^{1/2}} \tag{9.52}$$

Of course, to be able to cause the maximum disruption of communications, the jammer must know the ratio of received energy per bit to effective jammer noise PSD, E_b/N_j, at the input to the receiver and also be able to adjust his duty cycle ratio ρ to the optimum value, while maintaining constant average power. This may not always be possible; but since the designer of the communication system cannot predict exactly what the capability of the jammer may be in every situation, the designer must assume that the system will be facing the worst possible jammer. To maximize effectiveness, the jammer must turn the jamming signal on and off at unpredictable times, otherwise the communicator can simply avoid transmissions during the intervals when the jammer is on.

Now that we have analyzed the performance of the system when the communicator does not use coding, we consider the case in which a block code and a hard-decision decoder are used in the communication system. In this case, the development we have given above applies to the error probability of code symbols. A jammer who is free to adjust ρ to maximize the error rate for the code symbols will choose $\rho = \rho_{\text{opt}}$, as given in Equation 9.51 with E_b replaced by E_s, the energy per code symbol. This is given by

$$\rho_{\text{opt}} = \frac{N_j}{2E_s} \tag{9.53}$$

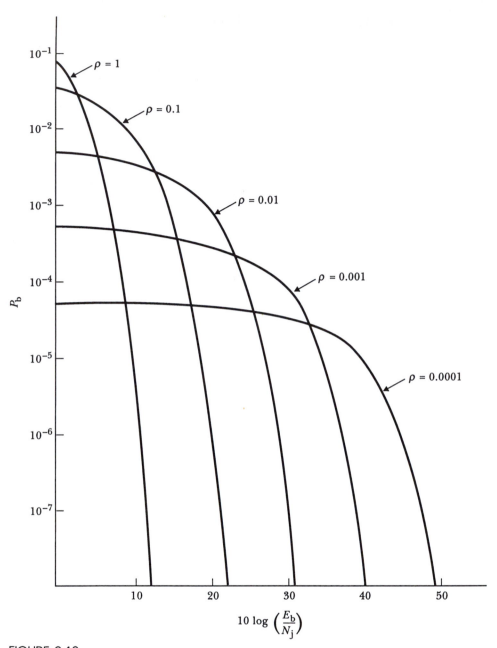

FIGURE 9.12
Bit-error probability versus signal-to-jammer ratio for various values of jammer duty ratio.

The resulting maximum symbol-error probability is given by Equation 9.52, again with E_b replaced by E_s, resulting in

$$P_{s\,max} = \frac{\mathcal{N}_j}{2E_s(2\pi e)^{1/2}} \qquad (9.54)$$

The word-error probability is approximately given by Equation 9.25, with P_s replaced by $P_{s\,max}$:

$$P_w \simeq \binom{n}{t+1} P_{s\,max}^{t+1} \qquad (9.55)$$

Notice that for this last result to be valid, errors in the various symbols of a code word must be independent. This is not the case if the successive symbols of a code word are transmitted one after another because if the jammer is on for a given symbol, it will also be likely to be on for the following symbol. Therefore, errors in symbols that are transmitted close together in time will not be independent. To insure that symbol errors are not concentrated in a few code words, the communication system must be designed to *interleave* the symbols from many code words. If this is properly done, the jammer can cause only independent errors in the symbols of a code word. Ideally (from the communicator's point of view), the interleaving should be done on a pseudorandom basis so that the jammer cannot learn the interleaving pattern. This prevents the jammer from adjusting its transmissions so that the symbol errors are concentrated in some of the code words, thereby defeating much of the error correction power of the code.

The bit-error probability is related to the word-error probability by Equation 9.27. Substituting Equation 9.55 into 9.27, we find that

$$P_b \lesssim \frac{d_{min}}{n} \binom{n}{t+1} P_{s\,max}^{t+1} \qquad (9.56)$$

The maximum symbol-error probability is found in terms of E_b by substituting Equation 9.22 into 9.54. The result is

$$P_{s\,max} = \frac{\mathcal{N}_j}{2R_c E_b(2\pi e)^{1/2}} \qquad (9.57)$$

EXAMPLE 9.4

Compute the bit-error probability of a PRK communication system operating in the presence of a jammer with $E_b/\mathcal{N}_j = 10$ if (a) the jammer is *on* continuously and the communicator does not use coding, (b) the jammer optimizes his duty ratio ρ, and the communicator does not use coding, (c) the jammer optimizes his duty ratio, and the communication system uses a $(5,1)$

repetition code with an interleaver to spread the errors and a hard-decision decoder.

SOLUTION

(a) When the jammer is *on* continuously, the error probability is given by Equation 9.48 with $\rho = 1$. Thus, we have

$$P_{\text{bon}} = Q\left[\left(\frac{2E_b}{N_j}\right)^{1/2}\right] = Q[(20)^{1/2}] = 3.88 \times 10^{-6}$$

(b) When no coding is used, the optimum duty ratio for the jammer is given by Equation 9.51, which is

$$\rho_{\text{opt}} \simeq \frac{N_j}{2E_b} = 0.05 = \frac{1}{20}$$

Thus, when the jammer is pulsed, the optimum duty ratio for the given value of E_b/N_j results in having the jammer on only 5% of the time. However, when the jammer is on, its power is 20 times its average power. This strategy concentrates the jamming signal, causing the maximum degradation of communication performance. The bit-error probability when no coding is used is given by Equation 9.52, which is

$$P_{\text{b max}} \simeq \frac{N_j}{2E_b[2\pi e]^{1/2}} = 0.0121$$

Thus, by concentrating the jamming into short, higher-power pulses, the jammer can increase the error probability by more than three orders of magnitude.

(c) When the communication system uses an $n = 5$, $k = 1$ code, which is capable of correcting up to $t = 2$ errors per code word, the ratio of symbol energy to jammer PSD is

$$\frac{E_s}{N_j} = \frac{R_c E_b}{N_j} = 2$$

Under these conditions the optimum jammer uses a duty ratio given by Equation 9.53, resulting in

$$\rho_{\text{opt}} \simeq \frac{N_j}{2E_s} = 0.25$$

Thus, when the communicator reduces the energy per binary symbol by using coding, the worst case jammer responds by reducing its power in the *on* state and remaining *on* for a higher fraction of the time to disrupt more symbols. The resulting symbol-error rate is given by Equation 9.57, which is

$$P_{\text{s max}} \simeq \frac{N_j}{2R_c E_b (2\pi e)^{1/2}} = 0.0605$$

Notice that the symbol-error rate for the code symbols is five times higher than the uncoded bit-error rate. The code must more than overcome this increase if coding is to be of any advantage to the communicator. The bit-error probability at the decoder output is bounded by Equation 9.56:

$$P_b \lesssim \frac{d_{min}}{n} \binom{n}{t+1} P_{s\,max}^{t+1} = 1.95 \times 10^{-3}$$

Thus, the use of the simple repetition code regains much of the performance lost when the jammer employs pulsing.

The error probability can be computed for the three cases for different values of E_b/N_j. The results are plotted in Figure 9.13. Notice that at a bit-error probability of 10^{-6}, the (5,1) repetition code provides an impressive coding gain of 29.7 dB against the worst case pulsed noise jammer. Recall that this same repetition code provides no advantage for the AWGN channel. Thus, we see that the selection of a coding scheme can depend strongly on the application.

9.4
AUTOMATIC REPEAT QUERY TECHNIQUES

The *automatic repeat query* (ARQ) approach to error control in communication systems uses a code to detect the presence of errors in blocks of received symbols and a return link to request retransmission of any blocks found to have errors.

An advantage of ARQ, compared with forward error control techniques, is that error detection is much easier to accomplish than error correction. High rate ($R_c = k/n$ close to unity) codes exist that can reduce the undetected word-error probability to exceedingly low values. As we have seen in Section 9.1, error detection is accomplished for linear block codes by computing the syndrome of the received sequence and checking to see if it is zero. This is relatively simple compared with multiple-error correction for long codes.

Three variations of ARQ are possible. In the simplest system, known as *stop-and-wait ARQ*, the transmitter sends a block of coded data and then stops transmission and waits for a *positive acknowledgment* (ACK) or a *negative acknowledgment* (NAK) from the receiver. If the receiver does not detect an error in the received block, it returns an ACK, and the transmitter sends the next block of data. If the receiver detects an error, it returns a NAK, and the transmitter repeats the block. Thus, a particular coded block is repeated until an ACK is returned. Since the data rate on the return path is usually much lower than on the forward path, the return path can be designed so that a negligible number of errors occur in the return of an ACK or a NAK. A typical stop-and-wait ARQ transmission sequence is shown in Figure 9.14a. When the round-trip delay D from the transmitter to the receiver and back is very long,

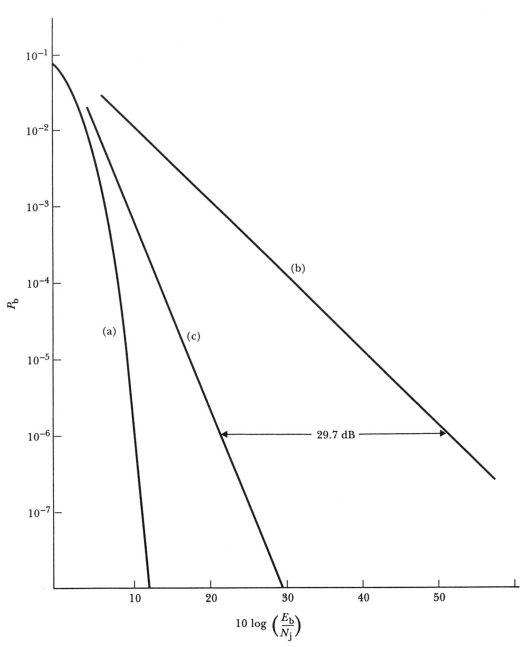

FIGURE 9.13
Bit-error probability versus E_b/N_j. (a) jammer *on* continuously and no coding, (b) worst case pulsed noise jammer and no coding, (c) worst case pulsed noise jammer and (5,1) repetition code.

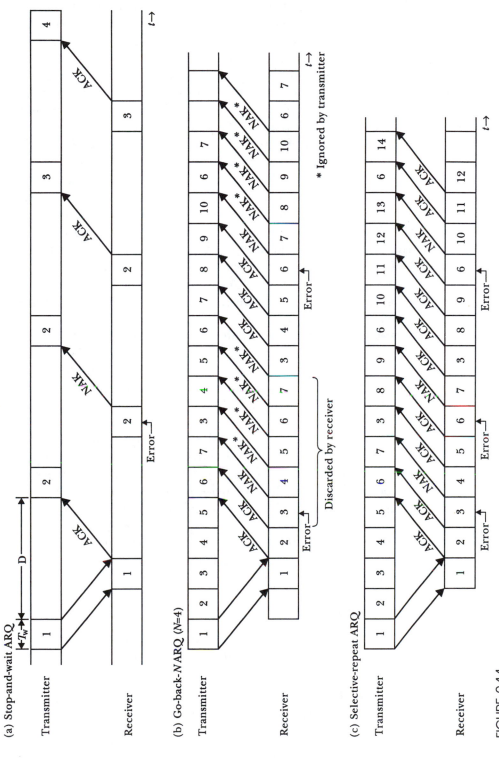

FIGURE 9.14
ARQ schemes.

the transmitter spends a large fraction of its time waiting for the return of an ACK or a NAK. As a result, the *throughput efficiency*, which is the ratio of the rate of the data bits successfully received to the data rate possible when data are transmitted continuously without coding, is reduced. This is a significant disadvantage of the stop-and-wait approach.

The *go-back-N* ARQ approach attempts to increase the throughput efficiency by continuing the transmission of code blocks without waiting for the return acknowledgment for each block. When the blocks are received without detected errors, the receiver returns an ACK for each block. As long as it continues to receive ACKs, the transmitter continues to transmit new blocks. When the receiver detects an error, it returns a NAK and ignores the next N received blocks. Upon receiving the NAK, the transmitter backs up N blocks in the sequence and retransmits the last N blocks. N is chosen large enough so that the block received in error is one (ideally the first) of the blocks retransmitted. A typical operating sequence for a go-back-N system is shown in Figure 9.14b. When the system is operating so that most of the blocks are received correctly and the round-trip delay is large, the go-back-N approach has a significantly higher throughput efficiency than the stop-and-wait approach. However, when the error probability becomes higher, the throughput efficiency of the go-back-N system eventually falls off.

The *selective-repeat* ARQ approach achieves even higher throughput efficiency by repeating only those blocks that are received in error. When the receiver detects a block with errors, it returns a NAK including an identification of the block that is in error. The transmitter continues to send new blocks and repeats only those blocks for which a NAK has been received. An illustration of the selective-repeat technique is shown in Figure 9.14c.

Of the three techniques, stop-and-wait is the least complex, and selective-repeat is the most complex. Selective-repeat attains the highest throughput efficiency, and stop-and-wait has the smallest. The throughput efficiencies of all three systems become small as the channel error rate becomes high. It is not unusual for the throughput to drop to unusable levels at high error rates because the transmitter spends most of its time repeating code groups. As the round-trip delay becomes large, the throughput efficiencies of both the stop-and-wait and the go-back-N systems decline. The efficiency of selective-repeat is not affected by the length of the round-trip delay.

Error probability relationships

Now we will develop expressions for the error probabilities and throughput efficiencies of these ARQ schemes. We begin by defining the following error probabilities:

P_c Probability that a given n-bit word experiences no errors during a single transmission through the channel

P_d Probability that a given word experiences a detectable error during a single transmission through the channel

P_e Probability that a given word experiences an undetectable error during a single transmission through the channel

P_w Probability that a given word is eventually accepted by the receiver with errors (possibly after repeated transmissions)

P_s Probability that a given bit of a code word experiences an error in transmission

P_b Probability that a given data bit delivered to the system output is in error

Clearly, when a word is transmitted through the channel, it is received correctly, received with a detectable error, or received with an undetectable error. Thus, we can write

$$P_c + P_d + P_e = 1 \qquad (9.58)$$

The probability of correctly receiving a code word of length n is the probability that each bit of the code word is received correctly. This is given by

$$P_c = (1 - P_s)^n \qquad (9.59)$$

The probability of receiving a word with a detectable error pattern is given by

$$P_d = \sum_{j=1}^{l} \binom{n}{j} (P_s)^j (1 - P_s)^{n-j} \qquad (9.60)$$

where, as before, l is the error detection capability of the code. Here we have assumed that the code can detect up to l errors and no error patterns of greater length. Actually, codes that are able to detect all errors of length l or less are usually also able to detect some longer error patterns as well. Thus, the actual performance of many codes is better than predicted by Equation 9.60 when l is taken as the guaranteed maximum error detection length. For example, a code with a minimum distance of 7 would be capable of detecting all error patterns of $l = 6$ errors or less, but it may also be able to detect some error patterns with 7 or more errors.

Similarly, the probability of an undetected error is given by

$$P_e = \sum_{j=l+1}^{n} \binom{n}{j} (P_s)^j (1 - P_s)^{n-j} \qquad (9.61)$$

When P_s is small, the summations of Equations 9.60 and 9.61 can be closely approximated by only the first term.

A given word may be incorrectly accepted with errors the first time it is sent, the second time it is sent if it is received with a detectable error pattern on the first transmission, or the third time it is sent if it is received with detectable error patterns on the first two transmissions, and so forth. In equation form, we have

$$P_w = P_e + P_d P_e + (P_d)^2 P_e + \cdots$$
$$P_w = P_e[1 + P_d + P_d^2 + \cdots]$$

In closed form, this equation becomes

$$P_w = \frac{P_e}{1 - P_d} = \frac{P_e}{P_c + P_e} \tag{9.62}$$

Most often when an undetected error occurs, the word contains $d_{min} = l + 1$ errors out of a total of n symbols in the code word. Thus, the probability that a given data bit is in error, given that an undetected word error has occurred, is approximately $(l + 1)/n$. Therefore, the probability of a bit error at the receiver output is given by

$$P_b \simeq \frac{l + 1}{n} P_w \tag{9.63}$$

Throughput efficiencies

Now, we develop expressions for the throughput efficiencies of the three ARQ approaches. First, we consider the selective repeat approach. The number of times that a given word must be repeated before it is accepted by the receiver is a random variable that we denote by m. The probability that a word is accepted on a given transmission is $P_c + P_e$, but for good error detection codes we have

$$P_c \gg P_e \tag{9.64}$$

Thus, we can assume that the probability of acceptance is almost exactly the same as P_c. Then the average value of m is given by

$$\bar{m} = P_c + 2P_c(1 - P_c) + 3P_c(1 - P_c)^2 + \cdots$$

It can be shown that this is equivalent to the simple closed form expression

$$\bar{m} = \frac{1}{P_c} \tag{9.65}$$

Thus, in the selective repeat approach, $n\bar{m}$ code symbols must be transmitted on average before k data bits are accepted by the receiver. The throughput efficiency, which is the ratio of the rate of data bits accepted to the code symbol rate for continuous transmission, is given by

$$\eta_{\text{sr}} = \frac{k}{n\bar{m}} = \left(\frac{k}{n}\right) P_{\text{c}} = R_{\text{c}} P_{\text{c}} \qquad (9.66)$$

where, as before, R_{c} is the rate of the error detection code. Note that as the error rate in the channel increases, P_{c} decreases and the throughput efficiency declines.

The average number of transmissions before a word is accepted for the go-back-N approach is given by

$$\bar{m} = P_{\text{c}} + (N + 1)P_{\text{c}}(1 - P_{\text{c}}) + (2N + 1)P_{\text{c}}(1 - P_{\text{c}})^2 + \cdots$$

In closed form, this equation is

$$\bar{m} = \frac{P_{\text{c}}(1 - N) + N}{P_{\text{c}}} \qquad (9.67)$$

Thus, the throughput efficiency for the go-back-N approach becomes

$$\eta_{\text{gbn}} = \frac{k}{n\bar{m}} = \frac{P_{\text{c}} R_{\text{c}}}{P_{\text{c}}(1 - N) + N} \qquad (9.68)$$

Notice that when P_{c} is close to unity, as it is under high SNR, the throughput efficiency of both selective-repeat and go-back-N are the same. When P_{c} becomes small, the throughput efficiency of the go-back-N approach is only $1/N$ times as large as for selective-repeat.

The throughput efficiency of the stop-and-wait scheme is reduced by the fact that the transmitter does not transmit continuously. In fact, the throughput efficiency of stop-and-wait is the same as that of selective-repeat times the duty ratio of the transmitter. If we denote the duration of a transmitted word as T_{w} and the round-trip delay from the transmitter to the receiver and back as D, the duty ratio of the transmitter is given by (see Figure 9.14a)

$$\frac{T_{\text{w}}}{T_{\text{w}} + D} \qquad (9.69)$$

Therefore, the throughput efficiency of the stop-and-wait technique is given by

$$\eta_{\text{sw}} = \frac{R_{\text{c}} P_{\text{c}} T_{\text{w}}}{T_{\text{w}} + D} \qquad (9.70)$$

EXAMPLE 9.5

Compare the throughput efficiencies of the selective-repeat and go-back-N ARQ schemes as a function of the channel error probability P_{s} for a communication link with a round-trip delay of $D = 500$ ms. (This rather long delay is typical for links that include satellite repeaters.) The system uses a

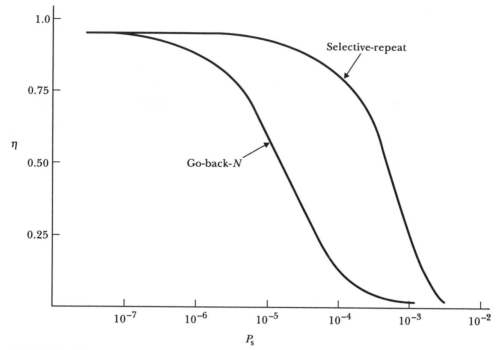

FIGURE 9.15
Throughput efficiency versus symbol error probability for Example 9.5.

block length of $n = 1000$ and $k = 950$ data bits per block. (The $n - k = 50$ bits are used for parity checks and control information.) The transmission rate for the channel is 100 kbit/s.

SOLUTION
The probability of receiving a given word with no errors is given by Equation 9.59. For this example, this becomes

$$P_c = (1 - P_s)^{1000}$$

The efficiency of the selective repeat scheme is given by Equation 9.66, which becomes

$$\eta_{sr} = 0.95P_c = 0.95(1 - P_s)^{1000}$$

This is plotted in Figure 9.15 as a function of P_s. Notice that at very low channel error rates, the efficiency for selective-repeat approaches 95%.

During a round-trip delay, the transmitter emits 50 coded blocks. Therefore, N must be at least 50 for the go-back-N scheme to be effective. Using $N = 50$ in Equation 9.68, we have

$$\eta_{gbn} = \frac{0.95P_c}{(50 - 49P_c)}$$

This is plotted versus P_s in Figure 9.15. Notice that at low error probability, both the go-back-N and the selective-repeat schemes provide high efficiency, but the go-back-N scheme deteriorates more quickly as the error probability increases. Of course, this difference is because the go-back-N scheme must repeat 50 blocks for each error whereas the selective-repeat scheme must repeat only the erroneous block.

EXAMPLE 9.6

Find the reduction in transmitter power that can be achieved by adding selective-repeat ARQ to a system that uses PRK modulation if a (15,11) Hamming code is used for error detection. Assume that the number of control bits required is negligible. The system is to operate with a bit-error probability of 10^{-6} at the receiver output. Also compare the bandwidth of the system using ARQ with the bandwidth of a system that does not use coding if the data rate at the system output is to be the same for both systems.

SOLUTION
The output bit-error probability for the ARQ system is related to the word-error probability by Equation 9.63. For the (15,11) Hamming code with $d_{min} = l + 1 = 3$, this becomes

$$P_b \simeq \frac{3}{15} P_w$$

Thus, to achieve a bit-error probability of 10^{-6}, the word-error probability at the system output is

$$P_w = 5 \times 10^{-6}$$

The word-error probability is given in Equation 9.62 as

$$P_w = \frac{P_e}{P_c + P_e}$$

The probability P_e is very small compared with P_c for a system operating with a small bit-error probability, so we can use the approximation

$$P_w \simeq \frac{P_e}{P_c}$$

Now, P_c is given by Equation 9.59, and P_e can be approximated by the first term of Equation 9.61. Substituting into the last expression, we obtain

$$P_w = 5 \times 10^{-6} = \frac{\binom{15}{3} (P_s)^3 (1 - P_s)^{12}}{(1 - P_s)^{15}}$$

This equation can be solved for P_s:

$$P_s = 2.22 \times 10^{-3}$$

Actually, this is somewhat lower than the value of P_s required to achieve the desired system performance because Equation 9.61 overestimates P_e for the Hamming code.

The throughput efficiency can be determined by use of Equation 9.66. Using the value found above for P_s, we find from Equation 9.59 that

$$P_c = (1 - P_s)^{15} = 0.967$$

Then from Equation 9.66 we have

$$\eta_{sr} = \frac{k}{n} P_c = 0.709$$

The number of symbols that must be transmitted by the ARQ system is

$$f_s = \frac{f_b}{n_{sr}} = 1.41 f_b$$

Thus, the ARQ system must transmit 41% more symbols than a system using no coding. The bandwidth required for the ARQ system is therefore 41% higher if the same data rate is to be maintained at the output of both systems.

The energy received for each binary symbol in the ARQ system is

$$E_s = E_b \eta_{sr} = 0.709 E_b$$

The symbol-error probability for the ARQ system using PRK modulation is given by Equation 8.46 with E_b replaced by E_s. This is

$$P_s = Q\left[\left(\frac{2E_s}{N_0}\right)^{1/2}\right]$$

Substituting values found above results in

$$P_s = 2.22 \times 10^{-3} = Q\left\{\left[\frac{2(0.709E_b)}{N_0}\right]^{1/2}\right\}$$

This can be solved to find that for the ARQ system,

$$\frac{E_b}{N_0} = 2.86 \quad \text{or} \quad 10\log\left(\frac{E_b}{N_0}\right) = 4.6 \text{ dB}$$

is required.

For the uncoded system without the use of ARQ, we have

$$P_b = 10^{-6} = Q\left[\left(\frac{2E_b}{N_0}\right)^{1/2}\right]$$

This can be solved to find that

$$\frac{E_b}{N_0} = 11.3 \qquad \text{or} \qquad 10 \log\left(\frac{E_b}{N_0}\right) = 10.5 \text{ dB}$$

Thus, we find that in this case the use of ARQ allows a 5.9 dB reduction in transmitter power by the use of the simple (15,11) Hamming code. More complex codes provide a greater improvement.

SUMMARY

1. The key ingredients of error control coding are controlled redundancy and noise averaging. These ingredients are found in both block and convolutional codes.

2. Forward error correction (FEC) schemes use coding to correct errors that have occurred in the channel. Automatic repeat query (ARQ) schemes use coding to detect errors and then request retransmission of erroneous data.

3. Block codes transform a group of k data bits into an n symbol code word. Linear codes are particularly easy to generate and analyze due to the structure imposed by linearity. An important measure of the error detection and correction power of a block code is the minimum Hamming distance between code words. For linear codes, the minimum distance is the same as the minimum weight of the words in the code. The generator matrix and parity check matrix are useful in characterizing linear codes. The syndrome of a received sequence, obtained by matrix multiplication of the parity check matrix and the received sequence, can be used to determine the most likely error pattern of the received word.

4. Hamming codes are an important class of linear codes that are capable of correcting one error per code word.

5. Cyclic codes can be generated with feedback shift register circuits. The syndrome of the received sequence can also be computed with a feedback shift register. Many powerful cyclic codes are known. Error correction decoding can be accomplished by calculation of the syndrome followed by a lookup table to find the most probable error pattern. Other important decoding algorithms for linear cyclic codes are known but are beyond the scope of this text.

6. Convolutional codes are generated by transmission of data bits through a shift register, and code symbols are generated after each shift by

modulo-two addition of the content of particular shift register stages. Convolutional codes can be described either by a tree diagram or a trellis diagram. Optimum convolutional codes have been found by computerized searches to find the input connections for the modulo-two adders resulting in the largest minimum Hamming distance between code sequences.

7. The Viterbi algorithm decodes convolutional codes by finding the code sequence closest to the received sequence at each step of the trellis diagram. Decoding of a given data bit is complete when a single survivor path remains for that bit. An important advantage of Viterbi decoding of convolutional codes is that it can easily be modified to make use of soft decisions from the demodulator.

8. Performance improvements provided by coding techniques depend on the code used, the channel characteristics, the type of decoder in use, and the SNR. For AWGN channels, performance can be stated in terms of the coding gain, which is a measure of the transmitter power savings provided by coding. Coding gains typically range from 2 dB to perhaps 8 dB for the AWGN channel.

9. A jammer, often encountered by military communication systems, tries to disrupt communication by its adversary. A jammer with fixed average power can greatly increase its effectiveness by pulsing its output. A communication system can overcome much of the advantage lost when the jammer uses pulsing by the use of interleaving and error correction coding.

10. ARQ techniques use an error detection code to detect errors in code blocks and then request a retransmission of erroneous blocks. The main disadvantage of ARQ is the need for a return link. The main advantage of ARQ over FEC is the fact that highly reliable error detection is much easier to achieve than error correction.

11. Three types of ARQ are in use: stop-and-wait, go-back-N, and selective-repeat. Throughput efficiency is highest for selective-repeat and lowest for stop-and-wait. Selective-repeat is the most complex system and stop-and-wait is the simplest.

REFERENCES

G. C. Clark, Jr., and J. B. Cain. *Error-Correction Coding for Digital Communications*. New York: Plenum Press, 1981.

R. W. Hamming. *Coding and Information Theory*. Englewood Cliffs, New Jersey: Prentice-Hall, 1980.

S. Lin and D. J. Costello, Jr. *Error Control Coding: Fundamentals and Applications*. Englewood Cliffs, New Jersey: Prentice-Hall, 1983.

S. Lin, D. J. Costello, Jr., and M. J. Miller. "Automatic-Repeat-Request Error-Control Schemes." *IEEE Communications Magazine* (December 1984): 5–17.

PROBLEMS

1. Construct an (8,4) single error correcting code using a two-by-two array like Table 9.1. Give the equations for the parity check bits r_1, r_2, c_1, and c_2. Assuming that the code words are given in the order $d_1, d_2, d_3, d_4, r_1, r_2, c_1, c_2$, find the generator matrix and the parity check matrix. Prepare a partial list of the syndromes and the corresponding error patterns. List all of the code words and find the minimum weight of this code.

2. A systematic (7,3) code includes the code words:

$$0011010$$
$$0100101$$
$$1001111$$

   (a) If the code is linear, list all of the code words. What is the minimum distance of this code?

   (b) List the words of a nonlinear code that includes the words given.

3. Give a parity check matrix for a systematic Hamming code with four parity checks. For your choice of the parity check matrix, what will happen if the decoder tries to correct a double error in the first two code symbols? Give the generator matrix corresponding to your parity check matrix.

4. List the error correction and error detection options for a code with a minimum distance of seven Hamming units. For each option, indicate the minimum number of code-symbol errors that results in undetected errors at the decoder output.

5. A (7,3) code is produced by the shift register encoder shown in Figure P9.1 by shifting the data bits into the register with the switch in position B. The switch is then moved to position A, and the register is shifted seven times to produce the code word. Assuming that the shift register starts in the state 101, list the successive states of the register and notice that the register returns to the original state after seven shifts. Because the register cycles through all of the possible non-zero three-bit patterns before returning to the original state, this is known as a *maximal length code*. Give the generator matrix and the parity check matrix for this code. What is the minimum Hamming distance of this code? How many errors can it detect if only error detection is desired? How many errors can it correct?

6. Assume that the data sequence 0110 is loaded into the shift register encoder of Figure 9.2a. Find the transmitted code word at the output of the coder. Demonstrate that the final content of the syndrome register of Figure 9.2c is 000 if this code word is received without errors. Then assume that the code word is

FIGURE P9.1

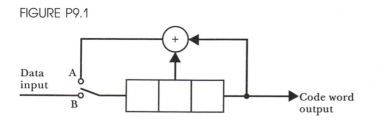

received with an error in the second code symbol and demonstrate that the decoder properly corrects the error.

7. Show that the (5,2) code with the code words listed below is a nonlinear code. What is the minimum distance between any pair of code words? What is the minimum weight of a nonzero code word? Notice that for this nonlinear code the minimum weight and the minimum distance are different. Give a different choice for the last code word in the list so that the code becomes linear.

$$00000 \quad 01111 \quad 10101 \quad 11101$$

8. As we have seen, one way to implement a decoder is to compute the syndrome of the received sequence and then use the syndrome as the address of the corresponding error pattern, which is stored in a ROM. A Golay code is a (23,12) code with a minimum distance of 7 that can correct all triple-error patterns. How many words does the syndrome ROM contain, assuming that all syndrome patterns are possible? How many bits are stored at each memory location if it is only desired to correct data bits?

9. The Hamming code of Example 9.1 is used in a communication system having a symbol-error probability denoted by p. The code is used with a decoder designed to correct single errors. Find an expression for the probability that the word at the decoder output is in error. Evaluate for $p = 0.1$, 0.001, and 10^{-6}. Hint: The decoder output is in error unless either there are no symbol errors or one symbol error in transmission of the code word.

10. An odd minimum distance linear code can be *extended* by adding another redundant bit at the end of each code word that is the modulo-two sum of all of the symbols in the original code word. Thus, the weight of all of the original code words with odd weight is increased by one, and the minimum distance of the code is increased by one. List all of the code words of the code obtained by extending the Hamming code of Example 9.1. Notice that the minimum distance of the extended code is 4. Find the parity check matrix and the generator matrix of the extended code. Is there any advantage in extending a code with even minimum distance? Explain your answer.

11. Prepare a tree diagram and a trellis diagram for the convolutional encoder shown in Figure P9.2.

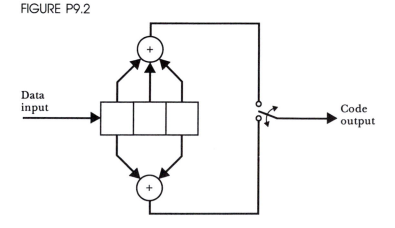

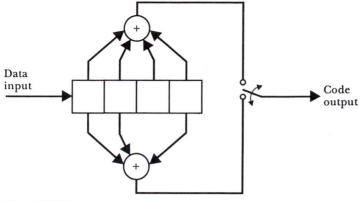

FIGURE P9.3

12. Use the Viterbi decoding algorithm to find the survivor paths and their accumulated distances for the following received sequence if the convolutional code of Problem 11 is in use:

$$10 \quad 10 \quad 11 \quad 00 \quad 00 \quad 00 \quad 00 \quad 00 \quad 00 \quad 00 \quad 00$$

Resolve any ties that occur in favor of the lower path.

13. Prepare a tree diagram and a trellis diagram for the convolutional encoder shown in Figure P9.3.

14. As we have noted in the text, a repetition code with PRK modulation results in the same transmitted signal as when no coding is used. Since the soft-decision decoder finds the closest signal in signal space, the soft-decision decoder with a repetition code provides the same performance as the optimum PRK receiver when no coding is used. When a hard-decision decoder is used with a repetition code, the performance is worse than if no coding were used. Assuming that $E_b/N_0 = 10$, compute the bit-error probability at the output of a soft-decision decoder and at the output of a hard-decision decoder if a $(7,1)$ repetition code is used with PRK modulation.

15. Three codes with the parameters shown in the following table are under consideration for use in an FEC system. Estimate the required value of E_b/N_0 to achieve $P_b = 10^{-6}$ if a soft-decision decoder is used and if a hard-decision decoder is used. Prepare a table showing the estimated coding gain of each combination compared with an uncoded system.

n	k	$d_{\min}$
31	26	3
31	21	5
31	16	7
255	123	39

16. Consider the Hamming code with three parity checks of Example 9.1. Prepare a table showing the weights and number of code words with each weight of all of the nonzero code words. This table is known as the weight structure of a code and

is known for only a few useful block codes. Use this weight structure to find an expression for the exact probability of an undetected error P_e, in terms of the symbol error P_s, when the code is used for error detection. Compare this exact expression with the approximation of Equation 9.61.

17. The (23,12) Golay code is one of the few useful codes for which the weight structure (see problem 16) is known. The weights and numbers of code words with each weight are the following:

Weight	Number of code words
0	1
7	253
8	506
11	1288
12	1288
15	506
16	253
23	1

If this code is used for error detection when $P_s = 0.001$, find values for P_c, P_d, and P_e that are correct to two significant figures. Compare these values with the bounds of Equations 9.60 and 9.61. If the full error correction capability of the code is used, estimate the probability of a word error at the decoder output for $P_s = 0.001$, and compare it with the value of P_e found for the error detection case. Comment.

18. Repeat Example 9.6 if a (255,223) block code with $d_{min} = 9$ is used.

19. ARQ is under consideration for a line-of-sight radio communication link with a one-way distance of 50 miles. PRK modulation is to be used with a (31,26) Hamming code. The system is required to have a data rate of 100 kbit/s and a bit-error probability of 10^{-6} at the receiver output. Determine if each of the ARQ techniques is suitable for use in this system. Find the throughput efficiency for each of the suitable techniques. What are the relative bandwidth and average transmitter power requirements of the suitable approaches?

10 INFORMATION THEORY

Information theory is concerned with characterizing the information content of analog messages or digital data streams and finding the ultimate performance limits of systems for storing or transmitting that information. In some circumstances, the theory promises attractive performance limits but does not indicate how these limits can be attained. The theory can be applied to discrete sources and systems as well as continuous ones. We will give only a brief introduction to this subject with an emphasis on the simpler discrete cases.

In developing a measure of information, we will not be able to take value judgements or aesthetics into account. We are concerned only with the engineering aspects of data transmission and storage. Thus, the term *information* has a meaning that is slightly different from common usage.

In the first section of this chapter, we develop a measure for the information content of discrete sources. In the second section, we demonstrate how this measure of information leads to a coding scheme for representing the source output that approaches the maximum theoretical efficiency. In the next section, we show how the channel models considered in this chapter relate to the physical channel and the digital modulation schemes considered in Chapter 8. Finally, we develop a measure of the information carried by each symbol transmitted through a channel. This leads to the important concept of channel capacity. We conclude the last section with a derivation of the channel capacity for the binary symmetric channel.

10.1
MEASURE OF INFORMATION

Information sources

We want to develop a means for measuring the information content of a source. Practical sources in communication systems can be either analog or discrete. Examples include a microphone, a television camera, a keyboard, and a computer. Although the theory can be extended to continuous sources, we concentrate on measuring the information content of the output of discrete sources.

Sources can be categorized as having memory or being memoryless. For a source with memory, a current symbol depends on the previous symbols. An English-text source has memory: A period followed by a space indicates that the next character probably starts a sentence and is therefore upper case; a word that starts with the letter q always has a u as the second letter; and so forth.

A memoryless source is one for which each symbol emitted is independent of the previous symbols. A *discrete memoryless source* (DMS) can be characterized by listing the symbols, giving the probability of each symbol, and specifying the symbol rate of the source. For example, we could have a source that emits the symbols A, B, C, and D with probabilities $P(A) = \frac{1}{2}$, $P(B) = \frac{1}{4}$, $P(C) = \frac{1}{8}$, and $P(D) = \frac{1}{8}$, with a rate of 1000 symbols per second. We will develop a measure for the information rate of such DMSs.

Information content of a discrete memoryless source

Intuitively, we can equate the information content of a symbol to our surprise when it becomes known to us or to the difficulty of predicting its content. When we know that the next symbol from some source is going to be the letter C, we are not surprised by it and do not derive any information from it. As another example, suppose that we have a source with two symbols A and B with probabilities $P(A) = 0.999$ and $P(B) = 0.001$. We can almost always predict that an A will occur, and we are not surprised when it does. Thus, a highly probable symbol carries very little information. On the other hand, when B occurs, we are surprised and presumably derive more information than when A occurs. It is very difficult to predict when symbols with a low probability will occur; therefore, they convey relatively large amounts of information. We are led to define the information content of a symbol from a DMS as a function of its probability.

486

Suppose that we have a DMS, denoted by X, with symbols $x_1, x_2, \ldots, x_M$. We denote the information content of symbol x_i by $I(x_i)$. We have already established that $I(x_i)$ should be a function of the probability of occurrence of x_i:

$$I(x_i) = f[P(x_i)] \tag{10.1}$$

Also, we expect more information from a less probable symbol; in equation form,

$$I(x_i) < I(x_j) \qquad \text{if } P(x_i) > P(x_j) \tag{10.2}$$

In addition, we expect that the information content of the joint occurrence of independent symbols is the sum of the information in the individual symbols. For example, if the first symbol from a source is x_i and the second symbol is x_j, the information content of the two symbols should be

$$I(x_i x_j) = I(x_i) + I(x_j) \qquad \text{if } x_i \text{ and } x_j \text{ are independent} \tag{10.3}$$

Intuitively, we also require the information content of any symbol to be equal to or greater than zero because it does not seem possible for a symbol to convey negative information:

$$I(x_i) \geq 0 \tag{10.4}$$

It can be shown that the only definition of information that satisfies the requirements we have stated in Equations 10.1 through 10.4 is

$$\blacksquare \qquad I(x_i) = \log \frac{1}{P(x_i)} = -\log P(x_i) \text{ bits} \tag{10.5}$$

where we assume a base-2 logarithm. If the base of the logarithm is taken as 10, the units of information become Hartleys, and if the natural logarithm is used, the units are nats. We will use base-2 logarithms unless stated otherwise, and the units of information will be bits.

The use of the term *bit* in information theory is different from the common usage of the term by electrical engineers. In common usage, a binary symbol, which can be a 0 or a 1, is called a *bit*. In information theory, a binary symbol carries one bit of information only if it has a probability of one-half. In this chapter, we reserve the term *bit* for a measure of information and use the term *binary symbol* to indicate a symbol that can take one of two different values.

EXAMPLE 10.1

Find the information content of each symbol of a DMS having the symbols A, B, C, and D with probabilities $P(A) = \frac{1}{2}$, $P(C) = \frac{1}{4}$, and $P(C) = P(D) = \frac{1}{8}$ in bits, Hartleys, and nats.

The information content of each symbol can be computed using Equation 10.5 using base-2 logarithms for the results in bits, base-10 logarithms for Hartleys, and natural logarithms for nats. The results are as follows:

Symbol	Probability	Information content		
		Bits	Hartleys	Nats
A	$\frac{1}{2}$	1	0.301	0.693
B	$\frac{1}{4}$	2	0.602	1.386
C	$\frac{1}{8}$	3	0.903	2.079
D	$\frac{1}{8}$	3	0.903	2.079

Note: One bit is equivalent to 0.301 Hartleys or 0.693 nats.

Average information rate of a discrete memoryless source

In a practical communication system, we usually transmit long sequences of symbols from an information source; therefore, we are more interested in the average rate that a source produces information rather than the information content of a single symbol. The information carried by each source symbol is

FIGURE 10.1
Average information content of a binary source versus the probability of one of the symbols.

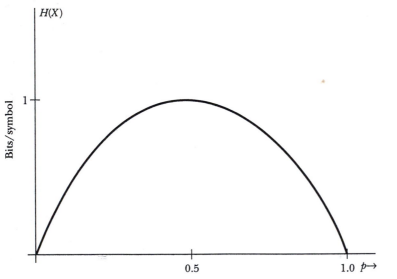

a random quantity since the symbols are produced at random, and the information content may be different for each symbol. The average information per symbol can be found by multiplying the information content of each symbol by its probability and summing. In equation form, the average information per symbol for a source X is given by

$$\blacksquare \qquad H(X) = \sum_{i=1}^{M} P(x_i) \log \left[\frac{1}{P(x_i)} \right] \text{ bits/symbol} \qquad (10.6)$$

where we have assumed that the source can emit M different symbols.

If the time rate at which the source emits symbols is denoted as r_s, the information rate of the source is given by

$$\blacksquare \qquad R = r_s H(X) \text{ bits/s} \qquad (10.7)$$

EXAMPLE 10.2

Find the average information per symbol and the information rate of the source of Example 10.1 if the symbol rate of the source is 5000 symbols per second.

SOLUTION

The average information per symbol can be computed from Equation 10.6, which for the present case takes the form

$$H(X) = P(A) \log \frac{1}{P(A)} + P(B) \log \frac{1}{P(B)} + P(C) \log \frac{1}{P(C)} + P(D) \log \frac{1}{P(D)}$$

$$= \left(\frac{1}{2} \right) 1 + \left(\frac{1}{4} \right) 2 + \left(\frac{1}{8} \right) 3 + \left(\frac{1}{8} \right) 3$$

$$= 1.75 \text{ bits/symbol}$$

If the symbol rate is 5000 symbols per second, the average information rate of this source is

$$R = 1.75 \times 5000 = 8750 \text{ bits/s}$$

It is interesting and instructive to consider the information rate of a binary source as a function of the probability of one of the symbols. Assume that we have a binary source X with symbols x_1 and x_2 with probabilities $P(x_1) = p$ and $P(x_2) = (1 - p)$. The average information per symbol for this source can be found from Equation 10.6 as

$$\blacksquare \qquad H(X) = -p \log p - (1 - p) \log (1 - p) \qquad (10.8)$$

This is plotted as a function of p in Figure 10.1. Notice that the maximum average information is produced by a source with equal probabilities $(p = \frac{1}{2})$. As p approaches either zero or one, the average information tends toward zero.

For example, consider these two binary sources with the characteristics:

	Source $X1$	Source $X2$
Symbols	A and B	C and D
Probabilities	$p = P(A) = \frac{1}{2}$	$p = P(C) = \frac{1}{32}$
	$P(B) = \frac{1}{2}$	$P(D) = \frac{31}{32}$
Information	$I(A) = 1$ bit	$I(C) = 5$ bits
	$I(B) = 1$ bit	$I(D) = 0.0458$ bit
Average information	$H(X1) = 1$ bit/symbol	$H(X2) = 0.201$ bit/symbol

Notice that a sequence from source $X1$ is, on the average, the most difficult to predict, indicating that it produces the largest average amount of information. Source $X2$ is quite predictable since symbol D occurs most of the time, and every time it occurs it carries only 0.0458 bits. When symbol C occurs, it is quite a surprise and carries 5 bits of information. It would be most difficult to predict when C might occur, and that is why its occurrence carries such a large amount of information. Even though one of the symbols from source $X2$ carries the most information of the four symbols, the average information rate of $X2$ is less than that of $X1$. Also, notice that information is produced at a uniform rate by $X1$ because every symbol carries one bit, whereas source $X2$ produces information in bursts when the symbol C occurs. For source $X2$ most symbols are Ds, which carry only a small amount of information, whereas occasionally a C occurs carrying a burst of 5 bits.

For a DMS with M symbols, it can be shown that the greatest average amount of information per symbol occurs when the symbols are equally likely. In this case, the average amount of information is $\log M$ bits per symbol.

10.2
SOURCE CODING

Now that we have developed a measure for the information output of DMSs, we will consider how to devise efficient codes to represent the information. Consider a source X with an average information rate of $H(X)$ bits per symbol for which we want to devise an efficient binary code. An efficient binary code consists of a code word, consisting of binary symbols, for each source symbol (or each group of source symbols), which is devised so that the average number of binary symbols per code word is minimized. A binary symbol is capable of conveying a maximum of one bit, and the average information per source symbol is $H(X)$. Therefore, we anticipate that it should be possible to devise a code with an average number of binary symbols per source symbol equal to the value of $H(X)$. We will shortly show how this is accomplished.

First we give a quantitative definition of the coding efficiency for representing the symbols produced by a discrete source. Let L represent the average length of the binary code words needed to represent a string of n source symbols. The efficiency of the code is defined by

$$\blacksquare \qquad \eta = \frac{nH(X)}{L} \qquad\qquad (10.9)$$

The Shannon-Fano code

An efficient code can be devised by listing the source symbols in order of decreasing probabilities and then assigning binary code symbols to divide this list as nearly as possible into equally likely groups and subgroups until finally only one symbol remains in each subgroup. This procedure is known as the *Shannon-Fano algorithm*. For example, consider a source with the symbols and probabilities shown in Table 10.1. The code words of the table have been obtained by first dividing the four symbols into two equally likely groups: A by itself in one group and B, C, and D in the second group. The first binary symbol of each code word indicates into which of these groups each source symbol falls. Since A falls into a group by itself, the first bit is all that is needed to represent it. The second group, containing three symbols B, C, and D, is then divided into two equally likely subgroups. Thus, BCD is divided into B and CD, and the second symbol of the code words indicates to which subgroup each of these source symbols is assigned. After these two divisions, B is in a subgroup by itself and needs only the two symbol code word to represent it. Finally, group CD is divided into C and D, and the last symbols of their code words distinguish them.

The source indicated in Table 10.1 was considered in Examples 10.1 and 10.2, in which we found that the average information per symbol is 1.75 bits.

TABLE 10.1
A discrete memoryless source, its probabilities, and the code obtained by use of the Shannon-Fano algorithm

Symbols	Probability	Code words
A	$\frac{1}{2}$	0
B	$\frac{1}{4}$	10
C	$\frac{1}{8}$	110
D	$\frac{1}{8}$	111

TABLE 10.2
A binary source with probabilities and code words

Symbols	Probabilities	Code words
A	0.9	0
B	0.1	1

The average length of the code words of Table 10.1 can be computed as

$$L = 1P(A) + 2P(B) + 3P(C) + 3P(D)$$
$$= 1.75 \text{ binary symbols per code word}$$

Therefore, the efficiency of the code can be found from Equation 10.9 as

$$\eta = \frac{nH(X)}{L} = 100\%$$

Since one source symbol is encoded by one code word, we have used $n = 1$. In this particular example, the efficiency of the code is 100%, indicating that each binary symbol of the code carries the maximum possible information (one bit per binary symbol). This is possible because the probabilities of the original source symbols allow subdivision into subgroups that are exactly equally likely. This is not always possible, as we will show by another example.

Consider a binary source with the two symbols and probabilities shown in Table 10.2. With only two symbols, the coding procedure is very simple. A single symbol code word is assigned to each source symbol. The average amount of information per source symbol for this source is given as

$$H(X) = -P(A) \log P(A) - P(B) \log P(B)$$
$$= 0.469 \text{ bits per symbol}$$

The average length of the code words of Table 10.2 is clearly one. Therefore, the efficiency of the code is given by

$$\eta = \frac{nH(X)}{L} = 46.9\%$$

where we have used $n = 1$ because only one source symbol is encoded per code word. This low value of efficiency is due to the fact that the source symbols cannot be subdivided into equally likely subgroups.

Extended sources

A solution for this problem is to represent a string of source symbols by a single code word. For example, Table 10.3 shows all of the two-symbol strings of symbols possible for the source of Table 10.2. When strings of two symbols

TABLE 10.3
Second-order extension of the source of
Table 10.2 with probabilities and code words
derived by the Shannon-Fano algorithm

Strings of source symbols	Probability	Code words
AA	0.81	0
AB	0.09	10
BA	0.09	110
BB	0.01	111

are considered, we call the source of the pairs of symbols the *second order exten-sion* of the original source. Before assigning code words, the strings are ordered by decreasing probability as shown in the table. The code words are assigned to each string by dividing the list of source symbols into two equally likely (as nearly as possible) groups and then assigning the binary code symbols to in-dicate into which group each string falls. Thus the first code symbol divides the list into a group containing only AA and a second group containing AB, BA, and BB. The second symbol of each code word subdivides this second group. The groups of strings of source symbols are subdivided, and a code symbol is assigned at each subdivision until only a single string remains in each subgroup. The resulting code is shown in the table.

The average length of the code words of Table 10.3 can be found as $L = 1.29$. Then the efficiency of the code can be computed as

$$\eta = \frac{nH(X)}{L} = 72.7\%$$

In this case, we have taken $n = 2$ because two of the original source symbols are encoded in each code word. When we encoded the symbols of Table 10.2 directly, we found an efficiency of only 46.9%, so a considerable improve-ment has resulted from encoding two symbol strings of the source symbols.

For this particular example, a further improvement can be obtained by considering the third-order extension of the source. Strings of three source symbols, their probabilities, and the resulting code words are shown in Table 10.4. As before, the code words are assigned by dividing the ordered list of strings of source symbols into the most nearly equally likely groups and assign-ing a code symbol to indicate each division. The average length of the code words is 1.598, and the efficiency is 88.0%.

It can be shown that it is possible to devise a coding scheme for a DMS that approaches 100% efficiency as closely as desired by encoding sufficiently long strings of source symbols. However, it should be noted that the improve-ment in efficiency is not always uniform as longer strings are considered. For

TABLE 10.4
Third-order extension of the source of
Table 10.2

Strings of source symbols	Probabilities	Code words
AAA	0.729	0
AAB	0.081	100
ABA	0.081	101
BAA	0.081	110
ABB	0.009	11100
BAB	0.009	11101
BBA	0.009	11110
BBB	0.001	11111

example, it is possible that the code obtained from the third-order extension is less efficient than the second-order extension.

Uniquely decipherable codes

The code obtained by the use of the Shannon-Fano algorithm has the property of being *uniquely decipherable*. This means that even if the code words are written in a string without an indication of the beginning and end of each code word, the string can be deciphered into the original symbols without ambiguity. For example, consider the source symbols and codes shown in Table 10.5. Code 1 is uniquely decipherable and code 2 is not. If the code sequence 010111110 is decoded using code 1, the only possible sequence of source symbols is ABDC. If the same code sequence is deciphered using code 2, several source sequences are possible, including ABABBBBBA and DDBBBC. Of course, it is important to use a uniquely decipherable code when strings of symbols are to be transmitted.

TABLE 10.5
Source symbols and codes

Source symbols	Code 1	Code 2
A	0	0
B	10	1
C	110	10
D	111	01

Note: Code 1 is uniquely decipherable; code 2 is not.

Source coding of practical sources

Practical sources are rarely as simple as the DMS we have studied. However, the process of source coding to obtain efficient codes for storing and transmitting information is still an important consideration. Many information sources such as English text, voice signals, and television signals contain redundancy. Letters that are missing from English text can usually be guessed by a reader, indicating that many of the letters supply little information. A television signal is often identical or nearly identical from one picture frame to another; only on scene changes is a new frame grossly different from the last. Similarly, voice signals have considerable redundancy. Many studies have been undertaken to code such practical sources efficiently; but because of the complexity of these real-world sources, the methods used for practical efficient source coding have been mainly empirical rather than theoretical.

10.3
COMMUNICATION CHANNELS

A communication channel is the path through which the symbols or waveforms of a transmitter flow to the receiver. A communication channel may either be discrete or continuous. A discrete channel is one through which the transmitter sends discrete symbols that are received (sometimes in error) by the receiver. A continuous channel carries a continuum of values or waveforms.

For analysis, we often find it convenient to locate the input and output of the channel at different points in a communication system as we consider its various aspects. For example, consider the baseband antipodal binary communication system shown in Figure 10.2. The information source emits binary symbols at a rate denoted by f_b. For each of the two possible symbols 0 and 1, the modulator selects one of two possible pulses (one is the negative of the other), which is transmitted to the receiver. The transmitted pulses are assumed to be given by

$$p(t) = A \operatorname{sinc} (f_b t)$$

FIGURE 10.2
Communication system example for discussion of channel models.

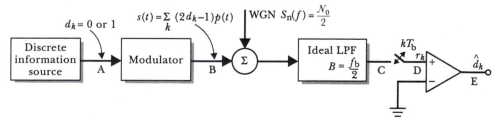

and the signal emitted by the modulator is

$$s(t) = \sum_k (2d_k - 1)p(t - kT_b)$$

where $T_b = 1/f_b$ is the interval between source symbols, and $d_k = 1$ or 0 are the source symbols.

In transmission, white gaussian noise is added to the signal, and the sum is passed through an ideal lowpass filter with a cutoff frequency of $B = f_b/2$. This filter is the matched filter for the transmitted pulse shape just chosen. The output of the matched filter is sampled at multiples of the symbol interval T_b, and the sampled values are compared to zero by the comparator to produce a series of decisions about the symbols that were transmitted.

If we choose to define the channel for the system of Figure 10.2 as starting at point A and ending at point E, a discrete channel results. The input to this discrete channel is the sequence of source symbols and the output is the sequence of decisions. For this system, there is no intersymbol interference at the output of the matched filter at the sampling instants, so each decision depends only on one source symbol. (Intersymbol interference or ISI was discussed for a system similar to this in Section 6.3.) Thus, between points A and E, we have a discrete channel with input symbols of 0 and 1 and output symbols of 0 and 1. This channel is said to be memoryless because each output decision depends only on the current input symbol. This channel can be characterized by the symbol error probability. An expression was given by Equation 8.46 for the error probability of an antipodal signal set in the presence of AWGN and is repeated here for convenience.

$$P_s = Q\left[\left(\frac{2E_s}{\mathcal{N}_0}\right)^{1/2}\right]$$

A simplified model for the channel between points A and E of Figure 10.2 is shown in Figure 10.3. The symbols input into the channel by the source are shown on the left side of the diagram, and the output symbols are shown on the right. The values shown on each of the transitions are the conditional probabilities of receiving the output symbol, given the input symbol. These are called *transition probabilities*. The simple channel modeled by Figure 10.3 is called a *memoryless binary symmetric channel*.

Another type of channel with discrete inputs and outputs would result between points A and E if the two-level comparator was replaced with a decision device capable of more kinds of decisions. For example, we could have a decision device that decides that the transmitted symbol is a 1 if the sample of the matched-filter output is large and positive, decides in favor of a 0 for large negative values, and gives a noncommital decision, known as an *erasure*, if the matched filter output is in a symmetrical zone around zero. This is the channel model shown in Figure 10.4, known as an *erasure channel*.

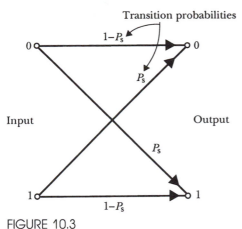

FIGURE 10.3
Model for binary symmetric channel.

Another variation of the channel model would result if, for example, the filter were changed so that the sample values show ISI. In this case, the decisions would depend not only on the current symbol but possibly on past symbols as well, and a discrete channel with memory would result. Then it would not be possible to model the channel by the simple transition diagram that we used for the memoryless channel. We will not pursue the subject of channels with memory any farther; we want only to indicate the possibility for memory in the channel.

If we were to consider the channel input to be at point A of Figure 10.2 and the channel output to be at point D, then we would have a channel with a discrete input $(d_k = 0$ or $1)$ and a continuous random variable r_k as the

FIGURE 10.4
Model for symmetric erasure channel.

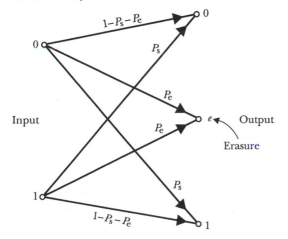

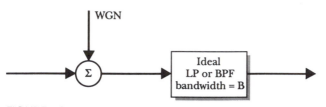

FIGURE 10.5
Bandlimited gaussian noise channel model.

output. This channel could be characterized by the conditional probability density functions for r_k given the input symbol.

In another case, we might be interested in the channel between points B and C. Then the input to the channel can be any waveform, and the output is a filtered version of the sum of the input and a sample function from a white gaussian noise process. This is a *bandlimited gaussian noise waveform channel*, modeled by the block diagram of Figure 10.5.

Obviously, many channel models can be envisioned. Channels with inputs and outputs that are discrete symbols, continuous random variables, or waveforms are of interest. Channels may or may not have memory. In this brief introduction to information theory, we concentrate on discrete memoryless channels (DMCs) such as those of Figures 10.3 and 10.4 and on the gaussian noise waveform channel of Figure 10.5.

10.4
INFORMATION TRANSFER AND CHANNEL CAPACITY
Mutual information for a discrete memoryless channel

Now we consider the amount of information transferred when symbols from a DMS are transmitted through a DMC. We denote the source by X, inputting symbols $x_1, x_2, \ldots,$ and x_M into the channel. The a priori probabilities of these source symbols are assumed to be known. The output of the channel is denoted by Y, and the output symbols are $y_1, y_2, \ldots,$ and y_N. Notice that the number of output symbols can be different from the number of source symbols. The channel is characterized by its transition probabilities $P(y_j|x_i)$. The source, channel, and output are depicted by the diagram in Figure 10.6.

Before a symbol is transmitted through the channel, $H(X)$ gives the average uncertainty of the receiver about the source symbol. If the receiver learns the symbol, this uncertainty is resolved, and the receiver has obtained $H(X)$ bits of information on the average. For many communication channels, the uncertainty of the receiver about the source symbol is not resolved when the symbol is received because of the possibility for errors in the channel. The uncertainty associated with symbol x_i before the output of the channel is

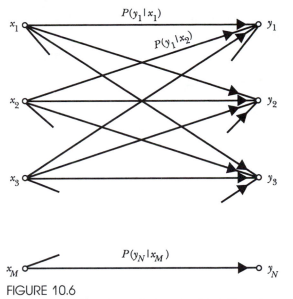

$P(y_1|x_1)$

$P(y_1|x_2)$

$P(y_N|x_M)$

FIGURE 10.6
Discrete memoryless channel with M input symbols and N output symbols.

learned is given by $-\log P(x_i)$. (This is the same as the amount of information conveyed when it is learned that x_i was transmitted.) When the output symbol y_j is learned, the probability of x_i changes from $P(x_i)$ to $P(x_i|y_j)$. Therefore, the uncertainty concerning x_i has changed from $-\log P(x_i)$ to $-\log P(x_i|y_j)$. The average uncertainty concerning the input symbol when the corresponding output symbol is known is denoted by $H(X|Y)$ and is given by

$$H(X|Y) = \sum_{i=1}^{N} \sum_{j=1}^{M} -P(x_i, y_j) \log P(x_i|y_j) \qquad (10.10)$$

Note that this is the uncertainty of x_i, given y_j, averaged over all combinations of input and output symbols.

$H(X)$ represents the average uncertainty about the source symbol before the output symbol is known and $H(X|Y)$ represents the average uncertainty after the output symbol is learned. Therefore, the average amount of information gained about the source symbol by learning the output symbol is given by

$$I(X,Y) = H(X) - H(X|Y) \qquad (10.11)$$

EXAMPLE 10.3

Find the average amount of information transferred by each symbol transmitted for the channel shown in Figure 10.7 if the source symbols are equally likely.

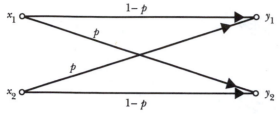

FIGURE 10.7
Binary symmetric channel.

For a binary source with equally likely symbols, the average amount of information conveyed by a source symbol is

$$H(X) = 1 \text{ bit/symbol}$$

The uncertainty remaining about the source symbols after the output symbol is known is given by Equation 10.10. To evaluate Equation 10.10, we need the joint probabilities $P(x_i y_j)$ and the conditional probabilities $P(x_i|y_j)$. These can be found as follows:

$$P(x_1 y_1) = P(y_1|x_1)P(x_1) = \frac{1}{2}(1-p)$$

Similarly, we find

$$P(x_2 y_2) = \frac{1}{2}(1-p)$$

$$P(x_1 y_2) = P(x_2 y_1) = \frac{p}{2}$$

To find the conditional probabilities, we have

$$P(x_1|y_1) = \frac{P(x_1 y_1)}{P(y_1)} = \frac{P(y_1|x_1)P(x_1)}{P(y_1 x_1) + P(y_1 x_2)} = \frac{\frac{1}{2}(1-p)}{\frac{1}{2}(1-p) + \frac{1}{2}(p)} = 1 - p$$

Similarly, we find that

$$P(x_2|y_2) = 1 - p$$

$$P(x_1|y_2) = p$$

$$P(x_2|y_1) = p$$

Evaluating Equation 10.10, we eventually obtain

$$H(X|Y) = -p \log (p) - (1-p) \log (1-p)$$

Then the average information transferred per symbol is

$$I(X,Y) = H(X) - H(X|Y)$$
$$= 1 + p \log (p) + (1-p) \log (1-p)$$

Notice that if $p = 0$ or 1, the channel transforms input symbols into output symbols in an unambiguous manner, and so the input symbol is known with certainty when the output symbol is learned. In this case, evaluation of $I(X,Y)$ shows that one bit of information is transferred with each symbol transmitted. On the other hand if $p = \frac{1}{2}$, knowledge of the output symbol does not give any indication of what the input symbol is, and evaluation of $I(X,Y)$ gives zero, as expected.

Channel capacity

The *capacity* of a communication channel is a measure of the maximum rate at which information can be transferred through the channel. For a DMC, the capacity is found by finding the maximum value of $I(X,Y)$. The maximization is with respect to the probabilities of the source symbols. Thus, the channel capacity is given as

$$C = \max_{P(x_i)} [I(X,Y)] \qquad (10.12)$$

For the binary symmetric channel of Example 10.3, the maximum information transfer occurs when the source symbols are equally likely. This can be anticipated due to the symmetry of the channel. Therefore, the expression found for $I(X,Y)$ in the example is also the capacity of the binary symmetric channel, given by

$$C = 1 + p \log (p) + (1 - p) \log (1 - p) \qquad (10.13)$$

This is plotted versus the error probability of the channel p, in Figure 10.8. Notice again that the capacity of the channel is one bit for each symbol transmitted if $p = 0$ or 1. If $p = \frac{1}{2}$ the capacity of the channel is zero.

EXAMPLE 10.4

Find the optimum source symbol probabilities and the channel capacity of the channel shown in Figure 10.9.

SOLUTION
Due to symmetry, we expect that the optimum source probabilities are $P(x_1) = P(x_4) = P$ and $P(x_2) = P(x_3) = Q$. Clearly, the probabilities of the source symbols must add to unity, so we have

$$2P + 2Q = 1 \qquad \text{or equivalently,} \qquad Q = \frac{1}{2} - P$$

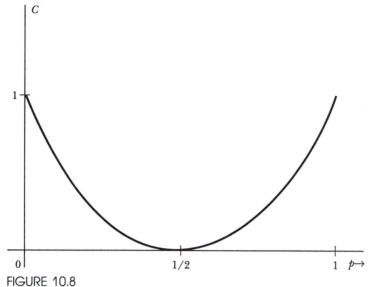

FIGURE 10.8
Capacity of the binary symmetric channel versus symbol error probability.

The entropy of the source can be computed using Equation 10.6, resulting in

$$H(X) = -2P \log P - 2Q \log Q$$

Using Equation 10.10, we eventually find

$$H(X|Y) = -2Q(p \log p + q \log q)$$

FIGURE 10.9
Channel for Example 10.4.

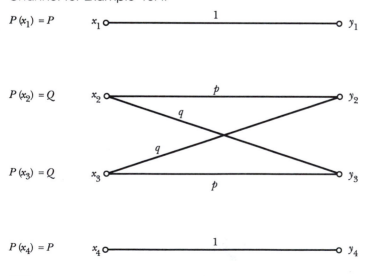

$P(x_1) = P$

$P(x_2) = Q$

$P(x_3) = Q$

$P(x_4) = P$

The information transferred per symbol is then

$$I(X,Y) = H(X) - H(X|Y)$$

$$= -2P \log P - 2\left(\frac{1}{2} - P\right) \log \left(\frac{1}{2} - P\right)$$

$$+ 2\left(\frac{1}{2} - P\right)(p \log p + q \log q)$$

where we have made use of the fact that $Q = \frac{1}{2} - P$.

The value of P resulting in the maximum transfer of information can now be found by differentiating the expression for $I(X,Y)$ with respect to P and equating to zero. Thus, we have

$$\frac{d}{dP}[I(X,Y)] = 0$$

$$-2P\left(\frac{1}{P}\right) - 2 \log P + 2 + 2 \log \left(\frac{1}{2} - P\right) - 2(p \log p + q \log q) = 0$$

Solving, we obtain

$$\log \left(\frac{\frac{1}{2} - P}{P}\right) = \log \frac{Q}{P} = p \log p + q \log q$$

This expression can be used with the requirement that $2P + 2Q = 1$ to find the optimum values of the source probabilities. The channel capacity can then be found by evaluating $I(X,Y)$ for the optimum values of P and Q.

It is interesting and instructive to consider several special cases. First, if $p = 1$ and $q = 0$, the channel transforms the input symbols into output symbols without ambiguity. In this case, the above expression for the optimum source probabilities gives $P = Q = \frac{1}{4}$. Thus, the source symbols are chosen to be equally likely. In this case, $H(X) = 2$ bits per symbol. Also $H(X|Y) = 0$ because there is no uncertainty about the source symbols when the output is known. Therefore, we have $C = \max[H(X) - H(X|Y)] = 2$ bits per symbol transmitted.

Consider another special case, with $p = q = \frac{1}{2}$. The optimum probabilities are $P(x_1) = P(x_4) = P = \frac{1}{3}$ and $P(x_2) = P(x_3) = Q = \frac{1}{6}$. Note that when y_2 or y_3 is received, it is known that x_2 or x_3 was transmitted, but there is no indication of which of the two source symbols occurred. Thus, the optimum strategy is to treat x_2 and x_3 as a single character. The three source characters x_1, x_4, and the paired symbol created from combining x_2 and x_3 are chosen to be equally likely to maximize the information transferred. The resulting channel capacity is $\log 3$, which is the average information content of symbols from a source with three equally likely characters.

The channel capacity theorem and its proof for the binary symmetric channel

The capacity of other DMCs can be found by a procedure similar to the previous example. The significance of channel capacity is due to a theorem first proven by Shannon. This important theorem states that in principle it is possible to transfer information through a channel at a rate not exceeding the capacity of the channel with as close to zero error probability as desired. Thus, channel capacity sets a limit on the rate at which information can be transferred but not on the accuracy.

Now we give a proof of this theorem for the binary symmetric channel shown in Figure 10.7. Consider encoding length-k blocks of the source symbols into length-n code words, which are transmitted through the channel. At the receiving end, each received block is compared with the list of code words to find the one closest in Hamming distance, as discussed in Chapter 9. Eventually, in the proof, we take the limiting case as k and n approach infinity. Assuming that the source symbols are equally likely, each source symbol carries one bit of information. Thus, the amount of information per code symbol is k/n (which is the rate of the code). The theorem we want to prove is that coding schemes exist for which the error probability approaches zero as k and n approach infinity, provided that k/n is less than the channel capacity C.

In the proof of this theorem, we make a number of approximations. To illustrate the validity of the approximations and provide a clearer picture of the concepts involved, we carry a numerical example along with the proof.

EXAMPLE 10.5

Find the maximum value of k/n permitted by the channel capacity theorem for error-free communication over a binary symmetric channel having a symbol error probability of $p = 0.05$.

SOLUTION

The channel capacity of the binary symmetric channel is given by Equation 10.13. Evaluating for $p = 0.05$, we obtain

$$C = 1 + p \log p + (1 - p) \log (1 - p)$$
$$= 0.7136 \text{ bits/symbol transmitted}$$

Thus, as k and n approach infinity, the error probability can approach zero as long as k/n is less than 0.7136. Assuming that $n = 20,000$ is large enough to provide nearly theoretical performance, a value of k less than about $20,000 \times 0.7136 = 14,270$ provides a very low error rate. In the series of examples to follow, we illustrate that for $k = 13,000$ the error rate can be forced very low.

In proving the channel capacity theorem, we will not use any of the codes discussed in Chapter 9. Instead, we assume that the code words are obtained by a random choice of the symbols. Each code symbol is independent and is equally likely to be either of the two possible transmitted symbols. Conceptually, the code word that represents a given data sequence could be obtained by tossing a coin n times and assigning one symbol for a head and the other symbol for a tail. Of course, the decoder must have a list of the code words to be able to perform the decoding operation. Notice that it is possible for the same code word to represent different data sequences. If this happens, the decoder will not be able to decide which data sequence is correct. This does not prevent the error probability from approaching zero as n approaches infinity, however, because as k and n approach infinity, the probability of such an event becomes vanishingly small.

To obtain expressions for the decoder error probability, we need to consider the statistics of the number of symbol errors in each received word. Consider a set of random variables $z_1, z_2, \ldots, z_n$, where $z_j = 0$ if the jth symbol of the code word is received correctly and $z_j = 1$ if the jth symbol of the code word is received in error. Clearly, the total number of symbol errors in the received word is given by

$$
\mathcal{N}_e = \sum_{j=1}^{n} z_j \tag{10.14}
$$

Now we consider the statistics of $\mathcal{N}_e$. Since we will eventually allow n to approach infinity, we expect from the central limit theorem that the cumulative distribution function for $\mathcal{N}_e$ approaches that of a gaussian random variable. Therefore, only the mean and variance of $\mathcal{N}_e$ is needed to characterize its statistics.

The average number of symbol errors per code word is given by

$$
E(\mathcal{N}_e) = E\left(\sum_{j=1}^{n} z_j \right) = nE(z_j) \tag{10.15}
$$

where the last expression follows because the expected value of a sum is the sum of the expected values. Furthermore, the expected values of each of the z_j random variables are the same and are given by

$$
E(z_j) = 1P(z_j = 1) + 0P(z_j = 0) = p \tag{10.16}
$$

Substituting this result into Equation 10.15 results in

$$
E(\mathcal{N}_e) = np \tag{10.17}
$$

The variance of the number of errors per code word is given by

$$
\text{var}\,(\mathcal{N}_e) = \text{var}\left(\sum_{j=1}^{n} z_j \right) \tag{10.18}
$$

The random variables $z_1, z_2, \ldots, z_n$ are independent because we have assumed that the channel is memoryless. The variance of the sum of independent random variables is equal to the sum of the variances. Since the variance of each z_j is the same, we can write

$$\operatorname{var}(\mathcal{N}_e) = n \operatorname{var}(n_j) \tag{10.19}$$

Now the second moment of z_j is given by

$$E(z_j^2) = 1^2 p + 0^2(1 - p) = p \tag{10.20}$$

The variance of z_j is given by

$$\operatorname{var}(z_j) = E(z_j^2) - [E(z_j)]^2 \tag{10.21}$$

Using Equations 10.16 and 10.20 to substitute into Equation 10.21, we find

$$\operatorname{var}(z_j) = p - p^2 \tag{10.22}$$

Substituting into Equation 10.19, we find

$$\operatorname{var}(\mathcal{N}_e) = n(p - p^2) \tag{10.23}$$

Thus, we have established that the number of errors per code word approaches a gaussian random variable with mean given by np and variance given by $n(p - p^2)$.

EXAMPLE 10.6

Find the statistics of the number of errors per code word if $p = 0.05$ and $n = 20,000$. Estimate the probability that the number of errors exceeds 1300 in a given code word.

SOLUTION

The mean value of the number of errors is given by Equation 10.17. Substituting and evaluating, we find

$$E(\mathcal{N}_e) = np = 20,000 \times 0.05 = 1000$$

The variance is given by Equation 10.23, which results in

$$\operatorname{var}(\mathcal{N}_e) = n(p - p^2) = 950$$

The standard deviation of the number of errors is the square root of the variance or 30.82. Due to the central limit theorem, the distribution of the number of errors approaches a gaussian distribution as the code words become very long. A sketch of the probability density function of $\mathcal{N}_e$ is shown in Figure 10.10. The probability of the number of errors exceeding 1300 is the shaded area in the figure and is given by

$$P(\mathcal{N}_e \geq 1300) = Q\left[\frac{1300 - E(\mathcal{N}_e)}{\sigma_{\mathcal{N}_e}}\right] = 1.1 \times 10^{-22}$$

Notice that the probability that the number of errors exceeds $K = 1.30$ times the mean is very small.

Notice in Equation 10.17 that the mean value of $\mathcal{N}_e$ increases directly with n. In Equation 10.23 we see that the standard deviation of $\mathcal{N}_e$ increases only as the square root of n. Thus, as n approaches infinity, the number of errors becomes more and more likely to be close to the average value. In the limit, the probability that the number of errors exceeds the average number by an arbitrarily small percentage tends to zero. This is evident in Example 10.6. Thus, the number of errors is nearly always less than Knp when K approaches unity as n approaches infinity. In equation form we can write

$$\lim_{n \to \infty} P(\mathcal{N}_e > Knp) = 0 \qquad \text{for } K > 1 \qquad (10.24)$$

Since $\mathcal{N}_e$ is nearly always less than Knp, the decoder can assume that any code word within a Hamming distance of Knp of the received word is likely to be the transmitted code word. Certainly for sufficiently large n we have seen that the correct code word is almost always within Knp distance units of the received sequence. However, since code words are selected at random, it is also possible for a code word other than the one transmitted to fall within Knp of the received sequence. When this happens, it is likely that the decoder will select the incorrect word. However, we will now show that if k/n is less than the channel capacity, then the probability of another code word falling within the decoding distance tends to zero as n becomes large.

FIGURE 10.10
Probability density function for the number of errors per code word of Example 10.6.

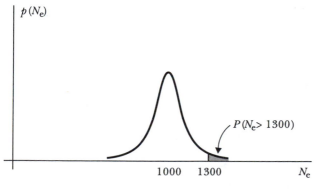

Consider the probability P_2 that a particular code word other than the one transmitted falls within Knp of the received sequence. Since the code words have a length of n symbols, the total number of potential code words and received sequences is 2^n. The number of sequences within Knp of the received sequence is given by

$$\sum_{j=0}^{Knp} \binom{n}{j} \tag{10.25}$$

where we have assumed that Knp is an integer. The first term in the summation is unity, representing the received sequence. The second term is n and accounts for the sequences that differ from the received sequence in only one symbol. The third term accounts for the sequences that differ from the received sequence in exactly two symbols, and so forth. For large n and when $Knp \ll n$, the summation is dominated by the last term. Since we are interested in cases for which p is small and K eventually approaches unity, the condition $Knp \ll n$ applies. Thus, the number of sequences within Knp of the received sequence can be approximated by

$$\binom{n}{Knp} \tag{10.26}$$

EXAMPLE 10.7

Find the ratio of the last two terms of the summation of Equation 10.25 for the case where $n = 20{,}000$ and $Knp = 1300$.

SOLUTION
The ratio of the last two terms is

$$\frac{\binom{20{,}000}{1300}}{\binom{20{,}000}{1299}} = \frac{\dfrac{20{,}000!}{1300!\,18{,}699!}}{\dfrac{20{,}000!}{1299!\,18{,}700!}} = \frac{18{,}700}{1300} = 14.4$$

Thus, for this case, the summation is dominated by the last term.

Now we can find that the probability that a given code word other than the one transmitted falls within Knp of the received sequence is the ratio of the number of sequences within Knp of the received sequence to the total number of sequences. Thus, we have

$$P_2 = \frac{\displaystyle\sum_{j=0}^{Knp} \binom{n}{j}}{2^n} \tag{10.27}$$

Approximating the summation by the last term, we find

$$P_2 \simeq \frac{\binom{n}{Knp}}{2^n} \tag{10.28}$$

$$P_2 \simeq \frac{n!}{(Knp)!(n - Knp - 1)!2^n} \tag{10.29}$$

When n is very large and K is slightly more than unity, the correct code word nearly always falls within Knp of the received sequence. Then the probability that the decoder makes an error in decoding a given received sequence is overbounded by the probability that at least one code word other than the one transmitted falls within Knp of the received sequence. The probability that a given code word falls in this region is P_2, and there are $2^k - 1$ code words other than the correct one. Thus, the probability of a decoding error P_e is overbounded by

$$P_e \leq (2^k - 1)P_2 \simeq 2^k P_2 \tag{10.30}$$

Using the approximation of Equation 10.29 in 10.30, we have

$$P_e \lesssim \frac{n!}{(Knp)!(n - Knp - 1)!2^{n-k}} \tag{10.31}$$

To proceed, we use Stirling's formula for large factorials, which is given by

$$x! \simeq x^x \exp{(-x)}(2\pi x)^{1/2} \tag{10.32}$$

This can be rewritten as

$$x! \simeq 2^{(x \log x - x \log e + 1/2 \log 2\pi x)} \tag{10.33}$$

where the logarithms are base 2.

EXAMPLE 10.8

Evaluate the error probability given in Equation 10.31 for the case in which $n = 20{,}000$ and $Knp = 1300$, and for various values of k.

SOLUTION
Substituting values into Equation 10.31 and using Equation 10.33 to approximate the large factorials, results in

$$P_e \lesssim 2^k 2^{-13{,}052}$$

For k somewhat less than 13,052, the error probability becomes very small. For example, if $k = 13{,}000$, the error probability is less than 2.22×10^{-16} and for $k = 12{,}950$, the error probability bound is 1.97×10^{-31}.

Thus, for $n = 20,000$, a channel error probability of 0.05, and random selection of code words, we have seen in Example 10.6 that the received sequence is almost never (probability of 1.11×10^{-22}) more than 1300 Hamming units from the transmitted sequence. In this example, we have found that a code word other than the one transmitted almost never falls within 1300 units of the received sequence as long as k is somewhat less than 13,000. Thus, we can guarantee a very low error probability with any code rate less than about $13,000/20,000 = 0.65$. The theoretical channel capacity of the channel was found in Example 10.5 to be 0.7136 bits per symbol. According to the channel capacity theorem, it should be possible to achieve as small an error probability as we desire as long as the code rate is less than the channel capacity and n is chosen large enough. From our sequence of examples, we see that n must be quite large for the error rate to become very small with randomly selected codes. In fact, the values of n required are so large that the use of random codes is far from practical.

Now we continue with our proof of the channel capacity theorem for the binary symmetric channel. Since we are interested in factorials of very large numbers, the second two terms in the exponent of Equation 10.33 will become insignificant, and we can use the approximation

$$x! \simeq 2^{x \log x} \tag{10.34}$$

Substituting Equation 10.34 into 10.31, we have

$$P_e \lesssim \frac{2^{n \log n}}{2^{Knp \log (Knp)} 2^{(n - Knp - 1) \log (n - Knp - 1)} 2^{n-k}} \tag{10.35}$$

Now we take the limit as n approaches infinity. As we have noted earlier, K approaches unity as n becomes large. Also $n - Knp - 1$ can be replaced by $n - np$. Making these changes in Equation 10.35 and rearranging, we find

$$P_e \lesssim 2^{n[k/n - p \log p - (1 - p) \log (1 - p) - 1]} \tag{10.36}$$

Notice that P_e can be made as small as desired by increasing n, as long as the coefficient of n in the exponent is negative. Thus, for error free transmission, we must have

$$\frac{k}{n} - p \log p - (1 - p) \log (1 - p) - 1 < 0 \tag{10.37}$$

Rearranging, we obtain

$$\frac{k}{n} < 1 + p \log p + (1 - p) \log (1 - p) = C \tag{10.38}$$

Notice that the right-hand side of this inequality is the capacity of the binary symmetric channel as given in Equation 10.13.

Thus, we have shown that it is possible to find coding schemes for the binary symmetric channel that provide an arbitrarily low error probability as long as the code rate is less than the channel capacity. Another important result, which we simply state without proof, is the converse of the channel capacity theorem. This theorem states that it is not possible to find any code with a rate greater than the channel capacity that can provide an arbitrarily low error probability.

We have proved the channel capacity theorem by considering the use of randomly selected codes. We certainly expect that codes exist that provide much better performance than randomly selected codes. Also, we considered infinitely long code words. In the numerical examples, we have illustrated that good performance with randomly selected codes having rates closely approaching the channel capacity requires codes of impractical length. The real problem of error-control engineering is to find codes with good performance that are short and contain structure that can be exploited to allow practical decoding. We have demonstrated how this can be achieved in Chapter 9.

Channel capacity results similar to the derivation we have done for the binary symmetric channel can be found for other communication channels of interest. For example, we have stated the capacity of the additive gaussian noise channel in Chapter 1 and compared practical digital modulation formats with this capacity bound in Chapter 8.

Proofs of capacity for the various channels of interest are existence arguments often involving impractical choices of codes or signal sets. Therefore, the capacity theorems and their proofs do not illustrate how the promised performance can be achieved with a practical approach. Nevertheless, the capacity provides a goal for the communication designer and a reference for the amount of improvement that is possible.

SUMMARY

1. Information sources may be continuous or discrete and have memory or be memoryless. Practical sources usually have memory with a complex dependence between successive symbols or amplitudes.

2. A discrete memoryless source can be characterized by listing the symbols and their probabilities.

3. The information content of a symbol from a discrete memoryless source is a function of the probability of the symbol. Symbols with lower probability convey more information. The information associated with a given symbol is defined as the negative of the logarithm of its probability.

4. The average information conveyed by a symbol from a source, denoted by $H(X)$, is important in characterizing an information source. This quantity is also called the entropy of the source. The time rate of information

produced by a discrete source is the product of the source entropy and the symbol rate of the source.

5. The average information conveyed by a symbol from a discrete memory-less source is maximized with respect to the source probabilities when the symbols are equally likely.

6. Source coding is the process of coding the information content of a source. The efficiency of a binary source code is defined as the ratio of the average amount of information in bits represented by a code word to the average number of code symbols per word. The most efficient binary code results in one bit of information per code symbol.

7. The Shannon-Fano algorithm is a method for devising an efficient code for a discrete memoryless source. The nth order extension of a discrete source is obtained by grouping n successive symbols from the source and treating each group as a new set of source symbols. A source coding efficiency as close to 100% as desired can be obtained by use of the Shannon-Fano algorithm to encode an extension of high enough order.

8. Practical sources such as voice, text, and television signals are known to contain a large amount of natural redundancy. Due to the difficulty of modeling these sources, source coding for them is usually done on an ad hoc basis.

9. Many models for communication channels are of interest. The inputs and outputs can be discrete symbols, continuous values, or waveforms. The definition of the channel for a given system can be different depending on which aspect of the system is under consideration.

10. A discrete channel has a discrete set of input symbols and a discrete set of output symbols. If each output symbol depends only on one input symbol, a discrete channel is said to be memoryless. A discrete memory-less channel (DMC) is characterized by the list of input symbols, the list of output symbols, and the transition probabilities. For example, a DMC results when antipodal signals are used with additive white gaussian noise and the receiver makes hard decisions.

11. The average uncertainty associated with an input symbol of a DMC after the output symbol is known can be found by averaging $-\log P(x_i|y_j)$ over all combinations of input and output symbols. The difference between the original uncertainty associated with the input source and the uncertainty after the output symbol is known is a measure of the information transferred by each symbol transmitted.

12. The information transferred through a DMC by each symbol is a function of the probabilities of the source symbols. The capacity of such a channel is the maximum value of information transferred per symbol. (The maximization is with respect to the source probabilities.)

13. The channel capacity theorem states that the capacity of a channel sets a limit on the rate at which information can be transferred through the channel but not on the accuracy. In principle, coding schemes exist that enable information to be transferred at rates up to the channel capacity with as small an error probability as desired. The converse of the channel

10 / INFORMATION THEORY

capacity theorem states that no coding schemes exist that enable information transfer at a rate greater than the channel capacity with vanishing error probability.

14. Proof of the channel capacity theorem is an existence argument that does not show how the capacity can be realized with practical coding schemes.

REFERENCES

R. E. Blahut. *Principles and Practice of Information Theory*. Reading, Massachusetts: Addison-Wesley, 1987.

R. W. Hamming. *Coding and Information Theory*. Englewood Cliffs, New Jersey: Prentice-Hall, 1980.

K.S. Shanmugam. *Digital and Analog Communication Systems*. New York: Wiley, 1979.

PROBLEMS

1. The probability of a particular message is $\frac{1}{16}$. Find the information associated with this message in bits, nats, and Hartleys.

2. A discrete memoryless source emits the symbols A, B, C, D, and E with probabilities $\frac{1}{2}$, $\frac{1}{4}$, $\frac{1}{8}$, $\frac{1}{16}$, and $\frac{1}{16}$, respectively. Find the information content of each symbol and the average information content of a symbol from this source.

3. When it is raining, it is cloudy 99.9% of the time. On the other hand, the sun shines during 0.1% of the occasions that it rains. Given that you know that it is raining, how much additional information do you receive by learning that it is (a) cloudy, (b) sunny?

4. Two dice are available, a red one and a green one. One of the dice is selected at random and tossed. How much information do you receive when you learn which color die was selected? How much information do you receive when you learn how many spots turn up on the toss? How many outcomes does this experiment have? What is the probability of each outcome and the information received by learning which outcome occurred? Notice that the information obtained in learning the outcome is the sum of the information obtained by learning the color and the information obtained by learning the number of spots.

5. A card is drawn from a deck of playing cards. How much information is gained by learning only the suit of the card drawn? How much information is gained by learning only the value (i.e., ace, two, . . . , queen, king) of the card drawn? How much information is gained by learning both the suit and the value?

6. A monochrome television picture consists of 525 lines, each containing approximately 700 picture elements (pixels). Thirty pictures are transmitted per second. Assuming that each pixel can take on 256 brightness levels, what is the maximum information rate of a monochrome television signal in bits per second? (Actually, television signals are very redundant, with one picture often very similar to the last and with large areas of a picture at the same brightness. Therefore, the required information rate for typical program material is much less than the maximum value.)

7. Consider an information source with three symbols x_1, x_2, and x_3 with probabilities p_1, p_2, and $1 - p_1 - p_2$. Write an expression for the average information content of this source. Find the values of p_1 and p_2 that maximize the average information content by setting the partial derivatives of $H(X)$ with respect to p_1 and p_2 equal to zero and solving.

8. Use the Shannon-Fano algorithm to encode the symbols of Problem 2. Find the efficiency of the resulting code.

9. Consider a source having symbols A, B, and C with probabilities $\frac{3}{4}$, $\frac{1}{8}$, and $\frac{1}{8}$. Use the Shannon-Fano algorithm to encode these source symbols and find the efficiency of the code. Repeat for the second-order extension of the source.

10. If the average information per symbol for a discrete source is $H(X)$ and successive symbols from the source are independent, show that the average information per combined symbol for the nth-order extension of the source is $nH(X)$.

11. An alternative source coding scheme is to arrange the source symbols in the order of decreasing probability and assign 0 to the first (most probable) symbol, 10 to the second symbol, 110 to the third, and so on. The last code word is M ones, where M is the number of source symbols. Use this coding scheme to encode the source of Problem 9. Find the efficiency of the resulting code. Repeat for the second-order extension of the source. Show that a code generated in this manner is uniquely decipherable.

12. Consider the DMC depicted in Figure P10.1a. If the source symbols A and B have probabilities p and $1 - p$, respectively, find an expression for the information

FIGURE P10.1

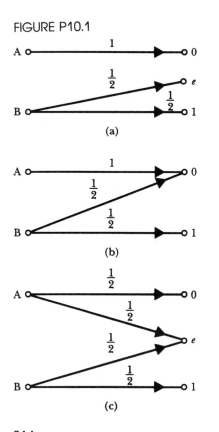

(a)

(b)

(c)

transferred per symbol transmitted through the channel. Find the value of p that maximizes the information transfer and the resulting channel capacity. Repeat for the channels of Figures P10.1b and c.

13. Find the capacity of a channel that has outputs independent of its inputs.

14. Estimate the information rate of spoken English by assuming a rate of 2 words/s, 6 letters/word, and that the letters are independent and equally likely.

15. The minimum channel requirements for transmission of intelligible speech are approximately a 3.5 kHz bandwidth and a received SNR of 30 dB. What is the capacity of this channel? Compare this with the result of the previous problem and comment.

16. A certain start-up company has made an announcement that it has made a breakthrough in data communications. It will soon market a modem (modulator-demodulator) capable of transmitting data at the rate of 100 kbit/s over channels such as ordinary voice grade telephone lines. Voice grade telephone lines are typically limited to a bandwidth of 3500 Hz and have a received SNR of 30 dB. The company claims to have demonstrated their new modem in the laboratory for a gaussian noise channel limited to a bandwidth of 3500 Hz and having a SNR at the channel output of 30 dB. Do you recommend buying stock in this company? Why or why not?

17. A voice signal is to be sampled, converted to digital form, and transmitted using PCM. If the signal is limited to a highest frequency of 5 kHz, what is the minimum sampling rate? If the ratio of peak to rms values of the signal is ten, how many bits per sample must be used to represent each sample if a uniform quantizer is used and a SNR of 30 dB is wanted after the signal is reconstructed? (Hint: See Chapter 6.) Neglect the effect of any bit errors in transmission. What is the data rate of the resulting data stream? If these data are to be transmitted through a channel having a bandwidth of 2 kHz, what is the SNR at the output of the channel if the channel is operating at capacity?

18. A continuous channel is modeled in Figure P10.2. If the signal power is 1 W and the noise PSD is 10^{-3} W/Hz, plot the capacity of the channel as a function of the filter bandwidth.

19. The error probability of a binary-symmetric channel is 0.1. Find the probability of one or more errors in a five symbol code word. Repeat for 10 or more errors in a 50 symbol code word. Comment.

20. A (15,10) binary code is to be devised by selecting the symbols of the code words at random. Find or estimate the probability that two or more code words are the same. Repeat for a (150,100) code.

FIGURE P10.2

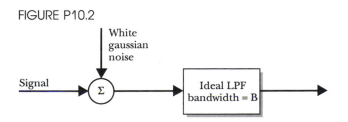

11

SPREAD-
SPECTRUM
COMMUNICATIONS

So far we have taken the attitude that the signals used in a communication system should occupy the minimum possible bandwidth because spectral space is limited, and many demands are made for it. Now we consider spread-spectrum techniques, which purposefully expand bandwidth to the maximum. This spreading process is accomplished by the use of codes in such a way that it can be reversed by the receiver. In military applications, the code is chosen so that it is difficult to predict and cannot be easily despread (reversed) by unintended receivers.

Two types of spread-spectrum techniques are in wide use and will be discussed in this chapter. They are *direct sequence* (DS) [also known as *pseudonoise* (PN)] and *frequency hop* (FH) spread spectrum. Other types of spread spectrum, known as time hop, chirp, and hybrid combinations of techniques, exist but will not be discussed here.

In a DS system, a narrowband data signal is phase modulated by a very high rate pseudorandom code sequence. The data signal can be any of the digital or analog modulation signals discussed earlier in the book. The phase modulation by the PN code is typically PRK, QPSK, or OQPSK. Due to limited space, we will confine our discussion to the case in which both the data modulation and spreading code modulation are PRK.

Frequency-hop spread spectrum is accomplished by using the spreading code to select the carrier frequency. The resulting carrier frequency rapidly switches over a very large number of values spread over a very wide range. The data or message signal is modulated onto this frequency-hopped carrier. Various types of data modulation are possible, but we will confine our discussion to the case in which binary or M-ary FSK is used.

The potential benefits of spread-spectrum techniques include:

1. Antijamming (AJ), by means of which a jammer is forced to use a higher power signal to disrupt communication
2. Low probability of detection (LPD), in which an unauthorized receiver cannot reliably detect even the presence of the communication signal
3. Code division multiple access (CDMA), in which many transmitters can simultaneously use the same part of the spectrum without cooperation or mutual interference
4. High-resolution ranging, in which the distance from the transmitter to the receiver can be accurately determined and, if a number of transmitters at known locations are in use the receiver can be located
5. Rejection of interfering multipath signals
6. Accurate universal timing

Although spread spectrum can improve performance against fixed power interfering signals such as multipath or intentional jammers, no improvement against white noise is achieved because the power of white noise in the band of the signal increases with bandwidth.

In Chapter 8, we saw that signals can be represented as vectors in signal space. Spread-spectrum techniques for the purpose of antijamming or low probability of detection use a signal set that occupies a very large number of dimensions in signal space. The location of the signal varies from one signaling interval to another in a pseudorandom way that is predictable by the intended receiver but not by the jammer or unintended receiver. Forcing the jammer to expend energy to jam a large number of unused signal-space dimensions for every signaling interval minimizes the amount of jammer energy that falls in the portion of signal space occupied by the signal. In the case of LPD, the unintended receiver is forced to accept many noise components from potential signal-space dimensions, which reduces the SNR for the unintended receiver and makes the determination of signal presence difficult. Thus, a key feature of spread-spectrum signals for these applications is to potentially occupy a very large number of dimensions in signal space.

Consider a set of signals that have a duration of T seconds and have most of their energy confined to a bandwidth of B Hz. It can be shown that the number of signal-space dimensions occupied by this signal set is approximately

$2BT$ (Wozencraft and Jacobs, 1965, p. 196). Thus, to have a large number of available dimensions for signals of fixed duration, it is necessary to use a large bandwidth. This is why modulation techniques for AJ and LPD use a large bandwidth. It also is important to notice that the mere use of large bandwidth is not sufficient for AJ or LPD purposes. If signals with large bandwidth lie in predictable regions of signal space, the jammer or unintended receiver can concentrate on this region. The important thing is to use large bandwidth in such a way that any of the $2BT$ dimensions can be used unpredictably.

11.1
PSEUDORANDOM SEQUENCE GENERATORS

In either DS or FH spread-spectrum techniques, a pseudorandom code spreads the signal at the transmitter and reverses the spreading at the receiver. Codes that are well suited for this application should have the following properties:

1. They are easy to generate.
2. They appear to be random to an unintended receiver.
3. They are either aperiodic or have very long periods.

Linear feedback shift register sequences

Linear feedback shift register (**LFSR**) sequences exist that come close to meeting these requirements. A generator for these sequences is shown in Figure 11.1. The tap coefficients $a_0, a_1, a_2, \ldots, a_r$, are either 0 or 1. (Actually, a_0 is associated with the feedback path and is not a tap coefficient. In our discussion, a_0 will always be a 1, so we could have dispensed with its use. We have included it

FIGURE 11.1
Linear feedback shift register. If $a_j = 1$, the jth stage is connected to the summer. If $a_j = 0$, the jth stage is not connected. Note: a_0 will always be 1.

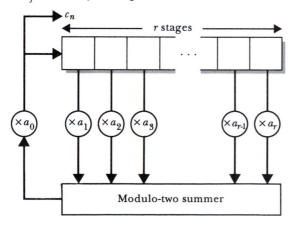

because it occurs in a more detailed discussion of LFSRs based on the theory of polynomials with binary coefficients. Inclusion of a_0 in our discussion should give the reader a smoother transition to the more advanced literature.)

In Figure 11.1, the content of the shift register stages that have a coefficient of 1 are modulo-two added and the result is fed back to the input of the shift register. The successive symbols of the code, denoted by c_n, are given in terms of the earlier symbols by the recursion relation

$$c_n = \sum_{k=1}^{r} a_k c_{n-k} \tag{11.1}$$

Notice that the content of the shift register can only be one of 2^r possible states. If an LFSR reaches the all-zero state, it stays in the that state indefinitely. Starting from some state other than all zeroes, the register must eventually return to a previous state after some number of shifts. Then the shift register cycles through a periodic sequence of states time after time.

EXAMPLE 11.1

Write the recursion relation and find the periodic sequences of states for the LFSR of Figure 11.2a.

SOLUTION
The LFSR has $r = 3$ and $a_0 = a_1 = a_2 = a_3 = 1$. Therefore from Equation 11.1, the recursion relation for the code sequence is

$$c_n = c_{n-1} + c_{n-2} + c_{n-3}$$

For this short register, we can find the cycles simply by considering the result of loading the register with a bit pattern and then shifting to find the

FIGURE 11.2
Example LFSR with $r = 3$ and $a_0 = a_1 = a_2 = a_3 = 1$, including its state cycles.

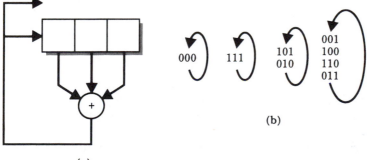

(b)

(a)

resulting cycle. As for any LFSR, if the register is loaded with the all-zero pattern, it remains in that state as indicated in Figure 11.2b. If the register is loaded with the pattern 001, it cycles through three additional states before returning to the 001 state. If the register is loaded with 101, it cycles to 010 and then back to 101. Finally, if loaded with the all-ones pattern, this particular register remains in that state. Thus, as indicated in Figure 11.1b, the register has four cycles of states. Two cycles are of length one, one cycle is of length two, and one cycle is of length four. Thus, the register can produce any of the following output sequences:

$$00000000 \cdots$$
$$11111111 \cdots$$
$$10101010 \cdots$$
$$11001100 \cdots$$

The particular binary sequence generated depends on the initial state of the register.

Maximal-length linear feedback shift register sequences

Certain selections of the tap coefficients of an LFSR lead to a sequence of register states containing all of the possible nonzero states. The binary sequence produced is known as a *maximal-length* sequence or simply as an m-sequence. The length of an m-sequence is $2^r - 1$ where r is the register length. (The m-sequence cannot contain the all-zero state because the register stays in the all-zero state.) M-sequences can be shown to have the following properties:

1. The sequence contains 2^{r-1} ones and $2^{r-1} - 1$ zeroes. Thus, for long shift registers, the sequence contains nearly half zeroes and half ones.
2. The modulo-two sum of any m-sequence and a time-shifted version of the same sequence is the same sequence with a third time shift. (The time shift must be a multiple of the symbol interval.)
3. If an r-symbol window is slid along the sequence, we observe each of the possible r-symbol patterns (except the all-zero pattern) exactly once.
4. A run is a sequence of identical symbols. An m-sequence has
   (a) one run of ones of length r,
   (b) one run of zeroes of length $r - 1$,
   (c) one run of zeroes and one run of ones of length $r - 2$,
   (d) two runs of zeroes and two runs of ones of length $r - 3$,
   (e) four runs of zeroes and four runs of ones of length $r - 4$,
   $$\vdots$$
   (k) 2^{r-3} runs of zeroes and 2^{r-3} runs of ones of length one.

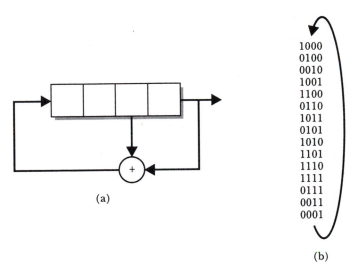

(a)

1000
0100
0010
1001
1100
0110
1011
0101
1010
1101
1110
1111
0111
0011
0001

(b)

FIGURE 11.3
A four-stage maximal-length LFSR and its cycle.

EXAMPLE 11.2

Demonstrate the stated properties for the sequence generated by the maximal-length register of Figure 11.3a.

SOLUTION
The sequence of register states for this LFSR is shown in Figure 11.3b. If the output sequence is taken from the right-hand stage, we obtain the sequence

$$\underbrace{000100110101111}000100 \cdots$$

one period

We can see that the sequence repeats after $2^4 - 1 = 15$ symbols and that each period contains 8 ones and 7 zeroes.

The shift-and-add property can be illustrated as follows:

$$00010011010111110001001101 \cdots$$
$$0001001101011110001001 \cdots$$
$$100010011010111110001000100 \cdots$$

The first line is the original sequence, the second line is the sequence shifted three symbols to the right, and the bottom line is the symbol-by-symbol modulo-two sum of the top two lines. The bottom line can be recognized as the original sequence shifted four symbols to the right.

Examination of the 15-symbol sequence shows that, as required by the run property, it has one run of ones of length four, one run of zeroes of length three, one run of zeroes of length two, one run of ones of length two, two runs of zeroes of length one, and two runs of ones of length one.

Notice that this distribution of runs is similar to what we would expect on the average for a truly random sequence. About half of the symbols are zeroes and about half are ones. A 1 is followed by a 0 half of the time, a 1 is followed by a 1 half of the time, and so forth.

M-sequences have many of the properties desired for a spreading code. Certainly, they are easy to generate. They approximate many of the properties of a truly random code with respect to run lengths and the average number of ones and zeroes. They are periodic, but the period can easily be made very long. For example, if a 100-bit shift register is used and the shift rate is 1 GHz, there are $2^{100} - 1 \simeq 1.27 \times 10^{30}$ symbols in the code and the period is approximately 4×10^{13} years!

Even though m-sequences approximate many of the properties of a random sequence, they have one major weakness. An entire LFSR sequence can be predicted from knowledge of only $2r$ symbols where r is known to be the longest possible value of the register. This is because the recursion relation of Equation 11.1 can be solved for the tap coefficients. Then, once the feedback connection is known, the remainder of the sequence can be predicted.

EXAMPLE 11.3

Suppose that the sequence 11001000 is produced by an LFSR of length not more than four. Find the values of the feedback coefficients and predict the remainder of the sequence.

SOLUTION

The shift register is known to be of the form shown in Figure 11.4a. The recursion relation of Equation 11.1 for $r = 4$ is

$$c_n = a_1 c_{n-1} + a_2 c_{n-2} + a_3 c_{n-3} + a_4 c_{n-4}$$

The first five symbols in the sequence given are 11001, where the last symbol is produced from the first four by the recursion relation. Thus, we can write

$$1 = 0a_1 + 0a_2 + 1a_3 + 1a_4$$

Using the second through fifth symbols, we can write

$$0 = 1a_1 + 0a_2 + 0a_3 + 1a_4$$

Similarly, using the next five consecutive symbols, we have

$$0 = 0a_1 + 1a_2 + 0a_3 + 0a_4$$

11 / SPREAD-SPECTRUM COMMUNICATIONS

The last five symbols result in

$$0 = 0a_1 + 0a_2 + 1a_3 + 0a_4$$

Thus, we have obtained four linear equations for the four tap coefficients. These equations can be easily solved to find

$$a_1 = 1, \qquad a_2 = 0, \qquad a_3 = 0, \qquad a_4 = 1$$

Thus, the LFSR takes the form shown in Figure 11.4b. As the last of the given symbols is being emitted, the state of the shift register is the preceding four symbols in reverse order. Therefore, the successive states can be predicted as shown in Figure 11.4c. Thus the sequence generated is

given symbols↘   ⌐predicted symbols

1100100011110101100 · · ·

during this symbol the register is in state 0010

It turns out in this case that the sequence is an m-sequence. (If only a three-stage register had been necessary to generate the given sequence, we would have found that $a_4 = 0$.)

This example shows that for an r-stage LFSR we can find the feedback connections and predict the entire sequence by knowing only a short segment ($2r$ symbols) and solving a set of r linear equations. Even for a long shift register

FIGURE 11.4
Four-stage LFSR of Example 11.3.

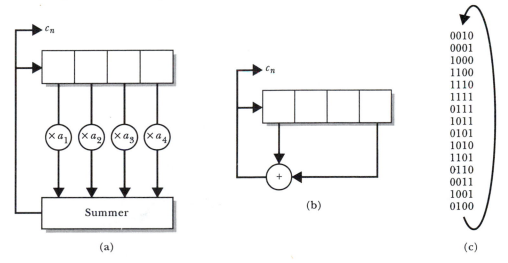

(a)

(b)

(c)

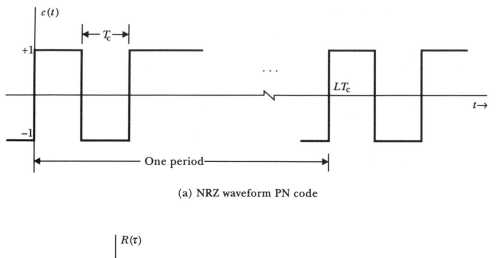

(a) NRZ waveform PN code

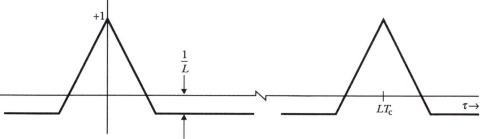

(b) Autocorrelation function

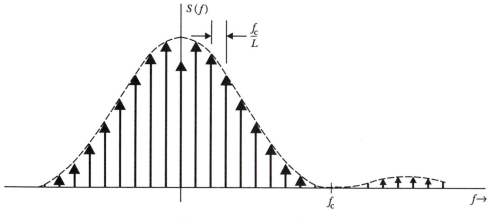

(c) PSD of maximal length PN Code.

FIGURE 11.5
Maximal-length PN code, its autocorrelation function, and its PSD.

of, say, 100 stages, this could be accomplished in a short time on a simple computer. Thus, an LFSR sequence may not always be a good choice for a military system, in which it is important to eliminate the possibility of an adversary predicting the code. It may be possible to avoid this problem by using nonlinear elements (AND gates) in the feedback path. No doubt, more information on this subject is available to those with access to the appropriate classified literature. Actually, since the signal is often available to an adversary only in a form with low SNR, it may be difficult for the adversary to obtain $2r$ consecutive correct code symbols.

Autocorrelation and the power spectral density of maximal-length sequences

Assuming that the logic levels of the maximal length code signal are ± 1 as indicated in Figure 11.5a, the autocorrelation function of the code signal is as shown in Figure 11.5b. Notice that this is similar to the autocorrelation function of an NRZ signal with random symbols as shown in Figure 7.20. There are two significant differences between the autocorrelation functions of an m-sequence and a random sequence. First, the autocorrelation function of the m-sequence is periodic because the m-sequence is itself periodic. The other difference is that the autocorrelation function of the m-sequence is not zero between correlation peaks. However, both of these differences become negligible as the register length becomes long.

The PSD of the m-sequence can be found by taking the Fourier transform of the autocorrelation function. Since the autocorrelation function is periodic, this is accomplished by first finding the Fourier series and then transforming the series term by term. The resulting PSD is given by

$$S(f) = \left[\sum_{m=-\infty}^{\infty} \delta\left(f - \frac{mf_c}{L}\right) \right] \frac{L+1}{L^2} \operatorname{sinc}^2\left(\frac{f}{f_c}\right) - \frac{1}{L}\delta(f)$$

where f_c is the symbol rate of the sequence and $L = 2^r - 1$ is the length of the sequence. The quantity f_c/L is the repetition frequency of the sequence. The PSD of m-sequence is shown in Figure 11.5c. Note that when the sequence is long, the frequency components become very close together. Thus, for very long periods, the autocorrelation function and PSD of an m-sequence approximate those of a random sequence.

Connection coefficients for maximal-length linear feedback shift register sequences

Selected connection coefficients for maximal-length shift registers of various lengths are shown in Table 11.1. Many additional connection coefficients are known and can be found in the literature.

TABLE 11.1
Connection coefficients of maximal-length shift register generators

Register length	Octal representation of connection coefficients[a]
2	[7]
3	[13]
4	[23]
5	[45], [75], [67]
6	[103], [147], [155]
7	[203], [211], [217], [235], [367] $\cdots$
8	[435], [453], [537], [545], [551] $\cdots$

[a] The coefficients are given in octal form (i.e., $0 = 000$, $1 = 001$, $2 = 010, \ldots$, $7 = 111$), a_r is the leftmost nonzero bit, and a_0 is the rightmost bit.

EXAMPLE 11.4

Use Table 11.1 to find a maximal-length shift register connection for $r = 6$ and verify that the length of the sequence is $L = 63$.

SOLUTION
From Table 11.1 we find that one choice of the connection coefficients in octal form is [103]. Converting each octal symbol to its binary equivalent gives (001 000 011), in which we have left the spaces for clarity. Since we are considering a six-bit register, seven coefficients are needed, a_0 through a_6. Thus, the leftmost pair of zeroes of the binary string are neglected, and then we have the following:

$a_6 = 1$ corresponding to the third bit from the left in the binary string
$a_5 = 0$ corresponding to the fourth bit in the binary string
$a_4 = 0$
$a_3 = 0$
$a_2 = 0$
$a_1 = 1$
$a_0 = 1$ corresponding to the rightmost bit in the binary string

Using these coefficients with the generic diagram of Figure 11.1 and noting that a zero coefficient can be accomplished by no connection whereas a one coefficient can be accomplished by a connection to the adder, we obtain the circuit shown in Figure 11.6a.

If this circuit is loaded with the pattern 000001, successive register states are as indicated in Figure 11.6b. It can be verified that the register returns to the original state after 63 shifts.

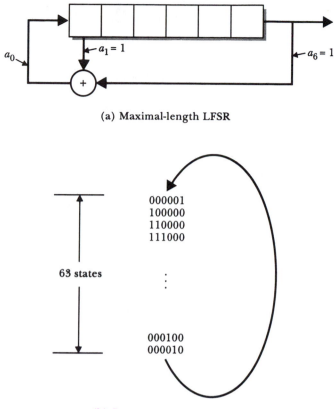

(a) Maximal-length LFSR

(b) Sequence of register states

FIGURE 11.6
Maximal-length LFSR based on the tap coefficients [103].

Thus, we have seen that maximal-length LFSRs provide a practical means for generating pseudorandom sequences with many properties that closely approximate those of a random sequence, provided that a long register length is used. However, we have also seen that these sequences can be predicted if an error-free segment of twice the register length is available.

11.2
DIRECT SEQUENCE SPREAD-SPECTRUM SYSTEMS

A PRK-PRK system

In this section, we discuss a direct sequence system that uses **PRK** for both the data modulation and the spreading-code modulation. We briefly illustrate how such a system can be used to achieve the benefits of spread spectrum listed in the introduction to this chapter.

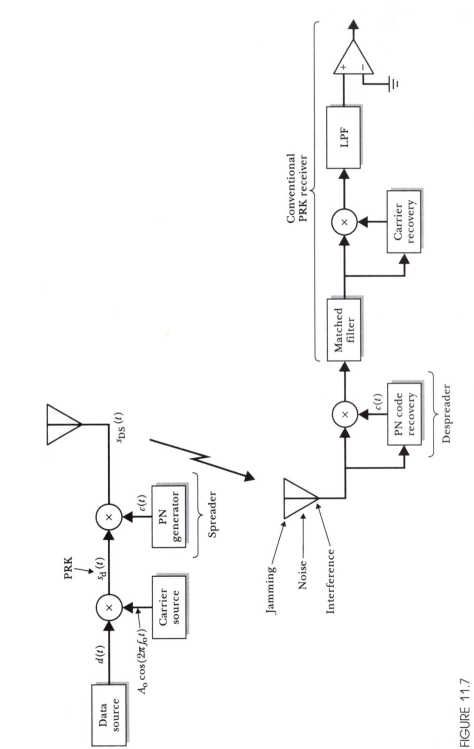

FIGURE 11.7
A direct sequence spread-spectrum system using PRK modulation.

The block diagram of the spread-spectrum system to be discussed is shown in Figure 11.7. The data source is assumed to produce a ± 1 NRZ waveform, denoted by $d(t)$ and illustrated in Figure 11.8a. We also assume that the data sequence is equally likely to be $+1$ or -1 and is independent from bit interval to bit interval.

FIGURE 11.8
Typical data and spreading-code waveforms for direct sequence spread spectrum.

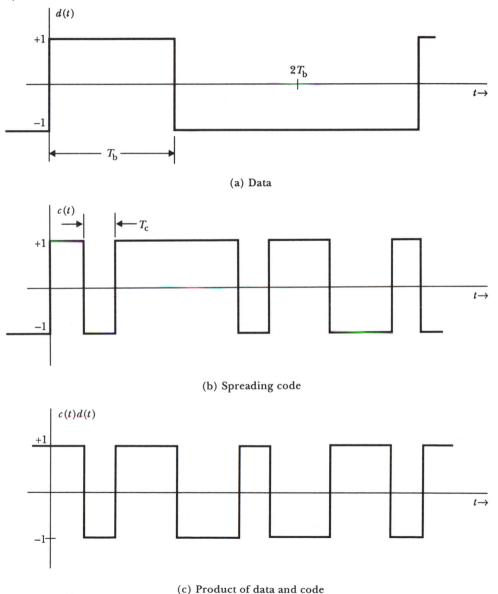

(a) Data

(b) Spreading code

(c) Product of data and code

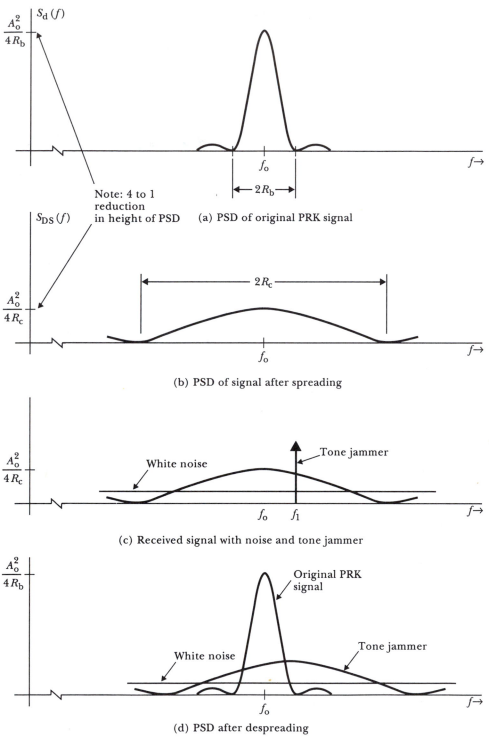

$\dfrac{A_o^2}{4R_b}$ $S_d(f)$

f_o

$\leftarrow 2R_b \rightarrow$

$f \rightarrow$

Note: 4 to 1
reduction
in height of PSD   (a) PSD of original PRK signal

$S_{DS}(f)$

$\dfrac{A_o^2}{4R_c}$

$\leftarrow \qquad 2R_c \qquad \rightarrow$

f_o

$f \rightarrow$

(b) PSD of signal after spreading

$\dfrac{A_o^2}{4R_c}$

Tone jammer

White noise

f_o f_1

$f \rightarrow$

(c) Received signal with noise and tone jammer

$\dfrac{A_o^2}{4R_b}$

Original PRK
signal

Tone jammer

White noise

f_o

$f \rightarrow$

(d) PSD after despreading

FIGURE 11.9
PSDs at various points in a direct sequence spread-spectrum system with a tone
jammer.

The data signal is PRK modulated onto the carrier $A_o \cos 2\pi f_o t$, resulting in the data-modulated signal given by

$$s_d(t) = A_o d(t) \cos 2\pi f_o t \qquad (11.2)$$

This PRK data-modulated signal has the PSD, denoted by $S_d(f)$, shown in Figure 11.9a. (Only one side of the PSD is shown because the PSD is symmetrical about $f = 0$.) Notice that this PSD has a null-to-null bandwidth equal to twice the data rate and a height inversely proportional to the data rate.

The PRK signal is then multiplied by the spreading code $c(t)$ as shown in Figure 11.7. The spreading code is assumed to take values of ± 1 with equal likelihood and independently in each T_c interval. Each symbol of the spreading code is called a *chip*. The symbol rate of the spreading code, or chip rate, is given by

$$R_c = \frac{1}{T_c} \qquad (11.3)$$

In practical spread-spectrum systems, the chip rate is typically several orders of magnitude larger than the data rate. The spreading-code waveform shown in Figure 11.8b has a chip rate equal to four times the data rate. We have used this low spreading rate simply to allow the waveforms and spectra of the signals before and after spreading to be drawn to the same scale so that the effects of spreading can be clearly illustrated.

The spread signal obtained by multiplying the PRK data-modulated signal by the spreading code is given by

$$s_{DS}(t) = A_o d(t) c(t) \cos 2\pi f_o t \qquad (11.4)$$

The product of the data signal and the spreading code, $d(t)c(t)$, has the same statistical characteristics as the spreading code by itself. Since we have assumed that the transitions of the data and the spreading code are aligned, the effect of multiplying the spreading code by the data signal is simply to invert some of the chips, as illustrated in Figure 11.8c. Therefore, the symbol rate of the product $d(t)c(t)$ is the same as the chip rate, and the symbols of the product are independent and equally likely to be ± 1. The spread signal is a carrier that has been PRK modulated by the product $d(t)c(t)$. Thus, the PSD of the spread signal is the same as the PSD of the data-modulated signal except that the data interval T_b is replaced by the chip interval T_c. The PSD of the spread signal is shown in Figure 11.9b for the case in which $T_c = T_b/4$. Notice that spreading has increased the null-to-null bandwidth and reduced the height of the PSD by a factor of four.

The received signal consists of an attenuated and delayed version of the transmitted spread signal plus noise, interference, and jamming signals added in the channel. For convenience, we neglect the attenuation and delay. In

Figure 11.9c, we show the PSD of the received signal including white noise and an impulse at a frequency denoted by f_1 representing a *tone jammer*. A tone jammer is designed to disrupt a communication system simply by transmitting a tone in the frequency band used by the system. Thus, the received signal, including the tone jamming and white noise, is given by

$$r(t) = A_o c(t)d(t) \cos 2\pi f_o t + A_j \cos 2\pi f_1 t + n(t) \qquad (11.5)$$

In the receiver, a locally generated spreading code is produced that is identical to the code used at transmitter. Phase-locked loop techniques are used to time align the local spreading-code generator output with the code imbedded in the received signal. This technique will be discussed further in Section 11.4. The product of the received signal and the locally generated spreading code is given by

$$r(t)c(t) = A_o c^2(t)d(t) \cos 2\pi f_o t + A_j c(t) \cos 2\pi f_1 t + c(t)n(t) \qquad (11.6)$$

Because the spreading code $c(t)$ takes values of ± 1, the square of the spreading code is unity at all times, and multiplication of the transmitted spread signal by $c(t)$ in the receiver reverses the spreading process. Thus, the first term on the right-hand side of Equation 11.6 reduces to the original PRK data-modulated signal. The action of the second multiplication by the spreading code is to despread the signal.

On the other hand, the jamming signal is spread by multiplication with the spreading code in the receiver. This is shown by the second term on the right-hand side of Equation 11.6.

The total power in the noise signal $n(t)$ is unaffected by multiplication with $c(t)$ since $c(t)$ takes only the values ± 1. Provided that two signals are independent, the PSD of their product can be shown to be the convolution of their individual PSDs. Since the convolution with a constant, such as the PSD of white noise, results in a constant, we conclude that the PSD of $c(t)n(t)$ is constant and has the same PSD as the input noise alone. The PSDs of the three components of the despreader output of Equation 11.6 are illustrated in Figure 11.9d. Notice that the desired signal power has been collapsed so that it falls into a bandwidth approximately equal to the data rate (the null-to-null bandwidth is twice the data rate), the tone jammer power has been spread over a bandwidth about equal to the spreading rate, and the white noise PSD has been unaffected by the despreader.

Processing gain against a tone jammer

When the ratio of the chip rate to the data rate is very large and the jammer frequency f_1 is close to the carrier frequency f_o, the PSD of the jammer contribution at the output of the despreader is approximately constant in the fre-

11 / SPREAD-SPECTRUM COMMUNICATIONS

quency range of the desired signal and is given by

$$S_j(f) \simeq \frac{A_j^2 T_c}{4} = \frac{P_j}{2R_c} \tag{11.7}$$

where P_j is the total received jammer power. The noise bandwidth of the matched filter for the desired despread PRK signal is R_b. Therefore, the despread jammer power that passes through the matched filter (see Figure 11.7) is

$$P_{jo} = S_j(f)\, 2\mathrm{R_b} \simeq P_j \frac{R_b}{R_c} \tag{11.8}$$

where the factor of two is necessary to account for the two-sided nature of the PSD. Thus, only a small fraction of the jammer power succeeds in passing through the matched filter to potentially disrupt the data decisions. This is due to the fact that in despreading the desired signal the jammer is spread, so only a small fraction of its power is in the frequency range of the desired signal. The ratio R_c/R_b is called the *processing gain* of the system. It gives a measure of the ability of the system to reject jammer signals.

Thus, we have demonstrated that direct sequence spread spectrum can help to reject the power of a tone jammer. We have also seen that white noise is not affected by the despreading process, so the concept of processing gain does not apply to it. Many other jamming signals are possible. For example, the jamming signal can be pulsed, multiple tones, or a noise-like signal. Various combinations of modulation formats for the data and spreading-code modulation are useful. The study of the noise rejection capability of all these systems is more extensive than space permits here.

Application to covert communication systems

Another application of spread-spectrum techniques is in *covert communications*, in which it is desired to carry on communication in such a manner that it is difficult for an unintended receiver to reliably detect the presence of the communication signal with a low false-alarm probability. A false alarm occurs when the detection receiver indicates the presence of a communication signal when in fact, none is present. Direct sequence spread spectrum is useful to the communicator in this situation because if a very high processing gain is used, the PSD of the spread signal at the input to the detection receiver can be reduced below the noise level. Thus, the signal is obscured by the noise. This is also the case for the intended receiver, but the intended receiver has knowledge of the spreading code in use and can despread the signal so that it rises above the noise level at the output of the despreader. This is illustrated in Figure 11.9, except that the figure is drawn for a processing gain of only four. When a processing gain of 1000 or higher is used, the peak value of the PSD of the spread signal can be much below the level of the white noise at the input to the receivers and much above the noise level after despreading. Of course, to be successful, the communicator must keep knowledge of the spreading code from the designer of the detection receiver. Ideally, the communicator should use a modulation format for spreading that results in a spread signal as nearly as possible indistinguishable from thermal noise.

Code division multiple access applications

Another application for direct sequence spread-spectrum techniques is code division multiple access (CDMA). In this type of system, multiple users can simultaneously transmit signals in the same frequency band without disrupting one another. If a PRK direct sequence spread-spectrum signal is used for this purpose, the received signal is given by

$$r(t) = \sum_{i=1}^{N} A_i d_i(t) c_i(t) \cos (2\pi f_o t + \theta_i) + n(t) \tag{11.9}$$

We assume that signals from N users are received. A_i is the amplitude of the ith signal, $d_i(t)$ is the data from the ith user, $c_i(t)$ is the spreading code used by the ith user, and θ_i is the phase of the ith signal. The thermal and receiver noise is represented by $n(t)$. Since this type of system is often intended for use without coordination among users, the transitions of the various code sequences are not assumed to be synchronized.

A receiver for this type of CDMA signal is shown in Figure 11.10, in which it is assumed that the desired signal is the jth signal in the summation of Equation 11.9. The despreader multiplies the received signal by the spreading code used by the desired transmitter. Of course, for proper despreading it is necessary for the despreading code to be time aligned with the code imbedded in the desired component of the received signal. Thus, the desired component at the output of the despreader is

$$A_j d_j(t)[c_j(t)]^2 \cos (2\pi f_o t + \theta_j) = A_j d_j(t) \cos (2\pi f_o t + \theta_j) \tag{11.10}$$

FIGURE 11.10
Receiver for direct sequence CDMA.

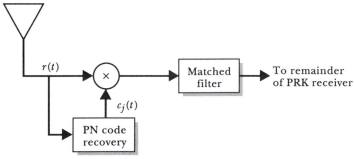

The right-hand side of this expression follows because of the fact that the square of the spreading code is unity. Thus, the despreader output contains the desired PRK-modulated data signal.

The output of the despreader due to one of the undesired signals is given by

$$A_k d_k(t) c_k(t) c_j(t) \cos\left(2\pi f_0 t + \theta_k\right) \tag{11.11}$$

Provided that the two pseudorandom spreading codes are independent, their product will be a waveform that switches pseudorandomly between $+1$ and -1 at the spreading rate (or even higher if, as is usually the case, the spreading-code transitions are not time aligned). Thus, the output of the despreader due to the undesired signals is still spread over a wide bandwidth, and most of its power is not passed by the matched filter. If a high processing gain is used and the spread signals are below the level of the noise, the small amount of power passed by the matched filter due to a single undesired user will not increase the error probability of the system significantly.

Of course, as more users are added to a CDMA system, the interference level becomes higher. Thus, there is a limit to the number of users that can be accommodated, just as there is in a frequency division multiple access (FDMA) system or in a time division multiple access (TDMA) system. An important basis for comparison of the various multiple access techniques is the number of users that can be simultaneously accommodated with a given bandwidth, transmitter power, and error performance. Even with the best-known spreading codes, CDMA seems to be significantly inferior to the other techniques in this respect unless time synchronization of the codes is employed.

One advantage of CDMA, compared with FDMA or TDMA, is that it is not necessary to coordinate the transmissions of the users. With FDMA or TDMA, a user must select a frequency or time slot that is not being used. With CDMA, provided that each user is assigned a different spreading code that is independent of the others, so that the code waveforms are essentially orthogonal, each transmitter can be turned *on* or *off* independently without coordination of time or frequency.

A significant problem with direct sequence CDMA, is the *near-far problem*. This comes about because, in many systems, the transmitters and receivers are geographically distributed. Furthermore, the desired transmitter may be much farther from the receiver than some of the undesired transmitters. Thus, the amplitude of some of the undesired signals can be much higher than the desired signal. This can be such a severe problem that the processing gain required for direct sequence CDMA is impractical.

Direct sequence CDMA is probably most useful when (1) the distances from the transmitters to the receivers are all the same so that the signal levels are nearly all equal, (2) many users want to have access to the system but only use the system a very small percentage of the time, and (3) some of the

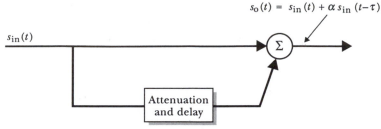

FIGURE 11.11
Simplified model of multipath communication channel.

other benefits (such as antijamming) provided by the use of spread spectrum are desired.

Application to multipath channels

Another problem that can potentially be solved by the use of spread-spectrum techniques is *multipath*. In some communication systems such as mobile radio in an urban environment, the signal travels from the transmitter to the receiver by a multiplicity of paths having different attenuations and delays. A model of a simple example of such a communication channel is shown in Figure 11.11. In the figure, the input signal travels to the output by two paths. We have assumed that a direct path exists with no delay or attenuation and that another path exists by which the signal arrives at the output with delay and attenuation. In the case of an urban mobile radio communication link, the direct path could result from line-of-sight transmission from the transmitter to the receiver, and the delayed path could result from reflection of the electromagnetic wave from a building. In this case, both the direct and reflected paths would encounter delay and attenuation. However, if we normalize to the amplitude and delay of the direct path, the model of the figure applies, but the attenuation factor α represents the relative attenuation of the indirect path and τ represents the relative delay.

The impulse response of the channel shown in Figure 11.11 is given by

$$h(t) = \delta(t) + \alpha\delta(t - \tau) \tag{11.12}$$

The transfer function of the channel can be found by taking the Fourier transform of this impulse response. The result is given by

$$H(f) = 1 + \alpha \exp(-j2\pi f\tau)$$

$$H(f) = 1 + \alpha \cos 2\pi f\tau - j\alpha \sin 2\pi f\tau \tag{11.13}$$

The magnitude of this transfer function is given by

$$|H(f)| = [(1 + \alpha \cos 2\pi f\tau)^2 + (\alpha \sin 2\pi f\tau)^2]^{1/2} \tag{11.14}$$

This is plotted for several values of α in Figure 11.12. Notice that when α is close to unity, the transfer function has a very small gain at certain frequen-

11 / SPREAD-SPECTRUM COMMUNICATIONS

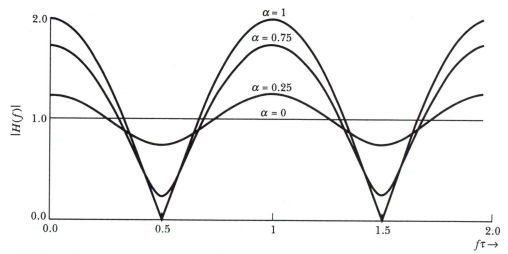

FIGURE 11.12
Transfer function of a simple multipath channel.

cies. Of course, this is because a transmitted signal at these frequencies is received from the two paths with nearly equal amplitude and out of phase, so
the received components nearly cancel. If a conventional narrowband communication system is used in a multipath environment and the frequency in
use falls at one of the nulls in the transfer function, the signal level can easily
fall below useable levels. This problem is particularly great in the case of urban
mobile radio, in which there are usually many paths due to reflections from
buildings. As the transmitter or receiver moves, the relative strengths and
delays of the various paths change, resulting in nulls in the corresponding
transfer function that move about in the frequency domain. Thus, the link
experiences deep fades from time to time. This can be a severe problem.

A narrowband signal is subject to fading caused by multipath because
its power is confined to a narrow range of frequencies. On the other hand, a
spread-spectrum signal has its power spread over a wide range; and even if a
deep fade falls in this range, part of the power will still be received. To illustrate
the advantage of direct sequence spread spectrum in the presence of multipath, assume that the direct sequence signal of Equation 11.4 is transmitted
through the channel defined by Equation 11.12. In this case, the received signal can be written as

$$r(t) = A_o d(t) c(t) \cos (2\pi f_o t)$$
$$+ \alpha A_o d(t - \tau) c(t - \tau) \cos (2\pi f_o t + \theta) + n(t) \qquad (11.15)$$

where θ represents the phase shift at the carrier frequency due to the delay,
and $n(t)$ is the noise. The first term on the right-hand side represents the direct
path; the second term represents a reflection.

If the received signal is despread by multiplication with a spreading code synchronized with the code imbedded in the direct path term, the direct path term is despread. The result of multiplying the delayed term by this despreading code is given by the expression

$$\alpha A_o d(t - \tau)c(t - \tau)c(t) \cos(2\pi f_o t + \theta) \qquad (11.16)$$

Now, if the relative path delay τ is greater than the chip interval of the spreading code, the product $c(t)c(t - \tau)$ will be equally likely to be ± 1 at every point in time, and it switches to new values at the chip rate or higher. In fact, as we pointed out in the first section of this chapter, if $c(t)$ is an m-sequence and if τ is a multiple of the chip interval (but not equal to the period of the code), the product is also a delayed version of the code. Thus, if the chip interval is less than the relative path delay, the direct path can be despread so that it passes through a relatively narrowband matched filter; but the delayed multipath signal component is not despread, and most of its power is rejected by the matched filter. Thus, the delayed path can be rejected by the processing gain of the system if a receiver is used that locks onto the (hopefully) strongest received component. This provides a solution to the multipath problem when the channel characteristics are stationary in time. A problem that remains to be solved for the mobile case is that each of the path attenuations changes as the transmitter or receiver moves. Thus, the receiver periodically experiences a loss of the signal to which it is locked. Perhaps the ultimate receiver for the mobile case would use multiple despreaders for each of received signal components and would select the strongest received path at any time to retrieve the data.

Time and position determination

Additional potential uses for direct sequence spread spectrum are dissemination of accurate time and determination of the position of a receiver. First, suppose that a receiver has a clock that is precisely synchronous with a clock at a transmitter and that the phase of the spreading code used by the transmitter at each point in time is predetermined and known to the receiver. Then the relative distance from the transmitter to the receiver can be deduced by comparing the phase of the spreading code imbedded in the received signal with the phase known to exist at the transmitter. For example, if the spreading rate is 10 Mchips/s, and the received code phase is 500 chips delayed from the phase known to exist at the transmitter, the distance between the transmitter and receiver can be computed as

$$\text{distance} = (500 \text{ chips}) \times (0.1 \times 10^{-6} \text{ s/chip}) \times (3 \times 10^8 \text{ m/s})$$
$$= 15 \text{ km}$$

We have assumed that free-space conditions exist and have taken 3×10^8 m/s for the speed of light.

If the ranges to three transmitters at known locations are determined in this manner, then the three-dimensional location of the receiver can be deduced. (Actually, two possible positions can result.) The distance from the first transmitter locates the receiver on the surface of a sphere centered at the first transmitter. Similarly, the range from the second transmitter locates the receiver on the surface of a second sphere. The first two ranges together locate the receiver on the circle where these two spheres intersect. The third range defines another sphere centered on the third transmitter, which potentially intersects the circle found from the first two ranges in two points. Thus, three ranges determine the position but with an ambiguity. Often, one of the potential positions can be eliminated because it is impossible, inside the earth, for example.

When the receiver does not have a clock of extremely high accuracy, a so-called clock bias term is added to the range from each of the transmitters. Thus, the receiver has four unknowns to resolve: three position coordinates, $x, y,$ and z, and the distance B corresponding to the receiver clock bias. When the receiver determines the distance to each of the transmitters using the received phase of the spreading code, it actually finds a pseudorange, which is the sum of the actual range and the clock bias. If the pseudoranges to four transmitters with known positions are measured, the position of the receiver and the clock bias can be determined. Thus, the receiver obtains accurate position and time information. The position coordinates, ranges, and clock bias are related by the equation set

$$R_1 = \rho_1 - B = [(X_1 - x)^2 + (Y_1 - y)^2 + (Z_1 - z)^2]^{1/2}$$
$$R_2 = \rho_2 - B = [(X_2 - x)^2 + (Y_2 - y)^2 + (Z_2 - z)^2]^{1/2}$$
$$R_3 = \rho_3 - B = [(X_3 - x)^2 + (Y_3 - y)^2 + (Z_3 - z)^2]^{1/2} \quad (11.17)$$
$$R_4 = \rho_4 - B = [(X_4 - x)^2 + (Y_4 - y)^2 + (Z_4 - z)^2]^{1/2}$$

where R_i is the true range to the ith transmitter, ρ_i is the pseudorange to the ith transmitter, B is the distance equivalent of the receiver clock bias, $x, y,$ and z are the coordinates of the receiver, and $X_i, Y_i,$ and Z_i are the (known) coordinates of the ith transmitter.

The receiver can measure the pseudoranges to the four transmitters by comparing the phase of the locally generated PN code with the PN codes imbedded in the four received signals. Then the receiver can use Equation set 11.17 to find its position and clock bias. Thus, the receiver can determine its position and the bias of its clock.

The Global Positioning System (GPS) is based on these principles of position and time determination. Eventually, GPS will consist of 18 transmitters

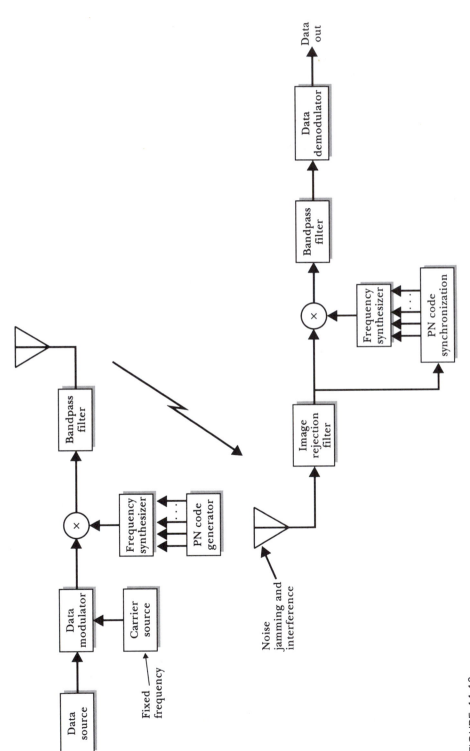

FIGURE 11.13
Block diagram of frequency hopped system.

in orbit above the earth. The transmitted signals are direct sequence CDMA signals capable of providing position information within tens of meters.

We have seen that expansion of bandwidth by the use of direct sequence spread spectrum can simultaneously provide many benefits. In the next section, we will see that similar results can be obtained with frequency-hop spread spectrum.

11.3
FREQUENCY-HOP SPREAD SPECTRUM

A typical frequency-hopped system

The block diagram of a frequency hopped (FH) system is shown in Figure 11.13. The data are first modulated onto a fixed-frequency carrier using any of a variety of modulation formats. The data-modulated signal is then mixed with a frequency-hopped sinewave produced by a frequency synthesizer. The sum frequency term at the output of the mixer is selected by a bandpass filter and transmitted. The frequency of the sinewave produced by the frequency synthesizer is determined by k bits produced by a PN code generator. The resultant carrier frequency is usually one of 2^k different frequencies uniformly spaced over the frequency range used by the system, with the spacing about equal to the bandwidth of the data-modulated signal. The k-bit output of the code generator changes periodically, so the carrier frequency of the signal transmitted hops pseudorandomly over a wide range of frequencies.

The received signal, consisting of the attenuated frequency-hopped signal plus noise and interference from various sources, is passed through an image rejection filter to a mixer that is driven by a frequency synthesizer identical to the one at the transmitter. The frequency of the synthesizer is determined by k bits from a PN code generator matching the transmitter code generator. A feedback control loop keeps the PN generator in the receiver synchronized with the received signal. (The synchronization problem will be discussed in more detail in the next section.) The output of the mixer is passed through a bandpass filter to select the difference frequency component between the received signal and the output of the local synthesizer. As a result, when perfect synchronism exists, the input to the demodulator is the same fixed carrier frequency signal that was produced by the data modulator in the transmitter.

Usually it is difficult to maintain an exact phase relationship between the two synthesizers as they are switched in frequency. Thus, each time the frequency changes, the phase of the input signal to the data demodulator changes to a new random value. If we were to attempt to use a coherent modulation scheme such as PRK, it would be necessary to wait for the carrier recovery loop to adjust to this new phase after each hop before data could be reliably demodulated. Usually, the desired hop rate is too high for proper

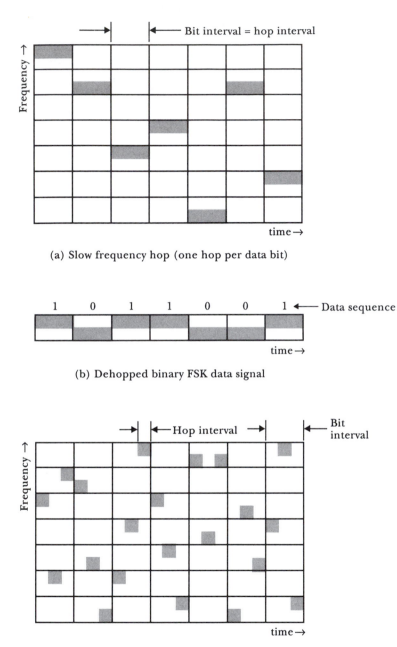

(a) Slow frequency hop (one hop per data bit)

| 1 | 0 | 1 | 1 | 0 | 0 | 1 | ◄── Data sequence |

time →

(b) Dehopped binary FSK data signal

(c) Fast frequency hop (three hops per data bit)

FIGURE 11.14
Frequency versus time for frequency hopped binary FSK.

phase settling on each hop, and coherent modulation schemes cannot be used. Therefore, noncoherent modulation schemes such as binary or *M*-ary FSK are usually used with FH spread spectrum. We will assume in the remainder of our discussion that binary FSK is used.

Slow versus fast frequency hop

Slow frequency hop results when the hop rate is less than or equal to the data rate, so that one or more bits are transmitted at each hop frequency. In *fast frequency hop*, the hop rate is greater than the bit rate. Figure 11.14a depicts frequency versus time for a typical slow FH signal, in which there is one hop per data bit. The corresponding binary FSK signal resulting from dehopping this signal is shown in Figure 11.14b. A fast FH signal with three hops per bit is shown in Figure 11.14c.

When slow FH is used with binary FSK modulation, the signal at the input to the data demodulator in the receiver is one of two frequencies (depending on the data bit) with a phase that, although random, is constant for the entire duration of the bit. The demodulator for this noncoherent binary FSK signal was discussed in Section 8.4 and is shown in Figure 11.15a. The filters are matched for segments of each of the two possible tones with a duration equal to the bit interval. These matched filters combine all of the energy for a given bit to produce a decision variable at each filter output at the end of the bit interval. The magnitudes of these decision variables are compared to make the data decision.

When fast FH is used, the input to the data demodulator is one of two frequencies as before, but the phase of this tone changes to a new random value for each hop interval. In this case, the receiver must be modified as shown in Figure 11.15b. The filters are now matched for a tone with a duration equal to the hop interval rather than the longer bit interval. This is necessary because, if the FSK signals with different random phases for each hop are combined in a filter matched to the bit interval, cancellation of energy from one hop with energy from another hop can occur when the segments are out of phase. A hard decision is made on the data bit for each hop interval, and these decisions are then used in a majority vote logic circuit to make the final data decision. On the other hand, the receiver for slow hop coherently combines all of the energy in the matched filter before the data decision is made.

Thus, in the fast hop receiver, data decisions are made for each hop interval, and these decisions are combined by majority logic to make the final decision. (Other receiver structures are possible. For example, the outputs of the envelope detectors for each hop could be added before a decision is made. The majority logic approach discussed here is advantageous when a partial band jammer tries to disrupt communications.) The segments of the signal for the various hops are said to be noncoherently combined because the phase information is lost in the envelope detectors before the segments for each hop

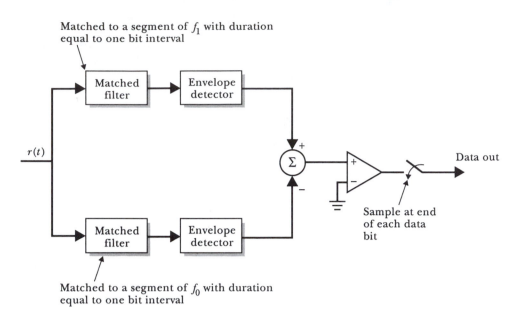

(a) Data demodulator for binary FSK used with slow frequency hop

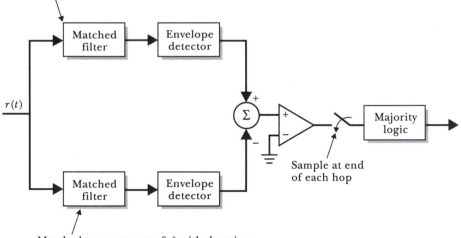

(b) Data demodulator for binary FSK used with fast frequency hop

FIGURE 11.15
Block diagrams of data demodulators for frequency hop.

are combined in the majority logic gate. For an AWGN channel, the performance of fast FH is poorer than that of slow FH. This loss in performance is called *noncoherent combining loss*. Nevertheless, as we will see shortly, fast FH has an advantage against certain types of jammers.

Application of FH for antijamming

A *broadband noise jammer* transmits a noise signal over the entire band used by the communication system. To maintain the same degradation when frequency hopping is used, the total jammer power must be increased by the factor that the bandwidth is increased by hopping. This is because the broadband noise jammer produces a constant noise PSD over the bandwidth used by the system. Therefore, the total jammer power is proportional to the system bandwidth. Thus, the concept of processing gain, which is the ratio of bandwidth after spreading to the bandwidth before spreading, applies when frequency hopping is used against a broadband noise jammer.

As we saw in Section 9.3, a jammer can greatly increase its effectiveness under certain conditions by jamming only part of the data being transmitted but with a higher jamming level. We considered a pulsed noise jammer that concentrates its energy in high power bursts and is off for the rest of the time. In the case of slow frequency hop, the jammer has an easily achievable method for concentrating its power: It needs only to jam part of the band used by the communication system. By concentrating its power in a smaller fraction of the band, the *partial band jammer* can achieve a higher noise PSD for those bits that fall in the jammed part of the band and thereby will cause a higher average error rate. Usually, the power levels in such a system are much above the natural noise level, and the error probability is negligible when the signal falls in the unjammed portion of the band. We saw similar effects for the pulsed noise jammer in Section 9.3 where we saw that coding and interleaving provide a means for the communicator to overcome much of the degradation caused by pulsing. Even a simple repetition code regained most of the advantage lost when the jammer employed pulsing. Fast FH provides a similar advantage against the partial band jammer. Each bit is divided into a number of segments that are hopped to different parts of the band in use. For example, if there are five hops per bit and the partial band jammer is jamming 20% of the band, we would expect only one hop in each bit to encounter jamming on the average. If the noise level in the jammed part of the band is very high, the error probability for the segments falling into the jammed band might approach one-half, and we would have a resulting error probability for each hop of 10%, which is the product of the probability of a hop falling into the jammed portion of the band and the error probability given that the hop falls in the jammed portion. A data-bit error occurs only when a majority of the hops are in error, and the probability of this occurrence can be made small by increasing the number of hops per bit. Thus, fast FH is useful against a partial band jammer.

Application of FH to CDMA

Frequency hop is also useful as a CDMA technique. Multiple users simultaneously use the same frequency band but with different pseudorandom hopping patterns. When the hops do not land on the same frequency, virtually no interference occurs because the transmitted signal for each hop is confined to a limited range of frequencies and the filters in the receiver can reject energy that falls on adjacent frequencies. When two or more transmitters choose the same frequency, a *hit* is said to occur, and an error is probable. Thus, frequency hop CDMA relies on having the transmissions of one transmitter avoid the transmissions of another. On the other hand, direct sequence CDMA relies on the use of pseudorandom codes that average to zero when integrated by a correlator or a matched filter. Thus, frequency hop CDMA can be termed an avoidance system and direct sequence CDMA an averaging system.

Because the energy of a given hop interval of an FH system is confined to a narrow frequency band that can be rejected by receivers centered on adjacent bands, the near-far problem is less significant for frequency hop CDMA than it is for direct sequence CDMA. Thus, frequency hop CDMA is better suited for an application such as mobile radio in which the desired signal may be much smaller than some of the interfering signals.

Application of FH to covert communications

When slow FH is used, the energy for one or more bits is confined to a bandwidth on the order of the data rate. This fact makes frequency hop less desirable for covert communication because an intercept receiver can use a bank of narrowband filters (with bandwidth about equal to the data rate) and detect the energy found at the output of each filter during each bit interval. Since the SNR in a bandwidth on the order of the data rate must be fairly high to provide low bit-error probability for the communication system, the intercept receiver, which may also have a path loss advantage, can easily detect the presence of the signal. On the other hand, with direct sequence, the energy of each bit can be spread over the entire band. Because the energy spectral density of a single hop time segment of the slow hop signal is confined to a narrow bandwidth, slow FH signals are said to have a narrow *instantaneous bandwidth*. Fast FH spreads the energy of each bit over a larger band; but since fast FH encounters noncoherent combining loss in the receiver, a higher power signal is required than for the coherent modulation schemes, which can be used with direct sequence. Thus, direct sequence spread spectrum seems to be a better choice than frequency hop for covert communications.

Application of FH to multipath channels

Frequency hop is also useful as an antimultipath technique. Usually in a multipath situation, the strongest component of the received signal results from the direct path between the transmitter and the receiver; therefore, it is the first to arrive. The other paths, such as reflections, arrive later and can add destructively to the main path signal. When FH is used, the hop rate should be high enough so that the main path signal and the receiver have hopped to a new frequency before the multipath components arrive. Thus, the FH system can avoid interference from multipath.

11.4
ACQUISITION AND TRACKING

To operate properly, a receiver in a spread-spectrum system must have its pseudorandom code properly time aligned with the code sequence imbedded in the desired component of the received signal. The initial coarse alignment of the local code phase within a fraction of a chip is called *acquisition*. Continuing fine adjustment of the code phase during operation is called *tracking*. Thus, on start-up, the receiver first searches for the correct code phase to within about one-half chip. Then the receiver enters the tracking phase in which a phase-locked loop structure maintains the correct code phase.

Acquisition in direct sequence systems

Two approaches to acquisition are possible: *serial search*, in which each potential code phase is tried one after the other, and *parallel search*, in which all possible code phases are tested during the same time interval. Most often, the serial search is used because a separate correlator is needed to test each code phase in the parallel approach, and the resulting complexity is too great for practical implementation. However, if the number of code phases to be searched is reasonably small and the system requirements dictate rapid acquisition, the parallel search method can be used. The disadvantage of serial search is that it can lead to very long acquisition times.

A block diagram of a typical system for serial code acquisition when direct sequence PRK modulation is used for the spreading code is shown in Figure 11.16. The received signal containing the desired direct sequence signal is first amplified and filtered to eliminate the noise and interference outside the spread bandwidth. The resulting filtered signal is then multiplied by the output of the local code generator. The product is applied to the input of a bandpass filter with a bandwidth about the same as that of the desired despread data modulated signal. The output of this filter is applied to a square-law device followed by an integrator. The integral of the squared filter output over an interval of time is the energy in the filter output for that interval. When the code generator is misaligned by more than one chip, the received signal is

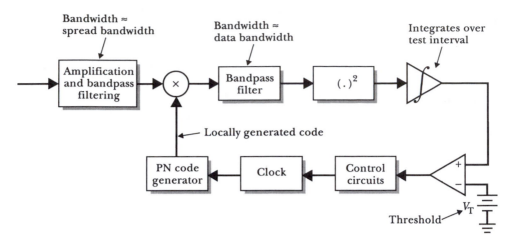

FIGURE 11.16
Serial search acquisition system for direct sequence with PRK spreading code
modulation.

not despread by multiplication with the local code, and very little of the signal
power passes through the bandpass filter following the multiplier. Thus, during
incorrect code phase alignment, the output of the integrator is mainly due to
noise. On the other hand, when the local code is phase aligned with the desired
signal, despreading occurs, and the desired signal power passes through the
bandpass filter, resulting in a larger output from the integrator. (Note that
if the alignment is not exact, only partial despreading occurs. For example,
if the codes are misaligned by one-half chip, the despread component has an
amplitude about half of its value for exact alignment because only half of each
chip is despread.)

A description of the operation of the acquisition system of Figure 11.16
follows. The phase of the local code generator is set to the first phase to be
tested, and the integrator computes the energy in the output of the bandpass
filter. At the end of the integration interval, the output of the integrator is
compared with a threshold. If the integrator output is less than the threshold,
the phase of the code generator is stepped to a new value, the integrator is
reset to zero, and the test is repeated. On the other hand, if the output of the
integrator exceeds the threshold, the system is automatically reconfigured, and
tracking is attempted. In the tracking mode, a lock detector determines if
proper tracking is taking place. If tracking is not successful, the system reenters
the acquisition mode.

A trade-off exists for the selection of the size of the steps by which the
code phase is changed for each new test. If the step size is small, the local code
phase eventually comes very close to the received signal code phase, resulting
in nearly complete despreading and the highest possible SNR at the output of
the integrator. On the other hand, small steps require more steps to be tested.

Ultimately, we want to select the system parameters that result in the minimum acquisition time. Typically, the code phase is stepped by one-half chip between tests.

When the acquisition circuits decide, because of noise and interference, that the local code phase is close to the correct value when, in fact, it is not close, a *false alarm* is said to occur. It is important to keep the probability of a false alarm at a low value because much time is lost when the system enters the tracking mode with the incorrect code phase. If a nearly correct code phase is tested and the system decides that it is not correct, a *miss* occurs. It is important to keep the probability of a miss very low (or, equivalently, the probability of detection very high) because many incorrect code phases usually must be tested before the correct one is again tested.

The integration interval should be long to provide a high SNR at the integrator output, resulting in high detection probability combined with low false alarm probability. However, long integration times potentially waste much time on incorrect code phases before getting to the correct phase. The system designer tries to find the best compromise, resulting in minimum acquisition time.

The threshold value used by the comparator also affects the probabilities of detection and false alarm. As the threshold is lowered, the probability of detection of the correct code phase becomes larger, but the false alarm probability also becomes higher. As with the other parameters, the best compromise is the one leading to minimum acquisition time.

The acquisition system we have described uses a *single dwell* energy detector, in which the same integration dwell interval is used in all decisions. A number of variations of the system are possible that use variable dwell intervals. One of these, known as a *double dwell* detector, first integrates the output of the squaring device for a short time and then compares the result with a threshold that is set low enough so the probability of detection is acceptably high. Since only a short integration interval is used, false alarms due to short bursts of high noise power occur often with the low threshold. Thus, when the threshold for this first integration is exceeded, a longer integration is performed to produce a high SNR at the integrator output, and the threshold is raised. After this second integration, the probability of detection is high, and the false alarm probability is low. Therefore, if the higher threshold is exceeded on the second integration, tracking is attempted. This double dwell approach results in faster acquisition than the single dwell approach because many of the incorrect phases can be dismissed after only the shorter first integration.

Acquisition in frequency-hopped systems

Acquisition systems similar to the one shown in Figure 11.16, with a frequency synthesizer added between the code generator and the mixer, are used for FH systems. Acquisition is less of a problem for slow frequency hop than for direct sequence systems with the same spread bandwidth because the code generator rate is much lower. Only k binary code symbols are needed for each data interval when using FH with one hop per data bit in order to cover a bandwidth of 2^k times the data bandwidth (assuming that the hop spacing is equal to the data bandwidth). On the other hand, with PRK direct sequence spread spectrum, 2^k binary code symbols must be produced for each data bit to attain the same amount of bandwidth expansion. The ability to more rapidly acquire code synchronization is a significant advantage of frequency hop spread spectrum as compared with direct sequence.

Code phase uncertainty

A factor that is very important with respect to acquisition time is the number of code chips of uncertainty that the receiver has concerning the correct code phase. If absolutely precise clocks were available and the range from the transmitter were known, the acquisition problem would not exist. The receiver and the transmitter could agree on the time for starting the code generators and on their starting states. Then, the receiver could generate the correct code phase without the acquisition process. It is because of time and range uncertainty that the receiver must search for the correct code phase. For a given code rate, the potential accumulated time offset and range inaccuracy can be converted to the number of chips of code phase uncertainty in a straight-forward manner.

A number of options are available to the system designer for minimizing the amount of code phase uncertainty. Very accurate clocks and various methods for estimating the range can be used. Sometimes, a common timing signal can be provided to both the transmitter and the receiver so that the timing uncertainty is very small. If a relatively short spreading code is used, the entire code can be searched in a short time. However, use of a short code may enable an adversary to detect or jam the signal more effectively. A short code can be used for acquisition and switched to a longer code for operation, though this presents a problem when the receiver fails to acquire during the short code interval unless a return link is available so that efforts can be coordinated.

A baseband code phase tracking system

Now we consider the tracking aspect of the code synchronization problem. Figure 11.17 shows a system, known as a *delay locked loop*, for synchronizing the phase of a local PN code generator with an input PN code signal. The input code $c(t)$ is assumed to be a ± 1-valued NRZ waveform produced by a maximal-length LFSR as described in the first section of this chapter. The local PN code generator produces three versions of this code. One version is intended to be synchronized with the input signal and is known as the on-time signal. It is denoted by $c(t + \tau)$ where τ is the time offset, which is intended to be forced close to zero by the control action of the tracking system. Another

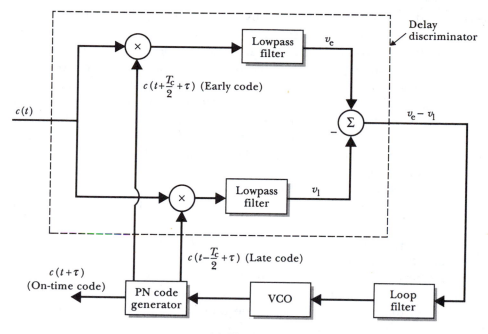

$c(t+\frac{T_c}{2}+\tau)$ (Early code)

$c(t)$

$c(t-\frac{T_c}{2}+\tau)$ (Late code)

$c(t+\tau)$
(On-time code)

(a) Block diagram

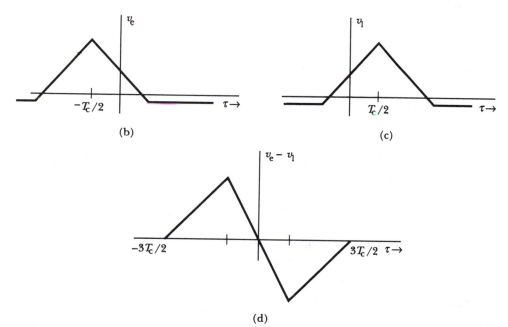

(b)

(c)

(d)

FIGURE 11.17
Baseband delay locked loop.

version of the local code is advanced by one-half chip from the on-time signal and the third version is delayed by one-half chip. These are known as the early and late signals. The early and late signals are multiplied by the input signal as shown in Figure 11.17. The outputs of the multipliers are passed through lowpass filters that pass only the dc component (time average) of the product signals. The time average of the product of a signal with a delayed version of itself is the autocorrelation function of the signal. Thus, the output of each filter is the autocorrelation function of the PN code except that the correlation peaks occur at $\tau = \pm T_c/2$ rather than at $\tau = 0$, due to the half-chip time offsets of the early and late signals. The resulting output voltages of the two lowpass filters are shown in Figure 11.17b and c as a function of the timing error τ. The difference of the filter outputs is shown in Figure 11.17d. This is the control voltage that is applied to the VCO to cause tracking of the phase of the input PN code. For example, if τ is $+ T_c/4$, indicating that the local code phase is $\frac{1}{4}$ chip advanced, the input to the VCO is a negative voltage that decreases the code rate until τ is returned to zero. Thus, the delay locked loop maintains the phase of the local code the same as the phase of the input code.

Notice from the error signal produced by the delay discriminator, shown in Figure 11.17d, that a corrective action exists only if the timing error τ is within $\pm 3T_c/2$ of the correct value. This is why the acquisition process is necessary in a spread-spectrum system. If the code is very long and the system is started in the tracking mode with a very large timing error, no corrective error signal is produced, and it could be a very long time before the code phases finally drift close enough so lock occurs.

Except for the more complicated delay discriminator, the delay locked loop can be seen to be a phase-locked loop much like those analyzed in Section 4.3. In fact, much of the analysis of the dynamic behavior developed there can be applied to the delay locked loop in a straight forward manner. As in the case of the phase-locked loops studied earlier, it is often desirable to include a loop filter ahead of the VCO so that desirable combinations of hold-in range and loop bandwidth can be attained. If the loop bandwidth is very narrow, code phase jitter due to noise is averaged out, but the loop is not able to track variations of the phase of the input due to such things as relative motion between the transmitter and receiver. Thus, the loop characteristics must be carefully chosen. The tracking dynamics of such loops is similar to the analysis we have presented in Section 4.3, and the reader can find material on the noise behavior of code tracking loops in the references.

A direct sequence delay locked loop

The input to the delay locked loop of Figure 11.17 was assumed to be a noise-free PN code. In a more practical case, the PN code is imbedded as the spreading code in the received signal. A delay locked loop suitable for use with a direct sequence signal with PRK spreading code modulation is shown in Figure 11.18.

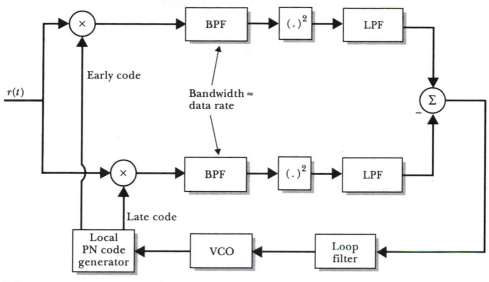

FIGURE 11.18
Noncoherent delay locked loop for PRK direct sequence signals.

In this circuit, the local spreading-code generator produces both early and late versions of the code which multiply the input signal. This results in partial despreading of the input signal when the relative delay between one of the local signals and the input signal is less than one chip. The outputs of the multipliers are applied to bandpass filters whose bandwidths are about equal to the bandwidth of the despread data-modulated signal. The outputs of the bandpass filters are squared and time averaged by lowpass filters. Notice that when $\tau = 0$, each multiplier despreads the same fraction of the input signal, the outputs of the lowpass filters are equal, and the output of the summer is zero. On the other hand, if τ becomes slightly different from zero, one of the despreaders produces a greater despread power, and the output of the summer has a value whose sign depends on the sign of τ. Thus, an error signal is produced at the input to the VCO, and tracking is possible.

Code tracking in frequency-hop systems

Figure 11.19a shows a tracking system for an FH input signal. The locally generated FH signal has the same hop pattern as the received signal but is lower in frequency by an amount equal to the IF. The received FH signal is mixed with this locally generated FH sinusoid, and the resulting dehopped difference signal is passed through a bandpass filter with a bandwidth wide enough to pass the data modulation. The output of this filter is applied to an envelope detector. Noise-free waveforms are shown in Figure 11.19b for the case in which the local generator lags the received FH signal. Notice that when the received signal hops to a new frequency, the output of the mixer

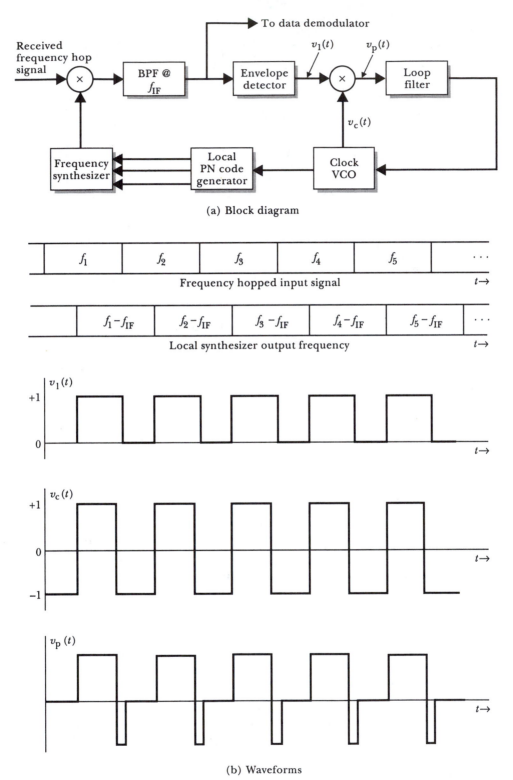

(a) Block diagram

(b) Waveforms

FIGURE 11.19
Tracking loop for frequency hopped signals.

has a difference frequency component, which no longer falls in the passband of the bandpass filter, and the output of the envelope detector falls to zero. When the local frequency hops to the correct value, the difference frequency component passes through the bandpass filter, and the output of the envelope detector is high. Thus, the output of the envelope detector is high when the local generator is on the correct frequency and low otherwise. The clock that drives the code generator also generates a square wave $v_c(t)$ that is $+1$ for the first half of each receiver hop interval and -1 for the second half of each hop, as shown in the figure. The clock square wave is multiplied by the output of the envelope detector, resulting in the product denoted by $v_p(t)$. Notice that the average value is positive for the case shown, in which the local generator lags the received signal. It can be verified that the average value of the product signal $v_p(t)$ is negative for the case in which the local generator leads the received signal by a fraction of a hop. Of course, no error signal is developed when the generator is out of phase by more than one hop. Thus, if the inital phase is within one hop, the circuit of Figure 11.19 is capable of phase-locked loop tracking of the received frequency hop signal.

SUMMARY

1. Spread-spectrum techniques include direct sequence, in which the spreading code phase modulates the carrier, and frequency hop, in which the spreading code selects the carrier frequency.

2. Spread-spectrum techniques are useful for antijamming, low probability of detection, code division multiple access, high-resolution ranging, accurate time dissemination, and rejection of multipath interference. Spread-spectrum techniques offer no improvement for additive white gaussian noise.

3. A linear feedback shift register is formed by a modulo-two adder that adds the content of some of the shift register stages and returns the resulting sum to the input of the shift register. As the register is shifted, a cycle of states occurs. Certain connections lead to a single cycle containing all of the nonzero states, and the binary sequence generated is known as a maximal-length sequence.

4. Ideal pseudorandom codes for spread spectrum are easy to generate, are very long, appear to be random, and are difficult to predict by anyone who does not know the details of the structure of the code generator. Maximal-length linear feedback shift registers have all of these desirable characteristics except that they are easy to predict from a short segment.

5. Long maximal-length LFSR sequences have run properties, autocorrelation function, and power spectral density closely approximating those of a random sequence.

6. A PRK data-modulated signal can be spread by multiplying it by a ± 1 NRZ spreading waveform. The ratio of the spread bandwidth to the bandwidth of the data-modulated signal before spreading is equal to the ratio of the symbol rate of the spreading code to the data rate. This ratio is called the processing gain.

7. A tone jammer tries to disrupt communication by transmitting an unmodulated carrier in the frequency band of the communication system. PRK direct sequence spread spectrum reduces the effect of a tone jammer by spreading the jammer while despreading the desired signal in the receiver. Most of the power of the jammer then falls outside the band of the despread data-modulated signal. The ratio of the power in the data band to the total power of the tone jammer is the same as the processing gain. Thus, the effect of a tone jammer can be greatly reduced by the use of direct sequence spread spectrum with a high processing gain. Similar improvements are possible for other types of direct sequence spread spectrum, as well as frequency hop, against various types of jammers.

8. Covertness of a communication signal is increased by direct sequence spread spectrum because the peak PSD is reduced by spreading. If the PSD of the signal is reduced far below the noise floor of the unauthorized receiver, the signal is difficult to detect. Frequency hop reduces the long time average PSD, but short segments of the signal can have their energy confined to a much narrower bandwidth. As a result, frequency hop is less useful than direct sequence for covert applications.

9. CDMA is accomplished by using a different spreading code for each of the users. For the direct sequence case, the power of other users is partially rejected when the desired signal is despread. For frequency hop, interference occurs only when several users hop to the same frequency. Direct sequence averages out the interfering signals whereas frequency hop relies on avoidance. Thus, frequency hop systems are less affected by the near-far problem than direct sequence systems. Without coordination among users, CDMA can support significantly fewer users than other multiple access techniques. The chief benefits of CDMA are that coordination is not necessary and that the other advantages of spread spectrum can be attained while providing multiple access.

10. Multipath interference can be reduced by spread-spectrum techniques in the same way that jamming signals or CDMA signals are rejected.

11. The range from a transmitter to a receiver can be accurately determined if the phase of the PN code at the transmitter is known and is compared with the phase of the received signal. When the transmissions of four or more transmitters at different known locations are compared with a local PN generator, accurate determination of position and time can be attained.

12. Frequency hop systems usually use noncoherent modulation schemes because it is difficult to maintain a known phase between a local synthesizer and the transmitter when hopping to a new frequency. This results in a loss of communication efficiency.

13. Alignment of the local code phase with the code imbedded in the received signal is a significant problem for spread-spectrum systems. Usually serial search is used to align the code phase to within a fraction of a chip, and then a phase-locked loop technique is used to maintain alignment.

REFERENCES

G. R. Cooper and C. D. McGillem. *Modern Communications and Spread Spectrum.* New York: McGraw-Hill, 1986.

J. K. Holmes. *Coherent Spread Spectrum Systems.* New York: Wiley, 1982.

R. L. Pickholtz, D. L. Schilling, and L. B. Milstein. "Theory of Spread Spectrum Communications—A Tutorial," *IEEE Trans. Commun.* COM-30, (May 1982): 855–884.

M. K. Simon, J. K. Omura, R. A. Scholtz, and B. K. Levitt. *Spread Spectrum Communications, Vols. I, II, and III.* Rockville, Maryland: Computer Science Press, 1985.

J. M. Wozencraft and I. M. Jacobs. *Principles of Communication Engineering.* New York: Wiley, 1965.

R. E. Ziemer and R. L. Peterson. *Digital Communications and Spread Spectrum Systems.* New York: Macmillan, 1985.

PROBLEMS

1. Consider the set of orthonormal basis functions consisting of nonoverlapping rectangular pulses as shown in Figure P11.1. Find the energy spectral density of

FIGURE P11.1

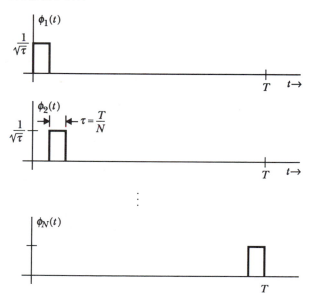

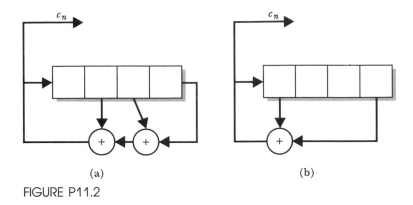

(a)   (b)

FIGURE P11.2

each of the basis functions. Find the bandwidth B of the energy spectral density from dc to the first spectral null. If total duration of the signal set is T seconds, find the maximum number of basis functions in the set in terms of B and T. What bandwidth is required if the total duration is 1 ms and a signal space of 5000 dimensions is required?

2. Write the recursion relation and find the cycles of states for each of the LFSRs shown in Figure P11.2.

3. It is not possible for an LFSR with an odd number of tapped stages to be a maximal-length sequence generator. Why not?

4. Find the connections for an LFSR of not more than four stages that produces the following sequences, and predict the future symbols of each sequence: (a) 00010011, and (b) 10101010

5. From Table 11.1 we find that one of the sets of connection coefficients for a five-stage maximal-length generator is [75]. Draw the diagram for this LFSR, and find the sequence generated if the register is initially loaded with the all-ones pattern. Verify that the sequence generated has the number of ones and zeroes and obeys the run properties given in Section 11.1.

6. The entry in Table 11.1 for the connection coefficients of a four-bit maximal length generator is [23]. The corresponding binary pattern is 010011. Since this is for a register of length four, the leading zero is dropped, resulting in 10011. If this pattern is reversed, we obtain 11001. Draw the diagram of the LFSR based on this reversed pattern, and verify that it produces an m-sequence that is a time-reversed version of the sequence generated by the original pattern. Similarly, a generator producing a time-reversed sequence exists for each of the generators specified by Table 11.1.

7. Consider a direct sequence spread-spectrum signal with PRK modulation for both the data and the spreading code as discussed in Section 11.2. Assuming a data rate of 100 kbit/s, a spreading rate of 500 kchips/s, and a carrier frequency of 1 MHz, sketch a typical NRZ data waveform, a typical NRZ spreading waveform, and the resulting spread signal waveform as given by Equation 11.4.

8. A communication system operates in the presence of a jammer, and the received jammer power is ten times higher than the power of the desired signal. The signal and jammer power are high enough so that the effects of thermal noise can be neglected. The jammer is a noiselike signal with a constant PSD over the

band from 2300 to 2301 MHz. (a) Consider the case in which the communication system uses PRK modulation with no spreading and a carrier frequency of 2300.5 MHz. The data rate is 200 kbit/s. Sketch the PSD of the desired signal and of the jammer to scale versus frequency. Estimate the error probability of the system assuming that the effect of the jammer is the same as that of white gaussian noise with the same PSD in the signal band. (b) Now consider the case in which the communication system uses PRK direct sequence spreading with a spreading rate of 100 MHz. Again sketch the PSD of the jammer and the received spread signal to scale. Sketch the PSD of the desired signal and the jammer at the output of the despreader. Estimate the error rate of this system, again assuming that the effect of the noise is the same as WGN with the same PSD in the despread signal band.

9. A covert communication system is designed to operate with a range from the transmitter to the receiver of 10 km. The system uses direct sequence spread spectrum with PRK modulation for both the data and the spreading code. The transmitted power is adjusted so that the bit-error probability for the intended receiver is 10^{-4}. The data rate is 10 kbit/s, and the chip rate of the spreading code is 100 Mbit/s. Assume line-of-sight free-space propagation conditions and that the noise level is the same for both the intended receiver and the detection receiver. The antenna gains for both receivers are the same. If the detection receiver can detect any signal with a peak PSD that is higher than 10 dB below the noise level, how close to the transmitter must the detection receiver come to detect the signal?

10. A number of users want access to a communication channel. The data rate for each user is 100 kbit/s, so a reasonable bandwidth to allot for each user is 200 kHz. Suppose that a total bandwidth of 50 MHz is available. The received signal power from each transmitter is the same.
    (a) If FDMA is used, how many users can be accommodated assuming that only one user is assigned to each channel?
    (b) Suppose that CDMA is used and each user spreads his power uniformly over the entire band. At the despreader, the desired signal is collapsed to the 200 kHz band whereas the undesired signals remain spread uniformly over the 50 MHz band. What is the resulting SNR if the same number of users as in part (a) are transmitting simultaneously? Consider only the interference due to other users and neglect thermal noise. How many users can be accommodated at one time if a SNR of 13 dB is required in the 200 kHz band at the despreader output?
    (c) Suppose that the probability that a given transmitter is *on* at any instant is 0.01. How many users can be accommodated so that when the average number of transmitters are *on*, the SNR is 13 dB?

11. A certain communication channel suffers from multipath, in which the direct path is always present and a number of other paths occasionally occur. The smallest relative delay for these added paths is 0.5 μs. If PRK direct sequence spread spectrum is to be used, what is the minimum spread rate that will enable the despreader to fully discriminate against the additional paths?

12. A system for locating a receiver and accurately determining time is depicted in Figure P11.3. Two transmitters, T_1 and T_2 are located at the known locations of the xy plane as shown, and the receiver is at an unknown location on the positive x axis. The transmitters use clocks that can be considered to be perfectly accurate to generate (different) PN codes at a rate of 3 Mchips/s. The receiver uses its inaccurate clock to generate the same PN codes. When the receiver clock is at

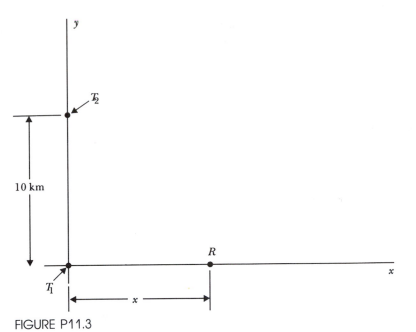

FIGURE P11.3

$t = 200$ s, the phase of the locally generated PN code 1 is leading the received signal from transmitter T_1 by 120 chips and leading the code from transmitter T_2 by 161.4 chips. Find the location of the receiver and the true time when the receiver clock is at $t = 200$ s. Assume that the speed of propagation of the radio signals is exactly 3×10^8 m/s for the purposes of this problem.

13. Two frequency synthesizers are to switch from a frequency of 1.5 GHz to a frequency of 2.0 GHz. Assume that the outputs of the two synthesizers are exactly in phase before switching. What tolerance is allowed for the switching time of one synthesizer relative to the other if the phase between the signals is to be maintained to within 10°?

14. Compare the ratio of the received energy per bit to noise power spectral density (E_b/N_o) required to obtain a data-bit error rate of 10^{-4} for the binary FSK receiver used with slow FH with the value required for the fast FH receiver. Assume five hops per data bit for the fast hop system. The receivers are shown in Figure 11.15a and b, respectively. The difference in performance is known as the noncoherent combining loss of the fast hop receiver. Hint: The error probability for the noncoherent FSK receiver was given in Equation 8.95 as

$$ P(E) = \frac{1}{2} \exp\left(-\frac{E_b}{2N_0}\right) $$

This equation also applies to the fast FH receiver for the error probability of the decision after each hop if E_b is replaced by the received energy per hop.

15. Suppose that a communication system uses slow FH with noncoherent binary FSK data modulation. A partial band noise jammer tries to disrupt this system by transmitting noise in part of the band used by the communication system. The fraction of the band that is jammed is ρ and the (two-sided) noise PSD produced by the jammer in this portion of the band is $P_j/(2B\rho)$, where B is the total

bandwidth of the system and P_j is the received power due to the jammer. When a bit falls into an unjammed portion of the band, assume that the error probability is negligible. The effect of the jammer is the same as white noise when the bit falls into the jammed portion of the band. Also assume that if a hop falls into the jammed portion of the band, the jammer noise covers the entire data band for that hop (i.e., no partial jamming of bits). Find the value of ρ that maximizes the average bit-error probability and the resulting error probability.

16. Sketch to scale the output of the delay discriminator of Figure 11.17 versus τ if the early code leads the on-time code by one chip and the late code lags the on-time signal by one chip. For this case, how close must the initial code phase be aligned before an error signal is developed for tracking?

17. A certain direct sequence spread-spectrum system operates with a data rate of 100 kbit/s and a spreading rate of 100 Mchips/s. The maximum uncertainty of the local clock at the receiver is ± 1 ms, and the maximum range uncertainty is ± 100 km. A single-dwell serial search is used with a dwell time equal to the duration of ten data bits. Find the amount of time required to test all of the possible code phases in half-chip steps. Assuming that the probability of a missed detection is negligible, this is the maximum acquisition time.

12

OPTICAL FIBER
COMMUNICATION
SYSTEMS

An optical fiber communication system consists of a light source that can be modulated by a data signal, a glass or plastic fiber that guides the modulated lightwave to the destination, and a receiver that converts the lightwave to an electrical signal and recovers the data. The light source usually consists of either a light-emitting diode (LED) or a laser diode. Modulation is often of the on-off keying type in which the light source is turned on and off by a binary digital data signal, but many other analog or digital modulation techniques can be used. The glass fiber is made of purified silica that is doped to create a core with a slightly higher refractive index than the cladding. Because of these variations in the refractive index, light that enters a fiber under the proper conditions can be confined to the core and guided to a destination. (A typical fiber, including a protective plastic coating, has approximately the same external appearance as six-pound test monofilament fishing line.) The receiver typically consists of a photodiode in which each photon of light creates a hole–electron pair, giving rise to electrical current. This current is amplified, filtered to minimize noise effects, and used to retrieve the original signal in a manner similar to other systems.

Each of the elements that we have encountered previously, such as carrier sources, modulators, and demodulators, has its counterpart in optical communication systems. Actually, a lightwave is an electromagnetic wave just as a microwave is, but the extremely high frequency of the lightwave carrier leads to quite different characteristics for lightwave components compared with the corresponding conventional electrical components.

Fiber optic communication systems using silica glass fibers to guide lightwaves were first seriously proposed in 1966 by Kao and Hockham and also by Werts. The potential for these systems was recognized by many workers, and the field has progressed at an astounding rate. The first commercial systems were put in place in the late 1970s using an infrared lightwave carrier with a free-space wavelength of about 800 nm (corresponding to a carrier frequency of about 375,000 GHz). These first systems used fibers with an attenuation of about 4 dB/km and achieved data rates on the order of 10 Mbit/s over links a few kilometers in length. Recently, systems have shifted to longer wavelengths in the range from 1.3 to 1.5 μm to take advantage of lower attenuation and dispersion possible for silica glass fibers in this region. As of 1989, experimental systems have achieved operation at several gigabits per second over links longer than 100 km. A transatlantic fiberoptic cable, known as TAT-8, went into service in December 1988 and can carry 40,000 telephone conversations—more than all of the previously existing transatlantic cables and satellite links. The field has been characterized by simultaneous and rapid development of all of the system components: lightwave sources, modulators, fibers, splicing technology, connectors, optical detectors, and the associated electronics. The outlook is for even greater improvements in system performance and reduction in cost.

Optical fiber systems have a number of advantages over conventional electrical cable systems. The bandwidth already achieved by optical fibers greatly exceeds the bandwidth of the best coaxial cables and the potential for increased bandwidth is very great. The region of low attenuation in silica, 1.3 to 1.5 μm, corresponds to a bandwidth of 30,000 GHz. Thus, the potential bandwidth of a single fiber is not only larger than that of a cable it is also about three orders of magnitude larger than the entire spectrum used for all types of radio communication. Work is currently under way to find practical ways to exploit this bandwidth.

Optical fibers have a much smaller size and weight than waveguides or cables of similar capacity. This is an advantage in cities where duct space is at a premium and the demand for increased communication capacity is great. Small size and weight are also advantageous in applications such as those on aircraft.

Optical fibers do not conduct electrical current, which is an advantage when electrical isolation of the source and the destination is important, as in control signals in an electrical power substation. Furthermore, since the fibers are nonconductors, they are not subject to interference from coupled electromagnetic fields. Properly designed optical cables experience virtually no crosstalk between fibers as electrical cables do. Another application of fibers based on their immunity to electromagnetic fields is in military systems that would have to function after they were exposed to the extreme electromagnetic pulses created by a high-altitude nuclear blast.

Low transmission loss is another advantage of optical fibers, which have been fabricated with an attenuation as low as 0.2 dB/km. This leads to wider spacing between repeaters in a long-haul telecommunication system.

Optical fibers potentially have lower cost than electrical cables because the glass is derived from sand, which is an abundant resource. This advantage has not yet been fully realized, but in the near future it can be expected that fiber optic cables, in addition to providing their other advantages, will also be less expensive than any type of electrical cable.

Even though optical fibers are constructed from a seemingly fragile material, small, flexible, and extremely rugged cables result when fibers are properly combined with strength members and protective coverings.

As with other topics covered in this book, we provide only an introduction to the most important aspects of fiber optic communication systems. In the first section of this chapter, we review the ray theory approach to optics and then discuss some of the more important characteristics of several types of fibers. In the second and third sections, we discuss the characteristics of optical sources and detectors. In the fourth section we consider the sources of system noise and the ultimate performance limits for several digital modulation techniques and receiver types. Finally, we consider the complete system design and analyze an example system.

12.1
OPTICAL FIBERS

Ray optics

Before the characteristics of optical fibers can be discussed, the behavior of lightwaves striking the boundaries between optically dissimilar materials must be considered. Light is an electromagnetic wave of very high frequency. It can be described by Maxwell's equations, but for our purposes the simpler ray theory will be sufficient. The ray theory treats light as rays that travel in straight lines but undergo *refraction* (bending) or *reflection* when they strike the interface between different materials. This is illustrated for several cases of interest in Figure 12.1.

A transparent material can be characterized by its *index of refraction*, which is the ratio of the speed of light in a vacuum to the speed of light in the material. For silica glass, the index of refraction is about 1.5, but the exact value depends on the precise composition of impurities in the glass. The index of refraction of air is almost exactly unity. Note that this implies that light travels significantly slower in glass than in air.

When a light ray traveling in a material of higher index of refraction n_1 strikes an interface with a material of lower index of refraction n_2, the ray

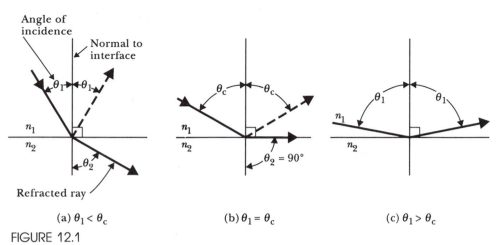

Angle of
incidence

Normal to
interface

Refracted ray

(a) $\theta_1 < \theta_c$ (b) $\theta_1 = \theta_c$ (c) $\theta_1 > \theta_c$

FIGURE 12.1
A light ray striking the interface between materials in which $n_1 > n_2$.

experiences reflection or refraction. If the angle of incidence θ_1 is smaller than the critical angle θ_c, the ray will be partially reflected back into the first material and partly refracted. The refracted ray will be bent away from the normal to the interface at the point where it enters the second material. This is illustrated in Figure 12.1a. If the angle of incidence θ_1 is increased, eventually the *critical angle θ_c* is reached at which the refracted ray travels parallel to the interface as indicated in Figure 12.1b. Finally, if the angle of incidence is greater than the critical angle, the ray experiences total reflection into the material with the higher index of refraction as illustrated in Figure 12.1c.

Snell's law gives the relationship between the angles of incidence and refraction and the indices of refraction as

$$n_1 \sin \theta_1 = n_2 \sin \theta_2 \qquad (12.1)$$

where the angles are defined in Figure 12.1a. The critical angle of incidence occurs when the angle of refraction θ_2 is 90°. Substituting $\theta_2 = 90°$ into Snell's law and solving, we find

$$\sin \theta_c = \frac{n_2}{n_1} \qquad (12.2)$$

Notice that the angle of reflection is exactly equal to the angle of incidence. For the case in which the angle of incidence is zero, so that the ray is normal to the surface, the ratio of the power carried by the reflected ray to the power carried by the incident ray is given by

$$\frac{P_r}{P_i} = \left(\frac{n_1 - n_2}{n_1 + n_2} \right)^2 \qquad (12.3)$$

where P_r is the power carried by the reflected ray, and P_i is the power carried by the incident ray. This partial reflection is called a *Fresnel reflection*. For the

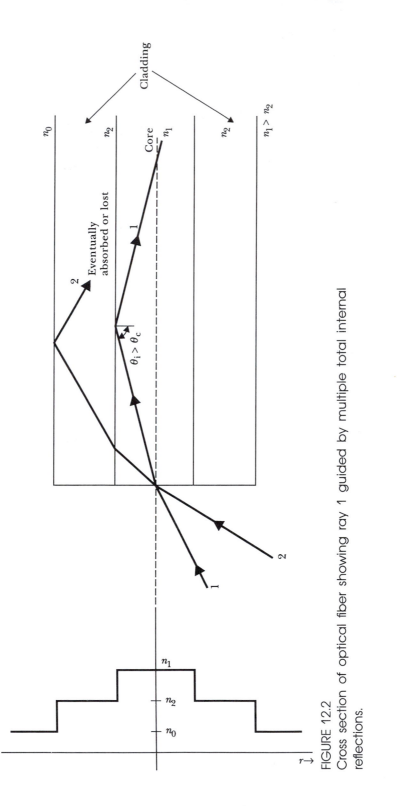

FIGURE 12.2
Cross section of optical fiber showing ray 1 guided by multiple total internal reflections.

12 / OPTICAL FIBER COMMUNICATION SYSTEMS

case in which the angle of incidence is greater than the critical angle, almost all of the incident power is reflected, and only a very small amount of absorption occurs at the interface between the materials.

Step index multimode fiber

Figure 12.2 shows a cross section along the length of a cylindrical fiber having a *core* material with slightly higher index of refraction than the outer *cladding*. This type of fiber is known as a *step index fiber* because of the step change in the index of refraction at the interface between the core and the cladding. Later, we will consider *graded index fibers*, for which the index of refraction varies continuously over the cross section of the fiber. When the diameter of the core is much greater than the wavelength of the lightwave to be propagated, the fiber is called a *multimode fiber*, and ray theory can be used to analyze the propagation of light down the fiber. When the core diameter is smaller, comparable to the wavelength of the light, it turns out that ray theory is not an adequate means of describing light propagation in a fiber, and electromagnetic field theory must be used. This theory shows that under certain conditions there is in effect only a single ray path through the fiber. In this case, the fiber is said to be a *single mode* fiber.

Figure 12.2 also illustrates the light guiding mechanism of the multimode step index fiber. A light ray that enters the end of the fiber in such a way that it arrives at the interface between the core and the cladding at greater than the critical angle will be totally reflected back into the core and will thus be guided along the fiber. This is illustrated by the ray labeled 1 in the figure. On the other hand, a ray entering at an angle so that it is incident on the core-cladding interface at less than the critical angle will escape from the core and will eventually be absorbed by the cladding, which usually has a much greater loss than the core material. Thus, the rays that enter the face of the fiber at a small enough angle to the center line are guided through the core of the fiber by multiple total internal reflections.

The rays shown in Figure 12.2 are called *meridional rays* because they pass through the center line of the fiber. Other rays known as *skew rays*, which do not pass through center line, can also be guided along the length of the fiber. The paths followed by skew rays resemble helices. We will confine our attention to the meridional rays.

The maximum angle of incidence at the face of the fiber that results in total internal reflection at the core cladding interface is known as the *angle of acceptance* of the fiber. It can be shown that the acceptance angle θ_a for a step index fiber is related to the fiber parameters by

$$n_0 \sin \theta_a = [(n_1)^2 - (n_2)^2]^{1/2} \qquad (12.4)$$

The term $n_0 \sin \theta_a$ is called the *numerical aperture* of the fiber. When the medium surrounding the fiber is air, so that n_0 is unity, the numerical aperture is simply

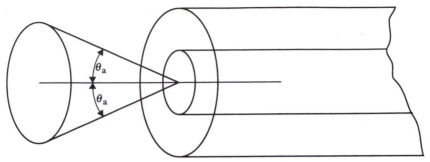

FIGURE 12.3
Cone of acceptance defined by the acceptance angle θ_a.

the sine of the acceptance angle. Notice that all rays that arrive from directions within a cone defined by the acceptance angle are guided by the fiber. This *cone of acceptance* is illustrated in Figure 12.3. A large angle of acceptance makes it easy to couple light into the fiber from a source, but as we shall see, also leads to lower bandwidth for the fiber.

Dispersion effects in fibers

Since rays that are at different angles to the center line travel different distances through a fiber, a short pulse of light containing rays at different angles will spread out in time as it travels. A ray traveling down the center line will take the least time to reach the destination. Considering only meridional rays, the ray that enters the fiber at the maximum acceptance angle and thus strikes the core cladding interface at the critical angle will take the longest time to reach the destination. Thus, a narrow pulse entering the fiber will become spread out in time. This effect is called *modal dispersion* and leads to bandwidth limitations for the modulating signal. Modal dispersion does not occur in single mode fibers because only a single ray propagates through the fiber. However, other characteristics of the fiber can also lead to dispersion. Before discussing this, we will consider an example of a typical multimode step index fiber.

EXAMPLE 12.1

A typical multimode step index fiber has a core diameter of 62.5 μm, an outside cladding diameter of 125 μm, a core refractive index of $n_1 = 1.47$, and a cladding refractive index of $n_2 = 1.46$. Compute the critical angle of incidence for the interface between the core and cladding, the angle of acceptance, and the range of distances traveled by the meridional rays for a 1 km length of fiber. Also find the difference in transit time between the fastest and the slowest meridional rays.

SOLUTION

The critical angle is related to the indices of refraction by Equation 12.2, which is

$$\sin \theta_c = \frac{n_2}{n_1} = \frac{1.46}{1.47} = 0.9932$$

Solving this for the critical angle, we find

$$\theta_c = 83.31°$$

Thus, the meridional rays travel at a maximum angle of $90° - 83.31° = 6.69°$ to the center line of the fiber.

The acceptance angle is related to the indices of refraction by Equation 12.4. Substituting the given values (with $n_0 = 1$ since we are assuming that the fiber is immersed in air) and solving, we find

$$\theta_a = 9.86°$$

Figure 12.4 shows the fiber and the meridional rays taking the shortest and the longest paths. The shortest distance traveled is the length of the fiber l and the longest distance is given by

$$l_2 = \frac{l}{\sin \theta_c} = \frac{n_1 l}{n_2}$$

For a fiber length of 1 km, we find $l_2 = 1.00685$ km. Thus, the longest path is $\Delta l = 6.85$ m longer than the shortest path. The velocity of light in the core is the velocity of light in a vacuum divided by the index of refraction of the core. This is given by

$$v_c = \frac{3 \times 10^8}{n_1} = 2.04 \times 10^8 \text{ m/s}$$

FIGURE 12.4
Multimode step index fiber and the meridional rays with the longest and shortest paths (not to scale).

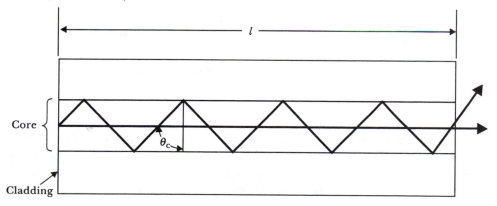

Thus, the difference in delay for the two rays is

$$\Delta t = \frac{\Delta l}{v_c} = 33.6 \text{ ns}$$

A short pulse of light entering the fiber would be dispersed and appear at the output as a broadened pulse due to this delay difference.

Another mechanism also leads to pulse broadening in optical fibers. This is called *material dispersion* and is due to the fact that lightwaves of different wavelengths travel at different velocities through the core. Equivalently, we can say that it is because the index of refraction of the core material varies with wavelength. If the light were all of the same frequency, material dispersion would not be a problem. However, the light sources currently available do not emit a single wavelength but instead emit light over a range of wavelengths. The width of this range is called the *linewidth* of the source. Broader linewidths lead to greater material dispersion and research is underway to produce light sources with narrow linewidths. Of course, when the lightwave is modulated, sidebands are produced that spread the power of the lightwave carrier over a range of frequencies. Therefore, the modulation type and data rate eventually set a lower limit on the linewidth (or equivalently, the bandwidth) of the light-waves to be conveyed by the fiber. However, with sources currently used in fiber optical systems, the linewidth is determined by the source rather than the modulation.

Figure 12.5 shows the dispersion of silica, normalized to the path length and the linewidth of the source, versus wavelength. Positive values of dispersion indicate that the longer wavelength components have longer time delays whereas negative dispersion occurs when the shorter wavelengths are delayed more.

Another source of dispersion is called *waveguide dispersion*. Like material dispersion it results from the differing velocities of the components of the light-wave with different wavelengths. Waveguide dispersion is due to the interaction of the lightwave with the guiding structure of the fiber. A quantitative discussion of this effect would consider the lightwave as an electromagnetic wave and use field theory to consider the interaction of the field with the core boundary.

By proper selection of the core diameter and doping profile, the wave-guide dispersion can be adjusted so the total dispersion due to material effects and waveguide effects can pass through zero at a wavelength in a range about 1.3 μm. Thus, the wavelength for zero net dispersion due to material and wave-guide effects can be shifted by the design of the fiber. Ideally, a fiber would have very low dispersion over a fairly wide range of wavelengths.

12 / OPTICAL FIBER COMMUNICATION SYSTEMS

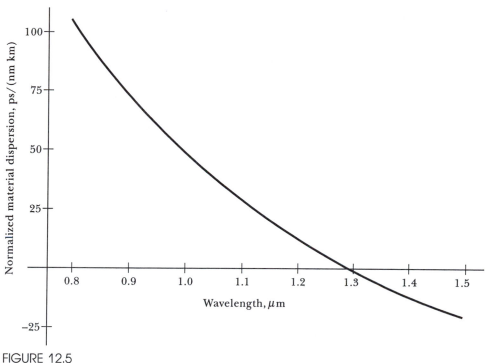

FIGURE 12.5
Normalized material dispersion for silica glass.

The combined effect of material dispersion and waveguide dispersion is referred to as *chromatic dispersion* because both effects depend on the spread of wavelengths of the light propagating through the fiber.

Modulation bandwidth

Of course, the total dispersion due to all sources sets a limit on the modulation rate, just as bandwidth limits any type of system. Figure 12.6a shows the amplitude of a lightwave that has been modulated by a return-to-zero binary signal using on-off keying. Figure 12.6b shows the amplitude of the resulting output pulses after a moderate amount of dispersion, and Figure 12.6c shows the output after a large amount of dispersion. Notice that the original data can still be reliably recovered if the dispersion is not too great, but if the dispersion is severe, the data will become difficult to retrieve due to intersymbol interference. Thus, dispersion sets a limit on the maximum data rate for digital systems and on the modulation bandwidth of analog modulation schemes such as amplitude modulation.

It is important to notice that the bandwidth limitation set by dispersion in an optical fiber is on the bandwidth of the modulating signal and is not a

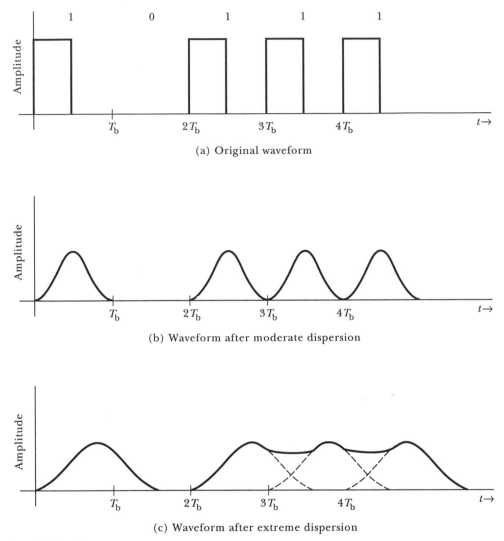

(a) Original waveform

(b) Waveform after moderate dispersion

(c) Waveform after extreme dispersion

FIGURE 12.6
Effect of dispersion on a return-to-zero on-off keyed light wave.

limit on the carrier frequency. In principle, it is possible to send several modu-lated lightwaves through a fiber if their frequencies do not overlap. This is the familiar idea of frequency division multiplexing except that in optical systems it is usually called *wavelength division multiplexing*. If we say that the bandwidth of an ordinary electrical cable is 100 MHz, for example, this means that the cable acts as a lowpass filter with a cutoff frequency of 100 MHz. Thus, the cable does not pass signals in excess of 100 MHz. On the other hand, when we say that the modulation bandwidth of an optical fiber is 100 MHz, we mean that dispersion sets an effective limit on the highest usable modulation fre-quency of 100 MHz. The fiber is capable of conveying many lightwave carriers,

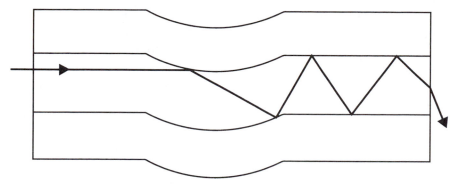

FIGURE 12.7
Fiber bends convert fast rays into slow rays and vice versa.

each carrying a 100 MHz modulation, provided that the linewidths of the sources are narrow enough so that the modulated lightwave signals do not overlap in the frequency domain.

So far, we have implied that the amount of dispersion in an optical fiber is proportional to the length of the fiber. In fact, this is true for chromatic dispersion but is an approximation for modal dispersion. The reason for this is called *mode mixing*, which is illustrated in Figure 12.7. Recall that modal dispersion is caused by the differing distances traveled by the various rays. As illustrated in the figure, bending of the fiber core can convert fast rays to slow rays and vice versa. Thus, as the fiber is increased in length, the rays tend to change speeds, so they all tend toward an average velocity and the dispersion does not continue to increase at the same rate as for a short length of fiber. Nevertheless, dispersion is often assumed to be proportional to fiber length because this usually leads to a conservative system design.

As we will see later in this chapter, the modulation bandwidth of a fiber is inversely proportional to the dispersion. Therefore, the fiber bandwidth is approximately inversely proportional to fiber length. Thus, fibers are often stated to have a certain bandwidth–length product. A fiber with a bandwidth–length product of 50 MHz km would have a bandwidth of 50 MHz if the length is 1 km and a bandwidth of only 5 MHz for 10 km.

A point of possible confusion with respect to bandwidth is that it can be measured in two ways that give different results. The *optical bandwidth* applies to power fluctuations of lightwaves traveling down the fiber. Since the light sources used in fiber optic systems convert an electrical current to optical power and optical detectors convert optical power into an electrical current, the *electrical bandwidth* is specified for an overall system including the source and detector. Thus, electrical bandwidth assumes that an electrical current is linearly converted to light power, which is then conveyed by the fiber to a detector that linearly converts light power to an electrical current. The optical bandwidth of a fiber is greater than the electrical bandwidth. Electrical bandwidth is the value that applies in most system studies since sources and detectors that make approximately linear conversions between electrical current and optical power are most often used. In the last section of this chapter, we will give quantitative relationships between dispersion and the two bandwidths.

Attenuation

Attenuation of the lightwave as it travels through the fiber is another characteristic of optical systems. Ordinary window glass appears to be transparent but in fact has very high attenuation if light must travel through several kilometers of the material. This loss in ordinary glass has been found to be mainly due to impurities such as copper. The production of fibers with low loss has become possible by purifying the silica. It is necessary to control the concentrations of certain impurities such as copper and chromium to less than one part in 10^{10}. Hydroxyl (OH) ions are responsible for a loss peak at wavelengths in the vicinity of 1380 nm. Certain impurities added to the fiber core produce the index of refraction variations needed to create a light-guiding structure, but these are selected to cause little increase in attenuation.

An unavoidable loss mechanism is *Rayleigh scattering*, caused by random density fluctuations frozen into the glass. Light traveling through the fiber is scattered by the resulting variations in the index of refraction and is lost from the fiber. The loss due to Rayleigh scattering is inversely proportional to the fourth power of wavelength, so it is much less important at longer wavelengths. This fact, plus the lower chromatic dispersion in the vicinity of 1300 nm, has resulted in the shift away from the 800 nm region used by the first fiber optic systems. In the region around 1300 nm, the lower limit on loss imposed by Rayleigh scattering is approximately 0.3 dB/km.

Another loss mechanism is scattering due to bends in the fiber. *Microbending* refers to small imperfections that occur in the straightness of the core during manufacture. *Macrobending* is due to bending the fiber around obstructions during installation. Both types of bending cause reflections at angles that lose light rays from the fiber core. In general, it is found that loss due to bending increases at longer wavelengths for a given fiber. Figure 12.8 shows the total loss of a typical optical fiber versus wavelength.

Losses also occur when fibers are spliced and when connectors are used to join fibers or connect fibers to sources or detectors. Splices are made by fusion welding and by mechanical means. In either case, the ends of the fibers to be joined are cleaved to produce flat surfaces that are butted together and carefully aligned. In a fusion splice, the fibers are then welded with an electrical arc. In a mechanical splice, the space between the fibers is filled with an index-matching gel to reduce loss from Fresnel reflections, and the fibers are held in position mechanically. A connector is similar to a mechanical splice but is designed for easier and quicker dismounting and reconnection. For multimode fibers, splice losses are in the region of 0.2 dB. Single mode fiber

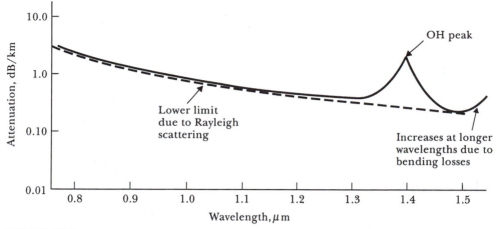

FIGURE 12.8
Typical attenuation versus wavelength for a silica glass fiber.

requires more careful alignment due to smaller core diameter, and splices typically have losses in the region of 0.5 dB. Splicing and connector technology have received considerable attention. No doubt, steady improvement in performance and reduction in cost will continue in the future.

Graded index fibers

We introduced our discussion of fibers by consideration of step index multimode fibers because they are the simplest to understand and were the first type to be used. However, the modal dispersion due to the varying distances traveled by the rays imposes a narrow modulation bandwidth with step index fibers. One approach to reduction of the amount of dispersion in multimode fibers is the graded index fiber. In a graded index fiber, the index of refraction varies continuously as a function of distance from the center line of the core. A particularly good profile for the index of refraction is the parabolic profile shown in Figure 12.9a.

In a graded index fiber, refraction is a continuous process. The index of refraction varies continuously, bending the rays gradually in smooth curves back toward the center of the core (Figure 12.9). This is in contrast to the step index fiber, in which rays travel in straight lines that bend sharply due to reflection at the interface between the core and the cladding. Perhaps the paths followed by the rays in a graded index fiber can be better understood by considering a ray traveling through layers of material with successively lower indexes of refraction, as in Figure 12.10. At the first several interfaces, the ray is refracted so that the angle of incidence is greater at each successive interface. Finally, the critical angle is exceeded, and the ray is reflected downward. If the layers become thinner and the change in the index becomes less at each

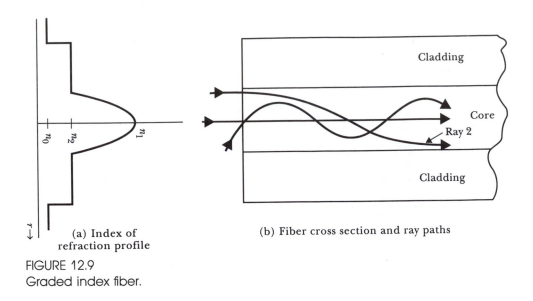

FIGURE 12.9
Graded index fiber.

(a) Index of refraction profile

(b) Fiber cross section and ray paths

interface, the sequence of straight line segments becomes closer to a smooth curve, as in the ray paths for the graded index fiber shown in Figure 12.9b.

The time dispersion of a graded index fiber can be much less than the dispersion of a step index fiber. The distances traveled by various rays differ in the graded index fiber just as in the step index fiber. However, the velocity of some rays varies in the graded index fiber because of the variations in the index of refraction in the core. For example, the ray that travels down the

FIGURE 12.10
Bending of a light ray in a series of layers of glass.

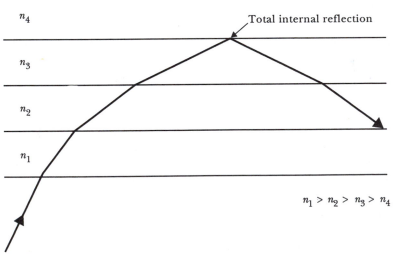

$n_1 > n_2 > n_3 > n_4$

center of the fiber travels the least distance; but since the index of refraction is highest at the center of the core, the average velocity is least for this ray. On the other hand, the ray labeled ray 2 in Figure 12.9b travels the greatest distance but has the highest average velocity because it is in the low index region close to the cladding much of the time. By properly tailoring the index of refraction profile, modal dispersion can be minimized. The profile that approximately achieves this is the parabolic profile shown in Figure 12.9a. Typical length–bandwidth products for graded index fibers are on the order of several hundred megahertz·kilometers whereas step index fibers have products of only several tens of megahertz·kilometers.

Single mode fibers

As we have noted, electromagnetic field theory is necessary for a more exact description of lightwave propagation in fibers than ray theory can give. When the core diameter is small enough, field theory shows that, in effect, only a single ray propagates through the fiber. As a result, a single mode fiber does not suffer from modal dispersion; only material and waveguide dispersion apply. It is possible for these effects to cancel at one or more wavelengths in the vicinity of 1.3 μm, so the single mode fiber operated close to this point can have a very large length–bandwidth product. The length–bandwidth product is highly dependent on the linewidth of the source because only chromatic dispersion effects occur in single mode fibers.

A typical single mode fiber has a core diameter of 8 μm whereas a multimode fiber usually has a core diameter of 62.5 μm. Furthermore, a single mode fiber usually has a smaller difference in index of refraction between core and cladding than a multimode fiber, resulting in a smaller acceptance angle. These facts make it more difficult to couple light into the fiber and to splice the fiber.

12.2
LIGHT SOURCES

The light sources most often used in optical fiber communication systems are light emitting diodes (LED) and laser diodes. Both of these devices can be designed to emit light at the wavelengths most useful in fiber systems, but there are several differences in their characteristics, which we will describe in this section.

Interaction of lightwaves with matter

When energy in the form of a lightwave is emitted or absorbed by matter, it is quantized and occurs in integer multiples of the frequency of the light times Planck's constant. Thus, when light is absorbed or emitted it is appropriate

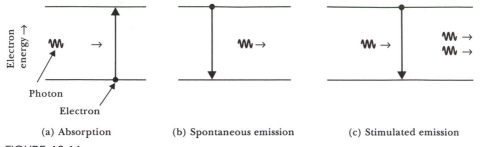

(a) Absorption   (b) Spontaneous emission   (c) Stimulated emission

FIGURE 12.11
Interaction of photons with matter.

to view the lightwave as a stream of particles known as *photons*. The energy carried by each photon is

$$E = hf \text{ J} \tag{12.5}$$

where $h = 6.63 \times 10^{-34}$ J s is Planck's constant, and f is the frequency of the lightwave.

The electrons surrounding the nucleus of an isolated atom or molecule, as in a gas, are allowed to take on only certain values of energy known as energy levels. Various interactions can occur between a photon of light and the electrons in such an atom. If the energy of the photon matches the difference between a higher energy state and the energy of one of the electrons in a lower state, the photon can be absorbed raising the electron to the higher energy state. Similarly, an electron in a higher energy state can spontaneously fall to a lower empty state emitting a photon. Finally, if an electron is in an energy state above an empty state and if a photon with an energy matching the difference between the electron energy and the empty state interacts with the atom, the electron can be stimulated to emit a second photon of the same frequency as, and in phase with, the original photon. These interactions, known as *absorption*, *spontaneous emission*, and *stimulated emission*, are illustrated in Figure 12.11.

In a solid material, the energy levels become bands of allowed energies. In certain solids, all of the interactions we have described are possible. A semiconductor is a material with a characteristic energy band structure. It has a valence band that is filled with electrons in their lowest energy states, as they would be at absolute zero temperature. Above the valence band is the forbidden gap, which is a band of energy with no allowed states. Above the forbidden gap is the conduction band, which would be empty for a pure semiconductor at absolute zero temperature. This band structure is illustrated in Figure 12.12. At normal temperatures, a few electrons move from the valence band to the conduction band, leaving holes in the valence band. The holes can be considered as positively charged particles that are free to move through the material. The electrons in the conduction band are not bound to a partic-

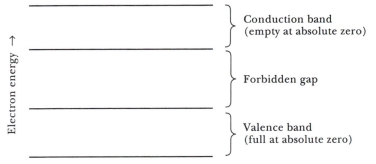

FIGURE 12.12
Energy band structure of a pure (intrinsic) semiconductor.

ular atom and can also move freely through the material. Thus, electrical conduction is due to the free electrons in the conduction band and the holes in the valence band.

When a photon of light of the proper energy strikes a semiconductor, it can be absorbed, raising an electron into the conduction band. This is the principle of operation of a photoconductor. Light striking the material creates more free electrons and holes making the material a better conductor as the light intensity increases. In some semiconductors, known as *direct bandgap* materials, the processes of spontaneous and stimulated emission of photons are possible when electrons fall from the conduction band to the valence band.

Impurities can be added to semiconductors to alter the population of free electrons and holes. Certain types of impurities produce a material with a large number of free electrons and very few holes. This type of material is known as *n*-type semiconductor. On the other hand, other impurities can be added to create *p*-type material, in which there are a large number of holes and only a few free electrons.

Light emitting diodes

A *pn* junction is created when the semiconductor on one side of a junction is *p*-type and the material on the other side is *n*-type. The energy level diagram of a *pn* junction is shown in Figure 12.13a. Notice that an energy barrier exists at the junction because the valence energy levels are higher on the *p* side than they are on the *n* side. This energy barrier is formed by a flow of charge from one side of the junction to the other when the junction is formed. The result is a layer of net negative charge on the *p* side and a layer of net positive charge on the *n* side. It is the electric field directed from the *n* side to the *p* side at the junction that is responsible for the energy barrier holding the free electrons on the *n* side and the holes on the *p* side.

When an external voltage source is connected to the *pn* junction with the positive end of the source connected to the *p* side and the negative side of the

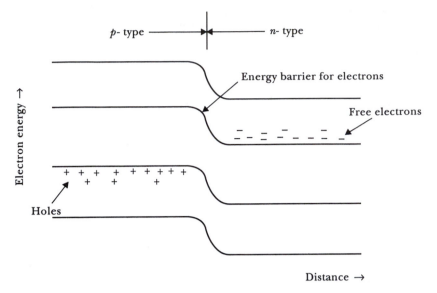

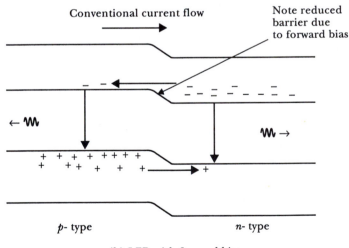

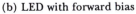

(a) *pn* junction with no bias applied

(b) LED with forward bias

FIGURE 12.13
Energy diagrams for an LED.

source connected to the *n* side, the height of the energy barrier is reduced, resulting in the flow of electrons and holes across the junction. This application of external voltage is said to *forward bias* the junction. If the semiconductor is a direct bandgap type, such as gallium arsenide (GaAs), the electrons and holes that have crossed the junction can recombine, emitting photons of light.

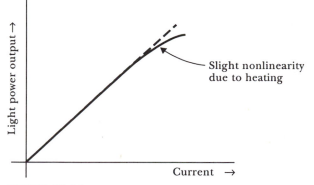

FIGURE 12.14

Optical power output of a typical LED versus forward current.

Then a light-emitting diode (LED) results. The energy diagram of an LED under forward bias conditions is shown in Figure 12.13b.

In an LED, the number of photons of light emitted is nearly proportional to the number of free charge carriers crossing the junction. (Not all of the recombinations result in emission of photons.) The optical power output is proportional to the rate at which photons are emitted. Thus, the optical output power is almost directly proportional to the forward current flowing in the LED. This is illustrated in Figure 12.14. The slight departure from linearity is due to second-order effects related to heating of the junction by the higher current levels.

The LED we have described suffers from some characteristics that are not desirable for a source in a fiber optical communication system. The electrons and holes that cross the junction diffuse into the material for some distance before recombination occurs, and the emitted photons travel in all directions instead of being confined to a small area and range of directions that can be easily coupled to a fiber. This simple junction is called a *homojunction* because it is the junction between doped semiconductor of the same material on both sides of the junction. A *heterojunction* occurs when a junction between different materials is formed. By proper selection of materials, a heterojunction LED can be formed in which the charge carriers are confined to a thin active layer, thereby confining the light emitted to a smaller region. The energy level diagram of such a heterojunction under forward bias conditions is shown in Figure 12.15.

One practical form of LED structure for optical fiber communication is the *etched-well* or *Burrus* diode construction shown in cross section in Figure 12.16. In this structure, light emission is limited to the active layer by the choice of materials and doping, which create a energy band structure confining the carriers. The electrical contact to the bottom surface is confined to a small area directly below the etched well so that the current is confined

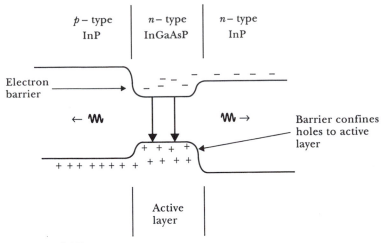

p – type
InP

n – type
InGaAsP

n – type
InP

Electron
barrier

Barrier confines
holes to active
layer

Active
layer

FIGURE 12.15
Energy diagram of heterojunction LED under forward bias conditions.

FIGURE 12.16
Typical etched well LED (not to scale).

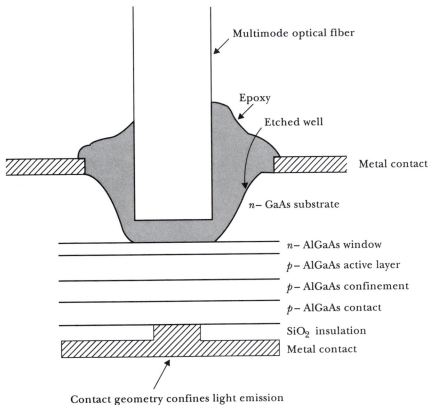

Multimode optical fiber

Epoxy

Etched well

Metal contact

n– GaAs substrate

n– AlGaAs window

p– AlGaAs active layer

p– AlGaAs confinement

p– AlGaAs contact

SiO₂ insulation

Metal contact

Contact geometry confines light emission
to the region under the fiber

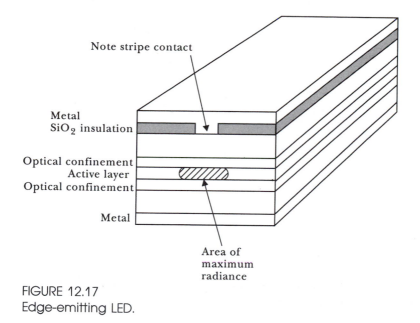

Note stripe contact

Metal
SiO$_2$ insulation

Optical confinement
Active layer
Optical confinement

Metal

Area of
maximum
radiance

FIGURE 12.17
Edge-emitting LED.

to the region of the active layer near the fiber. Unfortunately, even with a well-designed structure such as this, only a fraction of the light emitted by the diode is coupled to the fiber because only light that is emitted in a direction within the acceptance cone of the fiber is coupled. Nevertheless, the etched well heterojunction LED can be a effective source.

Another commonly used structure is the edge emitting LED illustrated in Figure 12.17. Again, the recombination of carriers is confined to an active layer by the choice of materials. Current is confined to a narrow stripe by the electrical contacts. Finally, if the proper dopants are added to the layers directly above and below the active layer, the active layer can have a slightly higher index of refraction than the adjoining confinement layers, resulting in a light-guiding structure. The active layer with its higher index of refraction and the confinement layers with their lower index of refraction form an optical slab waveguide that confines much of the emitted light to the active layer by repeated internal reflection, just as light is guided in the core of a fiber. Thus, light is emitted from a bright spot on the edge of the diode. The angular distribution of the optical power emitted is more tightly confined in the vertical plane than in the plane of the active layer, as illustrated in Figure 12.18. This is due to the light-guiding action in the active layer. The end of the diode opposite the desired point of emission can be coated with a reflective material to enhance the efficacy of the device.

If a pulse of current is applied to an LED, the light output will take some time to build up after the current is applied because a time interval is required for the concentration of charge carriers to build up in the active layer. After

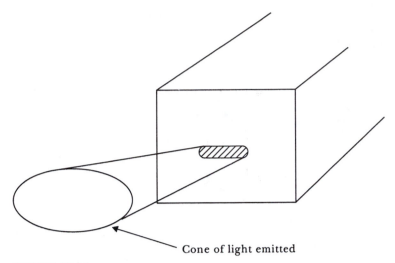

FIGURE 12.18
Radiation from an edge-emitting LED spreads more rapidly horizontally (in the plane of the active layer) than vertically.

the current drops to zero, light is emitted for a short period until the carriers all have recombined. The result is a smeared and broadened pulse of light. This imposes a limit on the rate at which the light output can be modulated by fluctuation in the drive current. Thus, carrier lifetime in the device leads to a bandwidth restriction.

The light emitted by an LED is spread over a range of wavelengths because a range of energy transitions from levels in the conduction band to the valence band is possible. The relative power output versus wavelength for a typical LED is shown in Figure 12.19. Recall that when light is spread

FIGURE 12.19
Typical power spectral density versus wavelength for an LED with an FWHM line-width of 0.12 μm.

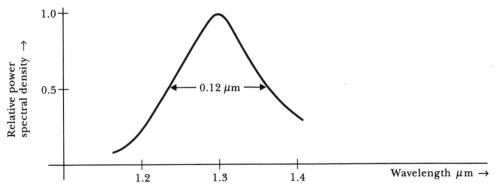

over a range of frequencies, chromatic dispersion occurs in the fiber due to material and waveguide effects. Since the linewidth of an LED is fairly wide, the use of an LED accentuates the problem of chromatic dispersion. As we will see, the linewidth and therefore the chromatic dispersion is much less when laser diodes are used.

The line shape of a typical LED can be modeled by a gaussian distribution, given by

$$P(\lambda) = \exp\left[-\frac{(\lambda - \lambda_c)^2}{2\sigma^2}\right] \tag{12.6}$$

Where $P(\lambda)$ is the normalized spectral power density of the LED output as a function of wavelength λ. The wavelength of peak output is λ_c. The linewidth is determined by the parameter σ, which is known as the *rms linewidth*. As indicated in Figure 12.19, linewidth can also be specified as the *full width half maximum* (FWHM) width denoted by Δ. It can be shown that the rms linewidth and the FWHM linewidth are related by

$$\blacksquare \qquad \Delta \simeq 2.36\sigma \tag{12.7}$$

Laser diodes

The term *laser* was originally an acronym for *light amplification by stimulated emission of radiation*. However, the term now usually refers to a light-amplifying material combined with a positive feedback structure resulting in an optical oscillator.

Certain materials have an energy structure in which the application of energy in an appropriate form can create a *population inversion* in the electron energies, a condition in which many electrons are in high energy states and many empty levels exist at lower energies. The population inversion is created and maintained by the application of external energy, known as *pumping*. Pumping can be done by external light from flash tubes around the active material or in some cases by passing an electrical current through the active material.

A photon traveling through a material with a population inversion can cause additional photons to be emitted by stimulated emission. This effectively increases the power of the original lightwave. Thus, the original lightwave is amplified by passing through the material. If the lightwave is reflected back into the material by mirrors, the lightwave can grow in power, limited by the ability of the pumping to maintain the population inversion.

A laser diode as used in fiber optic communications is very similar to the edge emitting diode shown in Figure 12.17. In the case of the laser diode, the front and back faces are cleaved to produce two parallel and highly reflective surfaces. In some cases, the back face may be coated with a partially or totally reflective material. The population inversion is created by forward current in

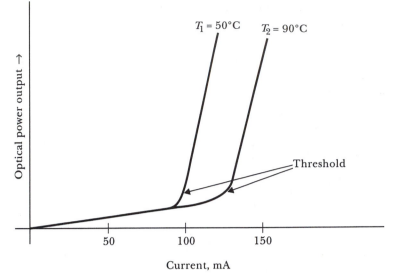

FIGURE 12.20
Power output for a typical laser diode versus current. Note the change in threshold current with temperature.

the diode, which brings free electrons and holes together in the active layer of the diode. For small amounts of forward current, the degree of the population inversion is not great enough to overcome the losses in power at the faces of the device. In this case, a self-sustaining, oscillation does not occur, though some small amount of light is emitted due to spontaneous emission. As the current increases, the degree of population inversion increases to the point at which the optical gain of the device overcomes the losses, and a self-sustaining lightwave oscillation occurs. Thus, the device reaches a *threshold* at which lasing commences. As the forward current is increased beyond threshold, the ability to maintain the population inversion increases, and the optical power output increases approximately in proportion to the current. Plots of output power versus current for a typical laser diode are shown in Figure 12.20 for several temperatures.

As we have seen, a laser source consists of a material with optical gain and positive feedback provided by mirrors at the ends of the material. In a solid material, optical gain is provided over a range of wavelengths due to the band structure of the solid; in other words, stimulated emission can occur at any frequency that corresponds to the energy difference between a free electron and a hole. One prerequisite for optical oscillation is that the gain must be large enough to overcome the losses. A second requirement is that the phase of the reflected lightwave should match the original phase so that the lightwave fed back adds constructively to the accumulated lightwave. These requirements for oscillation are the same as for an ordinary electronic circuit oscillator.

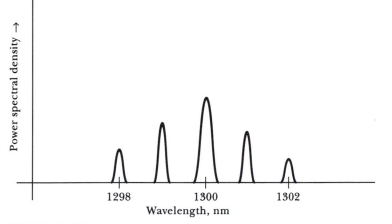

FIGURE 12.21

Power spectral density of a typical multimode laser diode.

Because of the phase-shift requirement, it is necessary for the round-trip distance through the material to be an integer number of wavelengths. In equation form, we have

$$\frac{\lambda}{n} = \frac{2L}{m} \tag{12.8}$$

where λ is the free-space wavelength of the lightwave, n is the index of refraction of the active material, L is the one-way distance between the reflectors, and m is an integer. Note that λ/n is the wavelength in the active material. (We have assumed that the active material fills the region between the reflectors.) Thus, optical oscillation occurs at specific frequencies and only if the optical gain is large enough. Typically in laser diodes, the conditions are met for several wavelengths, and oscillations occur simultaneously at each wavelength. In this case, the laser is called a *multimode laser*. The output power of a typical laser diode is shown versus wavelength in Figure 12.21.

Since stimulated photons are emitted with the same frequency, phase, and polarization as the incident photon, any modulation present on the incident lightwave is also present on the output. Thus, the potential exists for using this effect for amplifying modulated lightwave signals. However, this potential has not yet been fully realized in practical form for communication system applications, and lasers are usually used as oscillators.

Now that we have described the principles of laser diodes and LEDs, we will give a brief discussion and comparison of their characteristics as they relate to fiber optical communications. In general, LEDs are cheaper and simpler to operate than the more complex laser diode, which requires more complicated driver circuitry. Thus, LEDs are used where they allow the desired system requirements to be met. Laser diodes are reserved for certain higher performance applications.

The size and divergence of the light beam produced by a source are important characteristics relevant to coupling the optical power into a fiber. Since the beam size and divergence are smaller for laser diodes, it is easier to couple the optical power from a laser diode into a fiber than it is for an LED. As a result, the power coupled into a fiber can be considerably more for a laser than for an LED. This difference is most important for single mode fibers in which the core diameter and acceptance angle are smaller than they are in multimode fibers. Typical power levels coupled into multimode fibers from LEDs are on the order of 100 μW. The power levels for laser diodes are on the order of several milliwatts for either multimode or single mode fibers.

As we have noted, the linewidth of the output of LEDs is considerably wider than for laser diodes. This results in greater chromatic dispersion for lightwave carriers produced by LEDs, and a lower modulation bandwidth results. Furthermore, the optical output power for laser diodes changes faster in response to changes in the drive current than it does for LEDs. Thus, in general, laser diodes can achieve higher modulation rates than LEDs. However, for a multimode fiber operated in the low material-dispersion vicinity around 1300 nm, total dispersion is dominated by modal dispersion, and there is no bandwidth advantage to be gained from the narrower linewidth of a laser for this particular combination of parameters.

As indicated in Figure 12.20, the threshold current for laser diodes varies significantly with temperature. Therefore, to obtain a constant optical power output as temperature changes, it is necessary to adjust the drive current. Often, the optical output from the rear face of the laser diode is applied to an optical detector that can be included in the same package to monitor the laser output. A feedback control circuit can then be used to adjust the drive current to the laser to produce constant optical output. The optical power output of an LED is not such a strong function of temperature, so a simpler drive circuit can be used.

Initial attempts at laser diodes resulted in devices with lifetimes of only a few minutes. A large number of failure modes have been found and remedied. The lifetimes have improved to the point at which laser diodes have expected useful lifetimes on the order of 10 years. Nevertheless, LEDs are still more reliable.

The system designer must select a light source and compatible fiber to achieve the desired system requirements with minimum cost and complexity. For a short link of about several hundred meters and a low desired modulation bandwidth of several tens of megahertz, an LED operating in the vicinity of 850 nm with a step index multimode fiber would provide a good solution. For a link of several kilometers and a modulation bandwidth of several hundred megahertz, we could operate in the vicinity of 1300 nm using an LED and a graded index fiber. The graded index fiber and lower dispersion at 1300 nm would allow increased bandwidth. For a system with a modulation bandwidth of several gigahertz and a link approaching 100 km, a single mode laser diode and a single mode dispersion shifted fiber operating in the vicinity of 1500 nm would be used.

12.3
PHOTODETECTORS

The detector in an optical communication system converts the received lightwave into an electrical current that is amplified and used to recover the message signal or data. Although other devices such as phototransistors and photomultiplier tubes can be used, the most commonly used devices are the *pin diode* and the *avalanche photodiode* (APD). In this section, we will discuss the principles of photodetection in semiconductor junction diodes and the special features of the *pin* diode and the APD.

Figure 12.22 illustrates a *pn* junction under reverse bias. The applied voltage aids the electric field in the depletion region directed from the *n* side to the *p* side. This electric field holds the electrons on the *n* side and the holes on the *p* side of the junction, so very little current flows in the absence of light. When a photon of light strikes the depletion region, it can be absorbed, creating a hole–electron pair. These carriers are then pulled apart by the electric field, resulting in a short pulse of current in the external circuit. Photons that strike

FIGURE 12.22
Photodetection in a *pn* junction.

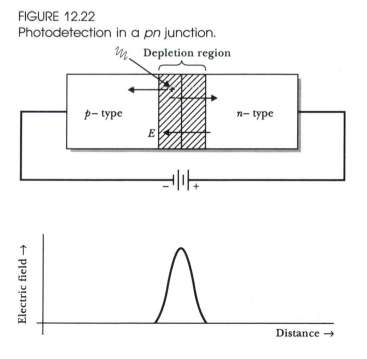

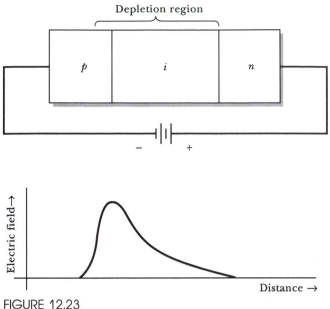

FIGURE 12.23
The *pin* diode photodetector.

the material outside the depletion region can also create hole–electron pairs, but these carriers are not quickly separated by an electric field. Instead, they move by diffusion and may eventually move into the depletion region where they are swept across by the electric field, resulting in current in the external circuit. On the other hand, charge carriers liberated away from the junction may recombine before crossing the junction without contributing to current in the external circuit. Thus, to achieve maximum speed of response and effectiveness, it is important for all of the photons to be absorbed in the depletion region. In an ordinary *pn* junction, the depletion region is not thick enough to ensure that all of the photons are absorbed before moving through the region.

The *pin* diode illustrated in Figure 12.23 is constructed with a layer of very lightly doped or nearly intrinsic material on the *n* side of the junction. With sufficient applied reverse bias, the depletion region extends all the way across the intrinsic layer. Most of the photons are absorbed in this thick depletion layer. The charge carriers liberated are quickly separated by the field in the depletion region and contribute to external current. A cross section of a typical surface-illuminated *pin* diode is shown in Figure 12.24. In this surface-illuminated device, the *p* layer is thin, so most of the incident photons travel through it to be absorbed in the intrinsic layer.

The *quantum efficiency* η of a *pn* junction used as a photodetector is the ratio of the rate of light-generated charge carriers crossing the plane of the junction, denoted by r_c, to the rate of photons incident on the device, denoted

590

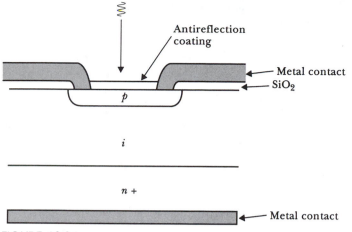

FIGURE 12.24
Typical surface illuminated *pin* photodiode structure.

by r_{p}. In equation form, this is

$$\eta = \frac{r_{\mathrm{c}}}{r_{\mathrm{p}}} \text{ charge carriers/photon} \tag{12.9}$$

The quantum efficiency is often expressed as a percentage and is a function of the wavelength of the light, the material used, and the physical construction of the device. Typical quantum efficiencies of devices used in fiber optical communication systems are in the vicinity of 75%.

The energy carried by each photon is hf. Therefore, the rate of arrival of photons at the junction is related to the optical power falling on the junction by

$$r_{\mathrm{p}} = \frac{P}{hf} \text{ photons/s} \tag{12.10}$$

The photocurrent is the product of the rate of charge carriers crossing the plane of the junction and the charge carried by each carrier. Of course, the carrier charge is the charge of an electron, $q = 1.59 \times 10^{-19}$ C. In equation form, we have

$$I_{\mathrm{p}} = qr_{\mathrm{c}} = \eta q r_{\mathrm{p}} \text{ A} \tag{12.11}$$

From Equations 12.9 through 12.11, we can find the following relation

$$I_{\mathrm{p}} = \frac{\eta q}{hf} P \text{ A} \tag{12.12}$$

The *responsivity* of a photodiode is the ratio of the photocurrent to the incident optical power. From Equation 12.12, we find

$$R_{\mathrm{d}} = \frac{I_{\mathrm{p}}}{P} = \frac{\eta q}{hf} \text{ A/W} \quad \text{(for a } pin \text{ diode)} \tag{12.13}$$

In addition to the photocurrent, the normal reverse current, caused by thermally generated carriers, crosses the junction. Since this current flows even in the absence of light, it is called the *dark current*. Dark current is an undesirable effect because it varies randomly due to shot noise effects and can obscure the current variations due to modulation of the received lightwave. Dark current is higher for materials with smaller bandgaps because a smaller bandgap leads to easier thermal generation of charge carriers.

To create charge carriers, a photon must have enough energy to raise an electron from the conduction band to the valence band. Thus, the quantum efficiency is effectively zero for light with such a long wavelength that the photon energy is less than the bandgap. It is therefore necessary to select a material with a bandgap less than the photon energy. However, as we have noted, the bandgap should be large to minimize the dark current. Thus, the best materials for a given wavelength have a bandgap slightly less than the photon energy. At 850 nm, silicon is a good choice, resulting in high quantum efficiency and low dark current. Beyond about 1100 nm, photon energy is insufficient to create charge carriers in silicon, and other materials such as germanium must be used. Alloys such as indium gallium arsenide have a bandgap that depends on composition, so a material optimized for a given wavelength can be produced.

In a power-limited system, it is desirable to be able to reliably detect the presence of only a few photons in a bit interval. Such small electrical currents are difficult to amplify without adding noise in the amplifier circuits. The APD avoids some of this difficulty by amplifying the detected charge carriers in the device. An illustration of an APD is shown in Figure 12.25. It consists of a layered *pn* junction. When a reverse bias is applied to the device, the width of the depletion region at the *pn* junction increases. For a sufficiently high reverse bias, the depletion region reaches through both the *p* layer and the intrinsic layer. A strong electric field is created at the *pn* junction and a weaker electric field also exists in the intrinsic layer.

When a photon creates a hole–electron pair in the intrinsic absorption layer of the APD, the electric field causes the free electron to drift toward the junction. The field at the junction is strong enough to cause the electron to gain enough energy so that additional hole–electron pairs are created by collisions with the neutral atoms in the crystal lattice. These carriers in turn generate more carriers. Thus, the effect of a single photon can be multiplied by a large factor, and many charge carriers flow in the external circuit as a result. This multiplication effect is one of the causes of reverse breakdown in ordinary *pn* junctions. The responsivity of an APD can be obtained by including a gain factor M in Equation 12.25, resulting in

$$\blacksquare \qquad R_{\mathrm{d}} = \frac{I_{\mathrm{p}}}{P} = \frac{\eta M q}{hf} \; \mathrm{A/W} \qquad \text{(for an APD)} \qquad (12.14)$$

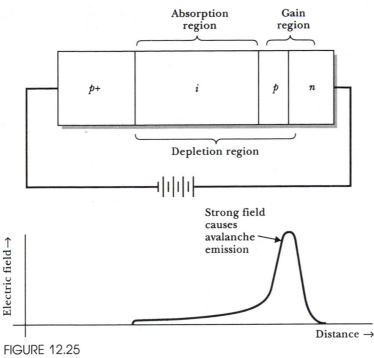

FIGURE 12.25
Avalanche photodiode.

For typical APDs used in fiber optical communication systems, the gain factor M is on the order of 100.

Unfortunately, the avalanche gain mechanism is random. The number of electrons resulting from each photon is a random variable. Thus, the photocurrent fluctuates randomly, which can obscure the data when the received power is small. Nevertheless, the APD usually gives better overall system noise performance than the *pin* diode. The noise contributed by the amplifier is more significant for the smaller current produced by the *pin* diode than it is for the internally amplified current of the APD.

12.4
SOURCES OF SYSTEM NOISE AND PERFORMANCE LIMITS

In this section, we will consider the sources of system noise and evaluate the ultimate error performance for several digital modulation techniques. First, we will briefly consider the sources of noise in a practical system. Then, we will consider a noncoherent receiver for an on-off keyed lightwave and derive an expression for the ultimate lower limit on bit-error probability set by the

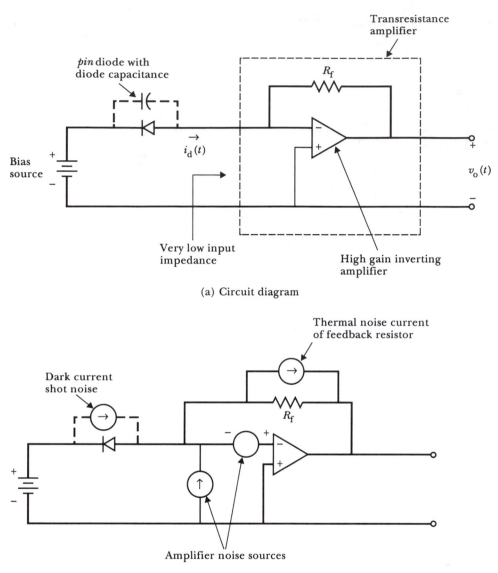

Transresistance amplifier

R_f

pin diode with diode capacitance

$i_d(t)$

Bias source

+
−

Very low input impedance

High gain inverting amplifier

+

−

$v_o(t)$

+

−

(a) Circuit diagram

Thermal noise current of feedback resistor

Dark current shot noise

R_f

+
−

−

+

−

+

Amplifier noise sources

(b) Circuit diagram with noise sources.

FIGURE 12.26
Commonly used receiver configuration.

quantum nature of lightwave detection. Finally, we will consider coherent receivers for PRK modulated lightwaves.

The receiver circuit and noise sources

A commonly used noncoherent optical fiber receiver configuration is shown in Figure 12.26a. A constant voltage source reverse biases one end of a photodiode. The opposite end of the diode is connected to the input terminal of a high-gain inverting amplifier. A feedback resistor is connected from the input terminal of the amplifier to the output. The combination of the feedback resistor and the amplifier is known as a *transimpedance amplifier* or *current-to-voltage converter*. This is a well-known operational amplifier circuit, though most integrated circuit operational amplifiers do not have sufficient bandwidth for high data rate applications. If the open loop gain of the amplifier is very large, the input impedance of the transresistance amplifier is very small, approaching a short circuit. The photocurrent flows through the feedback resistor and the resulting output voltage is given by

$$v_o(t) = -R_f i_d(t) \qquad (12.15)$$

where $i_d(t)$ is the total diode current consisting of the sum of the dark current and the photocurrent.

The receiver of Figure 12.26a is called a noncoherent receiver because the photodetector does not respond to the phase of the received lightwave. The photodetector responds only to the received optical power variations. Later we will consider coherent receivers, in which a locally generated lightwave is added to the received lightwave before it falls on the photodetector. Then the detected current depends on the phase of the received light.

One advantage of the circuit of Figure 12.26a is that the reverse bias voltage across the photodiode is fixed by the bias source and changes little as the light falling on the diode changes. This is because only a small voltage is developed at the amplifier end of the diode due to the low input impedance of the transresistance amplifier configuration. Thus, the device can be biased for its best operating point.

Another advantage of the transresistance amplifier is that the capacitance of the diode does not need to be charged and discharged by the signal because the voltage across the diode is nearly constant. Therefore, the circuit bandwidth is not limited by the diode capacitance. An alternative viewpoint is that the input impedance of the transresistance amplifier is very low, so the circuit time constant is short, resulting in wide bandwidth.

Figure 12.26b shows the receiver circuit including noise sources. A noise current source is shown in parallel with the photodiode to account for the shot noise component of the dark current. The noise current source and noise voltage source at the input terminals of the amplifier account for the noise added by the amplifier. The noise current source in parallel with the feedback resistor accounts for the thermal noise generated in the resistor. When an APD is used, an additional noise source can be included to account for the random nature of the internal gain mechanism of the device. These noise sources combine to degrade the error rate of a digital system or the SNR of analog systems, but space does not permit a full analysis of these noise effects here. Instead, we will only investigate the ultimate limits on performance set by the discrete nature of light absorption for a few digital modulation techniques.

The quantum limit

Ideally, the noise sources shown in Figure 12.26b would be zero, but even if this were the case, the system performance would not be perfect because of the discrete nature of the lightwave itself. For example, when a digital signal modulates the carrier by the use of on-off modulation, it is possible for no photons to be received during a bit interval even when light is being transmitted. This is due to the random nature of absorption of photons by matter. The probability of absorbing k photons in a bit interval T_b is given by the Poisson distribution of Equation 7.16. If the average rate of photon absorption (and hence electron generation) when the lightwave is in the *on* state is denoted as r_c, the probability of k electrons in a bit interval, given that the data bit is a 1, is

$$P(k|d(t) = 1) = \frac{(r_c T_b)^k \exp(-r_c T_b)}{k!} \qquad (12.16)$$

Assuming that there is no illumination when the data bit is a 0 and that there is no dark current, no electrons are liberated when a 0 is transmitted. Thus, a 1 bit results in a random number of electrons with probability given by Equation 12.16. A 0 bit always results in no electrons. In the ideal noise-free case, the optimum receiver decides that a 1 was transmitted if one or more electrons are liberated during the bit interval, and the receiver decides in favor of a 0 only if no electrons are liberated. Under these assumptions, errors never occur when a 0 is transmitted. When a 1 is transmitted, an error occurs if by chance no photons are detected. Assuming that ones and zeroes are equally likely to be transmitted, the average error probability is

$$P(E) = \frac{1}{2} P(k = 0|d(t) = 1) = \frac{1}{2} \exp(-r_c T_b) \qquad (12.17)$$

The product $r_c T_b$ is the average number of photons detected, given that a 1 is transmitted. Since the number of photons is zero when a 0 is transmitted, the average number of photons for a data stream with equal numbers of ones and zeroes is given by

$$p = \frac{1}{2} r_c T_b \qquad (12.18)$$

Substituting Equation 12.18 into 12.17, we have

$$\blacksquare \qquad P(E) = \frac{1}{2} \exp(-2p) \qquad \text{(for noncoherent on-off keying)} \qquad (12.19)$$

where p is the average number of detected photons.

Equation 12.19 is called the *quantum limit* because it is the lower limit on the error probability of a noncoherent, on-off keyed, optical communication system. This limit is set by the discrete nature of light absorption. Optical fiber communication system requirements are often compared for an error probability of 10^{-9}. From Equation 12.19 we can find that the average number of detected photons needed to achieve this error probability is about $p = 10$.

Actual optical fiber communication systems are not capable of operating at the quantum limit due to the noise sources in the receiver. After all, in deriving the quantum limit, we assumed that a single electron liberated in the detector could be distinguished from no electrons liberated. Achievement of the quantum limit with the noncoherent receiver structure of Figure 12.26 would require some very low noise electronics indeed. Well-designed non-coherent fiber optical systems operate with received power levels from 10 to 20 dB higher than the quantum limit. Thus, typical noncoherent systems need on the order of a thousand photons per bit incident on the photodetector for an error rate of 10^{-9}.

Notice that in an optical fiber communication system, the fundamental limit on error performance is set by the particle nature of lightwaves when they interact with matter. In a microwave radio system, the limit is set by thermal noise emitted by the background. Microwaves are also quantized, but the energy per photon at these very low frequencies (compared with light-waves) is so small that quantization effects are seldom of importance in communications. In an optical fiber system, background illumination is seldom a problem. When the light source is *off*, no photons emerge from the receiving end of the fiber.

Coherent receivers

In principle, it is possible to greatly improve the sensitivity of the optical receiver by the use of coherent techniques. A coherent receiver is shown in Figure 12.27. The coherent receiver differs from the noncoherent receiver we

FIGURE 12.27
Coherent receiver adds a locally generated lightwave to the received lightwave before detection.

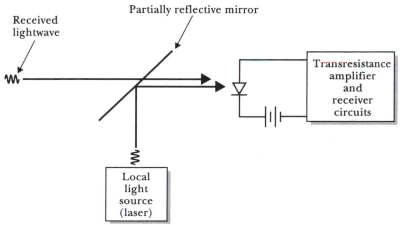

considered earlier in that a locally generated lightwave is added to the received lightwave before detection in the *pin* diode. Because the sum of the received and local lightwaves depends on their phase relationship, the coherent receiver is sensitive to the phase of the received signal. In the noncoherent receiver, in which no reference lightwave is added, absorption of photons from the received lightwave is not affected by phase.

We will only consider the case in which the received lightwave is PRK modulated by a bipolar NRZ data signal, denoted by $d(t)$, although coherent techniques are also useful with other modulation techniques.

In the analysis of the coherent receiver, it is necessary to consider the electromagnetic field nature of the lightwaves. We will assume, for ease of analysis, that the received and locally generated lightwaves are plane waves that are normally incident on the surface of the photodetector. Thus, the electric field intensity for the received lightwave component at the detector surface is given by

$$E_r(t) = E_{rp}d(t) \cos (\omega_r t) \tag{12.20}$$

where E_{rp} is the peak value of the electric field, $d(t) = \pm 1$ is the NRZ data waveform, and ω_r is the optical frequency of the received lightwave. As usual, we assume that $d(t) = +1$ when the data bit is a binary 1 and $d(t) = -1$ for a binary 0. For a normally incident plane wave, the electric field is constant over the surface of the detector, which has an area denoted by A. The locally generated field is also a plane wave, producing an electric field at the surface of the detector given by

$$E_l(t) = E_{lp} \cos (\omega_l t) \tag{12.21}$$

where E_{lp} is the peak field intensity and ω_l is the optical frequency.

When the locally generated lightwave has the same frequency as the received lightwave, the receiver is known as a *homodyne receiver*. We will also consider the *heterodyne receiver*, in which the locally generated wave is at a different frequency from the received wave. In this case, we will see that a difference frequency component is generated in the photodetector. This is similar to the familiar superheterodyne radio.

We also assume that the fields are linearly polarized in the same direction, so the total field is the algebraic sum of the received field and the locally generated reference field. Thus, the total field is given by

$$E_t(t) = E_{lp} \cos (\omega_l t) + E_{rp}d(t) \cos (\omega_r t) \tag{12.22}$$

The photodetector responds by producing a current proportional to the power in this combined wave. The power incident on the surface of the detector is given by

$$P_t(t) = \frac{A}{R_i} [E_t(t)]^2 \tag{12.23}$$

where A is the area of the photodetector and R_i is the intrinsic impedance of the material above the surface. For air, $R_i \simeq 377\ \Omega$. When $E_t(t)$ is squared, terms with frequencies of $2\omega_r$, $2\omega_l$, and $\omega_r + \omega_l$ result, but the photodetector does not respond to these optical frequencies. Thus, we retain only the low-frequency terms resulting in

$$P_t(t) = \frac{A(E_{lp})^2}{2R_i} + \frac{AE_{lp}E_{rp}}{R_i}\, d(t)\, \cos\,[(\omega_r - \omega_l)t] + \frac{A(E_{rp})^2}{2R_i}\, d^2(t) \quad (12.24)$$

The first term on the right-hand side is the average power of the locally generated wave. The middle term is the difference frequency component. The last term is constant and represents the average power of the received lightwave component, given by

$$P_r = \frac{A(E_{rp})^2}{2R_i} \quad (12.25)$$

where we have used the fact that $d^2(t) = 1$ always. Now we will consider several combinations of local optical frequency and relative field strength.

The first case we will consider is a homodyne receiver $(\omega_l = \omega_r)$ in which the peak field strength is the same for both fields $(E_{lp} = E_{rp})$. In this case, we find from Equations 12.24 and 12.25 that the total power incident on the detector is

$$P_t(t) = 4P_r \qquad \text{when } d(t) = 1$$

and

$$P_t(t) = 0 \qquad \text{when } d(t) = -1 \quad (12.26)$$

In this case, the total field at the detector is an on-off modulated signal. The total field is zero when the local wave and the received wave are out of phase so they add destructively. On the other hand, when the fields add in phase, the power at the detector is four times the average received power.

Recall that in the case of the noncoherent receiver with on-off modulation, the power at the detector for the *on* state was only twice the average received power. Thus, provided that the peak strengths of the received and local fields are equal, PRK modulation and homodyne detection has a factor of two advantage over on-off keying and a noncoherent receiver.

As before, we denote the average number of detected photons per data bit in the received wave as p. Recall that the rate of photon absorption by the photodetector is proportional to the total optical power at the detector. Thus, in the present case, the number of photons absorbed per data bit is $4p$ when the data bit is a 1 (because the power at the detector when a 1 is transmitted is four times the average received power). When the data bit is a 0, no photons are absorbed. If we assume that the system is noiseless except for quantization noise of the lightwave, no errors occur when the data bit is a 0

$[d(t) = -1]$. Furthermore, errors only occur for $d(t) = 1$ when by chance no photons are absorbed. When the data bit is a 1, the number of photons is Poisson distributed with an average of $4p$. We assume that $d(t) = 1$ half of the time. Thus, the average error probability is half the probability of no photons being absorbed given that the data bit is a 1. The probability of no photons when the data bit is a 1 is found from Equation 12.16 by setting $k = 0$ and $r_c T_b = 4p$. Then the average error probability is

$$P(E) = \frac{1}{2} P[k = 0 | d(t) = 1] = \frac{1}{2} \exp(-4p) \qquad (12.27)$$

It is important to note that this result applies for homodyne detection with the local field strength exactly equal to the received field and assumes that no noise other than the inherent discrete absorption process is present. Thus, it applies only for some rather idealized conditions that are not achieved in practice. Nevertheless, the result indicates an important limit on performance. Comparison of this result with the quantum limit of Equation 12.19 for non-coherent on-off keyed signaling shows that equal field homodyne detection of PRK signals requires only half as many photons to achieve a given error probability. Thus, the limit of Equation 12.27 has been called the *super quantum limit*. For example, to attain an error probability of 10^{-9}, the average number of received photons needs to be only about five for the super quantum limit as compared to about ten for the quantum limit.

Now we will consider another case of homodyne detection of PRK modulated optical signals, in which the local field strength is much higher than the received field (i.e., $\omega_l = \omega_r$ and $E_{lp} \gg E_{rp}$). In this case, the last term on the right-hand side of Equation 12.24 for the power at the detector is much smaller than the other terms and can be dropped. Substituting the shortened version of Equation 12.24 into Equation 12.12, we find the following expression for the current in the photodiode

$$I_p(t) \simeq \frac{\eta q A (E_{lp})^2}{2hfR_i} + \frac{\eta q A E_{lp} E_{rp}}{hfR_i} d(t) \qquad (12.28)$$

The first term represents a large constant current in the photodiode due to the high power local field. We will denote this first term by I_{dc}. The second term is much smaller than the first and represents the detected NRZ data signal.

Therefore, we can write the diode current as

$$I_p(t) = I_{dc} + K d(t) \qquad (12.29)$$

where

$$I_{dc} = \frac{\eta q A (E_{lp})^2}{2hfR_i} \qquad (12.30)$$

and the amplitude of the NRZ data signal current is

$$K = \frac{\eta q A E_{lp} E_{rp}}{hfR_i} \tag{12.31}$$

Since the current I_{dc} represents charge carriers crossing the barrier of the *pn* junction, it will fluctuate randomly due to shot noise effects as discussed in Chapter 5. The rms value of this noise current was given in Equation 5.21 as

$$i_{sh} = (2qI_{dc}B_n)^{1/2} \tag{12.32}$$

where B_n is the noise equivalent bandwidth of the system used to measure the noise. The normalized power of this current is simply the square of the current. The power spectral density of the noise current can be found by dividing the power by the bandwidth. This results in

$$\mathcal{N}_0 = 2qI_{dc} \text{ W/Hz} \tag{12.33}$$

where, as usual, we have used the symbol $\mathcal{N}_0$ to indicate the single-sided PSD of white noise. In summary, the dc component of the diode current displays shot noise. This noise current is a white gaussian noise signal with the PSD given in Equation 12.33.

The amplitude of the detected NRZ data signal was denoted by K. Therefore the (normalized) power of the detected signal is K^2, and the signal energy per bit in the detected signal is given by

$$E_{bd} = K^2 T_b \quad \text{(homodyne detection)} \tag{12.34}$$

where we have used E_{bd} to denote the bit energy in the detected electrical signal component since we have reserved E_b to denote the received optical energy per bit. As usual, T_b is the time interval per data bit.

The effect of the dc component of the diode current can be eliminated by ac coupling in the amplifier. The remaining signal then consists of the NRZ data signal plus white gaussian noise. These are exactly the same as the received signal and noise components in the baseband communication system discussed in Section 8.3. Therefore, the data and clock signals can be recovered by the same methods discussed for baseband systems in Chapter 8. The resulting probability of error was given in Equation 8.46 as

$$P(E) = Q\left[\left(\frac{2E_{bd}}{\mathcal{N}_0}\right)^{1/2}\right] \tag{12.35}$$

where we have used the present notation for the bit energy in the detected electrical signal E_{bd} in place of E_b, which we are now using to denote the average received optical power per data bit. Now, if we substitute Equation 12.31 into 12.34 and the result into 12.35, and also substitute 12.30 into 12.33 and the result into 12.35, we obtain

$$P(E) = Q\left[\left(\frac{4\eta E_b}{hf}\right)^{1/2}\right] \tag{12.36}$$

Since hf is the energy per photon, and E_b is the received optical energy per bit, $E_b/(hf)$ is the number of received photons per bit. If we assume an ideal detector that detects all of the photons (i.e., $\eta = 1$), then Equation 12.36 can be written as

$$\blacksquare \qquad P(E) = Q\left[(4p)^{1/2}\right] \qquad \text{(homodyne detection of PRK)} \qquad (12.37)$$

Note that this result applies for homodyne detection in which the local field is much stronger than the received field. Also we have assumed that the only source of noise is the shot noise caused by the discrete detection process.

In the case of high local field homodyne detection, note from Equations 12.30, 12.31, and 12.32 that both the detected signal amplitude and the shot noise current are proportional to the strength of the local field. Thus, by using a high power local optical source, the signal and shot noise can be made large enough so that the effects of dark current and amplifier noise are negligible. Therefore, Equation 12.37 represents a level of performance that we can eventually hope to approach with a practical implementation. On the other hand, the quantum limit for noncoherent reception of on-off keyed modulation would be very difficult to approach because of the difficulty of amplifying the current of only a single electron without adding significant noise.

In summary, high field homodyne detection results in a detected signal that is much larger than for direct detection. This is due to the simple fact that the power of an electromagnetic plane wave field is proportional to the square of the field strength. Thus, when a small received field is added to a large locally generated field, the resulting power fluctuations of the combined field can be much larger than the power in the received field by itself. It can be expected that coherent detection will eventually be able to achieve practical operation with much smaller received power than direct detection. Unfortunately, this is not presently the case with optical fiber systems because of implementation problems, which we will detail later. However, the field is advancing rapidly, and practical systems that approach the ultimate performance limits are expected in the near future.

Now we want to consider the case of heterodyne detection (i.e., $\omega_1 - \omega_r = \omega_{if}$) in which the local field is much stronger than the received field. In this case, the diode current can be found to be

$$I_p(t) = I_{dc} + Kd(t)\cos(\omega_{if}t) \qquad (12.38)$$

Notice that if ω_{if} is zero, so that we have the homodyne case, this reduces to Equation 12.29. I_{dc} and K were defined earlier in Equations 12.30 and 12.31, respectively.

From Equation 12.38, we can see that the diode current contains a dc component that is the same as for the homodyne case. This dc current exhibits a shot noise component that can be modeled as a white gaussian noise signal just as before. The last term on the right-hand side of Equation 12.38 is a PRK modulated signal with its carrier frequency equal to the difference between the optical frequencies. After ac coupling to eliminate the dc component, the detected signal consists of a PRK signal plus WGN. The methods for carrier recovery and demodulation discussed in Chapter 8 apply directly to the heterodyne detected optical signal.

The bit energy contained in the PRK signal term on the right-hand side of Equation 12.38, is given by

$$E_{bd} = \frac{1}{2} K^2 T_b \qquad \text{(heterodyne detection)} \qquad (12.39)$$

Comparison of Equations 12.34 and 12.39 shows that there is only half as much bit energy in the heterodyne detected signal as in the homodyne case. As we have noted, the shot noise in the detected signal is the same for homodyne or heterodyne detection. The error probability of both the baseband signal obtained in homodyne detection and the PRK signal obtained in heterodyne detection is given by Equation 8.46. Therefore, heterodyne detection is less effective than homodyne detection by a factor of two. The resulting lower limit on error probability for heterodyne detection of PRK signals is given by

$$\blacksquare \qquad P(E) = Q\left[(2p)^{1/2}\right] \qquad \text{(heterodyne detection of PRK)} \qquad (12.40)$$

Recall that this result was derived under the assumption that the local field is much stronger than the received field.

Table 12.1 summarizes the results we have found for the digital modulation and receiver techniques considered. Noncoherent reception of on-off keying is the most commonly used technique for digital transmission at present (1989). The likelihood of approaching the maximum theoretical performance or quantum limit with this technique is deemed to be small because of the difficulty of reliably detecting a single photon. Thus, current system performance is typically 10–20 dB worse than the quantum limit.

TABLE 12.1
Summary of the performance limits for several modulation and receiver techniques

Modulation type	Receiver type	Theoretical error probability	Likelihood of approaching theoretical performance	Photons/bit needed for $P(E) = 10^{-9}$
OOK	Noncoherent	$\frac{1}{2}\exp(-2p)$	Poor	10
PRK	Homodyne $E_l = E_r$	$\frac{1}{2}\exp(-4p)$	Poor	5
PRK	Homodyne $E_l \gg E_r$	$Q[(4p)^{1/2}]$	Good	9
PRK	Heterodyne $E_l \gg E_r$	$Q[(2p)^{1/2}]$	Very good	18

Homodyne detection of PRK, in which the received and local field strengths are equal, has a theoretical performance limit that is 3 dB better than noncoherent on-off keying. This is similar to the difference we found between PRK and on-off modulation for conventional additive WGN channels. However, this result is not likely to be of practical importance because the photocurrents are very small. Dark current and amplifier noise will seriously degrade the performance just as they do for noncoherent on-off keying.

Homodyne detection with a strong local field has a theoretical performance slightly better than the quantum limit. Because the detected signal is proportional to the local field strength, it is possible to make the amplifier and dark current noise negligible compared with the detected signal. Thus, the prospects for approaching the theoretical limit for this technique are deemed good. However, many problems need to be solved before practical implementations are achieved. For example, the local field must be at the same optical frequency as the received field and must be in time phase. This can be achieved by the use of phase-locked loop carrier recovery techniques similar to those discussed in Chapter 8. Also, the local field must be of the same polarization as the received field. The polarization of the received field can change with time due to changes in stress or in temperature of the fiber. The alignment of the fields so that they have the same phase relationship over the active surface of the detector is required. Nevertheless, workers in the field are optimistic about eventually being able to meet these requirements.

Heterodyne detection of PRK with a strong local field has a theoretical performance limit that is 3 dB worse than homodyne detection. However, heterodyne detection should be easier to implement because a free-running local oscillator that is not locked in phase and frequency with the received signal can be used. Thus, the prospects for approaching the theoretical limit of performance are rated very good.

12.5
DESIGN OF OPTICAL FIBER LINKS

The design of a communication system using optical fibers begins with selection of potential system components such as the light source, optical cable, detector, and receiver circuits. The resulting prototype system is then analyzed to see that each of the system specifications is met. In cases in which the specifications are not met, another choice of one or more of the system components is made, and the analysis is again carried out. Even when the specifications are met by a combination of components, it is important to consider whether another combination that makes a better use of resources will also meet the specifications. Thus, design is partly a trial and error process. Sometimes even the least expensive components provide a system that meets the desired level of performance so the design process is easy. On the other hand, the specifications are sometimes difficult to meet, and only a carefully selected combination of components will provide a satisfactory solution.

A large number of factors must be considered in the design, including mechanical considerations such as minimum allowed bend radius of the cable, maximum tensile load allowed on the cable both during installation and during operation, and the crushing strength of the cable. Often the space available for the cables is severely limited and becomes a major design consideration. Provisions for possible future upgrading of the link should also be considered. However, in this section, we will confine our attention to the communication aspects of the link. These are the overall modulation bandwidth of the link and the received optical power level. The modulation bandwidth must be great enough so that the frequency components of the analog or digital signal to be conveyed are accommodated. The received optical power must be large enough so that the SNR of an analog system is above the minimum specification or so that the error probability of a digital system is sufficiently low.

System bandwidth considerations

The overall link bandwidth is determined by four factors: transmitter bandwidth, modal dispersion in the fiber, chromatic dispersion in the fiber, and the receiver bandwidth. Of course, modal dispersion does not occur if single mode fibers are in use. A model of an optical fiber link for the purposes of bandwidth analysis is shown in Figure 12.28. The input signal current is first passed through a lowpass filter that accounts for the bandwidth limitations of the light source and driver circuits. The filtered current is then converted linearly into transmitted optical power denoted by $P_t(t)$. As it propagates through the fiber, the optical power is lowpass filtered by the dispersion mechanisms. Notice that it is the optical power as a function of time that is considered as the input signal to the optical components. In an electrical component, it is the voltage or current signal that is filtered rather than the power. Thus, we treat the optical power in a manner analogous to the input voltage or current in an electrical system block. For example, given the input optical power as a function of time and the impulse response of a fiber, we could compute the optical power at the output as a function of time by convolving the input power and the impulse response. As an alternative, we could multiply the Fourier transform of the input power times the transfer function of the fiber to obtain the Fourier transform of the output power.

As indicated in Figure 12.28, the received optical power is linearly converted to current. This current is then passed through a lowpass filter that models the bandwidth limitations of the detector and amplifier circuits. Provided that the conversions of current to optical power and back again are indeed linear, the overall system is linear. The overall linear system can be characterized by either its impulse response or its transfer function.

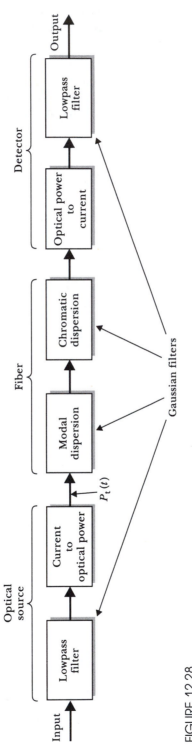

FIGURE 12.28
Model for bandwidth considerations.

The overall transfer function of the system is the product of the transfer functions of the individual blocks of Figure 12.28. (A constant multiplicative factor could also be included to account for attenuation, but instead we will treat this effect in a separate analysis.) However, the exact transfer functions of each of the filters of the figure are seldom available, so an exact analysis of the overall system is often impossible. Nevertheless, it is reasonable to model each of the filter blocks as having a gaussian pulse as its impulse response. Thus, we will assume that each of the filters has an impulse response of the form

$$h(t) = (2\pi\sigma^2)^{-1/2} \exp\left(\frac{-t^2}{2\sigma^2}\right) \tag{12.41}$$

This is exactly the same form as the gaussian probability distribution and is plotted in Figure 12.29a. Recall that σ is the standard deviation and is a measure of the width of the gaussian pulse. Since the standard deviation is the rms value of the variations of the random variable from its mean, σ is often referred to as the *rms pulse width* for a gaussian impulse response. In the case of an optical fiber, σ is a measure of the spreading or dispersion of a narrow pulse of light as it travels through the fiber. Each block in the system model as well as the overall system will have its own characteristic value for σ. As in the case of linewidth of a optical source, the width of an impulse response can also be measured at the half-maximum points as indicated in Figure 12.29a. In this case, we obtain the *full width half maximum* (**FWHM**) pulse width, denoted by Δ. It can be shown that, for a gaussian pulse,

$$\Delta = 2.355\sigma$$

We will use rms pulse width except where noted otherwise.

The transfer function corresponding to the gaussian impulse response is the Fourier transform of Equation 12.41. This transfer function is given by

$$H(f) = \exp\left[-2(\pi\sigma f)^2\right] \tag{12.42}$$

Note that this also has the same gaussian functional form as illustrated in Figure 12.29b. Note also that this is the transfer function of a lowpass filter.

The actual impulse response and transfer functions of the various filter blocks in the model of an optical fiber link are not gaussian filters, but we expect them to act as lowpass filters; so the performance predicted by assuming a gaussian filter with the same bandwidth is expected to be close to the actual system performance. Actually, we anticipate that many systems will have a nearly gaussian response overall. This is due to an argument based on the central limit theorem. Recall that according to the theorem, when many independent random variables with similar variances are added, the sum tends to become a gaussian random variable. Also recall that the overall probability density function is obtained by convolving the individual density functions.

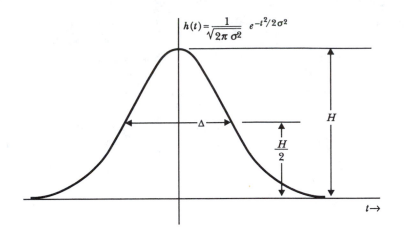

$$h(t) = \frac{1}{\sqrt{2\pi\,\sigma^2}}\,e^{-t^2/2\sigma^2}$$

H

$\frac{H}{2}$

Δ

$t \rightarrow$

(a) Impulse response illustrating FWHM pulse width Δ.
Note: $\Delta \approx 2.355\,\sigma$

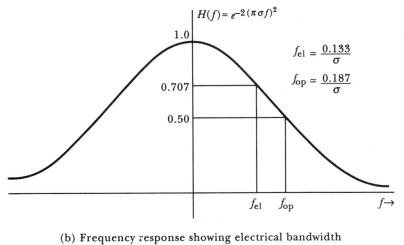

$$H(f) = e^{-2(\pi\sigma f)^2}$$

1.0

0.707

0.50

$f_{el} \approx \dfrac{0.133}{\sigma}$

$f_{op} \approx \dfrac{0.187}{\sigma}$

f_{el} f_{op} $f \rightarrow$

(b) Frequency response showing electrical bandwidth
f_{el} and optical bandwidth f_{op}

FIGURE 12.29
Gaussian filter characteristics.

Thus, the central limit theorem implies that when a large number of functions with similar variance are convolved, the result is gaussian. Therefore, since the overall impulse response of an optical fiber system is the convolution of the impulse responses of several blocks, we can expect the overall impulse response to be approximately gaussian. This will be especially true if the bandwidths of the individual blocks are similar so that a single block does not dominate the overall transfer function.

Often we find it convenient to express the magnitude of a transfer function in decibels. Usually, a transfer block has a current or voltage as its input and output, and we convert the transfer ratio to decibels by taking twenty times the logarithm of the magnitude. However, in the optical portion of the system, the input and output signals are power. Thus, the transfer functions in the optical system are power ratios and, as is standard practice for power ratios, are converted to decibels by taking ten times the logarithm of the magnitude.

The bandwidth of a system or component is usually specified as the 3 dB bandwidth. For a lowpass filter, this is the frequency at which the gain is 3 dB lower than the dc gain. For the gaussian transfer function of Equation 12.42, the dc gain is unity or zero decibels (regardless of whether we use ten or twenty times the logarithm). For an optical component, the 3 dB bandwidth is found from

$$-3 = 10 \log \left\{ \exp \left[-2(\pi \sigma f_{op})^2 \right] \right\} \tag{12.43}$$

where we have used the factor of ten for a transfer function relating optical power signals. Equation 12.43 can be solved for the *optical half-power bandwidth* f_{op} in terms of the rms dispersion σ, resulting in

$$\blacksquare \qquad f_{op} = \frac{0.187}{\sigma} \tag{12.44}$$

For an electrical component, we can find the 3 dB bandwidth from

$$-3 = 20 \log \left\{ \exp \left[-2(\pi \sigma f_{el})^2 \right] \right\} \tag{12.45}$$

Note that here we have used twenty times the logarithm to convert the transfer function to decibels because the transfer function is the ratio of currents or voltages. This can be solved for the *electrical half-power bandwidth*, resulting in

$$\blacksquare \qquad f_{el} = \frac{0.133}{\sigma} \tag{12.46}$$

An optical fiber can be considered as an optical component with the input and output signals being optical power. On the other hand, the conversions from current to optical power at the input and from optical power to current at the output can be included with the fiber, resulting in an electrical component. Thus, both an electrical and an optical bandwidth can be associated with the fiber. The dispersion parameter σ is the same in either case. Thus, by comparing Equations 12.44 and 12.46, we find the following relationship between the optical bandwidth and the electrical bandwidth:

$$f_{el} = 0.707 f_{op} \tag{12.47}$$

Actually, the electrical bandwidth is the more significant specification because the overall system has electrical input and output signals. The bandwidths of fibers are sometimes given as optical bandwidths, which can be converted to approximate electrical bandwidths by Equation 12.47.

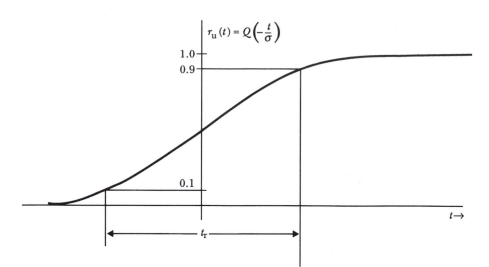

FIGURE 12.30
Step response of gaussian filter illustrating the 10 to 90% risetime. Note: $t_r \simeq 2.56\sigma \simeq 0.340/f_{el}$.

Often it is convenient to characterize the bandwidth limitations of a device by the 10 to 90% risetime of the response to a step input. The step response can be found from the impulse response by convolving a unit step function with the impulse response. The resulting step response for the gaussian impulse response of Equation 12.41 is

$$r_u(t) = Q\left(\frac{-t}{\sigma}\right) \qquad (12.48)$$

where the $Q(\cdot)$ function is defined in Equation 7.55. This step response is plotted in Figure 12.30, where the risetime is also indicated. It can be shown that the risetime is related to the rms dispersion parameter by

$$t_r = 2.56\sigma \qquad (12.49)$$

From Equations 12.46 and 12.49, we can relate the electrical half-power bandwidth to the risetime, resulting in

$$f_{el} = \frac{0.340}{t_r} \qquad (12.50)$$

Recall that we derived a similar relation between risetime and bandwidth in Equation 2.44, except that in that case the constant was 0.350 instead of 0.340. The difference is that in Section 2.5 we assumed a first-order system whereas we are now assuming a gaussian transfer function. The similarity of these results should give us additional confidence in the assumption of a gaussian response even though we know that the components in the actual system are not gaussian. Rarely will we try to design systems with such a small performance margin that the difference between 0.34 and 0.35 is significant.

The overall transfer function for two cascaded gaussian system blocks is the product of their individual transfer functions, given by

$$
\begin{aligned}
H_s(f) &= H_1(f)H_2(f) \\
&= \{\exp\,[2(\pi\sigma_1 f)^2]\}\{\exp[2(\pi\sigma_2 f)^2]\} \\
&= \exp\,\{2[\pi^2(\sigma_1^2 + \sigma_2^2)f^2]\} \\
&= \exp\,[2(\pi\sigma_s f)^2]
\end{aligned}
\tag{12.51}
$$

where we have defined the overall system rms dispersion factor σ_s in terms of the dispersion factors of the individual blocks by

$$
(\sigma_s)^2 = (\sigma_1)^2 + (\sigma_2)^2
\tag{12.52}
$$

Therefore, the overall system also has a gaussian transfer function, and the dispersion factors are combined by use of Equation 12.52. Of course, this development can be extended to any number of cascaded blocks, resulting in

$$
(\sigma_s)^2 = (\sigma_1)^2 + (\sigma_2)^2 + (\sigma_3)^2 + \cdots
\tag{12.53}
$$

where the number of terms on the right-hand side is the same as the number of cascaded blocks.

Equations 12.44 and 12.46 show that both optical bandwidth and electrical bandwidth are inversely proportional to the dispersion factor. Using this fact with Equation 12.53 results in the following equation for determining the overall bandwidth from the individual bandwidths of cascaded gaussian blocks

$$
(f_s)^2 = \frac{1}{(f_1)^{-2} + (f_2)^{-2} + (f_3)^{-2} + \cdots}
\tag{12.54}
$$

where the number of terms in the denominator matches the number of cascaded blocks. We could use either optical bandwidths or electrical bandwidths in this equation, but they should not be mixed. Since we are interested in the overall system with electrical input and output signals, we always use electrical bandwidths unless otherwise stated.

The system bandwidth required depends on the message and modulation technique in use. When the message is digital, the overall bandwidth sets a limit on the symbol rate that can be used effectively. For example, if we consider on-off keying by an NRZ waveform, the lowpass filtering of the overall system tends to round the individual pulses so that they spill over into adjacent bit intervals. This is intersymbol interference (ISI), discussed in Chapter 6. As we discussed in Section 8.3 and illustrated in Figure 8.21d and e, the eye pattern of a filtered digital signal shows the effects of ISI. When the bandwidth of the system is very wide, the eye pattern is open and the signal levels corresponding to 1 bits are clearly differentiated from the levels for 0 bits. As the bandwidth becomes smaller, the opening of the eye closes, showing that the 1-bit levels are not as far from the 0-bit levels as before. This closing of the eye pattern can be overcome by increasing the transmitted power level. Thus, the effects of lower bandwidth can be offset by using more power if the channel filtering is not too severe. When the system bandwidth becomes still smaller, the eye may completely close, indicating that some of the signal levels for 1 bits are lower than the levels for some of the 0 bits. In this case, increasing the transmitted power does not separate the 1-bit levels from the 0-bit levels. Thus, it is not possible to offset the filtering by additional transmitted power when the channel filtering is severe.

A commonly used rule of thumb for estimating the maximum data rate of on-off keyed NRZ optical fiber communication systems is

$$R_b \simeq \frac{0.3}{\sigma_s} \simeq 2.25 f_s \qquad (12.55)$$

where R_b is the approximate maximum practical data rate, σ_s is the rms dispersion of the overall system, and f_s is the overall electrical half-power bandwidth. Operation at this rate incurs a power penalty of about one to two decibels compared with a system without a channel bandwidth limitation. If the data rate is increased above the value given by Equation 12.55, the power penalty increases rapidly, and eventually it is not possible to obtain low error probabilities by increasing transmitted power. We will conclude our discussion of the bandwidth limitations of optical fiber systems with an example system calculation.

EXAMPLE 12.2

Find the overall electrical bandwidth for a 2.5 km optical fiber link constructed from the components listed below. Also estimate the maximum practical bit rate for on-off keying with NRZ pulses.

Source   LED operating at a wavelength of 830 nm with a FWHM linewidth of 40 nm. The 10 to 90% risetime of the source is 1.7 ns.

Fiber   Multimode graded index silica with an optical bandwidth of 300 MHz km when used with a laser source having negligible linewidth.

Detector   *Pin* diode and transresistance amplifier having a 10 to 90% risetime of 1 ns.

SOLUTION
First we find the electrical half-power bandwidth of each component and then use Equation 12.54 to find the overall bandwidth. Equation 12.55 is then used to estimate the maximum data rate.

The electrical bandwidth of the LED source can be estimated using Equation 12.50, resulting in

$$f_{\text{source}} = \frac{0.34}{t_{\text{r}}} = \frac{0.34}{1.7 \times 10^{-9}} = 200 \text{ MHz}$$

(Note that many engineers use $0.35/t_{\text{r}}$, which would result in 206 MHz.)

The fiber has two dispersion mechanisms that limit its bandwidth. These are modal dispersion due to the multiple ray paths allowed in a multimode fiber and chromatic dispersion due to the different velocities of the various wavelength components of the source. The 300 MHz km optical bandwidth specified for the fiber accounts only for the modal dispersion because it is specified for a narrow linewidth source in which chromatic dispersion does not occur. This optical bandwidth can be converted to electrical bandwidth by the use of Equation 12.47, resulting in

$$\text{electrical bandwidth} = 0.707 \times \text{optical bandwidth}$$

$$\text{electrical bandwidth} = 0.707 \times 300 \text{ MHz km}$$

$$= 212 \text{ MHz km}$$

For a 2.5 km length of fiber, the resulting electrical bandwidth of the modal dispersion block is

$$f_{\text{modal}} = \frac{212}{2.5} = 84.8 \text{ MHz}$$

The chromatic dispersion of the fiber is due to material dispersion and waveguide dispersion. At 830 nm, the material dispersion is the dominant effect and can be estimated from Figure 12.5, resulting in 95 ps/(nm km). The specified linewidth of the source is 40 nm. This can be converted to rms linewidth by the use of Equation 12.7, resulting in

$$\text{rms linewidth} = \frac{\text{FWHM linewidth}}{2.36}$$

$$= \frac{40 \text{ nm}}{2.36}$$

$$= 16.9 \text{ nm}$$

Now the rms linewidth and fiber length can be used to calculate the rms dispersion parameter associated with the material effects. The result is

$$\sigma_{\text{material}} = 95 \, \frac{\text{ps}}{\text{nm km}} \times 16.9 \text{ nm} \times 2.5 \text{ km}$$

$$= 4.01 \text{ ns}$$

The electrical bandwidth associated with the material dispersion effect can now be computed using Equation 12.46, resulting in

$$f_{\text{material}} = \frac{0.133}{\sigma_{\text{material}}}$$

$$= 33.2 \text{ MHz}$$

The bandwidth of the receiver can be estimated using the specified rise-time in Equation 12.50, resulting in

$$f_{\text{rec}} = \frac{0.340}{t_r} = 340 \text{ MHz}$$

Now we can estimate the overall system bandwidth by using Equation 12.54:

$$(f_s)^2 = \frac{1}{(1/f_{\text{source}})^2 + (1/f_{\text{modal}})^2 + (1/f_{\text{material}})^2 + (1/f_{\text{rec}})^2}$$

Substituting values and solving, we find that the system bandwidth is

$$f_s = 30.4 \text{ MHz}$$

Notice that the material dispersion is the most significant contributor to the system bandwidth. A significant increase in bandwidth could be achieved by using a laser diode with a much smaller linewidth.

Now the maximum practical data rate for on-off keying with NRZ pulses can be estimated using Equation 12.55, resulting in

$$R_b \simeq 68.5 \text{ Mbit/s}$$

Loss considerations

Now that we have considered the limitations on system bandwidth, we will consider the attenuation of the optical power in the system and the required power level at the receiver. Table 12.2 shows the loss budget for an optical fiber system constructed from the components of Example 12.2. The total LED output power is -3 dBm (0.5 mW) but only a fraction of this power can be coupled to the fiber because the light generated by the LED is emitted over an area that may be larger than the fiber core and at a range of angles greater than the acceptance angle of the fiber. The resulting coupling loss is 4 dB for the example illustrated. The coupling loss would be less if a laser diode with a smaller emitting area and smaller numerical aperture (resulting in a more collimated beam of light) were used. A fiber with a larger core and numerical aperture would accept more of the output of the LED. On the other hand, a single mode fiber with its small core diameter and small numerical aperture would accept much less.

TABLE 12.2
Power budget for the system of Example 12.2

LED power output	−3.0 dBm (0.5 mW)
Source coupling loss	4.0 dB
Connector loss	1.0 dB
Fiber loss (5 dB/km × 2.5 km)	12.5 dB
Splice losses	
(2 splices at 0.5 dB each)	1.0 dB
Connector loss	1.0 dB
Detector coupling loss	0.5 dB
Received optical power	−23.0 dBm (50 μW)

Often for convenience, a connector is used to attach a short piece of fiber from the source or detector (known as a pigtail) to the main fiber. Each of these connectors introduces some optical loss to the system. In the example illustrated in Table 12.2, two connectors, each with a loss of 1 dB, are assumed.

The loss of the main fiber depends on the fiber type and the wavelength of operation. The 5 dB/km figure used in the table is typical of multimode silica fibers operated at 830 nm. Simply shifting the system to a wavelength of 1300 nm could reduce the fiber loss to perhaps 2 dB/km.

Splices are necessary because fibers may not be available in the length required or it may not be possible to install the cable in a single piece even for short distances. In the example, two splices each with an attenuation of 0.5 dB are assumed.

Finally, some coupling loss occurs at the detector, but this is usually much less important than the coupling loss at the source.

The received optical power at the detector is found by subtracting the losses from the output power of the LED. In the example of Table 12.2, the resulting received power is −23 dBm or 50 μW.

The level of received power required for satisfactory operation depends on the type of modulation in use, the desired level of performance (SNR for an analog system or error probability for a digital system), and the noise performance of the receiver. When binary on-off keying is used and a bit-error probability of 10^{-9} is wanted, we have seen in Section 12.4 that a quantum-limited detector requires ten detected photons per bit on the average. Because of dark current, thermal noise in the detector load resistance, and amplifier noise, noncoherent receivers typically require many more photons than the quantum limit. If the receiver requires N photons per bit on average, the received optical power can be computed as

$$P_r = hfNR_b = \frac{hc}{\lambda} NR_b \qquad (12.56)$$

where, as usual, h is Planck's constant, f is the optical frequency, c is the speed of light, λ is the wavelength of the light, and R_b is the bit rate.

Notice from Equation 12.56 that the required received power is proportional to the data rate. Therefore, the received power sets a limit on the highest bit rate. When the system bandwidth is sufficient to allow operation at this limit, it is said to be a *power-limited system*. On the other hand, when the system bandwidth sets a lower limit on the data rate, it is said to be a *bandwidth-limited system*. It is important in the design process to determine which limit is in effect for a prototype system because different changes are called for to upgrade the system depending on the limiting factor. For example, in a bandwidth-limited system, increased bandwidth allows operation at a higher data rate. This might be achieved by changing to a source with a smaller linewidth to reduce chromatic dispersion, switching to a fiber with less modal dispersion, or using a source or detector with wider bandwidth to mention a few possibilities. On the other hand, if the system is power-limited, some of the effective changes would include the use of a higher power source, lower loss fiber, or a more sensitive (less noisy) detector.

EXAMPLE 12.3

Determine if the system of Example 12.2 and Table 12.2 is power-limited or bandwidth-limited if the receiver sensitivity is 20 dB above the quantum limit and the desired error probability is 10^{-9}.

SOLUTION
The quantum limit of Equation 12.19 indicates that about ten photons are required per bit to achieve the desired error rate with a noise-free receiver. Therefore, a receiver operating 20 dB above this limit will require $N = 1000$ photons/bit. Using Equation 12.56, we can solve for the maximum data rate allowed by the received power as

$$R_b = \frac{\lambda}{Nhc} P_r$$

Substituting values ($\lambda = 830$ nm, $N = 1000$, $c = 3 \times 10^8$ m/s, $P_r = 50\ \mu W$, and $h = 6.63 \times 10^{-34}$ J s), we find

$$R_b = 2.09 \times 10^{11} \text{ bits/s}$$

On the other hand, in Example 12.2, we found that the system bandwidth resulted in a limit on the data rate of 68.5 Mbit/s. Therefore, this system is bandwidth-limited, and improvement can be best obtained by changes to the system that increase the bandwidth. Because material dispersion is the most significant factor limiting the bandwidth, perhaps the most effective changes would be the use of a narrower linewidth source or changing to the low dispersion region for silica around 1300 nm.

SUMMARY

1. Optical fiber communication systems have a number of advantages over conventional electrical cable and waveguide systems. These include large bandwidth, low loss resulting in long distances between repeaters, smaller size and weight, electrical isolation between the source and destination, immunity to electromagnetic pulses or crosstalk, and potentially lower cost. On the other hand, a light source and detector must be provided, which is not necessary in an all-electrical system.

2. Optical fibers have of a core of higher index of refraction surrounded by a cladding of lower index. Several types of fiber are in use. They are step index multimode fiber, graded index multimode fiber, and single mode fiber. Fibers can be constructed of plastic, plastic-coated silica, or doped silica.

3. The important characteristics of optical fiber cables include their mechanical properties, attenuation of the lightwave as it travels through the fiber, and the modulation bandwidth.

4. Light is guided through a step index multimode fiber by multiple internal reflections at the core-cladding interface, so the ray paths consist of straight-line segments. In the graded index fiber, the light is continuously bent back toward the center line by refraction, and the light rays follow curved paths through the fiber. In single mode fiber, in effect only a single ray path propagates through the fiber.

5. The modulation bandwidth of a fiber is limited by dispersion effects. Modal dispersion results in multimode fibers due to differences in transit time between the various ray paths. Modal dispersion is most pronounced in step index fibers. Graded index fibers have much less modal dispersion, and single mode fibers are free from modal dispersion. Chromatic dispersion is due to waveguide effects and variations in the index of refraction of the core material with wavelength. Chromatic dispersion occurs in all types of fiber and is proportional to the linewidth of the source. Chromatic dispersion is a function of the wavelength of the source reaching a minimum in the vicinity of 1300 nm for silica fibers.

6. Attenuation of light in silica glass is very high unless certain impurity concentrations are reduced to very low levels. For very pure silica glass, Rayleigh scattering sets the lower limit on loss. This limit decreases with the fourth power of wavelength, so the lowest loss fibers operate in the vicinity of 1500 nm. At still longer wavelengths, losses increase due to bends in the fiber and intrinsic loss in the material.

7. In wavelength division multiplexing, several lightwave carriers with different wavelengths are carried simultaneously by the fiber, each with a different modulation.

8. Light energy is quantized when it interacts with matter. In interactions with matter, photons may be absorbed, spontaneously emitted, or emitted by stimulation from another photon. In stimulated emission, the emitted photon is coherent with the incident photon.

9. The two most important light sources for optical fiber systems are the LED and the laser diode. In an LED constructed from a direct bandgap material, forward bias results in spontaneous emission of photons when free electrons and holes recombine. The optical power output of the LED is almost directly proportional to the diode current. A heterojunction structure is often used in both devices to confine the recombination of carriers to a thin active layer, so the light emitted is confined to a small area and can be more easily coupled to the fiber.

10. A laser diode is similar to an LED except that reflectors are deliberately provided so that some of the light is returned to the material where stimulated emission takes place. When a sufficient population inversion exists, the positive feedback provided by the reflectors leads to self-sustained oscillations. Above this threshold point, the optical power increases linearly with additional forward diode current.

11. Laser diodes have much more temperature sensitivity and lower reliability than LEDs. On the other hand, laser diodes provide higher optical output power. Also, the beam is smaller and better collimated, so it is more effectively coupled to the fiber. The linewidth of the source is important because it determines the amount of chromatic dispersion that occurs in the fiber. Linewidths are much smaller for laser diodes than for LEDs.

12. The photodetectors most often used in optical fiber systems are either *pin* diodes or APDs. In the *pin* diode, incident photons create hole–electron pairs in the lightly doped depletion region. The ratio of charge carrier pairs produced to the number of incident photons is the quantum efficiency of the device. The diode is reverse biased, so the charge carriers result in current in the external circuit. The current is proportional to the incident optical power. APDs include a high field region where the detected charge carriers are multiplied by collisions with the crystal lattice. Because of this internal gain, the noise added in the receiver is less significant and the noise performance of systems using APDs is superior to those with *pin* diodes.

13. The detector is often used with a transresistance amplifier, which has the advantages of maintaining constant bias on the detector and wider bandwidth than other amplifier configurations.

14. Noise sources in an optical fiber system include the discrete nature of lightwave absorption, fluctuations in the dark current of the detector, amplifier noise, and thermal noise in the load resistor (or feedback resistor of the transresistance configuration). The discrete nature of light sets the ultimate lower limit on performance.

15. The theoretical lower limit on the error probability for noncoherent detection of on-off keyed modulation set by the discrete nature of light absorption is known as the quantum limit. This limit indicates that on average approximately ten photons per bit must be detected to achieve an error rate of 10^{-9}. Because of other noise sources in the receiver, noncoherent systems typically operate far from the quantum limit. It is

deemed unlikely that noncoherent receivers will be able to approach the quantum limit.

16. In a coherent receiver, a locally generated lightwave is added to the received lightwave before it falls on the detector. As a result, the detected current is affected by the phase of the received lightwave, so PRK modulation can be used. When the frequency of the local field is the same as the received field, the receiver is called a homodyne receiver. When the frequencies are different, a heterodyne receiver results, and a difference frequency component is produced in the detector. The theoretical performance limits of coherent receivers and the potential for realization of systems approaching these limits are summarized in Table 12.1.

17. Design of optical fiber communication systems is partly a trial-and-error procedure in which a prototype system consisting of a choice of modulation type, light source, modulator circuitry, fiber, connectors, splicing techniques, detector, and receiver circuitry is analyzed to determine if the desired system specifications can be met. The mechanical properties, modulation bandwidth, loss budget, and other factors of the fiber link must all be considered. Then appropriate adjustments are made to find a system that meets the desired objectives better. This process is continued until further improvements are judged unlikely to justify further design effort. As with all types of design, practice develops one's skill in choosing a good initial combination of parameters and in performing the required system analysis.

18. A number of factors contribute to the overall modulation bandwidth of a system. These are the modulation bandwidth of the source, modal dispersion in the fiber, chromatic dispersion, and the bandwidth of the receiver. Often, the specifications of the various components are given in different forms. The risetime of a component, electrical bandwidth, optical bandwidth, or a dispersion parameter may be available. In analyzing a system prototype, the designer must rely on conversion formulas based on appropriate models to combine these specifications.

19. System performance may be limited by either the received power or the modulation bandwidth. The modifications required to upgrade the system depend on the limiting factor.

REFERENCES

J. Gowar. *Optical Communication Systems*. London: Prentice-Hall, 1984.

W. B. Jones, Jr. *Introduction to Optical Fiber Communication Systems*. New York: Holt, Rinehart and Winston, 1988.

"Lightwave Technology." *AT&T Technical Journal* (January/February 1987).

R. A. Linke and A. H. Gnauck. "High-Capacity Coherent Lightwave Systems." *J. Lightwave Technol.* 6(11) (Nov. 1988): 1750–1769.

J. C. Palais. *Fiber Optic Communications*. Englewood Cliffs, N. J.: Prentice-Hall, 1984.

J. Salz. "Coherent Lightwave Communications." *AT&T Technical Journal* (December 1985).

J. Senior. *Optical Fiber Communications Principles and Practice*. London: Prentice-Hall, 1985.

D. J. Sterling, Jr. *Technician's Guide to Fiber Optics*. Albany, New York: Delmar Publishers, 1987.

PROBLEMS

1. The index of refraction of water is 1.33. What is the critical angle for rays striking an interface between air and water? Prepare a diagram showing ray paths for light emitted from an object at the bottom of a pool of water. Show several rays that strike the interface at a range of angles above and below the critical angle. Also prepare a diagram for the ray paths of light emitted by an object above the water surface.

2. Find the power loss in decibels due to Fresnel reflections for a connector that holds the ends of the fibers to be joined close together with a thin layer of air between them. Consider a light ray traveling down the axis of the fiber and assume that the index of refraction of the core material is 1.5. Repeat the calculation if the space between the fibers is filled with a material having an index of refraction of 1.45.

3. Consider a step index fiber with $n_1 = 1.45$ and $n_2 = 1.44$. Find the angle of acceptance and the range of distances traveled by the meridional rays for a 2 km length of fiber. What is the difference in transit times for the fastest and slowest meridional rays? Repeat for a second fiber having $n_1 = 1.45$ and $n_2 = 1.40$. Which fiber has the larger modulation bandwidth? Which fiber is it easier to couple light into?

4. Compute the bandwidth associated with the range of wavelengths from 1.30 to 1.35 μm. The approximate bandwidth of a telephone quality voice waveform is about 3.5 kHz. How many such voice signals can be accommodated in the wavelength range given if closely spaced frequency division multiplexing is used?

5. Compare the photon energy of a lightwave at a wavelength of 800 nm to the photon energy of a 10 GHz electromagnetic wave. Suppose that an optical digital communication system is operating at a data rate of 200 Mbit/s with a wavelength of 800 nm and an average of 10 photons are received per bit. What is the received power level of the system? How many photons would be received per bit on the average for a microwave system operating at 10 GHz with the same received power level?

6. Derive Equation 12.7 for the relationship between rms linewidth and the FWHM linewidth.

7. Assume that a perfectly monochromatic lightwave with a wavelength of 1500 nm is PRK modulated by a random binary data sequence having a data rate of 200 Mbit/s. Find the FWHM linewidth of the modulated lightwave. Compare the result with the typical linewidth of an LED.

8. The bandgap of lightly doped GaAs is approximately 1.40 electron volts. What is the approximate maximum wavelength of light photons emitted by a GaAs LED?

9. The distance between the reflecting surfaces of a GaAs laser diode is $L = 120 \ \mu m$. Find the free-space wavelength spacing of the modes of oscillation for the diode if it is emitting in the vicinity of 870 nm. The refractive index of GaAs is approximately 3.6. What is the frequency spacing of the modes?

10. A silicon *pin* diode has a dark current of 1 nA and a quantum efficiency of 75%. What optical power level at a wavelength of 830 nm is required so that the photocurrent is equal to the dark current? Find the responsivity R_d of this diode.

11. An APD has a quantum efficiency of 60%, operates at 1300 nm, and has an internal gain factor of 150. Find the responsivity of this diode.

12. List and briefly discuss the sources of noise in a noncoherent optical receiver for an on-off keyed signal. Which of these noise sources sets the ultimate lower limit on the error performance of the receiver?

13. In the text, we have considered coherent reception only for PRK modulation. Derive an expression for the lower limit of error probability for a coherent homodyne receiver for on-off keying in terms of the average number of received photons per bit. Assume that the local optical field is much stronger than the received field. Repeat for heterodyne detection. Compare the average number of received photons to achieve an error probability of 10^{-9} with the results found in the text for PRK.

14. Why is it unlikely that practical systems will be developed to operate at the super quantum limit for coherent detection of PRK with a local field equal in strength to the received field whereas prospects are good for approaching the theoretical limit with strong local fields?

15. An optical fiber link uses a source and driver circuitry having a 10 to 90% risetime of 2 ns, a fiber with an optical half-power bandwidth of 300 MHz km, and a receiver with a risetime of 3 ns. Find the overall electrical bandwidth of the link if the fiber length is 0.5 km. Repeat for a fiber length of 5.0 km.

16. A graded index multimode silica fiber has an optical bandwidth of 500 MHz km when used with a very narrow linewidth source at a wavelength of 830 nm. The fiber is to be used with an LED source with an FWHM linewidth of 20 nm. Find the optical bandwidth of a 5 km length of this fiber when used with the LED source. Find the corresponding electrical bandwidth.

17. Show that the overall 10 to 90% risetime of the cascade of two gaussian filters is given by the square root of the sum of the squares of the risetimes of the individual filters.

18. An on-off keyed optical fiber link is in operation with a detected signal amplitude at the amplifier output of 50 mV. A break occurs in the fiber and the (rather poor) splice made to repair the break results in an attenuation of optical power of 3 dB. What is the amplitude of the detected signal at the amplifier output after the break is repaired? By how many decibels has the electrical output decreased?

19. A certain on-off keyed fiber link has an average source power output of 1.0 mW, a source to fiber coupling loss of 5 dB, a fiber with a length of 5 km, and a fiber loss of 2.1 dB/km. Five splices, each with a loss of 0.6 dB, and a connector at each end with a loss of 1.2 dB each are used in the link. Coupling loss to the detector is 0.3 dB. Find the received optical power.

20. A certain on-off keyed optical fiber link has an average received optical power of -60 dBm and operates at a wavelength of 830 nm. Assume that the receiver requires 20 dB more power than predicted by the quantum limit to achieve a given error rate. The desired error probability is 10^{-9} or less. The overall electrical (half-power) bandwidth is 100 MHz. Estimate the maximum practical data rate for this system. Is the system power limited or bandwidth limited? Repeat if the received power is raised to -30 dBm.

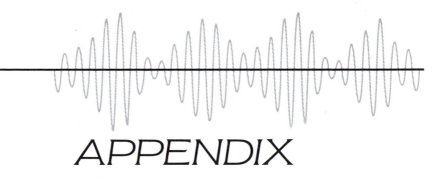

APPENDIX

Trignometric formulas

$$\exp{(\pm j\theta)} = \cos\theta \pm j\sin\theta$$

$$\sin\theta = \frac{\exp{(j\theta)} - \exp{(-j\theta)}}{2j}$$

$$\cos\theta = \frac{\exp{(j\theta)} + \exp{(-j\theta)}}{2}$$

$$\sin^2\theta + \cos^2\theta = 1$$

$$\cos^2\theta - \sin^2\theta = \cos 2\theta$$

$$\cos^2\theta = \frac{1}{2} + \frac{1}{2}\cos 2\theta$$

$$\sin^2\theta = \frac{1}{2} - \frac{1}{2}\cos 2\theta$$

$$\sin{(A \pm B)} = \sin A \cos B \pm \cos A \sin B$$

$$\cos{(A \pm B)} = \cos A \cos B \mp \sin A \sin B$$

$$\sin A \sin B = \frac{1}{2}\left[\cos{(A - B)} - \cos{(A + B)}\right]$$

$$\cos A \cos B = \frac{1}{2}\left[\cos{(A - B)} + \cos{(A + B)}\right]$$

$$\sin A \cos B = \frac{1}{2}\left[\sin{(A - B)} + \sin{(A + B)}\right]$$

Fourier transforms

See Table 2.2.

Fourier transform properties

See Table 2.1.

Binomial coefficient

$$\binom{n}{k} = \frac{n!}{(n-k)!k!}$$

Q Function

$$Q(x) = \int_x^\infty \frac{1}{(2\pi)^{1/2}} \exp\left(-z^2/2\right) dz$$

$$Q(x) \simeq \frac{1}{(2\pi)^{1/2}x} \exp\left(-x^2/2\right) \qquad \text{accurate for } x > 3$$

$$Q(-x) = 1 - Q(x)$$

Sinc function

$$\text{sinc}\ (x) = \frac{\sin\ (\pi x)}{\pi x}$$

GLOSSARY OF ABBREVIATIONS

A	Ampere
ac	alternating current
ACK	(positive) acknowledgment
A/D	analog to digital
AJ	antijamming
AM	amplitude modulation
AMI	alternate mark inversion
APD	avalanche photodiode
ARQ	automatic repeat query
AWGN	additive white gaussian noise
BER	bit error rate
BPF	bandpass filter
BPSK	biphase shift keying
C	Coulomb
CDF	cumulative distribution function
CDMA	code division multiple access
CPFSK	continuous phase frequency shift keying
CW	continuous wave
DAC	digital to analog converter
dB	decibel
dBK	decibels relative to unity Kelvin
dBm	decibels relative one milliwatt
dBV	decibels relative to one volt
dBW	decibels relative to one watt
dc	direct current
DEPSK	differentially encoded phase shift keying
DMC	discrete memoryless channel
DMS	discrete memoryless source
DMUX	demultiplex
DPSK	differential phase shift keying
DS	direct sequence
DSB-SC	double-sideband–suppressed-carrier
FDM	frequency division multiplexing
FDMA	frequency division multiple access
FEC	forward error control
FET	field effect transistor
FH	frequency hop
FM	frequency modulation
FSK	frequency shift keying
FWHM	full width half maximum
GaAs	gallium arsenide
GHz	gigahertz

GPS	Global Positioning System
Hz	Hertz
IF	intermediate frequency
ISI	intersymbol interference
J	Joule
K	Kelvin
kHz	kilohertz
km	kilometer
LED	light emitting diode
LFSR	linear feedback shift register
LNA	low noise amplifier
LO	local oscillator
LPD	low probability of detection
LPF	lowpass filter
m	meter
mA	milliampere
MAP	maximum a posteriori
MHz	megahertz
MSK	minimum shift keying
MUX	multiplex
mV	millivolt
mW	milliwatt
NAK	negative acknowledgment
NB	narrowband
NBFM	narrowband frequency modulation
NBPM	narrowband phase modulation
nm	nanometer
NRZ	nonreturn to zero
OOK	on-off keying
OQPSK	offset quadrature phase shift keying
PAM	pulse amplitude modulation
PBX	private branch exchange
PCM	pulse code modulation
PDF	probability density function
pF	picofarad
PLL	phase-locked loop
PM	phase modulation
PN	pseudonoise
PPM	pulse position modulation
PRK	phase reversal keying
PSD	power spectral density
PWM	pulse width modulation
QAM	quadrature amplitude modulation
QPSK	quadrature phase shift keying

RC	resistance-capacitance
RF	radio frequency
rms	root mean square
ROM	read only memory
RZ	return to zero
SNR	signal to noise ratio
SQPSK	staggered quadrature phase shift keying
SSB	single sideband
TDM	time division multiplexing
TDMA	time division multiple access
TV	television
V	volt
VCO	voltage-controlled oscillator
VSB	vestigial sideband
W	Watt
WGN	white gaussian noise
XOR	exclusive or

INDEX